Mechanical Life Cycle Handbook

MECHANICAL ENGINEERING
A Series of Textbooks and Reference Books

Founding Editor

L. L. Faulkner

*Columbus Division, Battelle Memorial Institute
and Department of Mechanical Engineering
The Ohio State University
Columbus, Ohio*

Additional Volumes in Preparation

Mechanical Engineering Software

Mechanical Life Cycle Handbook

Good Environmental Design and Manufacturing

edited by

Mahendra S. Hundal
University of Vermont
Burlington, Vermont

MARCEL DEKKER, INC.　　　　　　　NEW YORK · BASEL

ISBN: 0-8247-0572-6

This book is printed on acid-free paper.

Headquarters
Marcel Dekker, Inc.
270 Madison Avenue, New York, NY 10016
tel: 212-696-9000; fax: 212-685-4540

Eastern Hemisphere Distribution
Marcel Dekker AG
Hutgasse 4, Postfach 812, CH-4001 Basel, Switzerland
tel: 41-61-261-8482; fax: 41-61-261-8896

World Wide Web
http://www.dekker.com

The publisher offers discounts on this book when ordered in bulk quantities. For more information, write to Special Sales/Professional Marketing at the headquarters address above.

To all the people—past, present, and future—
who see it as their duty to help preserve our
natural environment for all the life on this planet.

Preface

Design for the environment (DFE), also called "green design" or "environmentally friendly design," is one of the greatest challenges facing engineers today and one that offers great potential benefits to society. DFE addresses the problem of waste and pollution at the design stage, where the potential for effecting results is the greatest. The decisions made at the design stage affect all phases of a product's life—manufacturing, transportation, operation, maintenance, and disposal. The traditional field of environmental engineering, on the other hand, deals with waste and pollution after the fact (i.e., after they have been generated) and with methods of mitigating environmental damage.

The need to develop products that minimize environmental damage has become increasingly evident. Products are important in fulfilling human needs, but the side effects of production—pollution and the depletion of natural resources—must also be of concern to designers and manufacturers. This handbook is divided into the following parts: General Concepts, Product Planning, Design and Manufacturing, Industrial Practice, and Management Aspects.

GENERAL CONCEPTS

Chapter 1 introduces DFE and the process of life cycle engineering (LCE). The product realization process (PRP) is described from the traditional perspective and from the perspective of the life cycle. The chapter also discusses consumption and waste of energy and material resources, as well as the environmental requirements of product development.

In Chapter 2, Conway-Schempf and Hendrickson present an overview of life cycle assessment (LCA), defining its terminology and providing a historical perspective on LCA stages and methods. Two case studies in the chapter show the results of using alternative materials in, respectively, drinking cups (paper vs. polystyrene bead) and pavement (asphalt vs. concrete).

Chapter 3, by Caudill et al., introduces an extension of LCA—multi–life cycle assessment (MLCA)—that focuses on quantifying the materials, energy, and environmental burdens associated with end-of-life options and on the value of returning parts and materials back into use. This chapter presents MLCA software as a tool for analyzing and comparing environmental impacts, energy consumption, and costs of various products.

In Chapter 4, Billatos focuses on design methodologies for the environment, providing an overview of the product life cycle, DFE drivers, and DFE guidelines. The author describes applications of DFE in the German automobile industry, consumer appliances, and packaging design.

Chapter 5 examines the laws and regulations that drive DFE in the United States. It describes major U.S. environmental regulations, their effectiveness, and their necessity, and considers the interaction of laws, economics, and product/process innovation.

In Chapter 6, Dammert et al. address the environmental laws and market-driven requirements in Europe, including those for the European Community in general, and those in Finland, Sweden, and Germany in particular. The authors also discuss the regulations for environmental management and environmental labeling and declarations, including an extensive list of European information sources and references.

PRODUCT PLANNING

Luttropp's Chapter 7 introduces a DFE method useful in the early product development stages. In this method, products are classified according to their degree of modularization and ease of disassembly. The author gives five steps in a green conceptual design process: market study, product planning, developing the module list, preparing a connection map of parts, and determining indices for sorting of parts and subassemblies.

Züst begins Chapter 8 by examining high resource use, environmental management, and early planning. He goes on to describe product life phases, environmental performance evaluation, an approximate life cycle model, and modeling of energy and material flows. The chapter's final section introduces a search heuristic for ecological improvement.

Chapter 9, by Wimmer, presents a procedure for evaluating designs for properties relevant to environmental impact. This procedure involves close scrutiny of the product's life cycle and its environment, with the goal of formulating targets for product improvement. The chapter also describes the environmental checklist method, which uses modules for part analysis, functional analysis, and product analysis.

DESIGN AND MANUFACTURING

In Chapter 10, Birkhofer and Grüner present a methodology for DFE that emphasizes marketing needs and computer support for the process. They present strategies and rules for DFE and examine conflicts between environmental and marketing requirements. The chapter also includes a discussion of procedures and computer support for an integrated product and process design that considers each life phase of the product—from raw materials extraction through disposal.

Chapter 11, by Anderl et al., shows the development of a model-based approach for assessment and optimization of a product based upon ecological considerations. The authors discuss the implementation of a three-tiered model in terms of information sources, system components, and system architecture. In order to create the underlying information model, the modeling tool integrates both process- and object-oriented modeling languages.

In Chapter 12, Sauer et al. investigate the significance of user behavior in ecological design and introduce an LCA-based environmental impact assessment. The authors present a model for understanding user behavior that focuses on the interaction of the user with the system. The chapter ends with an evaluation of trade-offs between ecological and commercial requirements.

Chapter 13, by Nicolaou et al., introduces a domain-independent mathematical model for design and manufacturing process analysis. The model employs multicriteria optimization and begins with a qualitative correlation between evaluation criteria (cost, reliability, environmental impact) and design decision variables (life phases, with emphasis on disposal alternatives.) Examples provided focus on computer products to facilitate product take-back, disassembly, and product reuse over multiple life cycles, and on the integration of statistical analysis of a machining process.

In Chapter 14, Sriram et al. propose a framework for collaborative design that addresses environmental concerns in the design phase. They review the extent of the pollution and waste problem, the steps in the product realization process, and the life phases of a product. The collaborative design system emphasizes access to environmental databases through an environmental data manager, which provides an intelligent interface between databases and applications.

INDUSTRIAL PRACTICE

Sullivan's Chapter 15 provides an in-depth discussion of life cycle assessment using the SETAC framework, which includes life cycle inventory, life cycle impact assessment, and life cycle improvement/interpretation. The author also addresses other approaches to LCA—namely, EPS, the eco-indicator 95 method, and the threshold inventory interpretation method (TIIM). The chapter presents applications of LCA in the production of aluminum; in the automobile industry; and, using TIIM, in a comparison of three types of fruit juice containers.

In Chapter 16, Gediga et al. examine life cycle assessment of nonferrous metals, emphasizing the effect of various technologies and geographic locations on life cycle inventory. The authors provide a model LCA of zinc production for two processes—electrolysis and smelting—and the resulting environmental profiles are compared. Environmental impacts considered in the chapter are primary energy demand, global warming potential, and heavy metals emissions.

Computer-based simulation and modeling of demanufacturing systems is analyzed by Caudill et al. in Chapter 17. A typical demanufacturing system involves inspection of collected products, staging of the workflow, disassembly of products, shredding of products or components, separation into bins, and shipment of the recovered materials and components for further processing or use. Computer simulation of such a system reveals system throughput, resource/worker utilization, bottleneck identification, and cost and profit assessment.

Chapter 18, by Vaajoensuu et al., presents a discussion of materials and manufacturing processes, with special emphasis on the production of electronic and electromechanical components. This discussion addresses active and passive components, cables, batteries, displays, printed circuit board, and joining materials. Environmental impacts of various metals are considered with regard to energy requirements, coating processes, and emissions to the environment. The chapter concludes with a brief presentation on recycling and reuse considerations in mechanical design.

The application of ecodesign in the electronics industry is described by Stevels in Chapter 19. General characteristics of environmental approaches are divided into defensive, cost-oriented, and proactive approaches.

In Chapter 20, Johnson and Wang discuss the economics of disassembling products for material recovery. The recovery process, disassembly sequence, choice of components recovered, and design characteristics for ease of disassembly are examined in detail. The authors present economic models of disassembly that consider the option of recovery, present disposal cost, and the costs of disposal and disassembly.

Disassembly and recycling of automobiles is addressed by Das in Chapter 21. The author looks at current recycling practice and particularly at the types of recyclers (scavengers and reclaimers) used. Disassembly economics is analyzed using, for example, the disassembly effort index and disassembly return on investment.

MANAGEMENT ASPECTS

Sarkis presents in Chapter 22 the requirements of management for environmentally friendly manufacturing. Included are an explanation of industrial ecology and an elaboration of the importance of green supply chains. The author describes the evolution of natural environment management theory, manufacturing strategy, and the linkage of functional strategies to manufacturing.

In Chapter 23, Stevels and Boks examine ways of dealing with a product at the disposal stage. The chapter is divided into three main parts: (1) development of end-of-life systems, (2) plan for end-of-life analysis at the product level, and (3) optimization of the product according to the end-of-life system. Quantitative evaluations of end-of-life systems are shown in the form of LCA, life cycle costs, and end-of-life costs.

In order for DFE to be most effective, it needs to be integrated into the complete business cycle, as shown by Stevels in Chapter 24. The management of the product creation process under DFE constraints is described in detail. The author suggests that the embedding of ecodesign in business should take place in three steps: green idea generation, green product creation, and the exploitation of the results in the marketplace.

Chapter 25, by Guide and Linton, focuses on minimizing the environmental impact of production by promoting an interaction between logistics planners and design engineers. The authors discuss management drivers, technological drivers, factors affecting product recovery, motivations for product recovery, and generalized design guidelines. Through a thorough understanding of reverse product flow, as well as current challenges, it becomes possible to better position a product for recovery as a postconsumer product.

Kaipainen and Ristolainen evaluate various DFE tools in Chapter 26. These tools, which are used for assessing the environmental impacts of product design, are divided into manual and software tools. The former include questionnaires, checklists, tables, and design guides; the latter comprise the software tools EcoScan and EIME. Such tools enable analyses of products, creation of databases, and calculation of environmental impacts.

CONCLUSION

The handbook is intended to be a concise and broad reference on the subject. Readers should find a wealth of further opportunities in the literature cited by the authors of various chapters; these sources vary in scientific rigor and thus require various degrees of technical sophistication from readers.

One of the objectives of a work such as this is that people in one industry learn from those in other industries and from active scholars. Most of the contributors to this volume are academics; others have backgrounds in industry. As such, each chapter herein expresses the richness of the author's background and contributes to an overall diversity of views.

ACKNOWLEDGMENTS

I would like to express my heartfelt thanks to all the contributors to this handbook. It has been a most rewarding experience working with such a distinguished group of scientists. And, for their help in the production of this book, I would like to thank the wonderful people at Marcel Dekker, Inc.—John Corrigan, acquisitions editor, and Paige Force and Michael Deters, production editors.

Mahendra S. Hundal

Contents

Contents

Contents

Contributors

Mauri Airila Laboratory of Machine Design, Department of Mechanical Engineering, Helsinki University of Technology, Helsinki, Finland

Robert H. Allen Department of Mechanical Engineering, University of Maryland, College Park, Maryland

Reiner Anderl Department of Computer Integrated Design, Darmstadt University of Technology, Darmstadt, Germany

Samir B. Billatos i2 Technologies, Inc., Irving, Texas

Herbert Birkhofer Institute for Machine Elements and Engineering Design, Darmstadt University of Technology, Darmstadt, Germany

Casper Boks Design for Sustainability Program, Delft University of Technology, Delft, The Netherlands

Reggie J. Caudill Multi-Lifecycle Engineering Research Center, New Jersey Institute of Technology, Newark, New Jersey

D. Navin Chandra* Carnegie Mellon University and TimeØ Inc., Cambridge, Massachusetts

Noellette Conway-Schempf[†] Green Design Initiative, Carnegie Mellon University, Pittsburgh, Pennsylvania

Taina Dammert Laboratory of Machine Design, Department of Mechanical Engineering, Helsinki University of Technology, Helsinki, Finland

Sanchoy K. Das Multi-Lifecycle Engineering Research Center, Department of Industrial and Manufacturing Engineering, New Jersey Institute of Technology, Newark, New Jersey

Bernd Daum Department of Computer Integrated Design, Darmstadt University of Technology, Darmstadt, Germany

Peter Eyerer Institute for Polymer Testing and Polymer Science, University of Stuttgart, Stuttgart, Germany

Harald Florin Institute for Polymer Testing and Polymer Science, University of Stuttgart, Stuttgart, Germany

Johannes Gediga Institute for Polymer Testing and Polymer Science, University of Stuttgart, Stuttgart, Germany

Chris Grüner Institute for Machine Elements and Engineering Design, Darmstadt University of Technology, Darmstadt, Germany

V. Daniel R. Guide, Jr. A. J. Palumbo School of Business Administration, Duquesne University, Pittsburgh, Pennsylvania

Chris Hendrickson Department of Civil and Environmental Engineering, Carnegie Mellon University, Pittsburgh, Pennsylvania

Jingjing Hu Multi-Lifecycle Engineering Research Center, New Jersey Institute of Technology, Newark, New Jersey

Mahendra S. Hundal Department of Mechanical Engineering, University of Vermont, Burlington, Vermont

---

Current affiliation: NovaSpike Inc., Boston, Massachusetts.
[†] *Current affiliation*: Automatika, Inc., Pittsburgh, Pennsylvania.

Jie Jin   Multi-Lifecycle Engineering Research Center, New Jersey Institute of Technology, Newark, New Jersey

Harald John   Department of Computer Integrated Design, Darmstadt University of Technology, Darmstadt, Germany

Michael R. Johnson   Industrial and Manufacturing Systems Engineering, University of Windsor, Windsor, Canada

Jukka Kaipainen   Institute of Electronics, Tampere University of Technology, Tampere, Finland

Markku Kuuva   Laboratory of Machine Design, Department of Mechanical Engineering, Helsinki University of Technology, Helsinki, Finland

Ketan Limaye   Multi-Lifecycle Engineering Research Center, New Jersey Institute of Technology, Newark, New Jersey

Jonathan D. Linton   Polytechnic University, Brooklyn, New York

Conrad Luttropp   Department of Machine Design, KTH Royal Institute of Technology, Stockholm, Sweden

Donna Mangun   Department of General Engineering, University of Illinois at Urbana-Champaign, Urbana, Illinois

Panicos Nicolaou   Department of General Engineering, University of Illinois at Urbana-Champaign, Urbana, Illinois

Christian Pütter   Department of Computer Integrated Design, Darmstadt University of Technology, Darmstadt, Germany

Eero Ristolainen   Institute of Electronics, Tampere University of Technology, Tampere, Finland

Bruno Rüttinger   Institute of Psychology, Darmstadt University of Technology, Darmstadt, Germany

Joseph Sarkis   Graduate School of Management, Clark University, Worcester, Massachusetts

Jürgen Sauer   Institute of Psychology, Darmstadt University of Technology, Darmstadt, Germany

Ram D. Sriram Department of Mechanical Engineering, University of Maryland, College Park, Maryland

Ab Stevels Philips Consumer Electronics, Eindhoven, The Netherlands

John L. Sullivan Ford Research Laboratory, Ford Motor Company, Dearborn, Michigan

Ying Tang Multi-Lifecycle Engineering Research Center, New Jersey Institute of Technology, Newark, New Jersey

Deborah L. Thurston Department of General Engineering, University of Illinois at Urbana-Champaign, Urbana, Illinois

Eero Vaajoensuu Laboratory of Machine Design, Department of Mechanical Engineering, Helsinki University of Technology, Helsinki, Finland

Michael H. Wang Industrial and Manufacturing Systems Engineering, University of Windsor, Windsor, Canada

Bettina S. Wiese Institute of Psychology, Darmstadt University of Technology, Darmstadt, Germany

Wolfgang Wimmer Institute for Engineering Design and Mechanical Handling, Vienna University of Technology, Vienna, Austria

Pingtao Yan Multi-Lifecycle Engineering Research Center, New Jersey Institute of Technology, Newark, New Jersey

MengChu Zhou Multi-Lifecycle Engineering Research Center, New Jersey Institute of Technology, Newark, New Jersey

Rainer Züst[‡] Leiter KTC, Winterthur, Switzerland

[‡]*Current affiliation*: Alliance for Global Sustainability, ETH-Zentrum, Zurich, Switzerland.

1

Introduction to Design for the Environment and Life Cycle Engineering

Mahendra S. Hundal
University of Vermont, Burlington, Vermont

1 INTRODUCTION

The need to develop products that minimize damaging effects on the environment has become increasingly evident. Manufacturers have realized that "end-of-the-pipe" solutions to pollution problems are not acceptable—neither from a legal nor from an economic viewpoint [1]. There is thus a need to apply knowledge and innovation right from the beginning of the product development process to create environmentally friendlier products and the associated processes. These processes include extraction of material, processing and manufacturing, transportation, and use and disposal of the product.

As environmental awareness has grown worldwide, customers are more willing to pay higher prices for "green" products. Companies that have innovated to be the first in the marketplace with new or improved products enjoy advantages of being the leaders. By early introduction, a product has a marketplace advantage by gaining early customers who lock on to it, develop loyalty, and are less likely to switch [2].

A manufacturer has four inputs, or resources, to put into the enter-prise: *raw material, energy, labor,* and *capital.* In the past a company with access to cheap inputs had a marketplace advantage. With the globalization of the economy, no single company, or indeed a country, has a lock on cheap resources. A company from a high-cost country can build a plant in a low-cost country. Rather, today, it is how a company uses its resources that makes it competitive. Porter and van der Linde [3] call this "resource pro-ductivity."

A competitive product is one that can be produced cheaper than simi-lar products or one that provides more value for which customers are willing to pay more. Competitiveness and environmental friendliness are not mutually exclusive. Through innovation, by making better use of resources, products can be made more competitive and environmentally friendly (Chapter 5).

2   DEFINITIONS

We begin with definitions of some of the terms used here. Design for envir-onment (DFE) is a design aim. It is one type of "design for X," where X may stand for M (manufacturability), C (cost), A (assembly), and so on. DFE aims to integrate environmental requirements in the design process. It thus seeks to minimize the environmental impacts of the product at each stage of its life cycle. The word "design," used here and elsewhere, is often used to describe the total product realization process (PRP).

Ecodesign or ecological design is a term often used in European lit-erature. It implies environmentally friendly design (i.e., low environmental burdens) and combines the ideas of DFE and life cycle design (LCD).

Green design is a term also used to denote environmentally friendly design (i.e., creating low environmental burdens). A report sponsored by the U.S. Office of Technology Assessment (OTA) [4] defines green design as addressing two goals: preventing waste and optimal use of materials.

The life cycle is when the life of a product begins with material extraction and proceeds through material processing, manufacturing, use, and its ultimate disposal. Each of these activities is referred to as a life stage. The term is generally qualified by a noun that indicates which aspect of the life cycle is addressed: A, assessment; D, design; E, engineering; I, inventory, and so on. Life cycle assessment is a method for assessing the environmental impact of a product over its life cycle (i.e., at each of its life stages).

LCD is a design methodology that considers all life stages of a product. Many authors regard LCD and DFE as interchangeable terms.

DFE may be regarded as one of the aims of LCD. Life cycle engineering (LCE) is concerned with the total product life, from raw material acquisition through material processing, manufacturing, use, and disposal. The aims of LCE may differ (e.g., low cost, long lifespan, minimizing resource use, etc.).

PRP consists of the steps of product planning, design, process planning, and manufacturing. The term "product design" is often used to indicate the complete PRP, although it is only one of the steps in the process.

Two approaches—DFE and LCE—have come to mean the same things. Once we begin to look at the complete life cycle of the product, the environmental impacts become important as a matter of course. The PRP/LCE follows the same steps as the traditional PRP, with the added considerations of environmental burdens integrated into each stage. Figure 1 shows the different life phases of a product and the PRP.

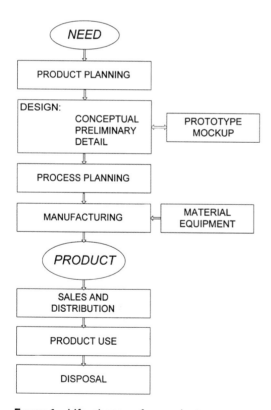

FIGURE 1 Life phases of a product.

3 STEPS IN THE PRP

The need for the product, whether real or imagined, must exist. This may come from external or internal sources. The external forcing for a new product may be due to an order directly from a customer, obsolescence of the existing product, availability of new technologies, or change in market demands. Internal to the company, new product ideas may come from new discoveries and developments within the company or need for a product identified by the marketing department.

Once the need has been established, the product has to be designed and manufactured. This process (PRP) includes the first four steps shown in Figure 1, viz., product planning; design; process planning; and manufacturing. The above steps appear in Figure 1 as sequential. This is the model of the so-called gateway project administration. In this process each step must be complete before the next step begins. This procedure can lead to delays, mistakes, poor quality, and high costs. A preferable product development procedure is one in which the activities in the successive stages are concurrent and partially overlap [5].

3.1 Product Planning

Product planning is the search, selection, and development of ideas for new products. A systematic approach to product planning will lead to a better meeting of the constraints of cost and time. Product planning activities include the following:

- Establishing product goals;
- Conducting market analyses;
- Detailing the benefits this product will provide the customer;
- Deciding on the features the product will have;
- Establishing product performance;
- Conducting an economic analysis and setting the cost target;
- Establishing the expected sales volume;
- Setting deadlines for completion of tasks, such as design, prototype; building, setting up manufacturing line, etc.

The two most important entities involved in making the product development decision are the company and the market. There are other secondary factors, such as government laws, economic policies, and state of the technology. This interaction is shown in Figure 2. The company needs to define its objectives and examine its capabilities. The types of strengths that a company has are its personnel, its facilities, and the financial situation. The personnel and facilities are distributed in various types of activities or departments (e.g., design, production, marketing, etc.) and the buildings

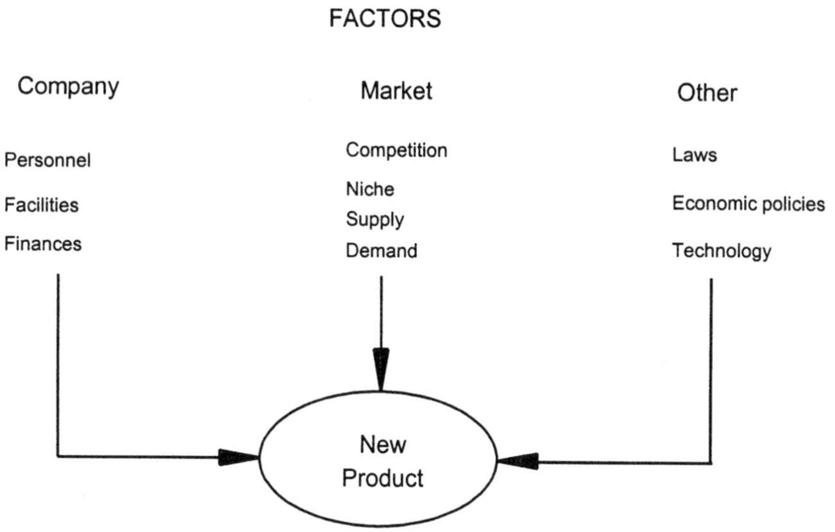

FACTORS

Company	Market	Other
Personnel	Competition	Laws
Facilities	Niche	Economic policies
	Supply	
Finances	Demand	Technology

FIGURE 2 Factors influencing new product development.

(e.g., design, test and production equipment, distribution systems, etc.). Such an evaluation will help focus the company on the type of products it should develop.

3.2 Design

Product design includes the activities of generating the requirements list for the product, developing ideas on what it should look like and how it should operate, and complete drawings and documentation. The latter contains the complete information on the product, from which it may be manufactured. Steps in design are

- Preparation of the requirements list;
- Conducting a technology survey: determine what is feasible at this time using feasible technologies;
- Conceptual design;
- Preliminary or embodiment design;
- Designing, building, and testing of a prototype; building a mockup if appropriate;
- Detail design;
- Preparing documents such as bill of materials, assembly, operating, and service instructions.

3.3 Process Planning

Also called production or manufacturing planning, process planning involves decisions on how the product is to be manufactured. For example, what steps are required to manufacture the product; which manufacturing processes, machines, tools are required; how the parts are to be assembled, and so on. Steps in process planning are producibility analysis, initial process design, vendor/sourcing selection, tooling design, and final process design.

3.4 Manufacturing

Under manufacturing we include materials handling, production of parts, assembly, quality control, and related activities. Many of the decisions regarding manufacturing have already been made during the embodiment design stage, knowingly or unknowingly. Steps in manufacturing are tool and equipment procurement, production line set-up, test runs, and production runs.

4 LIFE CYCLE ENGINEERING

LCD process considers a full range of environmental consequences and gathers information from all the stakeholders in the product—within the company and from the outside. The life cycle of the product includes not only the product but all activities associated with it, such as the manufacturing process, suppliers, and distribution.

Modern economies are material intensive. The United States extracts 10 tons of material (not including air, fresh water, stone, gravel, and sand) per year per person, of which only 6% is embodied in durable goods. The rest constitutes waste, according to a report by the OTA [4]. The overall picture of a product's life cycle and the materials flow associated with it are shown in Figure 3. This figure also illustrates the three terms frequently used in connection with DFE: *reuse, remanufacturing,* and *recycling.*

The material cycle begins at the extraction or acquisition of raw material and ends with the final disposal of the remains [6]. At each life stage there is a flow of materials, energy, and information through the pertinent system or subsystem. From the environmental viewpoint, at each stage there is energy input and waste is produced.

LCD aims to consider all elements of the product life in an integrated fashion. In particular it includes the following:

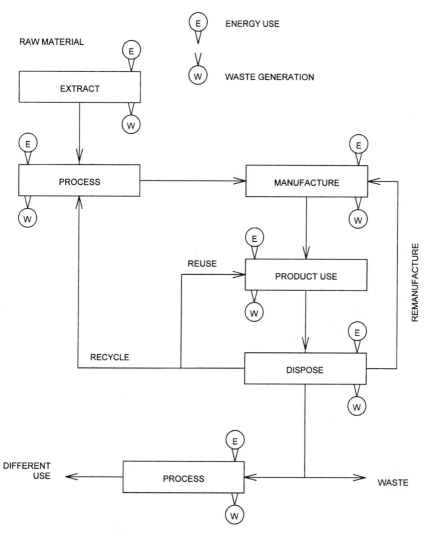

FIGURE 3 Product life stages and material flows.

- The product itself—all materials in the finished product, the raw materials used, energy used, waste produced;
- The manufacturing process—the processes, plant, equipment, and ancillary activities;
- The distribution—packaging, transportation (all materials), storage facilities.

Goals of LCD may be summarized into "reduce environmental burdens" and "find sustainable solutions." The environmental burdens consist of resource depletion, ecological effects, and human health effects. A product generates environmental burdens at each of its life stages. Different types of product generate varying amounts of burden in the different life stages. For example, a hand tool does so mainly during its production; a pump, on the other hand, requires most of the energy during its operation. Similarly, a piece of furniture creates environmental burden during its manufacture and disposal, whereas a conventional automobile uses fuel and produces air pollution during its use.

As an example we consider the life cycle of a plastic part [7]. The analysis includes the part itself, the manufacturing process, and the distribution of all the materials involved in the life cycle:

- For the *product* we consider the petroleum raw material, the plastic raw material, manufacture and use of the part, and the disposal of the remains.
- For the *process* we consider the equipment and energy requirements for drilling for petroleum, its processing to produce the plastic raw material, the molding and other processes in part production, processes during use of the part (e.g., cleaning), the processes for disposal.
- For the *distribution* we consider the transportation vehicles and equipment, packaging, containers, and storage facilities.

An integrated and interconnected product/process system can provide efficiencies beyond those in a "conventional" system. A simple example is a power plant with cogeneration and use of waste heat for district heating or agricultural purposes. An example of several industries colocated in an industrial park is from Kalundborg, Denmark [8]. An oil refinery, a plasterboard plant, a power plant, fish farming, and green houses share each other's input and output streams.

5   ENERGY AND MATERIAL CONSUMPTION AND WASTE

Sustainable development is defined as the ability to fulfill the present needs without compromising the future. The increasing world population requires ever greater use of materials and energy. The issue is further compounded by the fact that standard of living of most people around the world is improving, which translates into a greater per capita use of materials and energy and a greater rate of waste generation.

Figure 4 shows world power consumption and population growth. While the rate of population growth has increased somewhat, it is the

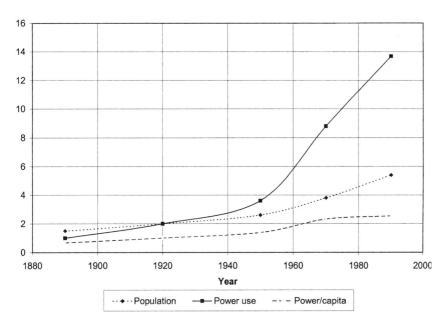

FIGURE 4 World population and power use. Population, billions; power use, trillion watt; power/capita, kW/capita. (From Ref. 9.)

power use that shows a dramatic increase. As the developing countries improve their standards of living, we should see an even sharper increase in world power use.

Figure 5 shows energy use per capita for different parts of the world. The disparity between the industrialized and developing nations is evident from this chart. It is also interesting to note that per capita energy use in Western Europe is lower than in the former Soviet Union, despite the higher standard of living in the west.

Figure 6 shows the generation of municipal solid waste per capita for some of the developed countries. The chart shows that in nearly all countries there has been an increase in solid waste generation. There is notable disparity between certain countries with comparable standards of living. There is a potential dependency of waste production in a country to resource consumption and the efficiency of resource use.

Although the most visible environmental impact of products is municipal solid waste (MSW), it is one of the smallest parts of the total waste generated during its life cycle. The difference between perception and reality

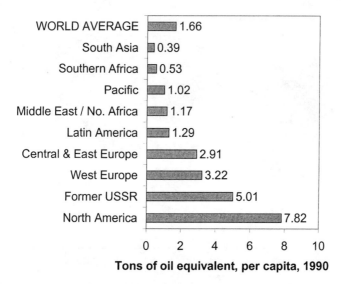

Tons of oil equivalent, per capita, 1990

FIGURE 5   Energy demand per capita. (From Ref. 10.)

Municipal Solid Waste Per Year

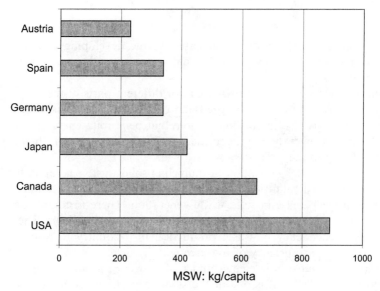

MSW: kg/capita

FIGURE 6   Worldwide generation of municipal solid waste. (From Ref. 11.)

is shown in Figure 7. The figure shows solid wastes produced annually, as defined by the Resource Conservation and Recovery Act (RCRA), from material extraction, to processing, manufacturing, transport, use, and finally disposal [4]. The total solid waste produced in 1988 in the United States was 11,700 million tons, of which 700 million tons was hazardous waste, as shown in Figure 7a. In Figure 7b the nonhazardous waste is shown divided up into its various components by source. We see that MSW is only 2% of the total nontoxic waste. An example of public's perception of waste is the belief that at fast food outlets it consists mainly of the containers. Actually, 80% of the waste occurs "behind the counter," not seen by the public. As shown in Figure 7 this is one part of the wastes produced in manufacturing that constitutes about one-half of the total wastestream and thus holds the potential of significant reductions by proper design and manufacturing practices.

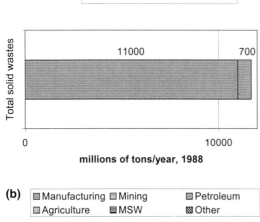

FIGURE 7   RCRA solid wastes. (From Ref. 4.)

6 PRODUCT DEVELOPMENT FROM A LIFE CYCLE PERSPECTIVE

Products are made of materials. They process one, two, or all three of the entities energy, material, and information (Figure 8.) We can thus summarize the core aim of engineering design as follows: Determine the shapes and arrangement of material that will provide the optimum flow of *energy*, *materials*, and *information* to fulfill the desired *requirements*.

The product development process begins with the need for the product and proceeds through product planning, design, and manufacturing. The life cycle development process looks at the potential environmental burdens: resource depletion, ecological effects, and effects on human health. Needs are analyzed at project initiation. Product planning identifies customers and their needs. An environmental assessment is made of the existing products.

Needs analysis is affected by life cycle principles (e.g., sustainable development and creation of low-impact products). The company should include environmental criteria in its product development process. It should not produce environmentally damaging product lines for which lower impact alternatives are available.

Project scope and purpose should be defined. System boundaries should be set: Which stage of life cycle? Which part of the product? At which time? Should the company focus on the complete life cycle, part of the life cycle, or only on certain stages or activities? The scope of the project will determine the extent of improvements that are possible. Define the level of analysis. How far upstream and downstream are the impacts to be evaluated? At the first level, look at material, energy, and labor. A deeper analysis will look at facilities and equipment needed to produce the items. Project schedule and budget should be established. Small firms cannot

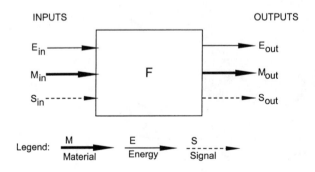

FIGURE 8   A function block.

afford full-scale life cycle assessment. Companies should consider long-term pay-off.

6.1 Baseline and Benchmark Environmental Performance

Baseline analysis yields the environmental profile of a product. It shows opportunities for improvement. Baseline analysis uses life cycle inventory analysis and audit team reports. A process flow diagram should be prepared. Baseline analysis helps formulate general design goals and detailed design requirements. Sources of environmental data include life cycle inventory, purchasing records, monitoring reports, quality assurance and control reports, legal department records, audit reports, compliance records, publicity records, and so on.

Benchmarking involves comparing one's own product to best products on the market, with the aim to target design improvements. It establishes reference points for improvement of the product. Life cycle assessment is one means of comparative assessment of product. It has been used, for example, to compare the impact of paper versus plastic cups.

Benchmarking and baseline analysis can reveal vulnerabilities associated with environmental risks or liability, performance, standards, cost, and so forth.

6.2 Developing Product Requirements

In the case of LCD, as indeed in traditional design, there are conflicting requirements. Whereas in traditional design we consider requirements based on performance, cost, and manufacturability, in LCD we must also consider the environmental objectives. The environmental objectives are achieved by reducing the environmental burdens.

As a means toward achieving the environmental objectives, all stakeholders should be involved in the design process. In traditional design we include the suppliers, consumers, and the marketing, planning, design, manufacturing, and costing groups. Additional stakeholders from the environmental viewpoint include the general public, the regulators, waste handlers, and resource recovery entities. The implementation follows the methods of concurrent engineering by forming cross-functional teams [5].

7 REQUIREMENTS LIST

The requirements list is a most crucial document, prepared at the start of the development cycle. It should be prepared by design and manufacturing (at the minimum) working together. It should define for the yet nonexistent product, its functions, and the constraints on it—not how the functions will be realized nor its physical features. The constraints should be real constraints, not those perceived by engineering or marketing. The requirements should be dictated by the customer's needs, expectations and problems, benefits to the customer, and expected improvements over existing products. The requirements list is not an engineering document per se nor the manufacturing plan. This is not the list of specifications that will appear with the product when it is marketed. It should be as abstract as possible, as far as its physical features are concerned and not rule out any options by containing false constraints. It should pay particular attention to the environment in which it will operate. For example, the design requirements on ovens for commercial and domestic use are different. Commercial units are designed for heavier duty cycles, greater reliability, use by professionals, and must conform to professional standards. Requirements are stated in the language of the designer. This is more precise than, say, the shop or colloquial language in which the problem may have been (and generally is) presented in the first place. The requirements are stated in "solution-neutral" terms, for instance, they should not restrict the product to a certain form unless it is specifically so desired. For developing a list of requirements, it is useful to categorize them, which can be done in several ways. The chief categories are as follows:

- The system interface: The system or product being designed almost always interfaces with some existing system or is to be located within it.
- Importance of the requirements: The requirements on a system vary in weight and priority. Some are more important than others. An important classification of requirements is as "demands" and "wishes."
- Life phases of the product and types of requirements: The requirements can be arranged in matrix form, as shown in Table 1.

7.1 Types of Requirements

Requirements on a product can be broadly classified as engineering (technical), economic, ergonomic, legal, environmental, and other. Before dealing with each of these in detail, it should be pointed out that the requirements in each category apply to the different life phases of the product. The life phases and requirement types can be thought of as forming a matrix, as shown in Table 1 [5]. Thus, for example, there are engineering requirements that pertain to manufacturing and use of the product. The same is true for economic and other categories. Environmental requirements for the manufacturing phase would call for the use of nonpolluting processes, easily separable inserts in molds, and so on. The most important requirements

TABLE 1 Matrix of Life Phases and Types of Requirements

Life phases	Types of requirements					
	Engineering	Economic	Ergonomic	Legal	Environmental	Other
Planning						
Design						
Manufacture						
Distribution						
Product use						

to be considered during conceptual design are the engineering requirements pertaining to product use (performance requirements) that must be met (i.e., demands).

The engineering (technical), economic, and ergonomic requirements have been discussed in detail by Hundal [5]. Legal aspects are given extensive coverage in Chapter 5.

7.1.1 Environmental Requirements

Environmental requirements are developed to minimize use of natural resources, energy requirements, waste generation, health and safety risks, and ecological degradation. These goals need to be translated to requirements.

To be effective, DFE must affect all stages of the materials cycle. The OTA report [4] defines green design as addressing two goals: preventing waste and optimal use of materials. Waste prevention is addressed by reducing the waste, toxicity, and energy use in the materials flow cycle and improving the product life. Use of less material reduces not only the product costs, as shown in [5], but also reduces the wastes, emissions, and energy use at each stage. For example, improvements in lead-acid battery design have led not only to the reduction in lead use by a third but also the elimination of arsenic/antimony in the batteries. Optimal use of materials addresses issues such as remanufacturing and recycling and more efficient disposal possibilities that facilitate energy recovery and composting. Extensive guidelines for developing environmental requirements and strategies for fulfilling these are given in references [7] and [12].

7.1.2 Other Requirements

Aesthetic appeal is an important consideration, especially in case of consumer products. Impacts on the society in various forms should also be taken into consideration. Certain organizational requirements and constraints of various types need to be considered. These include personnel considerations during design, manufacturing, servicing, customer contact, and times required and deadlines to be met in the various activities. Ethical, cultural, and political issues also need to be addressed. Shape, color, finish, material, and texture are properties that appeal differently to different people.

8   DESIGN SOLUTION AND IMPLEMENTATION

For design we begin with the definition and clarification of the problem. Next we look at the function(s) of the product with no consideration given to solutions or hardware at this stage. Once the functions have been identified, we consider the physical effects and processes that can be used to satisfy them. Only then we look for shapes and materials by which the physical phenomena can be utilized. The method therefore should tackle problems from a very fundamental level.

It is during the design phase that most properties of a product are set (e.g., functionality, manufacturability, reliability). This is especially true of costs. Yet the design phase itself constitutes one of the smallest portions of the cost. The major document coming out of the first step (i.e., the task clarification step) is the requirements list. The list includes the overall function of the device and any subfunctions that may be foreseen by the designer.

8.1   Conceptual Design

Conceptual design is the most important phase of design; it has the single largest influence on costs. It has been found that 70–80% of the product costs have been committed after only a small portion of the development resources have been expended in this phase. As an example consider the design of a heat exchanger. When the designer chooses the concept—parallel flow, counter flow, cross flow—he or she has already made the major decision. The concept determines the arrangement of elements through a function structure and thus the flows of energy, material, and signals in the system. Functional analysis is the basis of value analysis that is an accepted technique for product improvement. The physical effects used determine how subfunctions are realized. The solution principles used determine how complex the product will be. Major advances in technology are the results of new concepts rather than improvements in embodiment. Examples are internal combustion engine replacing steam engine, ballpoint pen replacing fountain pen, and fiberoptic cable replacing copper wire.

A function in its simplest form can be expressed as a verb (e.g., "sense," "amplify," "convey," etc.). A signal is the physical embodiment of information and can be considered to be energy at a low level. Figure 8 shows the general function block with its inputs and outputs. A function block is the basic element of a function structure [5].

Figure 9 shows the block diagram of a basic function block, used for the purpose of life cycle inventory analysis. The material and energy flows may be classified further as useful (u), waste (w), recycled.

The next step is to use the various effects to obtain solution possibilities in a sketch form (i.e., without considering the actual shapes which the parts might have). By considering different solutions for the various subfunctions and using systematic combination, a number of different solution concepts can be generated. The concept that best satisfies the specifications is chosen. In the conceptual design phase we look at the most important requirements (i.e., the functional requirements of the product). The functions are listed and complex functions are broken up into simpler subfunctions. These are arranged in the form of a function block diagram or function structure with inputs and outputs for the complete system and for the individual subfunctions. It is possible at this stage to form functional variants by rearranging, combining, and subdividing the functions. Up to this point little thought is given to the solutions, shapes, or the hardware that might be involved. The next step in conceptual design is to look for solutions for each subfunction. By considering different physical processes it is possible to obtain a number of solutions for each

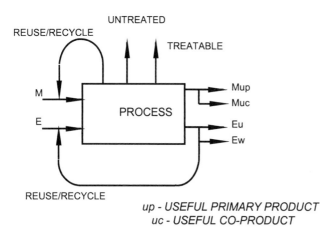

up - *USEFUL PRIMARY PRODUCT*
uc - *USEFUL CO-PRODUCT*

Fɪɢᴜʀᴇ 9   Basic function block with material and energy flows.

function. The subsolutions thus found are then combined in a systematic and rational way to obtain a number of solutions or concept variants for the task.

Implications of environmental factors at this stage are as follows. Since the product exists only at an abstract level, we have the greatest freedom in choosing what it might look like to fulfill the given functions. At the function structure level we should strive for function separation rather than integration. The resulting solution structures lead to modular design in which individual modules can be easily replaced and recycled and modernized with advancing technology. This is contrary to the rules for low-cost design but conforms to the methods of time-driven product development [5]. The chosen concept should use physical processes that call for fewer different materials and foresee jointings that are easy to separate.

8.2 Embodiment and Final Design

The preliminary or embodiment design phase (Figure 1) consists of determining shapes or surfaces, motions, if any, and principal material properties. The final or detail design phase leads to production documents. The final decisions on dimensions, arrangement, shapes of individual components, and materials are made. The design proceeds from a more abstract level at task clarification and takes on a more concrete form as it approaches this phase. In case of the heat exchanger mentioned above, these are the phases where the sizes of the tubes and cavities are optimized and materials and manufacturing methods are chosen.

Since it is at this stage that decisions regarding materials and shapes are made, this is the design stage that has the greatest influence on the materials flow cycle, production processes that are used, and many details. Design of the product in a modular form allows for easy disassembly. The arrangement of the overall layout can make it easier to perform the necessary steps in remanufacturing. The principle of division of tasks (function separation) used in embodiment design [5] should concentrate wear on parts that can be easily reconditioned. Surface treatments should be designed for the total life of the product, including remanufacturing. Otherwise they should be easily renewable during remanufacturing. Design should aim for extending product life, for instance, design for ease of maintenance— extend part life by providing easy lubrication, cooling, design for easy parts replacement, preferably by user. Use of same and similar parts leads not only to lower cost design but also to ease of remanufacturability. Types of information that can be useful to the designer at this stage are, for example, properties of joints.

8.3 Manufacturing

The most important points in manufacturing are the minimization of toxic byproducts and other nontoxic waste. The designer determines the type and amount of scrap by the choice of shapes and materials. Manufacturing technologies such as near-net-shape and high precision forging and casting, sheet metal stamping, and plastics molding are examples of low waste generation.

Optimum layout of stampings aims to minimize scrap. Fewer different materials used enable easier recycling.

9 EVALUATION OF DESIGN ALTERNATIVES

Evaluation of alternative designs may be performed at each stage of the process, for instance, at the requirements stage, concepts stage, or at final design. These alternatives must be evaluated and the best one selected before proceeding to the next stage. The earlier the design stage, the less information available to help with the decision making. Yet a decision made on a rational basis, even with the small amount of information, is better than one made spontaneously.

9.1 Initial Concept Selection

Figure 10 shows a scheme for selecting among concepts at early stages. This initial selection process requires the minimum of numerical data; rather, a judgement on part of the designer leads to rejecting the unsuitable concepts. The criteria used are fairly general and might seem so obvious that one might be tempted to say "those solutions should not have been brought this far." But this runs counter to the discursive process of design. All concepts ought to be considered, evaluated, and then rejected in a rational procedure if they violate these general criteria.

The evaluation criteria to be used in the above matrix would vary from problem to problem and the purpose for which it is used. For the initial selection among concepts, Pahl and Beitz [13] suggest the following:

A: Compatibility with the rest of the system or process.
B: Fulfills the "demands" in the requirements list.
C: Is basically realizable.
D: Can be accomplished within available resources.
E: Its safety is assured.
F: Is preferred in own company or field.

A similar selection matrix can be used for evaluating subsolutions in a concept.

ANALYSIS & EVALUATION	DECISION
Evaluation according to criteria: + yes - no ? need more information ! check requirements list ◄——— CRITERIA ———►	Mark decisions: + proceed further - solution unacceptable ? supply information ! check requirements list

A	B	C	D	E	F	CONCEPT/Remarks	
+	+	+	+	+	+	Concept # 1	+
+	+	+	+	+	+	Concept # 2	+
+	+	+	+	-	+	Concept # 3	-
+	+	+	?	+	-	Concept # 4	?
						....	

FIGURE 10   Initial concept selection matrix.

9.2   Other Methods

The method of weighted values, also called the decision matrix technique [14] can be used for a more detailed evaluation. Table 2 shows the decision matrix scheme for evaluating concept variants. Each criterion is assigned a weight and given a value for each concept variant. The product of the weight and the respective value produces a weighted value. The sums of the values or, more appropriately, the weighted values provide means for evaluating the concept variants.

The decision matrix can be prepared by using one of the criteria (e.g., technical, economic, etc.), or a combination of all, as a multidimensional matrix. Since not all criteria are equally important, they are assigned weights. For each concept variant, each criterion is assigned a value (e.g., on a scale of 1 to 10). The weighted value is then found by multiplying the value and the corresponding weight. The sums of the values and the weighted values are then found for each concept variant. The sums of the weighted values are generally normalized to 1.0 or 10.

While the actual criteria would vary from case to case, there are certain generic terms pertinent to most problems. A typical list is as follows:

Technical criteria: (a) sturdiness, (b) simple operation, (c) low wear, (d) safety, (e) robustness.
Economic criteria: (a) material costs, (b) number of parts, (c) assembly costs, (d) operating costs, (e) testing.

TABLE 2 Decision Matrix

Criteria	Weights	Concept variants						
		CV 1		CV 2		CV 3		...
		Value	W-V	Value	W-V	Value	W-V	
Criterion 1								
Criterion 2								
Criterion 3								
Criterion 4								
....								
Sum Weighted sum								

CV, concept variant; W-V, weighted value.

Environmental criteria:
  Materials used: renewable? recycled? where available? impact on other environmental criteria.
  Energy utilization: environmental impacts; where available? impact on other environmental criteria.
  Waste streams: type; hazardous? destination? reusability; treatment options.
  Health and safety: toxicity; effects on workers, users, public.
  Ecological impacts: type of impact; distribution of impact.

9.3 Environmental Evaluation Tools

For the evaluation of design alternatives from the viewpoint of environmental impact, a number of tools are available [7]. These range from life cycle assessment (comprehensive) to single metrics:

- Life cycle assessment tools, such as EPA/SETAC, DFEIS, EPS;
- Resource productivity index, waste per unit product;
- Specific metrics, such as energy consumed per unit activity, percent recycled content, percent recyclable content, etc.;
- Cost methods, such as life cycle costs, environmental accounting.

A framework for life cycle assessment consists of inventory analysis, impact assessment, and improvement assessment. Vigon et al. [15] show the details for performing life cycle inventory.

10 EXAMPLE: TELEPHONE TERMINAL

Keoleian et al. [7] described the development process of a new telephone terminal by AT&T, beginning in 1990. The 1990 environmental goals of AT&T are summarized as follows:

- CFC phase-out: 50% by 1991, 100% by 1994.
- Toxic air emissions decrease: 95% by 1995, 100% by 2000.
- Manufacturing process waste decrease: 25% by 1994.
- Reduce paper use and recycling.

Environmental requirements for the new phone terminal address the manufacturing, use/service, and end-of-life management. The aims included the following:

1. Manufacture
   a. Use recyclable materials
   b. Recycle molding scrap
   c. Maximize use of recycled materials
   d. Use nonozone depleting materials
   e. Eliminate toxic materials (e.g., Pb)
   f. Minimize defective products
2. Use/service
   a. Extend useful life through modular design with enough forward and backward capability
   b. Operate on-line power only
3. End-of-life management
   a. Reuse parts; standardize parts
   b. Components recyclable after use
   c. Open-loop recycling into fiber cables, spools, and reels
   d. Easy disassembly (no rivets, glues, ultrasonic welding, minimal use of composites)
   e. Easy sorting (shape, marking, etc.)

A comparison of features of the new telephone terminal to those in an older model shows a number of improvements:

- The high-gloss housing was replaced by textured housing, thus reducing molding waste.
- In the older model, feet were glued to the stand. This was changed to snapped-on feet. Thus rubber could be separated during disassembly.
- UL symbol on paper was substituted by symbol molded into the housing. Contamination was minimized.

- Foam piece glued to housing was replaced by foam piece press fit over speaker, also minimizing contamination of the housing during disassembly.
- The older model had a polycarbonate sheet glued to housing. In the new model no light diffuser was used, thereby also minimizing contamination.
- The housing material was not identified in the older model. In the new model, ISO plastic marking code was molded in. Thus the plastic will be identifiable by any recycling center.

Figure 11 shows material flow in the AT&T telephone business. Among others, it shows the advantage of leasing equipment for improved recycling.

11 SUMMARY

Environmental protection is a complex and controversial subject. Thus DFE inherits the conflicts inherent therein. Decisions on environmental issues require value judgements and involve public policy. Trade-offs that

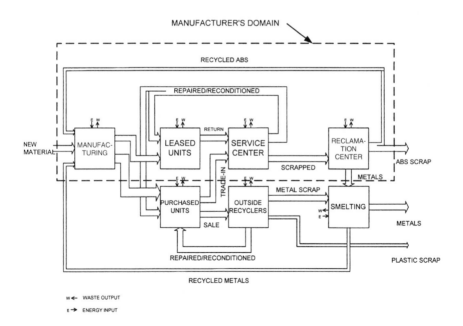

FIGURE 11 Material flow in telephone set manufacture, distribution, and disposal. (From Ref. 7.)

depend on economic values are influenced by a country's tax policies. The goal of engineering design has been and is still to design and produce products and systems to fulfill the needs of society. DFE enlarges the scope of the problem to include the natural environment.

Engineers and designers need information to operate successfully. There is need on the part of the governments to determine priorities (e.g., recycling vs. energy conservation) and relative risks of different materials to health and ecosystems. For an automobile's lifetime, 10% of the energy is used in manufacturing, another 10% in disposal, and the remaining 80% during its use. Thus the greatest potential in energy saving lies in making cars more efficient. A list of products and processes posing high risks is needed. There is a continuing need to find safer substitutes for hazardous substances.

DFE by itself does not make products cheaper nor more expensive. On the other hand, the overall costs, when one considers the environmental impacts, can be lowered. Environmental attributes must be considered at all stages of the design, as are other attributes (e.g., cost, manufacturability, quality, etc.). The application of DFE requires knowledge more than any other resource. Products designed for recyclability are more easily serviceable. Remanufactured products can be of as high quality as new products— and cheaper, as demonstrated at Xerox [4]. Products containing high-technology components likely to show consistent and frequent improvements in performance should be designed so that they may be easily disassembled. Such products cannot be designed for long life—they tend to quickly become obsolete. Thus they should be designed for easy renovation.

A number of problems in this field still require further study and research. Two of these areas are material flow and management and better design methods. Reliable data are needed on material balances and flow of wastes. All parties in the materials flow cycle (e.g., material processors, designers, manufacturers and waste handlers) require such information. Technologies for recycling and incineration need improvement. Typically, materials that are easy to recycle are the least toxic and cheapest to begin with. More information is needed on the compatibility of materials for recycling, including the suitability of adhesives, paints, and coatings.

Improvements in methods and knowledge are needed in the areas of design for recyclability, which includes design for disassembly. New fastening and joining techniques must be developed. The aims of the existing and well-advanced fastening and joining technology are, in general, to produce joints that stay intact. This technology must now be extended to develop joints that are strong yet can be easily separated. Information, such as that in the VDI Guideline 2243 [16], needs to be more widely disseminated and further enhanced. Since, in general, products will have longer life, material response to long-term loading needs to be investigated. Products that experience rapid technological change would need to be designed for easy disassembly, since they will have a short useful life.

Strategies for reducing environmental impact vary with the product system. Among proven approaches are the following (for example [5] and [7]):

- *Product life* can be extended by reducing wear and simplifying maintenance and service, allowing for low cost and easy repair.
- *Materials* should be selected for their recyclability and easy availability, thus conserving rare materials.
- *Processes* should aim for energy and material efficiency.
- *Distribution* should aim for minimizing packaging and reduced energy usage.

Recycling is not always the best strategy. An example is the snack bag, which consists of nine layers of five materials with a total thickness of 0.002 inch [4]. It is not possible to recycle it to its original use nor to separate it into its component materials. Yet such a snack bag conserves materials, energy, and lengthens shelf life of food, thus minimizing waste.

REFERENCES

1. R. E. Cattanch, J. M. Holdreith, D. P. Reinke, L. K. Sibik. Handbook of Environmentally Conscious Manufacturing. Chicago: Irwin, 1995.
2. P. G. Smith and D. G. Reinertsen. Developing Products in Half the Time. 2nd ed. New York: Van Nostrand Reinhold, 1998.
3. M. E. Porter and C van der Linde. Green and competitive: ending the stalemate. Harvard Bus Rev 73:120–134, 1995.
4. U.S. Congress, Office of Technology Assessment. Green Products by Design: Choices for a Cleaner Environment. Washington, DC: U.S. Government Printing Office, OTA-E-541, 1992.
5. M. S. Hundal. Systematic Mechanical Designing: A Cost And Management Perspective. New York: ASME Press, 1997.
6. M. E. Henstock. Design for Recyclability. London: Institute of Metals. 1988
7. G. A. Keoleian, J. E. Koch and D. Menerey. Life Cycle Design Framework and Demonstration Projects. Washington, DC: U.S. EPA. EPA/600/R-95/107, 1995.
8. T. E. Graedel. Streamlined Life-Cycle Assessment. Upper Saddle River, NJ: Prentice Hall, 1998.
9. J. P. Holdren. Energy in transition. Sci Am 263:157–163, 1990.
10. World Energy Council. Energy for Tomorrow's World. New York: St. Martin's Press, 1993.
11. Organization for Economic Cooperation and Development. Environmental Indicators: A Preliminary Set. Paris: OECD, 1991.

12. G. A. Keoleian and D. Menerey. Life Cycle Design Guidance Manual. Washington, DC: U.S. EPA. EPA/600/R-92/226, 1993.
13. G. Pahl and W. Beitz. Engineering Design—A Systematic Approach. Berlin, New York: Springer-Verlag, 1988.
14. R. C. Johnson. Mechanical Design Synthesis. Huntington, NY: Krieger Publishing Co., 1978.
15. B. W. Vigon, D. A. Tolle, B. W. Cornaby, H. C. Latham, C. L. Harrison, T. L. Bogoski, R. G. Hunt and J. D. Sellers. Life Cycle Assessment: Inventory Guidelines and Principles. Washington, DC: U.S. EPA, EPA/600/R-92/245. 1993.
16. VDI Guideline 2243: Konstruieren recyclinggerechter technischer Produkte. [Design for Recycling of Technical Products.] Duesseldorf: VDI Verlag, 1991.
17. University of Michigan's Center for Sustainable Systems (formerly National Pollution Prevention Center, NPPC.) http://www.umich.edu/~nppcpub/index.html

2

Life Cycle Assessment
A Synopsis

Noellette Conway-Schempf* and Chris Hendrickson
Carnegie Mellon University, Pittsburgh, Pennsylvania

1 INTRODUCTION

Life cycle assessment (LCA) is a powerful analytical approach developed to guide environmentally conscious product and process design and to help in the evaluation of the environmental implications of products, processes, and services. An LCA evaluates the relevant environmental, economic, and technical burdens associated with a material, product, process, or service throughout its life—from creation (material extraction) through manufacturing, use, and finally end of life (disposal or recycling, etc.), as shown in Figure 1. It is expected that the LCA approaches can guide product and process designers in assessing environmental issues associated with their products; guide decision makers in choosing more environmentally conscious products materials, components, products, and services; and guide consumers in making more environmentally conscious purchasing decisions. Some of the most powerful applications of LCA are internal use by companies and regulatory agencies to assess systemwide impacts of decisions and choices.

Current affiliation: Automatika, Inc., Pittsburgh, Pennsylvania.

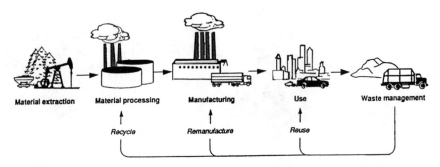

Material extraction   Material processing   Manufacturing   Use   Waste management

Recycle   Remanufacture   Reuse

FIGURE 1 Stages of a products life cycle. (From U.S. Congress, Office of Technology Assessment. Green Products by Design: Choices for a Cleaner Environment. OTA-E-541.)

2 DEFINITION

According to the U.S. Environmental Protection Agency (USEPA 1993a), LCA is "[a] concept and methodology to evaluate the environmental effects of a product or activity holistically, by analyzing the whole life cycle of particular product, process, or activity. The life cycle assessment consists of three complementary components—inventory, impact, and improvement, and an integrative procedure known as scoping." A commonly accepted international definition developed by the Society for Environmental Toxicology and Chemistry (SETAC) is as follows:

> Life cycle assessment is an objective process to evaluate the environmental burdens associated with product, process, or activity by identifying and quantifying energy materials used and wastes released to the environment, to assess the impact of those energy and material uses and releases to the environment, and to evaluate an implement opportunities to affect environmental improvements. The assessment includes the entire life cycle of the product, process, or activity, encompassing extraction and processing raw materials; manufacturing, transportation and distribution; use, reuse, maintenance; recycling and final disposal (SETAC 1993).

3 HISTORICAL PERSPECTIVE

Coca-Cola was one of the first corporations to attempt LCA when the company analyzed the environmental impacts of beverage containers in the 1960s. The energy analyses that were popular in the 1960s and early 1970s were preliminary LCA attempts to understand the environmental impacts of products and processes. The early 1990s brought about new corporate and government philosophies of waste minimization and pollution prevention and a renewed interest and demand for LCA approaches. In 1991, SETAC published the first guidelines for conducting an LCA. More recently, LCA criteria are beginning to find their way into environmental labeling schemes such as Germany's Blue Angel and the ISO 14000 environmental management standards (www.iso.ch). Simplified LCAs even pop up in popular magazines; for example, *Consumer Reports* occasionally comments on the environmental impacts of different product packaging types and chemicals.

An LCA can show the major environmental problems of a material, product, or process. The act of doing the assessment builds awareness about environmental impacts and focuses improvement efforts. This has led companies, such as AT&T and Volvo, to develop internal LCA tools for their product lines (Graedel and Allenby 1995) and government agencies, such as the EPA, to provide generic guidelines for conducting LCAs (USEPA 1993a). In fact, many proponents and users of LCA information suggest that the main role of LCAs should be to guide internal decision making rather than as a consumer marketing or information tool.

4   STAGES OF AN LCA

At its most basic, an LCA consists of three broad steps, which are described in detail in USEPA (1993a):

1. *Inventory analysis.* This step involves technical data gathering to support estimates of energy and raw material requirements, atmospheric emissions, waterborne emissions, solid wastes, and other releases for the product/process/activity life cycle. This aspect of LCA is well developed and the U.S. EPA even provides sample checklists data sheets for LCA inventory development (USEPA 1993a). Their sample checklist includes descriptive elements such as the purpose of the inventory, assumptions used, the boundary, the basis for comparative claims (if any), computational models used, quality assurance, peer review attempts, and details about the data presentation. The inventory checklist also includes quantitative elements, such as listings of materials used, energy use and water usage, product coproducts, air emissions, water effluents, solid waste, transportation, and personnel data (Figure 2).

2. *Impact analysis.* This stage involves technical, quantitative, and qualitative efforts to determine the effects or impacts of the environmental issues (resource use, emissions, etc.) determined in the inventory assessment. This is not a straightforward exercise. The impacts of many chemicals are unknown. Natural and biological systems are complicated, and it is difficult to predict the effects of combined groups of chemicals in different forms

---

LIFE-CYCLE INVENTORY CHECKLIST PART I—SCOPE AND PROCEDURES INVENTORY OF: _____

Purpose of Inventory: (check all that apply)

Private Sector Use
Internal Evaluation and Decision Making
☐ Comparison of Materials, Products, or Activities
☐ Resource Use and Release Comparison with Other Manufacturer's Data
☐ Personnel Training for Product and Process Design
☐ Baseline Information for Full LCA
External Evaluation and Decision Making
☐ Provide Information on Resource Use and Releases
☐ Substantiate Statements of Reductions in Resource Use and Releases

Public Sector Use
Evaluation and Policy-making
☐ Support Information for Policy and Regulatory Evaluation
☐ Information Gap Identification
☐ Help Evaluate Statements of Reductions in Resource Use and Releases
Public Education
☐ Develop Support Materials for Public Education
☐ Assist in Curriculum Design

Systems Analyzed
List the product/process systems analyzed in this inventory: _____

Key Assumptions: (list and describe)
_____
_____
_____

Define the Boundaries
For each system analyzed, define the boundaries by life-cycle stage, geographic scope, primary processes, and ancillary inputs included in the system boundaries.

Postconsumer Solid Waste Management Options: Mark and describe the options analyzed for each system.
☐ Landfill _____   ☐ Open-loop Recycling _____
☐ Combustion _____   ☐ Closed-loop Recycling _____
☐ Composting _____   ☐ Other _____

Basis for Comparison
☐ This is not a comparative study.   ☐ This is a comparative study.
State basis for comparison between systems: *(Example: 1000 units, 1,000 uses)* _____
If products or processes are not normally used on a one-to-one basis, state how equivalent function was established.
_____

Computational Model Construction
☐ System calculations are made using computer spreadsheets that relate each system component to the total system.
☐ System calculations are made using another technique. Describe: _____
Describe how inputs to and outputs from postconsumer solid waste management are handled. _____
_____

Quality Assurance: (state specific activities and initials of reviewer)
Review performed on: ☐ Data Gathering Techniques _____   ☐ Input Data _____
                    ☐ Coproduct Allocation _____   ☐ Model Calculations and Formulas _____
                                         ☐ Results and Reporting _____

Peer Review: (state specific activities and initials of reviewer)
Review performed on: ☐ Scope and Boundary _____   ☐ Input Data _____
                    ☐ Data Gathering Techniques _____   ☐ Model Calculations and Formulas _____
                    ☐ Coproduct Allocation _____   ☐ Results and Reporting _____

Results Presentation
☐ Methodology is fully described.
☐ Individual pollutants are reported.
☐ Emissions are reported as aggregated totals only.
    Explain why: _____
    _____
☐ Report is sufficiently detailed for its defined purpose.

☐ Report may need more detail for additional use beyond defined purpose.
☐ Sensitivity analyses are included in the report.
    List: _____
☐ Sensitivity analyses have been performed but are not included in the report. List: _____

FIGURE 2 Checklist of criteria with worksheet for life cycle inventory. (From USEPA 1993a).

entering the body or the ecosystem via a variety of pathways. The approach generally recommended (USEPA 1993a) is to determine links between the product/process/activity life cycle and potential impacts rather than actually quantifying a specific impact.

	Data Value[a]	Type[b]	Data[c] Age/Scope	Quality Measures[d]
LIFE-CYCLE INVENTORY CHECKLIST PART II—MODULE WORKSHEET				
Inventory of: _____ Preparer: _____				
Life-Cycle Stage Description: _____				
Date: _____ Quality Assurance Approval: _____				
MODULE DESCRIPTION: _____				

	Data Value[a]	Type[b]	Data[c] Age/Scope	Quality Measures[d]
MODULE INPUTS				
Materials				
Process				
Other[e]				
Energy				
Process				
Precombustion				
Water Usage				
Process				
Fuel-related				
MODULE OUTPUTS				
Product				
Coproducts[f]				
Air Emissions				
Process				
Fuel-related				
Water Effluents				
Process				
Fuel-related				
Solid Waste				
Process				
Fuel-related				
Capital Repl.				
Transportation				
Personnel				

(a) Include units.

(b) Indicate whether data are actual measurements, engineering estimates, or theoretical or published values and whether the numbers are from a specific manufacturer or facility, or whether they represent industry-average values. List a specific source if pertinent, e.g., "obtained from Atlanta facility wastewater permit monitoring data."

(c) Indicate whether emissions are all available, regulated only, or selected. Designate data as to geographic specificity, e.g., North America, and indicate the period covered, e.g., average of monthly for 1991.

(d) List measures of data quality available for the data item, e.g., accuracy, precision, representativeness, consistency-checked, other, or none.

(e) Include nontraditional inputs, e.g., land use, when appropriate and necessary.

(f) If coproduct allocation method was applied, indicate basis in quality measures column, e.g., weight.

FIGURE 2   Continued

3. *Improvement analysis.* The improvement analysis step aims to systematically evaluate the opportunities available to reduce environmental impacts and issues revealed in steps 1 and 2. This analysis step will be both quantitative and qualitative. An example would be efforts to reduce the emissions of toxic chemicals over existing levels quantified by the LCA.

4.1 Boundary of the LCA

A major component of the LCA, and one that is often dismissed or discussed only briefly, is determining the assessment boundary. Each of the three steps outlined above is limited by the boundary of the assessment—if the boundary is narrow (e.g., activities under the control of a particular corporation), many issues and impacts may be erroneously omitted from the assessment. In particular, the environmental impacts of the supply chain may be ignored. A thorough inventory and impact analysis, which is within too narrow a boundary, is not helpful. If the boundary is broad, data gathering may be too expensive or too time consuming—it may even be impossible! Most contradictory LCA results are due to the setting of different assessment boundaries by the analysts. Setting the boundary and determining the goals of the assessment are sometimes addressed by an informal additional step referred to as "scoping"; however, there are few specific guidelines for the scoping step of the assessment.

A number of efforts involving industry, government agencies, and universities are focusing on standardizing the analyses so that LCAs carried out by different organizations are comparable. LCAs have been criticized for a number of reasons:

- Environmental effects for all life cycle stages are not known.
- It is difficult to compare different types of effects—species extinction versus cancer incidence, for example.
- The amount of data required to analyze even simple products is enormous; the typical cost for a detailed LCA may exceed $200,000.
- Data gathering is difficult as many of the life cycle stages involve proprietary processes.
- It is difficult to know where to draw the boundary around the analysis—Should raw materials extraction be included? What about the life cycle impacts of the equipment used in the raw materials extraction? What about the environmental impacts of the detergent used to wash the clothes of the people mining the raw materials?

A number of controversies arising from LCAs, such as the debate over paper versus plastic bags at the grocery store or the use of cloth versus disposable diapers, illustrates some of these problems. The contrasting results of studies by different groups are generally the result of different assessment boundaries, demonstrating the importance of standard boundary-setting assumptions.

5 LCA METHODS

There are two main approaches to conducting LCAs—process based (SETAC) and input-output based.

5.1 Process-Based SETAC Approach

Most currently used LCA techniques are modifications of the approach developed by SETAC. The EPA guidelines on conducting LCAs (e.g., USEPA 1993a) are based on the SETAC approach. Guidelines are available for scoping, inventory development, assessment, and improvement. Example sheets provided by the EPA are shown in Figure 2. Practical use of the SETAC approach often involves drawing a boundary that may limit consideration to a few producers in the chain from raw materials to consumers. This is a problem if important stages or suppliers in the life cycle are excluded from the assessment. However, when the environmentally sensitive stages of the life cycle are well understood, the SETAC approach can allow for detailed process level examination of specific products, processes, or materials.

As an example of a SETAC-based LCA, Hocking (1991) carried out a study of the environmental implications of alternative hot drink cups—paper versus plastic (molded polystyrene bead foam). Hocking analyzed the materials and energy required for the fabrication of both cups and considered issues such as recycling, incineration, and biodegradation. His assessment followed process flow paths from material extraction to product disposal but did not consider the environmental impacts associated with suppliers to the main flow elements.

Hocking found that overall paper cups required about 2.5 times the material requirements of the polystyrene cups. Processing the paper cups required six times as much steam, 13 times as much electric power, and twice as much cooling water compared with the polystyrene cup (Table 1).

5.2 Input-Output Based

Researchers at the Green Design Initiative at Carnegie Mellon have found that limiting the boundary of the LCA to the major suppliers and manufacturing processes may lead to consideration of only a fraction of the total environmental discharges associated with the product or process. These researchers have developed an approach based on models of industrial activity and pollution discharge data (Cobas et al. 1995; Hendrickson et al. 1998; Lave et al. 1995, 1996). The resulting software tool, called economic input-output life cycle assessment (EIO-LCA), allows for economy-wide aggregate LCA (see www.eiolca.net for software).

TABLE 1 Paper Versus Polystyrene Cups—Raw Materials Requirements, Abbreviated Utility Requirements, and Emissions

	Paper cup (per cup)	Polystyrene cup (per cup)
Raw materials		
Total wood	~ 21 g	—
Total petroleum fractions	~ 1.8 g	~ 4.3 g
Other chemicals	~ 1.2 g	~ 0.08 g
	Paper cup (per 10,000 cups)	Polystyrene cup (per 10,000 cups)
Utilities		
Steam	~ 840 kg	~ 130 kg
Electricity	~ 78 kW	~ 6 kW
Cooling water	~ 4 m^3	~ 3 m^3
Water effluent		
Volume	~ 10 m^3	~ 0.05 m^3
Air emissions		
Chlorine	~ 0.02 kg	—
Chlorine dioxide	~ 0.02 kg	—
Reduced sulfides	~ 0.1 kg	—
Particulates	~ 0.2 kg	~ 0.008 kg
Pentane	—	~ 0.08 kg
Ethylbenzene/styrene	—	~ 0.05 kg
Carbon monoxide	~ 0.3 kg	~ 0.002 kg
Nitrogen oxides	~ 0.5 kg	~ 0.008 kg
Sulfur dioxide	~ 1 kg	~ 0.07 kg

Source: Hacking 1991.

5.2.1 EIO-LCA

Economic input-output analysis is a well-established modeling framework for tracing the flows of inputs and outputs in an economy. Input-output analysis is generally used for economic planning purposes, for example, calculating the resources needed to support an increase in the production of automobiles. The resulting estimates show the increases in production, both for automobiles and for the various sectors, which supply products directly or indirectly. For example, an expansion in automobile production would require steel, electricity, petroleum, plastics, and even additional automobiles. Most developed nations produce economic input-output tables that describe the interactions within their economy, although they do not all contain the same level of detail.

The green design researchers have used the 500-sector U.S. Department of Commerce input-output tables for the United States to develop an economy-wide LCA technique by linking the economic input-output tables with environmental databases (conventional pollutants, TRI, energy use, ore use, fertilizer use, global warming potential. and ozone depleting potential. etc.). The method is called EIO-LCA.

The EIO-LCA approach has been used to examine the environmental impacts of various industrial sectors (Cobas et al. 1995; Lave et al. 1995, 1998), automobile use (MacLean and Lave 1998), automobile components (Joshi 1997), and construction materials (Horvath and Hendrickson 1998a, b; Hendrickson et al. 1998).

The EIO-LCA tool has advantages over SETAC LCA approaches:

- EIO-LCAs can be used to examine the total direct and indirect economy-wide effects (effects of suppliers) on emissions and energy consumption resulting from changes in production.
- The EIO-LCA model uses the entire U.S. economy as the boundary for the analysis.
- The approach highlights priority areas for reduction in environmental impacts.
- Initial EIO-LCAs can be carried out at a fraction of the cost and time associated with SETAC LCAs.
- The EIO-LCA can be used in conjunction with current SETAC approaches, where the SETAC approach is used to analyze specific products and processes in detail and the EIO-LCA explores the indirect economy-wide effects of changes in product mix.
- The LCA is calculated by a transparent computer-based tool that is easily implemented as part of green design efforts.
- The LCA has underlying databases accessible by all environmental stakeholders.
- The method provides some environmental information for every commodity in the U.S. economy.
- The method allows for sensitivity analyses and scenario planning.
- The method is complete and the software is available either through the Green Design Initiative at Carnegie Mellon or via the World Wide Web at www.eiolca.net.
- Results are reproducible by environmental stakeholders.

The EIO-LCA approach also has important limitations. Even with 500 economic sectors, the amount of disaggregation may be insufficient for the desired level of analysis, for example, an assessment of one type of plastic versus another. EIO-LCA models include sectors of the economy rather than specific processes. Detailed analysis of the environmental impacts of the activities of the individual members of the supply chain will require the more traditional SETAC LCA technique described earlier. The use and disposal phases of certain products may be too difficult to analyze with EIO-LCA. In addition, the data in the model may reflect past practices and imports are treated as if they are U.S. products.

Example EIO-LCA: Asphalt Versus Steel Reinforced Concrete for Pavement Construction

Construction is a significant portion of U.S. economic activity. Construction and infrastructure development account for a disproportionate share of resource use. Understanding the resource use and emissions associated with construction related activities is an important component of developing sustainable industrial systems. Hendrickson et al. (1998) have used the EIO-LCA method to examine the resource inputs and emissions for aggregate U.S. construction sectors and found that construction accounts for a disproportionate share of several mineral and pollution emission categories.

Horvath and Hendrickson (1998b) used the EIO-LCA method to do specific comparisons of the environmental implications of asphalt and steel-reinforced concrete pavements. Asphalt and concrete (often reinforced with steel) are the most common materials used in pavement construction. They compared the resource requirements and emissions associated with constructing a new 1-km-long pavement section composed of either asphalt or concrete. Specific details about the roadway designs are provided in the source article (Horvath and Hendrickson 1998b). The resource inputs evaluated included consumption of electricity, fuels, ores, and fertilizers (Table 2). In addition, the EIO-LCA method allowed for the identification of inputs not directly related to pavement material manufacturing such as fertilizer demand. In this case, fertilizer consumption occurs with some of the agricultural suppliers of some of the direct suppliers to the asphalt and concrete sectors. Environmental effects include toxic chemical discharges to

TABLE 2 Summary of Resource Inputs for 1 km of Pavement (Asphalt vs. Steel-Reinforced Concrete)

Resource inputs	Asphalt	Concrete
Electricity (kWh M)	0.1	0.1
Coal (metric tons)	30	100
Other fuels (metric tons)	100	40

Electricity data from 1991; other data from 1987. The detailed listing of ores and fuels is found in the source publication.
Source: Horvath and Hendrickson 1998b.

air, water, land, and underground and transfers to off-site treatment plants, ozone depletion potential of chemical releases, hazardous waste generation, and conventional pollutant releases to air (Table 3). The environmental effects of both direct suppliers (such as the cement industry) and indirect suppliers (such as the agricultural sector) are included in the analysis.

The authors found that for the initial construction of equivalent pavement designs, asphalt appears to have higher energy input, lower ore and fertilizer inputs, lower toxic emissions, but higher associated hazardous waste generation than steel-reinforced concrete. Considering the uncertainties in the data, the environmental effects are roughly similar for the two materials. However, asphalt pavements have generally been recycled in larger quantities than concrete pavements, with consequent resource savings, suggesting that asphalt may be preferable from a sustainable development viewpoint, although other considerations may dictate the particular choice of material. However, steel-reinforced concrete pavement may last longer, further complicating the comparison.

6 ISO 14000 STANDARDS ON LCA

In September 1996, ISO, the International Organization for Standardization based in Switzerland, initiated the ISO 14000 series of environmental management system standards. These are a series of standards that deal with the components of an effective environmental management system along with guidelines for auditing, ecolabeling, environmental performance evaluation, and LCA (see Table 4 for a complete listing).

TABLE 3 Summary of Environmental Outputs for 1 km of Pavement (Asphalt vs. Steel-Reinforced Concrete)

Environmental outputs	Asphalt	Concrete
TRI total releases (kg)	30	80
TRI total releases and transfers (kg)	60	200
Ozone-depleting potential (kg CFC-11 equivalents)	0.05	0.06
RCRA generated hazardous waste (tons)	30	8
SO_2 (kg)	1000	4000
NOx (kg)	600	1000
VOCs (kg)	100	200

Data from 1993. A more detailed listing is given in the source publication.
Source: Horvath and Hendrickson 1998b.

TABLE 4   ISO 14000 Series of Environmental Standards (as of September 1999)

	Standards
ISO 14001	Environmental Management Systems—Specifications with Guidance for Use
ISO 14004	Environmental Management Systems—General Guidelines, Principles, Systems, and Supporting Techniques
ISO 14010	Guidelines for Environmental Auditing—General Principles on Environmental Auditing
ISO 14011	Guidelines for Environmental Auditing—Audit Procedures—Auditing of Environmental Management Systems
ISO 14012	Guidelines for Environmental Auditing—Qualification Criteria for Environmental Auditors
ISO 14040	Life Cycle Assessment—Principles and Guidelines.
ISO 14050	Environmental Management Vocabulary

Standards in draft or
committee discussion

ISO 14015	Environmental Aspects of Sites and Entities
ISO 14020	Environmental Labels and Declarations—General Principles
ISO 14021	Environmental Labels and Declarations—Environmental Labeling—Declared Environmental Claims—Terms and Definitions
ISO 14024	Environmental Labels and Declarations—Environmental Labeling Type I—Guiding Principles and Procedures
ISO 14025	Environmental Labels and Declarations—Environmental Labeling Type III—Guiding Principles and Procedures
ISO 14031	Environmental Performance Evaluation
ISO 14032	Environmental Performance Evaluation—Case Studies in the Use of ISO 14031
ISO 14041	Life Cycle Assessment—Goal and Scope Definition and Inventory Analysis
ISO 14042	Life Cycle Assessment—Impact Assessment
ISO 14043	Life Cycle Assessment—Interpretation

There are also additional ISO technical reports, which provide examples of applications and more specialized information for specific industries, such as forestry companies.

The "mother" standard is ISO 14001, which requires implementation of an environmental management system in accordance with defined internationally recognized standards (as set forth in the ISO 14001 specification—see www.iso.ch). The ISO 14001 standard specifies requirements for establishing an environmental policy, determining environmental aspects and impacts of products/activities/services, planning environmental objectives and measurable targets, implementation and operation of programs to meet objectives and targets, checking and corrective action, and management review. The other ISO 14000 series of standards support the environmental management system and can also be used as stand alone standards.

For our purposes here, the standards relating to LCA are of most interest (Table 5). These standards attempt to provide consistency among

TABLE 5   LCA Guideline Standards

Standard	Description
ISO 14040: Life Cycle Assessment—Principles and Framework. Published in 1997.	As it states, this standard describes the general principles, framework, and methodology for conducting LCAs of products, processes, and services.
ISO 14041: Life Cycle Assessment—Goal and Scope Definition and Inventory Analysis. Due to be published in 1999.	This international standard describes in addition to ISO 14040 the requirements and the procedures necessary for the compilation and preparation of the definition of goal and scope for an LCA by performing, interpreting, and reporting an LCI.
ISO 14042: Life Cycle Assessment—Life Cycle Impact Assessment. Due to be published in 1999.	This document offers guidelines for carrying out the assessment.
ISO 14043: Life Cycle Assessment—Life Cycle Interpretation. Due to be published in 1999.	This document discusses how to interpret the results.
TR 14048 and TR 14089	These technical reports (not standards) discuss ways of formatting data to support the development of LCAs (TR 14048) and provide examples that illustrate how to apply the 14041 standard.

Source: www.iso.ch

LCA efforts and ensure that all LCA practitioners are using similar tools and techniques.

7 ECOLABELS AND LCA

The U.S. EPA has also published guidelines on the use of LCA in independent-third-party environmental labeling (ecolabeling) programs (USEPA 1993b). Third-party environmental labeling programs highlight the environmental issues associated with product choices. The aim is to provide the consumer with environmental information regarding their product choices so that they will choose products with reduced environmental impacts. Ecolabels are generally "seal of approval" labels that identify products or services as being less harmful to the environment than similar products with the same functionality (e.g., the Green Seal program in the United States), "single-attribute certification programs" that indicate that a third party has validated a particular environmental claim (e.g., the Energy Star program in the United States), or "report card labels" that offer a listing of environmental information associated with the product (e.g., the Scientific Certification Systems Environmental Report Card in the United States).

The EPA (USEPA 1993b) carried out a study of the use of LCA in environmental labeling programs and found that, as of 1993, although most programs recognized that LCA was an important concept in developing labeling programs, few were using formal LCA methods. More recently, the advent of the ISO 14000 standards on environmental management (discussed earlier) which include LCA and ecolabeling suggest that LCA will be a more formal component of many ecolabel programs.

8 MOVING TO ASSESSMENT AND IMPROVEMENT

The assessment and improvement stages of LCA are not yet well developed. Assessment is probably the most difficult stage to accomplish. It is difficult to estimate all the possible environmental impacts of an activity product or process (health effects, ecological effects, etc.) as impacts are time and location dependent and may be synergistic or antagonistic. The health and ecological impacts of most chemicals and pollutants are not known; those that have been studied are usually analyzed in narrow contexts. However, carrying out rough approximations is probably sufficient for most applications; for example, assuming that large levels of toxic releases or other emissions are more environmentally damaging than lower levels is probably a safe assumption. Assuming that higher levels of energy use or raw materials requirements is also a good approximation. The focus can then shift to improving the situation—in this example, reducing the levels emissions or energy use. The important thing to remember is to keep the systems perspective and not to focus on one pollutant or environmental issue.

9 CONCLUSIONS

LCA is useful in guiding a wide range of decisions—from design choices to policy frameworks. LCA allows the user to take a "bird's eye" look at the situation rather than focusing narrowly on a narrow range of issues. The main contribution of LCA to engineering, design, and policy analysis is the systems perspective it provides. LCA can highlight priority areas for improvement and, if used correctly, can provide an objective means of comparing the environmental impacts of products, processes, services, and policies.

REFERENCES

E. Cobas, C. Hendrickson, L. B. Lave, and F. C. McMichael. Economic Input-Output Analysis to Aid Life Cycle Assessment of Electronics Products. Proceedings of 1995 International Symposium on Electronics and the Environment. Piscataway, NJ: Institute of Electrical and Electronics Engineers, 1995, pp. 273–278.

T. E. Graedel and B. R. Allenby. Industrial Ecology. New Jersey: Prentice Hall, 1995.

C. T. Hendrickson, A. Horvath, S. Joshi, and L. B. Lave. Economic input-output models for environmental life cycle assessment. Environ Sci Technol 184A–191A, 1998.

M. Hocking. Relative merits of polystyrene foam and paper in hot drink cups: implications for packaging. Environ Manage 15:731–747, 1991.

A. Horvath and C. T. Hendrickson. Steel vs. steel-reinforced concrete bridges: an environmental assessment. ASCE J Infrastruct Syst, 1998a.

A. Horvath and C. T. Hendrickson. A Comparison of the Environmental Implications of Asphalt and Steel-Reinforced Concrete Pavements. Washington, DC: Transportation Research Board Conference, January 1998, and Transportation Research Record, 1998b.

A. Horvath, C. Hendrickson, L. Lave, F. C. McMichael, and T.-S. Wu. Toxic emissions indices for green design and inventory. Environ Sci Technol 29:86–90, 1995.

S. Joshi. Comprehensive Product Life-Cycle Analysis Using Input-Output techniques. Ph.D. Thesis, Carnegie Mellon University, The John H. Heinz III School of Public Policy and Management, 1997.

L. B. Lave, E. Cobas-Flores, C. T. Hendrickson, and F. C. McMichael. Using input-output analysis to estimate economy-wide discharges. Environ Sci Technol 29:420A–426A, 1995.

L. B. Lave, E. Cobas-Flores, F. C. McMichael, C. T. Hendrickson, A. Horvath, and S. Joshi. Measuring Environmental Impacts and Sustainability of Automobiles. Sustainable Individual Mobility—Critical Choices for Government and Industry Conference, Zurich, Switzerland, Nov. 4–5, 1996.

W. Leontief. Environmental repercussions and economic structure: an input output approach. Rev Econ Stat, 1970.

W. Leontief. Input Output Economics. London: Oxford University Press, 1986.

H. MacLean and L. B. Lave. A life cycle model of an automobile: resourse use and environmental discharges in the production and use phases. Environ Sci Technol 322A–330A, 1998.

Society for Environmental Chemistry and Toxicology (SETAC). Guidelines for Life Cycle Assessment: A Code of Practice. Workshop held at Sesimbra, Portugal. March 31–April 3, 1993.

U.S. Environmental Protection Agency, Office of Research and Development, Risk Reduction Engineering Laboratory. Life Cycle Assessment: Inventory Guidelines and Principles. Cincinnati, OH: Prepared by Batelle and Franklin Associates, EPA/600/R/92/245, February 1993a.

U.S. Environmental Protection Agency, Office of Pollution Prevention and Toxics. The Use of Life Cycle Assessment in Environmental Labeling. EPA/742-R-93-003, September 1993b.

3

Multi–Life Cycle Assessment
An Extension of Traditional Life Cycle Assessment

**Reggie J. Caudill, MengChu Zhou,
Pingtao Yan, and Jie Jin**
New Jersey Institute of Technology, Newark, New Jersey

1 INTRODUCTION

With increasing product complexity, sophistication of manufacturing processes, range of environmental concerns, and other pressures on industry over time, environmental evaluation becomes increasingly difficult and a key issue in industrial ecology. Based on the U.S. Environmental Protection Agency's municipal solid waste (MSW) characterization and the durable goods fraction of MSW, the scrap electronics wastestream may be as high as 5 to 10 million tons per year. Consumer electronics, computers, and household appliances contribute significantly to the environmental burden placed on the municipalities across the nation. If discarded products and wastestreams such as those mentioned could be recovered and reengineered into valuable feedstreams, then we can break this trend and achieve sustainability. Over the last decade, efforts to reduce process wastes and design green products were made in America's manufacturing industry. Companies are pressured to increase the pace to develop more green products.

There are a number of different systems-based approaches to integrating environmental issues into industry, such as life cycle assessment (LCA), design for the environment (DFE), total quality environmental management, ecofusion, green supply chain management, and a number of national and international environmental standards (Curran 1996; Glantschnig 1994; Graedel and Allenby 1995, 1996; Hersh 1998; Vigon et al. 1993). DFE is a systematic process by which firms design products and processes in an environmentally conscious way. It requires environmental considerations over a complete product life cycle in the design process. Closely linking to DFE output, LCA is targeted at the analysis of designs and involves an examination of all aspects of product designs from the preparation of input materials to end use. This includes evaluation of the types and quantities of input materials such as energy, raw materials, and water; product outputs including atmospheric emissions, solid and aqueous waste, and the end product; and identification and evaluation of opportunities for reduction of the environmental impacts of processes and products. An LCA methodology has four interrelated components: definition of scope and boundaries, inventory analysis, impact analysis, and improvement analysis (Graedel and Allenby 1995).

Many LCA tools have been available for implementing a full LCA analysis. Some of the available LCA databases and software are ECO-it (1999), GaBi software (1997), TEAM (1998), and SimaPro 4.0 (1999). Since it is unlikely that one single LCA methodology can be optimal for all LCA analysis, differences can be found in these tools, depending on the boundaries set by the tool and the specific problems it is designed to solve. They may deal with energy at just one life cycle stage rather than the total life cycle. They may also vary in the type of databases of materials they use. This chapter extends the traditional LCA methodologies to multiple life cycle assessment (MLCA) and presents their applications to evaluate DFE guidelines and life cycle engineering design of telephones.

2 MLCA METHODOLOGY

As an extension to traditional LCA methodologies, multi–life cycle engineering (MLCE) is a new approach in today's environment (Zhou et al. 1996, 1999; Caudill et al. 1998). It is based on the principle of sustainable economy where competitiveness is balanced with environmental responsibility. MLCE takes a systems perspective and considers fully the potential of recovering and reengineering materials and components from one product to create another, not just once but many times. MLCA systematically considers and quantifies the consumption of resources and the environmental impacts associated with a product or process. It extends the structure of traditional LCAs to include explicit consideration of demanufacturing, remanufacturing, reengineering, and reuse-extending LCAs to the realm of MLCE. These end-of-life recovery processes have been modeled to account for material flows, energy usage, and environmental burdens associated with recovery and reprocessing of components and basic materials. The total life cycle engineering framework is presented in Figure 1.

Compared with the traditional life cycle model that consists of four life cycle stages—material manufacturing, product manufacturing, product use, and recovery—an MLCA frame refines some of these stages and introduces additional components:

1. Material production includes two stages—material extraction and material synthesis.
2. To quantify the materials used in a packaging process, methods of transportation, distance traveled, and energy and emissions associated with these processes, a packaging and distribution stage is separated from the production stage as an independent stage.

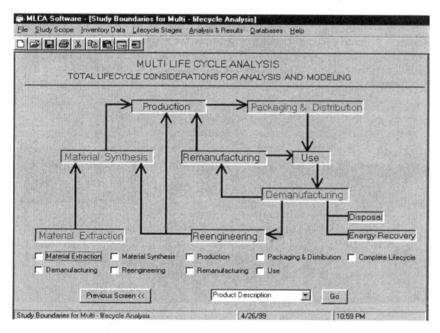

FIGURE 1   A total life cycle engineering framework.

3. The recovery and new life options of a product is the main point where MLCA differs from LCA. LCA specifies two types of recycling processes, open loop and closed loop. MLCA merges these two recycling options into one and throughout its life from raw material extraction to final disposal. MLCA calls this stage of a product life cycle demanufacturing. Its goal is to facilitate the multilife reuse of components and materials.

4. The reengineering stage acts as a link that closes the life cycle loop. It aims to supply the reusable materials to the production stage directly or recycled materials to the material synthesis step.

5. The remanufacturing stage is where the parts and subassemblies are reused. Those parts can be reused in new products at the production stage or as replacements at the use stage or can be sent back to demanufacturing.

3 MLCA TOOL: DESIGN AND IMPLEMENTATION

The MLCA software as a tool is developed for analyzing and comparing the environmental impacts, energy consuming, and cost of different products. With this software, one can practice a full life cycle analysis on products by considering all the effects and options based on product life cycle stages. The software tool is written for a personal computer Windows environment using Visual Basic with Microsoft Access as the database package. It aims to support the designers for implementation of the MLCA methodology; develop generic screens that can be used by most consumer electronic producers, with particular focus on telephones, computers, monitors, and televisions; have environmental information on products and processes readily accessible to designers, which simplifies the integration of DFE into a design process; help designer and decision makers track the environmental performance of products by using performance metrics and other indices such as Eco-Compass; and supply a cost model to put the cost issue on line.

The designs below present database, user interface, algorithm, analysis, and report design.

3.1 Database Design

MLCA, as a life cycle analysis tool, requires a large amount of data and information on product materials, manufacturing processes, energy, environmental, and cost issues. It allows designers to enter different types of materials they anticipate using in their product and can automatically generate a list of energy consumption and environmental burdens associated with these materials. The material price is the current market price. A set of databases coupled with MLCA software are designed to store all the data and information, which relate to each other in a complex way.

MLCA databases are divided into two classes: standard database with the verified data, and user-defined database where users can supply their own of different products. Users are allowed to use custom databases that address specific products and processes and large databases that describe generic processes, their emissions, and other effects.

3.2   User Interface Design

MLCA software supplies an easy Windows-based interface by using Visual Basic. It includes many advanced features used in Microsoft softwares, for instance, the ability to view the product by graphical tree structures. The MLCA software development is classified into three main stages: product description, life cycle stages, and analysis and reports.

3.2.1   Product Description Interface Design

Following the MLCA framework introduced in Section 2, to perform an MLCA evaluation on a product, we have to address the most important issue, for instance, the data collection. The product description interface is designed for product information gathering. Its structural design is shown in Figure 2. Four tabs are used to deal with different situations in a product's assembly:

1.   Subassembly with Part Tab: It contains subassemblies assembled by parts, or subassemblies assembled by other subassemblies in a product. Information such as material type, quantity, and weight of the subassembly or part is required to input into the subassembly or part section.
2.   Final Subassembly Tab: It contains subassemblies with no further parts or other subassemblies, but subassemblies may consist of more than one material. The information required in this tab relates to various materials and total weight of the subassembly.
3.   Parts Tab: It indicates parts that are not part of any subassembly with just one material.
4.   Circuit Boards Tab: It is designed to keep the information on the circuit board used in a product. The circuit board area and energy requirement for its manufacture are included.

3.2.2   Life Cycle Stage Interfaces Design

To get a complete analysis result on a product in the Analysis and Report stage, we design a life cycle stage interface for each specific life cycle stage.

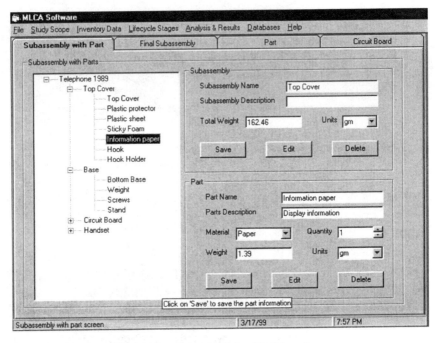

FIGURE 2 Product Description screen of the MLCA software.

Information is collected through these interfaces for later calculations and report.

Material Processing Stage. This stage focuses on material information of a product. A material inventory tree can be displayed clearly on the material processing screen, which is another way to show the product inventory information. All the materials used in a product are shown by material names as the parents in the tree structure and all the parts using the same material are shown under a material name as children. By clicking on one of the child nodes in the material inventory tree, users can also get information about a particular part. The material processing characteristics and activity-based cost factors are required from users. MLCA software can also summarize more material information of a product with a pull-down material information menu at the right bottom of the screen. The Material Processing screen is shown in Figure 3.

Production Stage. The production screen, shown in Figure 4, is designed for specifying the production processes that produce each part of a product. The present production process list has eight processes: extru-

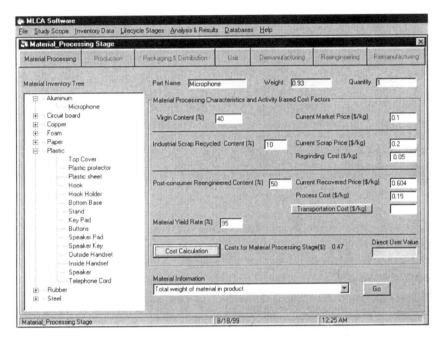

FIGURE 3   Material Processing screen of MLCA software.

sion, injection, thermoforming, stamping, milling, turning, semiconductor processing, and glass forming. If a process is companied with new materials to be consumed, users can input them during that process with a pop up frame of process materials. The production characteristics and activity-based cost factors for that particular process need to be specified as well. Finally, users should make a decision of the allocation of energy and environmental burdens. It can be allocated to the main product only, by mass ratio of product to coproduct, by the price market, or user defined.

Packaging and Distribution Stage.   The packaging and distribution interface is designed in MLCA software from which the user can input the specific information on the packaging material and transportation of that product. In a packaging material frame, information on the packaging material used for the product, the percentage of recycled material, material weight, volume of that material, and material market price can be filled in. Information on the yield rate and cost is also specified in the frame. Users can also quantify the transportation information of a product in Product Transportation screen. It includes the mode of transportation, actual load, load ratio, energy content factor, energy unit cost in dollar per kWh, and

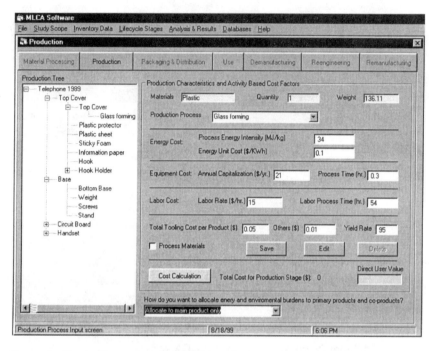

FIGURE 4   Production screen of MLCA software.

distance traveled. When users select a transportation mode, the corresponding default values of weight in pounds, maximum payload in pounds, and
the fuel efficiency in miles per gallon are displayed at the top right side of the
screen automatically. Users can edit the default ones by entering their values
in the textbox beside. Users can also assume the distance traveled to be
roundtrip or not. One can obtain the final report on transportation energy
consumption and environmental emissions at this stage. The details can be
seen in the thesis by Jin (1999).

Use Stage.   This stage relates to the energy consumed by a product in
different modes during its use stage. The four modes are active, idle, power
save, and off. In each mode a product consumes a certain amount of energy
that must be entered into this screen. The amount of time a product used in
each mode and the excepted lifespan of a product are required. Cost factors
in this stage include energy, supplies, product purchase price, update, and
maintenance/service. Users can access the calculation result by clicking on
the Calculate Use-Stage Energy Consumption button and the Cost
Calculation button.

Demanufacturing Stage. Demanufacturing is an important stage of the product multi–life cycle, as shown in Figure 5. The end fate of the various parts and subassemblies is determined here. The first step in analyzing a demanufacturing process is to quantify the facility structure used in terms of the size of the facility, yearly energy consumption, cost of energy, the volume of products handled per year, and the total disassembly time of a product. Then, users need to determine the end fate options for the product (Zussman and Zhou, 1999). MLCA software provides a list of end fate options: reuse a product, remanufacture parts and subassemblies, recover basic materials, and treat as remaining carcass.

PRODUCT REUSE. If the end fate chosen for a product is reuse, users are required to input its anticipated inspection pass rate, which means that this percentage of a generation of the product can be resold. The reselling price of a product is also required. Reuse of a product means taking the whole product back to its second life cycle without any additional costs and energy consumption. The benefit from reuse option can be obtained by multiplying the reselling price and the percentage of anticipated inspection pass rate, subtracting the cost that may be incurred at demanufacturing and remanufacturing stages.

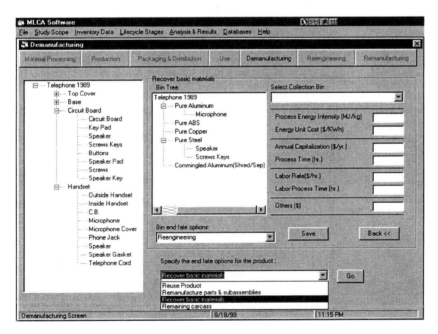

FIGURE 5   Demanufacturing stage of MLCA software.

PARTS AND SUBASSEMBLIES REMANUFACTURE. This option is chosen when some subassemblies or parts of a product can be remanufactured with an optimum value. Users need to add those subassemblies and parts into the remanufacture subassemblies and parts table provided in the screen. After the user specifies and selects the end fate option for each part or subassembly, that part or subassembly is deleted from the product tree automatically when its end fate is determined.

BASIC MATERIAL RECOVERY. Material can be recovered from parts and subassemblies of a product by disassembly or shredding. Users create a custom bin tree by selecting various materials from a collection bin list as the parents. This list includes various recovered materials with classification of pure and commingled. For example, if aluminum is part of the material in a product, then the collection bin list includes pure aluminum, pure aluminum shred, commingled aluminum, and commingled aluminum shred. The same method applies to all other materials in a product to be recovered. Parts and subassemblies made out of that material are selected from the production tree and dropped to the various recovered material nodes as the children of that material. Users then specify the four end fate options for each bin, for instance, reengineering, waste to energy, smelting, and landfill.

REMAINING CARCASS. The remaining parts and subassemblies of a product that do not apply to the above end fate options are entered in this stage. They cannot be reused anymore and are either smeltered, recovered for energy content, or put in a landfill.

Reengineering Stage. This stage specifies the process required to recover basic materials from selected materials in the demanufacturing stage and obtains the materials and energy and environmental burden information in each step of the reengineering process.

Remanufacturing Stage. This stage aims to quantify such data as material flow, energy consumption, environmental burdens, cost, and time requirement for each step to remanufacture a product, subassemblies, or parts. Users are required to specify the remanufacturing facility with such descriptions as facility size in square feet, volume of products handled per year, and total energy consumption and environmental burdens per year. Users then select a subassembly or part for remanufacturing from the given list that is taken from remanufacturing subassemblies and parts selected earlier in demanufacturing stage. The step by step remanufacturing process is shown in an interactive box. Finally, users select each remanufacturing step and input the required data in each popped up frame. Additional details can be seen in the thesis by Jin (1999).

3.3 Algorithm Design

Combined with product information stored in MLCA databases, MLCA software supplies the following algorithms to analyze the assessment result of a product or to compare different designs of a product or different products. To perform MLCA assessment on a product, five performance measures should be considered for each life cycle: environmental burden, energy consumption, material utilization, composite performance measure, and life cycle economics. The discussions below focus on the information of environmental burden and energy consumption for all life stages. Environmental burden is evaluated based on air emission, water effluents, and solid wastes.

3.3.1 Material Processing

The analysis of environmental burden generated from production of feedstock materials and emissions generated from power sources is performed in the material processing stage. Assuming that A_i is the total weight of material i used in a product, B_i is environmental burden generated from production of material i in unit gram, and C_i is environmental burden generated from power sources used during production of feedstock material i in unit gram, the environmental emissions of a product material generated from this stage is

$$\sum_{i=1}^{n} A_i \cdot (B_i + C_i)$$

3.3.2 Production Stage

Environmental burden in the production stage is generated from the use of electric power sources and processing of materials. Users are required to input information about environmental emissions from the production screen. The total environmental burden generated from this stage is the summarization of environmental burdens generated from each subassembly of a product.

3.3.3 Packaging and Distribution Stage

Environmental burden in this stage is from transportation tools used to transport a product. Environmental burden generated from production of packaging materials is not included in this study. Total emissions from product transportation can be calculated by

$$\sum_{i=1}^{n} (M_i \cdot N_i)$$

where M_i means environmental emission generated from transportation tool i per mile and N_i is traveled distance using tool i. Since environmental emission generated from the transportation tool is different, MLCA software owns different calculation engines on M_i (Ketan 1999).

3.3.4 Use Stage

Environmental emissions considered in the use stage are those generated from the use of electric power sources. The calculation is based on the MJ of energy consumed.

3.3.5 Demanufacturing Stage

Environmental emissions in this stage are those generated from electric power sources. The calculation is based on the MJ of energy consumed from demanufacturing facilities and operations.

3.3.6 Remanufacturing Stage

Environmental burdens from the remanufacturing stage are the sum of emissions generated from each remanufacturing process and remanufacturing facility power sources used. Users are required to input data in the remanufacturing life cycle screen.

3.3.7 Reengineering Stage

Environmental burden from the reengineering stage includes emissions generated from processes on each recovered material of a product. Users are required to input data in the reengineering life cycle screen at first.

The amount of energy consumption is also calculated over all the stages. For example, at the use stage, the four energy consumption modes are active, idle, power save, and off. Total energy consumption is

$$\sum_{i=1}^{4}(M_i \cdot N_i \cdot 24 \cdot E \cdot 365)$$

where M_i means energy consumption in unit MJ in each mode, N_i means utilization factor of each mode with $\sum_{i=1}^{4} N_i = 1$, and E is the expected lifespan of a product in years.

3.4 Analysis and Reports

With the data and algorithms required to evaluate a product, the analysis and report screen displayed in Figure 6 is designed to show product assessment results in different ways. Users can view their product analysis results by tables or graphics. They can focus on reports based on one or more life cycle stages of a product. The evaluation results using different methodol-

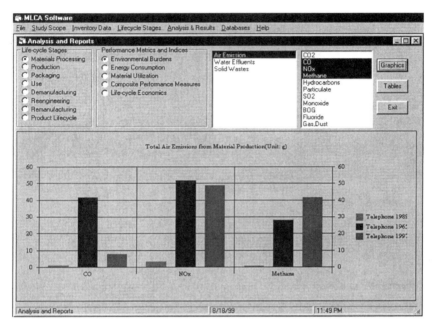

FIGURE 6 Analysis and Reports screen of MLCA software.

ogies such as eco-compass are also available. The detailed application procedures can be referred to (Jin 1999).

4 MLCA APPLICATION TO FOUR GENERATIONS OF TELEPHONES

We selected four generations of low-end business telephones used as an
MLCA software application example. Phones designed and manufactured
in 1965, 1978, 1989, and 1997 were available for study (Al-Okush et al.
1999), as shown in Figure 7. Based on the MLCA software, an MLCA
inventory database can be generated. Initially, a new project is created
with the project information screen and system boundary being defined.
Then, product information required for the product description level of
the MLCA software is entered. Finally, MLCA methodology is performed
on four generations of telephones by applying the analysis and report algorithms described in Section 3. As a result, evaluation on each generation of
telephone represents environmental performance of the business telephones.
    The inventory analysis generally includes energy consumption, waste
emissions, and process material requirements at each stage in the produc-

FIGURE 7 Four generations of office telephones. Top: 1965, left, and 1978, right. Bottom: 1989, left, and 1997, right.

tion of any raw material. In our study, process materials and energy required for the production of the materials under study are considered to be inside the system boundary. Energy consumption during the production and use of the telephones are also quantified. The energy consumption during the demanufacturing, reengineering, and remanufacturing were not quantified due to the unavailable resource data. The environmental burdens associated with the production, use, demanufacturing, reengineering, and remanufacturing were also unavailable and were not included in the present study. Materials used to fabricate fundamental equipment and tools and those indirectly consumed during the production and operation of a transportation vehicle remain outside the boundary. Finally, assumptions and adjustments were necessary to simplify the analysis. Materials considered for this study had to meet a threshold of being more than 2% by weight of the product or else they were excluded from the study, because their impact is considered to be negligible. Generally, the limits placed on the breadth and depth of LCA analysis can be classified as the restrictions on the life cycle boundaries of a system or the actual information collected, whether it is limited in its specificity or number of inventory categories. The discussion below takes the 1989 telephone as a specific example to show how to execute MLCA with MLCA software on a product.

4.1 Start of a New Project on Telephones

To perform an MLCA on the 1989 telephone, we need information on the product stored in the MLCA database. It can be obtained from users by the product description stage in the software. We start the MLCA software by double clicking the MLCA icon in the Windows program manager. The program will start up and display a toolbar with functions.

All work carried out in MLCA is stored as a *project* in the software. Users can enter new projects of their own by selecting the New Project option in the File menu at the top of the MLCA screen. The Project Information screen then appears. The project information already stored in the MLCA database is retrieved from the database automatically and showed in the Project Information table in the screen. Users can input the information about the 1989 telephone project in exact boxes such as "Telephone 1989" in the Project Name box, as shown in Figure 8. To review other project information already in the database, move the mouse to that project and click once to pop up the information in the corresponding boxes automatically by the software. After users enter the available data, they should click the Add button to save the new project information in the

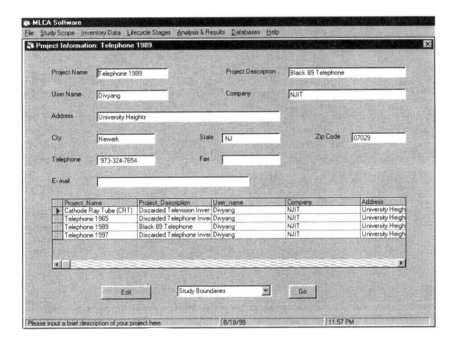

FIGURE 8 Telephone Project Information screen.

database. The new project added right now into the database appears in the project information table automatically. Figure 8 displays the detailed information.

4.2 Specifying the Study Boundary

It is important to define the study boundary of the product under study. It brings a clear purpose and process to carry out the MLCA. Users can access the study boundary stage by selecting the Study Boundary option in the bottom box of project information screen, shown in Figure 8, and then click the Go button beside the option box. According to the purpose of the study, users select different life cycle stages by marking the exact checkbox or the complete life cycle box to study all the life cycle stages. The selected stages are highlighted into blue color: material processing, production, packaging and distribution, use, and demanufacturing.

4.3 Product Description Stage

Users can input all the information from the Product Description screen. A detailed inventory table is shown of all the parts in 1989 telephone product, quantifying for each part the quantity, function, weight, material type, and market value. After it has been done, the inventory table can be displayed directly with a tree structure as shown in Figure 2.

4.4 Multi–Life Cycle Stages

Other information needed for the 1989 telephone assessment can be obtained from each of seven multi–life cycle stage screens.

4.4.1 Material Processing Stage

All the materials used in the 1989 telephone are shown in the material inventory tree in Figure 3. From that we know the material information of the 1989 telephone clearly. Clicking on each material mode, the tree is expanded to show all the parts using the same material. Detailed information such as total weight of selected material, number of parts made out of this material, total energy consumption, and environmental burdens of the material can also be accessed by selecting related topics in the material information box at the bottom of the screen.

4.4.2 Production Stage

Production processes of parts are specified. The database quantifies mainly materials, energy, and environmental burdens associated with those processes. Considering factors as yield rate, the way users allocate energy, and environmental burdens to primary products and coproducts, we get the energy consumption and environmental emissions of the 1989 telephone in the production stage. The production stage screen of the 1989 telephone is shown in Figure 4.

4.4.3 Packaging and Distribution Stage

Energy consumption and environmental burdens in this stage come from packaging materials and transportation tools used for transporting the 1989 telephone. For telephone packaging, papers and plastics are often used. The transportation tool is assumed to be a light duty truck. With these assumptions, we can obtain total energy consumption and environmental emissions at this stage.

4.4.4 Use Stage

According to the data entered in the use stage, we can obtain lifetime energy consumption in this stage. For the 1989 telephone, it works in active mode and stand-by mode only. The total energy consumption during the use stage can be achieved by the following assumption:

In active mode:

Energy consumption = 0.059 MJ.

Utilization factor = 3% per 24 hours (0.72 hours per day).

In stand-by mode:

Energy consumption = 0.02376 MJ.

Utilization factor = 97% per 24 hours (23.28 hours per day).

Expected lifespan = 7.5 years.

Total energy consumption during the use stage is 1630.49 MJ.

4.4.5 Demanufacturing Stage

A complete disassembly of the four telephones was conducted to establish an inventory database of materials, components, and assembly/disassembly techniques. The 1965 phone has 165 different parts held together with 140 screws and weighs 2325 g. The 1997 phone weighs 1085 g and has only 8 screws and 17 snap-fits to hold 32 parts together.

A technique developed at the New Jersey Institute of Technology (Das et al. 1998), the disassembly effort index (DEI), is used to quantify demanufacturability for the product and assign a score for each of the weighted performance parameters. DEI has six main parameters: disassembly time, fastener accessibility, tools required, part-hold device, force necessary, and instructions needed. These six parameters were integrated with another three—number of different types of materials, number of different types of fasteners, and stamping of material type into the part—to quantify revalorization. Revalorization, used for the eco-compass, expands DEI concepts. Table 1 highlights the major parameter values and disassembly comparisons. A higher DEI score reflects easier disassembly.

4.5 Analysis and Report

The analysis reports are obtained by summarizing information in each life cycle stage. One of the summarization reports on air emissions from material production of 1989 telephone is shown in Figure 6. We can follow the same path to get the assessment on the other three generations of telephones. Then, we can compare the four generations of telephones and show the result in graphic form to make the conclusions clear.

5 ASSESSING DFE GUIDELINES' IMPACT ON TELEPHONE DESIGNS

The first application of the MLCA tool is to assess the impact of DFE guidelines (Graedel and Allenby 1996) on four generations of telephone designs. These guidelines are generally divided into five major categories: material conservation, energy conservation, environmental burdens, service extension, and demanufacturability. For each of these categories, one or more performance metrics was defined, and data were collected to quantify these metrics for each of the four telephones. The data for the first three generations were used to establish a trend by regressing the data, in a least-square sense, to fit an exponential form, $\ln Y = aX + b$ ($Y = e^{ax+b}$). With coefficients a and b determined separately for each metric, the equation was used to forecast and estimate the expected metric value of a representative DFE 1997 telephone. Obviously, with only three data points to establish the trend for each performance metric, the results must be interpreted as yielding a better understanding of the issues of DFE effectiveness as opposed to a rigorous critical assessment.

5.1 Material Conservation

DFE: Can materials use be minimized by improved mechanical design (Graedel and Allenby 1996)?
This guideline reflects the concept of dematerialization, as quantified by the weight of the product and distribution of constituent material types. The impact of dematerialization is evident throughout the entire product life cycle from feedstock preparation and production to distribution and shipping and end-of-life recovery or disposition. The type and weight of materials directly affect other metrics such as energy consumption, environmental

TABLE 1 Disassembly Comparison of the Four Telephones

Telephones	Number of screws	Number of snap-fits	Total no. of fasteners	Number of parts	Time (sec)	DEI	Comments
1965	140	19	159	165	1902	62	Extensive use of screws and the large number of parts their complexity increased disassembly time and the ease disassembly.
1978	68	41	109	165	1640	71	Disassembly time reduced subassemblies are more accessible. Percentage of plastic increased by 30%, while metals used decreased by 89%
1989	12	12	24	43	354	93	Enormous change in design, number of parts reduced by 74% 1965. No problems faced disassembly.
1997	8	17	25	32	133	94	Handset disassembly time for the rest of phone decreased due to use of adhesives and snap-fits instead of screws.

Disassembly index includes time, tools, accessibility, force, part-hold, and instructions. Disassembly time for 1997 telephone excludes the handset disassembly, which is 180 seconds.
Source: Das et al. 1998.

burden, and demanufacturing. The complete material inventory is given in (Al-Okush 1999) for each generation of telephones. As shown in Figure 9, while dematerialization of the telephone has taken place, the material weight of the DFE-designed phone, 1085 g, is higher than the forecasted value of 633 g. Notice that the weight of the 1989 phone is less than the DFE phone. This is due primarily to the larger size and a heavier plastic stand attached to the base of the 1997 telephone. Although not shown here, the forecast value for each material category was lower than actual. It can be concluded that the desire for material conservation has given way to other design considerations; therefore, DFE guidelines did not affect dematerialization, making the concomitant gain in environmental benefits independent of DFE considerations.

5.2 Energy Conservation

The energy consumption of the telephones was measured over the following life cycle stages: feedstock materials, production, use, and recovery (including demanufacturing and recycling).

DFE: Is the product designed to minimize the use of materials whose extraction is energy intensive (Graedel and Allenby 1996)?
Data generated for the material synthesis stage is based on the energy required to produce the various materials used in the four telephones and

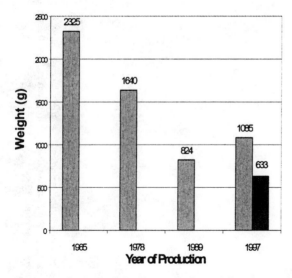

FIGURE 9   Dematerialization trends and forecast value.

calculated based on the energy intensity (MJ/kg) of each material in the MLCA tool. Table 2 shows that the actual energy value for feedstock materials in the DFE-designed phone is above the forecasted value. Therefore, DFE guidelines did not have a direct impact on feedstock energy conservation for the 1997 phone.

The entire production of the telephones is assumed to take place in a single plant with only the circuit board fabricated off-site. Based on annual energy and production volume data from a typical telephone manufacturing plant, the energy required to produce the 1997 phone was estimated to be 100 MJ (Al-Okush 1999), while the energy required to fabricate the 21,090-mm^2 1997 printed circuit board was calculated assuming an average of 0.002 MJ/mm^2 (Young 1995). Since 1965 there has been significant improvement in the energy efficiency of production of electronics products, with data from 1971 to 1985 showing that energy efficiency improved an average of 3% per year (Al-Okush 1999). Using this trend, production energy intensity (MJ/kg) was estimated for earlier generations of the telephone. Table 2 indicates that the production energy for today's telephone is 140 MJ, whereas the forecast value is 92 MJ.

DFE: Has the product been designed to minimize energy use while in service (Graedel and Allenby 1996)?
The energy consumption of the telephones during the use stage was defined into two categories: the in-use mode and the stand-by mode. The utilization factor for the office telephone during the 8-hour workday was estimated to be 9% based on a simple sampling of users. According to design specifications, the 1997 phone uses 5.5 W while in use and 2.2 W during stand-by. With a design life of 7.5 years, the total energy consumption during the use stage is 151 kWh$_e$ (1800 MJ assuming 30% electric power conversion rate). It is interesting to note that the energy consumption of the 1997 telephone during the stand-by mode is 14 times greater than the in-use mode over the telephone's lifetime. Therefore, designers should place emphasis on reducing the energy consumption of the telephone during the stand-by mode. The same assumptions have been made for the 1989 phone; however, the 1965 and 1978 phones do not have memory cells associated with the newer generations and are assumed to not draw power in the stand-by mode.

DFE: Is the product designed to minimize the use of energy-intensive process steps in disassembly? Is the product designed for reuse of materials while retaining their embodied energy (Graedel and Allenby 1996)?
The end-of-life recovery method is assumed to be basic materials recovery by a traditional shredder/separation process. Based on operational data for a material recovery facility designed for telecommunications equipment, the energy required is estimated to be 21.6 MJ/kg of product. The embodied

TABLE 2 Forecast and Actual Energy Consumption (MJ) over the Product Life Cycle

	Feedstock material	Production	Use	Recovery
Actual value (1997)	50	140	1800	−258
Forecasted value (1997)	28	92	N/A	−165
Percent difference	44.0%	34.0%	N/A	36.0%
Equation	$\ln y = -0.038x + 4.5$	$\ln y = -0.059x + 6.4$	N/A	$\ln y = -0.02x + 5.8$
R^2	0.68	0.99	N/A	0.35

energy of reclaimed materials, plastics only, is subtracted from total recovery process energy to give the net energy consumption during the end-of-life recovery. As indicated in Table 2, the DFE guidelines have been successful in retaining material embodied energy beyond the trends—the actual energy saved is 258 MJ compared with the forecasted value of 165 MJ.

Figure 10 summarizes the total energy associated with all considered life cycle stages of each telephone and shows that energy consumption during the use stage is by far the highest throughout a telephone life cycle.

5.3 Environmental Burdens

The analysis of air emissions, waterborne effluents, and solid wastes has been performed for each life cycle stage. With the exception of production of the feedstock materials, the only environmental burdens included are those associated with emissions from electric power generation during product manufacture, use, and recovery. For feedstock material synthesis, full environmental burdens are included. For electric power generation, the U.S. grid mix for 1990 was used as the default electric power mix, which is composed of 55% coal, 4% oil, 9% gas, 22% nuclear, and 10% hydroelectricity (Young and Vanderburg 1994). The DFE guidelines regarding reduction of air emissions, solid wastes, and waterborne effluents relate to life

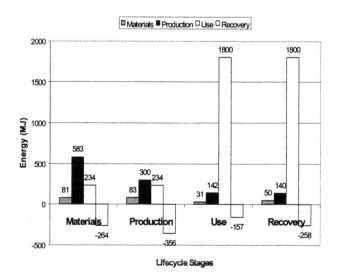

Figure 10 Total energy consumption over the product life cycle stages.

cycle environmental concern and strive to reduce impacts as much as possible.

DFE: Has manufacturing gaseous emissions been minimized to the greatest extent possible (Graedel and Allenby 1996)?
With growing concern over global warming and the effects of greenhouse gases, a focus has been on CO_2 emissions. Figure 11 displays the CO_2 emissions throughout the various life cycle stages of four telephones. The highest CO_2 emissions were during the use stage of the 1989 and 1997 telephones, correlating to the high use of electrical energy. Although not shown, similar results were found for CO, NO_x, and methane gas. Since most environmental burdens are tightly linked to the quantity and type of materials used and energy consumed, conclusions regarding the impact of DFE guidelines on environmental burdens are similar to those reached for material and energy conservation described above—forecast values are generally below actual values.

5.4 Service Extension

DFE: Are subassemblies designed for ready maintainability rather than solely for disposal after malfunction? Are modules designed for ready removal (Graedel and Allenby, 1996)?
These guidelines concentrate on service delivery to the product after it has been manufactured and during its use stage. Service extension is one of the

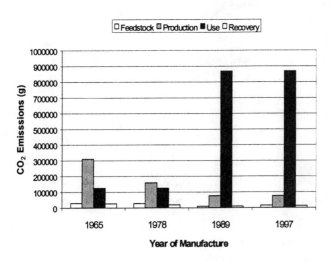

FIGURE 11 CO_2 emissions over the product life cycle.

six dimensions of the eco-compass (Fussler and James 1996). Three major metrics have been selected to quantify this dimension and assess the guidelines: commonality, upgradability, and modularity. Table 3 summarizes how each subassembly is defined as common, upgradable, or modular. It also shows for each of the above three variables the total number of subassemblies that meet their criteria. Defining whether a certain subassembly is modular, common, or upgradable is fairly subjective; however, as can be seen, commonality and modularity of subassemblies is evident for the 1997 DFE-designed phone.

5.5 Demanufacturing

Ease of disassembly and efficiency of demanufacturing is embodied in several DFE guidelines (e.g., service extension and energy conservation). Although not explicitly addressed in the general DFE guidelines, two metrics—disassembly time and number of parts—are presented here to assess inherent product demanufacturability. As indicated in Table 1, significant reduction in disassembly time and part count has occurred. Indeed, these metrics indicate that the DFE-designed phone performs better than forecasted. The extrapolated value for disassembly time is 270 seconds, while the actual time is 133 seconds. Also, the part count for the 1997 phone is 32, whereas the projected value is slightly higher at 36. The DFE guidelines do explicitly recognize the difficulty of identifying plastic materials so that they can be sorted efficiently.

DFE: Are all plastic components identified by ISO markings as to their content (Graedel and Allenby 1996)?
Parts should be stamped in a generic way that identifies the type of polymer used in the construction. Material stamping helps the demanufacturer identify the constituent material of each part so that higher purity of various materials can be recovered. Since the market value of pure materials is significantly higher than that of commingled or dirty materials, materials markings improve the economics of recovery. The 1997 telephone was the only telephone stamped with material type identification; consequently, it is evident that the DFE guidelines have had a direct positive effect on the product.

In summary, DFE guidelines have not had as great an impact on the environmental performance of the telephone as may have been desired. While many assumptions have been made in this study, two major concerns are that only four generations of telephones have been examined and that these four have been assumed as being functionally equivalent. From this standpoint, this study should be recognized as a beginning as opposed to a conclusion.

TABLE 3 Defining Subassemblies for each Service Extension Service

	Subassemblies						Total	Total subassemblies
	Top cover	Base	Handset	Circuit board	Key pad			
1997								
Upgradability	Y	Y	N	Y	N		3	5
Commonality	Y	Y	Y	Y	Y		5	
Modularity	Y	Y	Y	Y	Y		5	
1989								
Upgradability	N	Y	Y	Y	N		3	5
Commonality	Y	Y	Y	N	Y		4	
Modularity	Y	Y	Y	Y	N		4	
1978								
Upgradability	N	N	Y	Y	N	N	2	6
Commonality	N	N	Y	N	N	Y	2	
Modularity	N	N	N	Y	N	N	1	
1965								
Upgradability	N	N	Y	N	Y	N	2	6
Commonality	N	N	Y	N	Y	N	2	
Modularity	N	N	N	N	N	N	0	

6 INTEGRATED PRODUCT AND PROCESS DEVELOPMENT OF TELEPHONES

The MLCA tool can generate the necessary data to serve as a basis of integrated product and process development (IPPD) methodologies. This section presents the combination of such data and eco-compass concepts for life cycle engineering design of telephone designs. According to the IPPD method (Yan et al. 1998, 1999), different product development issues such as cost, benefit, and environmental impact can formally be described as constrained optimization problems. They are in turn solved via search algorithms over a life locus tree. In eco-compass, environmental impact is evaluated using six indices: mass intensity, energy intensity, health and environmental potential risk, revalorization, resource conservation, and service extension. Together with cost and benefit, an index vector of eight indices is set up to evaluate the performance of processes, life phases, and a product's different life loci. Development of business telephones is used to illustrate the application of the MLCA tool and the IPPD method (Yan et al. 1998, 1999).

6.1 Eco-Compass

LCA is a useful tool in analyzing the environmental impacts of a product by calculating the inputs and outputs of each stage of a product's life cycle (Caudill et al. 1998). However, LCA data are usually too complex and detailed to make sense to most business decision makers. Consequently, there is a need for a means of weighting the inputs and outputs to clarify important issues and make comparisons between options.

The eco-compass technique, developed at Dow Europe, is a comparative tool to evaluate one existing product with another or to compare a current product with new development options (Fussler and James 1996). The eco-compass has six dimensions, intended to encompass all significant environmental issues. Two are largely environmental: health and environmental potential risk and resource conservation. Four are of business and environmental significance: energy intensity, mass intensity, revalorization, and service extension. These six dimensions are explained as follows.

- *Mass intensity* reflects the change in the material consumption and mass burdens associated with the product over its life cycle.
- *Energy intensity* captures the change in the energy consumption associated with the product throughout its life cycle.
- *Health and environmental potential risk* detects the change in the environmental burdens associated with the product over its life cycle.
- *Revalorization* evaluates the ease with which remanufacturing, reuse, and recycling of the product can be carried out.

- *Resource conservation* detects the change in the conservation of materials and energy associated with the product over its life cycle.
- *Service extension* measures the extent to which service can be delivered to the product throughout its life cycle.

Using eco-compass, one of the products to be compared is chosen as the base case. The base case always scores a 2 in each dimension. The alternative product is then given a score relative to this base case on a scale of 0–5 in each dimension. The precise score depends on the percentage increase or decrease in performance. A diagram of the eco-compass is shown in Figure 12.

6.2   Integration of Eco-Compass into IPPD

Eco-compass uses six dimensions to evaluate significant environmental issues associated with a product over its life cycle. Therefore, we propose to use these six dimensions as six indices in the index vector **c** of our proposed IPPD methodology (Yan et al. 1998) to provide a more detailed and precise evaluation of a product's environmental impacts. Plus two other generally used indices (i.e., cost and benefit), the index vector **c** then consists of eight indices:

$$
\mathbf{c} = \begin{pmatrix} \text{Mass Intensity} \\ \text{Energy Intensity} \\ \text{H \& E Potential Risk} \\ \text{Revalorization} \\ \text{Resource Conservation} \\ \text{Service Extension} \\ \text{Cost} \\ \text{Benefit} \end{pmatrix}
$$

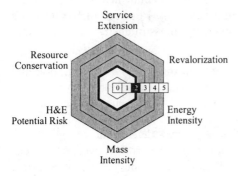

FIGURE 12   Eco-compass.

These eight indices are then used to evaluate the performance of processes, life phases, and a product's different life loci.

To apply this index vector **c** systematically in real product and process development, it is important to provide rules on how to evaluate each index.

Figure 13 shows a generic model of a life phase in a product's expected life cycle structure (ELCS). We characterize two inputs and three outputs for this life phase:

- I_1: Input from other life phases. In a simple ELCS (i.e., a sequence of consecutive life phases), this means input from the previous life phase.
- I_2: Input from sources other than life phases. It is divided into three parts: $I_2 = I_{2M} + I_{2E} + I_{2O}$.
  I_{2M}: Input of materials,
  I_{2E}: Input of energy,
  I_{2O}: Input of other forms.
- O_1: Output to other life phases. In a simple ELCS, this means output to the next life phase.
- O_2: Positive (good) output to places other than life phases. It is divided into three parts: $O_2 = O_{2M} + O_{2E} + O_{2O}$.
  O_{2M}: Output of materials,
  O_{2E}: Output of energy
  O_{2O}: Output of other forms.
- O_3: Negative (bad) output to places other than life phases.

Based on this generic model of a life phase, we can then evaluate each index in **c** as following:

- Mass intensity $= I_{2M}$,
- Energy intensity $= I_{2E}$,
- H & E potential risk $= O_3$,
- Revalorization $=$ Ease with which remanufacturing can be carried out,

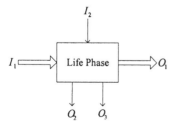

FIGURE 13   Generic model of a life phase.

- Resource conservation $= (I_{2M} - O_{2M}) + (I_{2E} - O_{2E})$,
- Service extension $=$ Ease with which service can be delivered to the product,
- Cost $= I_{2O}$,
- Benefit $= O_{2O}$.

The index of revalorization is only applicable to remanufacturing, recovery, or recycling-related processes and life phases, and the index of service extension is only applicable to processes and life phases related with product use.

From the evaluation equations above, it is noticed that the index vector **c** is only concerned with the interactions between the product's life phases and their surrounding environment. Flows inside or among the life phases are not explicitly considered in **c**. Process selections in the life phases reflect and determine the transformation of flows inside and among life phases. Using this index vector **c**, we can then follow the methods and procedures in the IPPD methodology (Yan et al. 1998, 1999) to search for an optimal life locus for a target product.

6.3 Business Telephones

We apply the extended IPPD methodology to a business telephone development. The MCLA tool is used to generate eco-compass LCA data.

6.3.1 Expected Life Cycle Structure

First, we need to set up an ELCS for a business telephone. For the illustration purpose of this case study, we apply a coarse granularity for life phases and expect a business telephone to have a typical life cycle structure as shown in Figure 14. It has five consecutive life phases: design, material extraction, telephone production, use, and recovery.

6.3.2 Processes

We then have to identify the possible processes in each life phase of the telephone's ELCS. A coarse granularity for processes in each life phase is applied.

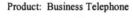

Product: Business Telephone

| Design | Material Extraction | Telephone Production | Use | Recovery |

FIGURE 14 An ELCS for business telephone.

In the design phase, there are four different telephone designs, D1, D2, D3, and D4, as shown in Figure 7. Their prototypes correspond to four generations of business telephones: D1 (1965), D2 (1978), D3 (1989), and D4 (1997).

In the material extraction phase, each telephone design may Di select a material extraction process from two possibilities, MEi1 and MEi2. They use different processing technologies and have different performances. Each process generates all the materials needed in a telephone (i.e., plastics, steel, aluminum, copper, etc.).

In the telephone production phase, each design has two alternative production processes, TPi1 and TPi2. They correspond to the use of different production equipment and therefore have different characteristics.

In the use phase, Di may be used in two different patterns, Ui1 and Ui2. They correspond to different utilization frequencies of the telephone and bear different characteristics.

In the recovery phase, two recovery processes are considered for each telephone design Di. Ri1 means shredding/separation of a telephone for material and energy recovery and Ri2 means landfilling.

6.3.3 Indices

Each process in a telephone's life cycle is characterized by eight indices (i.e., mass intensity, energy intensity, H & E potential risk, revalorization, resource conservation, service extension, cost, and benefit). Revalorization is only applicable to recovery processes Rij and service extension is only applicable to use processes Uij ($i = 1, 2, 3,$ and $4; j = 1$ and 2).

Applying the concept of tangible characteristics (Yan et al. 1998), we calculate the index values for each process. The data for processes MEi1, TPi1, Ui1, and Ri1 ($i = 1, 2, 3,$ and 4) are obtained through the MCLA tool. The indices for MEi2, TPi2, Ui2, and Ri2 ($i = 1, 2, 3,$ and 4) are calculated based on the following assumptions about each pair of processes:

- MEi1 and MEi2: The yield rate for material extraction process MEi1 is 95%. The yield rate for MEi2 is 98%, the energy consumption for MEi2 is 10% greater than MEi1, its H & E potential risk is 2% less, and its cost is 5% greater than MEi1.
- TPi1 and TPi2: Telephone production process TPi2 consumes 10% more energy than TPi1, its H & E potential risk is 10% greater than TPi1, and its cost is 20% less than TPi1.
- Ui1 and Ui2: The utilization factor of usage process Ui1 is 3%. Ui2 has a utilization factor of 12%. Consequently, Ui2 consumes more energy than Ui1, and incurs more H & E potential risk. The benefit of Ui2 is 3 times that of Ui1.

• Ri1 and Ri2: Ri1 means shredding/separation of the telephone. Metals in the telephone are recovered for reuse, and plastics are recovered for its embodied energy. Ri2 simply landfills the telephone. Therefore the energy consumption for Ri2 is 0; its revalorization index is 0 because remanufacturing, reuse, or recycling do not happen; and its benefit is also 0. The only cost for Ri2 comes from the cost of the space needed in the landfill and transportation. It is calculated based on the weight of the telephone. The H & E potential risk for Ri2 is much greater than Ri1, and it is also calculated based on the weight.

With these reasonable assumptions and the data calculated via the MLCA tool, we can summarize the index values for all the possible processes, as in Table 4. The following abbreviations for index names are used: MI (mass intensity), EI (energy intensity), H&E (H & E potential risk), RV (revalorization), RC (resource conservation), and SE (service extension). Indices RV, SE, and benefit are the higher the better; all the other indices are the lower the better.

6.3.4 Life Locus Tree

Based on the data and assumptions made, we can set up a life locus tree for the telephone as shown in Figure 15. For a clear illustration, on each level of the tree, details are only shown for one selected branch. There are 64 life loci in this tree, i.e.,

$$LL_k = \{Di, MEij_1, TPij_2, Uij_3, Rij_4\}$$

$$k = 1, 2, \ldots, \quad \text{and} \quad 64; i = 1, 2, 3, \quad \text{and} \quad 4; j_1, j_2, j_3, j_4 \in \{1, 2\}$$

$$k = 16(i - 1) + 8(j_1 - 1) + 4(j_2 - 1) + 2(j_3 - 1) + j_4$$

The index vector for a life locus is calculated as

$$c(LL_k) = c(Di) + c(MEij_1) + c(TPij_2) + c(Uij_3) + c(Rij_4)$$

6.3.5 Search

We can search in the life locus tree for an optimal life locus. In this case study, we set the optimization criterion as

$$C = w_1 MI + w_2 EI + w_3 H\&E - w_4 RV$$
$$+ w_5 RC - w_6 SE + w_7 Cost - w_8 Benefit$$

By selecting different weighting factors, we can obtain different optimal search results as discussed next.

TABLE 4 Indices for Processes in a Telephone's Life Cycle (unit of indices: $)

Process	MI	EI	H&E	RV	RC	SE	Cost	Benefit
D1	0	0	0	0	0	0	0	0
D2	0	0	0	0	0	0	0	0
D3	0	0	0	0	0	0	0	0
D4	0	0	0	0	0	0	0	0
ME11	13.20	2.25	6.28	0	15.45	0	20.00	0
ME12	12.90	2.48	6.15	0	15.38	0	21.00	0
ME21	13.40	2.31	3.57	0	15.71	0	15.00	0
ME22	13.00	2.54	3.50	0	15.54	0	15.75	0
ME31	11.80	0.86	1.25	0	12.66	0	10.00	0
ME32	11.40	0.95	1.23	0	12.35	0	10.50	0
ME41	12.50	1.39	1.95	0	13.89	0	5.00	0
ME42	12.10	1.53	1.91	0	13.63	0	5.25	0
TP11	0	16.20	36.23	0	16.20	0	10.00	0
TP12	0	17.82	39.85	0	17.82	0	8.00	0
TP21	0	8.33	18.63	0	8.33	0	8.00	0
TP22	0	9.16	20.49	0	9.16	0	6.40	0
TP31	0	3.94	8.85	0	3.94	0	6.00	0
TP32	0	4.33	9.74	0	4.33	0	4.80	0
TP41	0	3.89	8.70	0	3.89	0	4.00	0
TP42	0	4.28	9.57	0	4.28	0	3.20	0
U11	0	6.50	14.56	0	6.50	35.00	0	2100.00
U12	0	26.00	58.24	0	26.00	35.00	0	6300.00
U21	0	6.50	14.56	0	6.50	40.00	0	2100.00
U22	0	26.00	58.24	0	26.00	40.00	0	6300.00
U31	0	45.31	101.43	0	45.31	110.00	0	525.00
U32	0	51.20	114.62	0	51.20	110.00	0	1575.00
U41	0	45.31	101.43	0	45.31	130.00	0	525.00
U42	0	51.20	114.62	0	51.20	130.00	0	1575.00
R11	0	1.39	3.12	5.50	−7.55	0	0	0
R12	0	0	116.25	0	0	0	2.33	0
R21	0	0.97	2.20	17.00	−9.94	0	0	0
R22	0	0	82.00	0	0	0	1.64	0
R31	0	0.50	1.11	26.25	−4.38	0	0	0
R32	0	0	41.20	0	0	0	0.82	0
R41	0	0.64	1.46	36.50	−7.19	0	0	0
R42	0	0	54.25	0	0	0	1.09	0

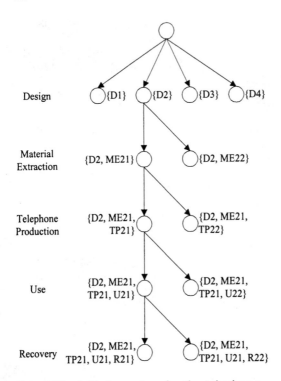

Design   {D1}   {D2}   {D3}   {D4}

Material
Extraction   {D2, ME21}   {D2, ME22}

Telephone
Production   {D2, ME21, TP21}   {D2, ME21, TP22}

Use   {D2, ME21, TP21, U21}   {D2, ME21, TP21, U22}

Recovery   {D2, ME21, TP21, U21, R21}   {D2, ME21, TP21, U21, R22}

FIGURE 15   A life locus tree for the telephone

6.4 Experimental Results and Discussions

The optimization criterion with different weighting vectors leads to different optimal life loci, for instance, optimal telephone design and its associated production, usage, and recovery processes. An exhaustive search algorithm is used. When more than one life loci correspond to the same optimal criterion value, we simply select the one that comes first in the search procedure. Table 5 provides some typical search results. They are based on different optimization criteria: some based on an individual index and others based on combinations of all eight indices.

For example, to minimize mass intensity, we should select D3, ME32, TP31, U31, and R31. The optimal life locus that has the best revalorization and service extension performance is D4, ME41, TP41, U41, and R41. The optimal life locus with minimum H & E potential risk is D2, ME22, TP21, U21, and R21. This life locus is also the optimal one when all six eco-compass indices are considered with equal importance, i.e., $\mathbf{w} = (1 \quad 1 \quad 1 \quad 1 \quad 1 \quad 1 \quad 0 \quad 0)^T$. However, if revalorization and service

TABLE 5   Optimal Life Loci Corresponding to Different Weighting Vectors

w	$(1\,0\,0\,0\,0\,0\,0\,0)^T$	$(0\,1\,0\,0\,0\,0\,0\,0)^T$	$(0\,0\,1\,0\,0\,0\,0\,0)^T$	$(0\,0\,0\,1\,0\,0\,0\,0)^T$
C	MI	EI	H&E	−RV
Optimal life locus	D3	D2	D2	D4
	ME32	ME21	ME22	ME41
	TP31	TP21	TP21	TP41
	U31	U21	U21	U41
	R31	R22	R21	R41
C^*	11.4	17.14	38.89	−36.5

w	$(0\,0\,0\,0\,1\,0\,0\,0)^T$	$(0\,0\,0\,0\,0\,1\,0\,0)^T$	$(0\,0\,0\,0\,0\,0\,1\,0)^T$	$(0\,0\,0\,0\,0\,0\,0\,1)^T$
C	RC	−SE	Cost	−Benefit
Optimal life locus	D2	D4	D4	D1
	ME22	ME41	ME41	ME11
	TP21	TP41	TP42	TP11
	U21	U41	U41	U12
	R21	R41	R41	R11
C^*	20.43	−130	8.2	−6300

w	$(1\,1\,1\,1\,1\,1\,0\,0)^T$	$(1\,1\,1\,2\,1\,2\,0\,0)^T$	$(0\,0\,1\,2\,1\,2\,1\,0)^T$	$(0\,0\,1\,0\,1\,0\,1\,1)^T$
C	MI + EI + H&E − RV + RC − SE	MI + EI + H&E − 2RV + RC − 2SE	H&E − 2RV + RC − 2SE + Cost	H&E + RC + Cost − Benefit
Optimal life locus	D2	D4	D4	D2
	ME22	ME42	ME42	ME21
	TP21	TP41	TP41	TP21
	U21	U41	U41	U22
	R21	R41	R41	R21
C^*	33.66	−100.39	−154.61	−6154.26

extension assume more significance, i.e., $\mathbf{w} = (1 \quad 1 \quad 1 \quad 2 \quad 1 \quad 2 \quad 0 \quad 0)^T$, the optimal life locus is D4, ME42, TP41, U41, and R41. When $\mathbf{w} = (0 \quad 0 \quad 1 \quad 0 \quad 1 \quad 0 \quad 1 \quad 1)^T$, the optimization criterion $C = I_2 - O_2 + O_3$ (referring to Figure 13), and the corresponding optimal life locus is D2, ME21, TP21, U22, and R21.

It needs to be mentioned that the validity of these search results depends on the validity of the assumptions made and the MLCA data obtained and the weighting factors used.

7 CONCLUSIONS

The LCA methodology has been extended to emphasize product end-of-life processes into a multi–life cycle framework. Software implementation is presented within the structure of a knowledge base management system. The models and databases are validated using a case study of four generations of office telephones and life cycle engineering design of telephones. These applications illustrate the potential of the developed MLCA methodology and tool. A major hurdle to the wide applications of LCA including MLCA methodologies lies in the buildup of a reliable and extensive database. The collection of environment-related data means tremendous work. The confidence level of the collected data's validity is an critically important factor in the application and interpretation of the assessment results. In addition, a significant amount of data is market sensitive, in particular, material prices and resale price of reuse parts and subassemblies. Demanufacturing and recycling technologies also play an important role in MLCA methodologies and tool.

ACKNOWLEDGMENTS

We acknowledge the support of the New Jersey Commission on Science and Technology through the Multi-Lifecycle Engineering Research Center at NJIT and the funding of Lucent Technologies. In addition, we recognize the stimulating technical discussions held with Joseph Moribito, John Mosovsky, David Dickerson, and Werner Glantznig of Lucent Technologies.

REFERENCES

H. Al-Okush. Assessing Design for Environment Guidelines: A Case Study of Office Telephones. M.S. Thesis, Industrial & Manufacturing Eng. Dep., New Jersey Institute of Technology, Newark, NJ, January 1999.

H. Al-Okush, R. J. Caudill, and V. Thomas. Understanding the Real Impact of DFE Guidelines: A Case Study of Four Generations of Telephones. In Proceedings of the 1999 IEEE International Symposium on Electronics and the Environment, May 1999, pp. 134–139.

R. J. Caudill. Multi-lifecycle Engineering and Demanufacturing of Discarded Electronic Products. Technical Report, Multi-lifecycle Engineering Research Center, New Jersey Institute of Technology, October 1998.

M. A. Curran. Environmental Lifecycle Assessment. New York: McGraw-Hill, 1996.

S. Das, R. Sodhi, Z. Ji, and R. J. Caudill. Disassembly Effort Index Calculator. Technical Report, MERC Center, NJIT, January 1998.

ECO-it. ECO-it: Eco-Indicator Tool for Environmentally Friendly Design. PRé Consultants BV, April 1999, http://www.pre.nl/eco-it.html

C. Fussler and P. James. Driving Eco Innovation: A Breakthrough Discipline for Innovation and Sustainability. London: Pitman Publishing, 1996.

GaBi. GaBi—the Software System for Life Cycle Engineering. IKP University of Stuttgart in cooperation with PE Product Engineering, October 1997, http://www.pe-product.de/

W. J. Glantschnig. Green design: an introduction to issues and challenges. IEEE Trans Components Packaging Manufact Technol Part A 17:508–513, 1994.

T. E. Graedal and B. R. Allenby. Industrial Ecology. Englewood Cliffs, NJ: Prentice Hall, 1995.

T. E. Graedel and B. R. Allenby. Design for Environment. Englewood Cliffs, NJ: Prentice Hall, 1996.

M. A. Hersh. A survey of systems approaches to green design with illustrations from the computer industry. IEEE Trans Syst Man Cybernet 28: 528–540, 1998.

J. Jin. Multi-Lifecycle Assessment Design Tools and Software Development. M. S. Thesis, Electrical and Computer Eng. Dept., New Jersey Institute of Technology, Newark, NJ, May 1999.

SimaPro. SimaPro 4.0—the Life Cycle Assessment Tool. January 1999, http://www.pre.nl/simapro.html

TEAM. Ecobilan Group's Life-Cycle Assessment: TEAM^{TM}. Ecobalance, Inc., July 1998, http://www.ecobilan.com/

B. W. Vigon, et al. Life-Cycle Assessment: Inventory Guidelines and Principles. Risk Reduction Engineering Laboratory Office of Research and Development, U.S. Environmental Protection Agency, February 1993.

P. Yan, M. Zhou, and D. Sebastian. A generic framework for integrated product and process development. Int J Environment Conscious Design Manufact 7:47–57, 1998.

P. Yan, M. Zhou, and D. Sebastian. An integrated product and process development methodology: a case study. In Proc. Rensselaer's Int. Conf. on Agile, Intelligent, and Computer-Integrated Manufacturing, Paper No. 46, Troy, New York, October 1998.

P. Yan, M. Zhou, and D. Sebastian. A methodology for integrated product and process development: concept formulation. Robot Comput Integr Manufact 15:201–210, 1999.

J. M. Young. Lifecycle Energy Modeling of Telephones. April 1995, http://sun1.mp-ce.stu.mmu.ac.uk

S. Young and W. H. Vanderburg. Applying environmental lifecycle analysis to materials. J Manufact 22–26, 1994.

M. C. Zhou, R. J. Caudill, D. Sebastian, and B. Zhang. Multi-lifecycle product recovery for electronic products. J Electron Manufact 9:1–15, 1999.

M. C. Zhou, B. Zhang, R. J. Caudill, and D. Sebastian. A Cost Model for Multi-lifecycle Engineering Design. In Proceedings of the 1996 IEEE International Symposium on Emerging Technologies and Factory Automation, Hawaii, November 1996, pp. 385–391.

4

Design Methodologies for the Environment

Samir B. Billatos

i2 Technologies, Inc., Irving, Texas

1 INTRODUCTION

Given the wide reporting of environmental concerns today, it is very difficult not to be pessimistic about the future of our planet. From the days of the Industrial Revolution, technology has spurred civilization on to new and staggering heights in very short spans of time. New technologies have made our lives much easier, but past designers did not always stop to consider undesirable side effects. The development of refrigerator compressors suitably illustrates this point: Early compressors used ammonia or sulfur dioxide, both of which are toxic chemicals that sometimes injured and even killed people. Then chlorinated fluorocarbons (CFCs) were developed and were hailed for their safety, low cost, and use in popular applications such as air conditioning. Only later was CFC use connected to phenomena like global warming and ozone layer destruction.

Humankind is always growing. By the year 2030, 10 billion people will probably live on this planet. Critical natural resources will be expended to support this population, and enormous wastestreams will be generated. Even today, each person in the United States alone produces a daily average of 4.5 pounds of solid waste. The result is approximately 180 million tons of municipal solid waste sent to landfills annually. By 2030, that figure could climb to 400 billion tons annually, which is "enough to bury greater Los Angeles 100 meters deep." With so much taken out and none being put back, it does not take much thought to conclude that we are on a collision course with environmental disaster. We must find efficient ways to recycle the wastestreams and put them back to work in the ongoing "industrial ecosystem."

In the light of this discussion, it is not difficult to see how design for the environment (DFE) can contribute toward minimizing the waste generated. Manufacturers produce goods for public consumption; because few mechanisms exist to recycle most products at the end of their useful life, landfills are put into service with the consequences outlined above. Thus, DFE dictates that designers and engineers should examine their products for environmental soundness, which encompasses an environmental "cradle-to-grave" snapshot across the entire product life cycle. The traditional product life cycle, as illustrated in Figure 1, must change to ensure that a DFE infrastructure is in place. The motivation for DFE should stem from more serious concerns than just reducing costs and minimizing waste. Fortunately, in many instances, the latter are also realizable, which makes DFE a methodology worth considering for more than purely altruistic reasons.

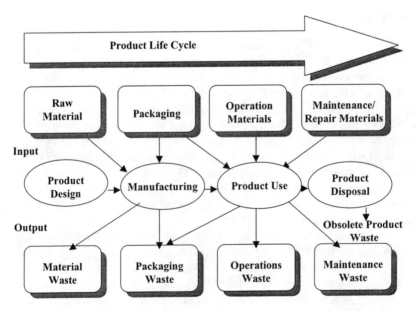

FIGURE 1   Material input and wastestreams in a traditional product life cyle.

Further, environmental legislation nationwide and worldwide is more stringent than it has ever been. Federal, state, and local statutes regulate air emissions, water discharges, occupational exposure, and the treatment and disposal of various hazardous chemicals. The 1990 Clean Air Act emphasizes waste minimization and emission monitoring and reporting and lists 189 compounds for regulation as air toxins. In addition, the internationally ratified Montreal Protocol regulates the production and use of halogenated organic compounds (CFCs are among them). With such severe legislative pressures being brought to bear upon manufacturers, the final outcome is clear: Changes in manufacturing and design philosophy must occur.

With this brief introduction, the stage for a closer look at the DFE process is now set. The "leading-edge" efforts of manufacturers today and highlights of the major challenges that DFE presents are examined in the following sections.

2 DFE EXPLORED

From concurrent engineering practice, it is known that approximately 75% of a product's total life cycle costs are determined in the design stage. Therefore, decisions made during the design phase profoundly impact the entire product's life cycle. This provides the incentive for "doing one's homework" very early on in the design process. To avoid environmental problems after the design phase, there is no apparent reason why environmental concerns could not also be addressed in the initial design stages. The overall costs associated with wastestreams can then be reduced.

Besides the reduction or elimination of product wastestreams from the manufacturing process, designers must examine the environmental impact of the design when it is being produced and disposed of or recycled, as described in Figure 2. Design for recyclability (DFR) can play a significant role in this effort. DFR is an infrastructure where products can be accepted at the end of their useful life by efficiently breaking them down and then recycling the individual products for use in other processes. Thorough materials research at the beginning of the design phase will go a long way in improving the recycling process. To enhance eventual breakdown and recycling, design for disassembly (DFD) principles must be used before the design is finalized. The times, techniques, and costs to disassemble a product are just as important as those factors that exist during its assembly. Since time is money, it is obvious that DFE cannot be implemented without added costs.

When environmental regulations become more strict, waste-associated costs increase. In addition, new studies and better chemical analyses can result in the recategorization of waste from nonhazardous to hazardous,

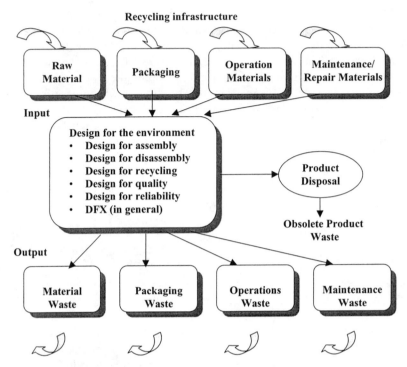

FIGURE 2 Material input and wastestreams in an ideal product life cycle.

"resulting in a tenfold increase in disposal costs." The obvious solution is to reduce waste in the first place but, more important, to ensure that any kind of waste (to include scrap, material rework, and large unused inventories) is minimized in all operations. Not surprisingly, this is exactly what just-in-time (JIT) and concurrent engineering propound; all production operations should be optimized to prevent waste and thus reduce costs. Sometimes, this optimization can involve the use of design for manufacturing and assembly (DFMA) principles; if a product takes less time to assemble, costs are further reduced. As recyclability of the product at the end of its life is a concern for DFE designers, they have the added problem of simultaneously designing for disassembly. Once again, if disassembly time is reduced, so are associated costs. Naturally, any cost saving over the entire product life cycle will benefit both the consumer and the producer.

DFMA and DFD philosophies may appear contradictory, but sensitivity to both aspects can ensure the design is sound for assembly and disassembly. For instance, dual-purpose snap fits can eliminate fasteners (a DFMA plus) and facilitate disassembly (a DFD plus). The Boothroyd and Dewhurst DFMA model may be used to predict times and cost for assembly. It is also important to recognize that the actual assembly sequence is not always the reverse of the disassembly sequence. Computer-aided design (CAD) systems can improve our understanding of this fact. For example, CAD systems have been developed with product assembly/disassembly planning functions that can be used in the earliest stages of design. The use of these functions in CAD software allow numerous changes and iterations and, "what if" analyses and provides the engineer with a tremendous amount of flexibility in design.

Another example of a DFD software design tool, which integrates DFE elements, is the Restar, developed by researchers at Carnegie-Mellon's Center for Integrated Manufacturing Decision Systems in Pittsburgh, Pennsylvania. Restar analyzes disassembly operations with regard to time involved, cost and effort of each step, value of recovered parts and materials, and any savings in energy and pollutant emissions. This tool quantifies the results of the DFE process for the designer's benefit and also shows how design changes may further improve those results. Other software tools covering material selection and compilation of a comprehensive database of the environmental impact of various material choices are in development.

Recent DFD seminars have focused on what industry can do upstream to reduce the enormous amount of plastic that are currently destined for landfills. The outcome of the seminar was three primary recommendations: source reduction, DFD, and use of recycled materials. It was emphasized that source reduction treats the symptom (full landfills) but not the cause (our inability to recycle certain materials and products); still, it finds use as a short-term strategy. It was also emphasized that reducing the total number of materials in the design is crucial to DFD. It was observed that the only answer is to overdesign the material performance in some parts to match materials with the parts requiring those additional characteristics. In so doing, the designer decreases separation time and increases the product's value to the recycler; costs may increase initially but will soon be canceled out by increasing disposal costs. Other material selection options include substituting equivalent materials to suit performance specifications or using two compatible materials that can be recycled together. It was stated that ideally, the goal of material selection in DFD is to eliminate the need for disassembly.

Specifying and using recycled materials in original designs will create and support the markets for these materials in the future. Unfortunately, many designers refuse to use recycled materials because they do not believe they are as good as virgin materials. Therefore, like GE Plastics of Pittsfield,

Massachusetts, even material suppliers involved in recycling put most of their development effort into virgin materials. This perception must change for DFE to succeed. Further, DFE is now seen to be so closely related to other design disciplines; so much that the whole "design for" realm is now increasingly referred to as DFX.

The DFE designer must ask and answer a long list of questions about the product and process: from start to end of the life cycle. The following questions are only a few that exists among a long list:

- What material properties are desirable? What strength and functional properties must the raw materials and the finished product possess?
- To conserve natural resources, can recycled materials be used instead of virgin raw materials? Will these materials provide the same "fit and function" to the product? If not, do other viable alternatives exist?
- What added costs (capital and operational) will be incurred if the new materials are used with (perhaps) different production processes?
- Has the production process been examined for waste generation? Are the solvents and fluids in use hazardous to humans or the environment? Are nontoxic substitutes available? Or can these solvents be eliminated from the picture by using a different manufacturing process? What added costs will be incurred?
- Has the product been designed for optimal DFMA and DFD? Are specified tolerances applied to individual parts? If so, will the completely assembled product meet the required overall tolerances?
- Have individual parts been marked with molded-in logos (to identify material types) and part separation points to facilitate breakdown and recyclability?
- Is a standardized industrial coding and information database available to track recycled material availability and breakdown and recycling facilities?

This is more in line with the all-encompassing "industrial ecosystem" and DFE infrastructure discussed earlier. For DFE to work, such a system will be absolutely vital. Fujitsu Ltd. of Japan claims its most significant achievement in the domain of environmental planning has been the development of an on-line database for environmental information. It is accessible via the company's personal computer network where data on properties, handling, hazards, and disposal is readily available. While this network may not appear very comprehensive, it certainly represents a step in the right direction.

The above list is not all inclusive, but several points become clear. First, not all the design considerations are necessarily environmental, but interestingly they fold in quite nicely with the other "optimal" design and manufacturing methods (DFMA, JIT, concurrent engineering, etc.) that fall under the umbrella of "continuous improvement." Second, industry must put out a lot of effort to establish a comprehensive nationwide and international DFE infrastructure. Effort equates to time and money, so while one can see this is not a trivial task, it is certainly within the realm of possibility. Finally, designers must use their creativity in several different ways to come up with efficient "green" product designs for future generations. Fortunately, the designers have several tools (such as CAD/CAM and DFMA software) at their disposal.

3 DFE PRESSURES

In this era of environmental awareness and (sometimes) activism, several diverse groups have pressured manufacturers to change their ways. Recyclers have tried for years to get manufacturers to use DFR. Legislation has been introduced in this country and abroad to get manufacturers to think in terms of environmental protection and recycling; the Europeans appear to be moving in a faster pace in this regard. In August 1992, Germany's environmental minister Klaus Topfer proposed a rule that required all automakers and their authorized representatives (authorized representatives include car importers, United States, Japanese, French, British, etc., and their suppliers) to take their cars back from consumers for recycling. The rule, which took effect in 1993, also mandated that automakers recycle plastics in cars starting with 20% by weight in 1996 and increase considerably in subsequent years. While the German auto industry already recycles 75% of its cars by weight, more is expected of it so the ramifications are global in scope. In the United States, the House of Representatives has considered amendments to the Solid Waste Disposal Act, which would require manufacturers to take back products that cannot be recycled. In 1990, the Energy and Commerce Committee introduced several bills addressing recycling of appliance components (H1810, H2845, H3735). Finally, consumers themselves have spoken through the use of surveys; 90% of German consumers said they "favored environmental protection measures and preferred clean green products."

And how have manufacturers and designers responded? Environmentally sound efforts abound in the United States and Europe. For example, Black & Decker of Canada has started a "take back" program for its cordless products and recyclable nickel–cadmium batteries. The consumers return the used products to Black & Decker authorized service centers where they receive a rebate toward their next purchase. In addition, Kodak's single-use camera is recycled after it is returned for film processing. Also, General Electric designed a refrigerator, called the Totem, that can be disassembled and recycled. As will be seen in the following sections, auto-makers in the United States and abroad and other large producers of consumer goods are incorporating DFE into the design and manufacture of their products and/or doing research to arrive at more reliable, practical, and recyclable plastics and nonhazardous cleaning solvents. The battle continues with new ground gained every day.

4 DFE GUIDELINES

Based on the above discussion, it is possible to summarize some of the general guidelines used in DFE. Again, an environmentally sound design (the flower in Figure 3) can grow in a "green" engineering environment (the sun) where the designer can understand the diverse nature of DFE and deal

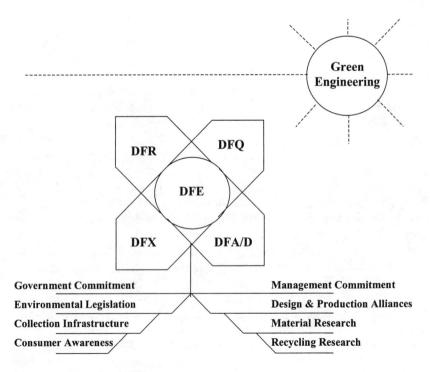

FIGURE 3 Growth of DFE

with it accordingly. In addition, there is no substitute for comprehensive knowledge, solid research, working partnerships, adequate funding, and a firm commitment from upper management.

With this understanding and the ideas presented above, some not-all-inclusive DFE guidelines may be stated as follows:

- Keep the design simple by using as few materials as possible. Also, incorporate as many functions as possible into any single part without compromising function. Avoid secondary finishes, toxic materials, and heavy metals that can contaminate the material.
- Find multiple or secondary uses for a product. Disposal will be less of a problem if a product has more "intrinsic value." As an example, a container protecting a product could also store the accessories that come with it.
- To ensure easier recycling, use materials that match each other closely or are of the same material. Look for ways to use recycled materials as starting compounds for a product.
- Modular design should be preferred whenever possible. This helps in maintenance and repair, a "black box" concept.
- Design for long product life and become more service oriented. If a manufacturer upgrades its product as technologies improve, a more loyal customer base is ensured.
- Ensure tracking mechanisms are available on a "cradle-to-grave" basis. Ensure up-to-date databases are available. Ensure parts are marked with logos to aid in recycling efforts.
- Examine those components in a design that may be reused upon failure or disassembly. This would reduce the need to recycle or dispose.
- Establish a network of producers and suppliers to form the beginnings of the "industrial ecosystem" and facilitate DFE efforts.
- Look for ways to reduce waste byproduct streams in manufacturing processes. Seek out nonhazardous solvents and cleaning materials. Reduce energy consumption by eliminating unnecessary manufacturing steps.
- Ensure a product "buy back" infrastructure is in place and well advertised to suppliers, producers, and consumers.
- Pay close attention to recyclability and reuse of packaging, shipping, and other peripheral requirements. Design reusable shipping vehicles.
- Whenever possible, attempt to incorporate a concurrent engineering philosophy (i.e., JIT, DFMA, DFD, etc.) to aid the overall DFE effort. Design for total ease of assembly, separation, handling, and cleaning.

- Apply tight tolerance design principles to reduce the use of fasteners and keep the separation process simple.

5 DFE APPLIED

As stated several times in this chapter, manufacturers in diverse industries are applying DFE principles very successfully today. It is their creative efforts that provide optimism for the future of our planet, both from a social and industrial perspective. The following industrial applications examine specific DFE efforts.

5.1 The U-Kettle

Polymer Solutions of Worthington, Ohio designed the U-Kettle for Great British Kettles, a United Kingdom–based appliance manufacturer. Polymer Solutions is a joint venture between GE Plastics and Fitch Richardson Smith, an industrial design firm. Besides a metal heating element, the electric U-Kettle consists of six parts injection molded of Noryl-modified polyphenylene oxide supplied by General Electric Plastics of Pittsfield, Massachusetts and two parts molded of GE's Lomod copolyester elastomer. The six parts are the reservoir, base, lid handle, cross-brace, an on/off button, a handle grip, and a lid grip. Mack Molding of Arlington, Vermont is the appliance molder; an initial run of 500,000 units required about 400,000 pounds of plastics, since each kettle consists of 21.5 ounces of Noryl and 2.6 ounces of Lomod.

In this design, dual-purpose snap-fits were used to eliminate fasteners and enhance disassembly at the same time—a good example of DFMA and DFD principles used in concert. Molded-in break points were also designed in without any loss of performance during normal operation. Another area of concern during product prototyping was the handle of the assembled kettle, which proved to be unsteady when filled; sonic welding of two points took care of the problem without affecting disassembly. Finally, to aid recycling, molded-in logos clearly identified material type. To aid disassembly, molded-in instructions clearly showed part separation points.

The U-Kettle is a classic example of what a DFE effort should represent. The partnership between an industrial design firm, a plastics producer, and an appliance manufacturer effectively married three critical areas of DFE: design, material selection/properties, and the manufacturing process. The "experts" in each area were able to put their heads together to arrive at a producible environmentally sound design. They recognized the lack of and need to push for a recycling infrastructure and ensured both thermoplastics used could be recycled together. Also, the U-Kettle is reported to consume less electricity than any other electric cookware item, which is another victory in the fight to conserve resources. Finally, the designers and manufacturers worked together using DFMA, DFD, and DFR creative design solutions to overcome practical problems. Their "trial-and-error" solutions have blazed a trail for future DFE designers and manufacturers.

Even though a lot remains to be done, several shining DFE successes in the auto industry stand out. The German manufacturing arm of General Motors (GM), Adam Opel AG, plans to use 22 million pounds per year of recycled plastics in more than 20 automotive parts. Furthermore, the company plans to recycle nearly 100% of the plastics used in its new models, called Astra and Calibre. Opel expects plastics to comprise about 15% of a car's weight by the year 2000, up from 10% in 1985. Fuel efficiency should improve through lightweight components and aerodynamics. Also, Opel made a conscious DFE decision not to use tough-to-recycle reinforced sheet molding compound in the design of its new cars. However, Opel executives expect renewed use of sheet molding compound when a new polyurethane recycling program is introduced.

5.2 Opel's DFE Program

Opel's DFE program has other significant elements. First, Opel designers will use recyclable material wherever possible. The material of choice is polypropylene, which costs less, is highly durable, recyclable, and has noise-reduction properties. Excluding fibers, polypropylene accounts for 68% of the plastics in the Astra model. Second, Opel will attempt to phase out polyvinyl chloride. This effort is driven by the refusal of municipal incinerators in Austria to accept vinyl waste. Third, Opel is committed to DFD principles. A new GM clamp uses a wedge catch and tooth strip for the rear bumper mount, enabling easy access and disassembly in 20 seconds. As much as possible, assemblies are made of like materials, without metals, to be fed directly into the recycling process, a classic DFR consideration. For example, a rack-and-pinion assembly previously made from steel and rubber is now made from polypropylene elastomer. Nylon and polybutylene (PBT) elastomer are used in a new gearbox design. Finally, as mentioned earlier, recycled materials are used whenever possible. Old polypropylene battery casings are now widely used by Opel in wheel arch fender liners; compounds are upgraded with bumper production scrap and virgin polypropylene. Polyurethane seat foam is used in a sound-proofing mat in dashboards, whereas industrial plants, dealers, and scrap yards supply scrap for these efforts.

Opel's experience points to the fact that a "green" vision is essential for DFE to succeed. Once long-range plans have been developed, all capabilities and research efforts must be brought to bear in an effort to implement those plans. Opel used classic DFE guidelines with regard to material selection, DFMA, DFD, and DFR.

5.3 DFE in the Auto Industry

Other examples of DFE include the efforts of Mercedes-Benz to implement a total vehicle recycling program with two main elements: vehicle design and vehicle recycling. Highlights of the Benz design effort include choosing environmentally compatible and recyclable materials for components, reducing the volume and variety of plastics used, marking plastic parts with logos, and avoiding composite materials as much as possible. Plastic parts are used only when practical advantages are offered by their use, and DFD and parts sorting upon breakdown are also given full consideration. Mercedes-Benz started taking scrap cars back in 1991. Once dismantling and shredding has been completed, the material is placed in a smelting reactor in a process Benz calls "metallurgical recycling." Obviously, such an extensive DFE program must be extremely well thought out, planned, and implemented from the very beginning. Such is the challenge for the DFE engineer.

BMW of North America recently announced a pilot program to test the feasibility of nationwide recycling of BMW automobiles: Because of tough German environmental laws, the company already recycles cars in Europe. Targeting three U.S. cities, BMW will give owners a $500 credit toward the purchase of a new or used BMW for turning in a car to a dismantling center. Using DFD principles and more recyclable components in the original design, BMW hopes to increase the percentage of recycled car weight from the present 75% to 90% in the future.

The U.S. auto industry is also looking at ways to break down polymeric materials into their original chemical state through hydrolysis, methanolysis, and pyrolysis. During hydrolysis, superheated steam is used to depolymerize resins, resulting in reduction to monomer. The monomer can be recycled when the plastic is repolymerized. During methanolysis, methanol is substituted to depolymerize resins. During pyrolysis, plastics and rubber are broken down to fuel-grade oil and gas by heating in an oxygen-deficient environment, resulting in usable products.

For the present and future, U.S. automakers are focusing on many areas to recover automotive plastics such as batteries, where 150 million pounds of polypropylene is now annually recycled from 95% to 98% of all batteries for reuse in new batteries and other products; reaction injection molding auto parts, which are currently landfilled; radiator end caps, where an infrastructure to collect the caps, recycle, and perhaps reuse them in other applications is being established; and sheet metal composite scrap, where new product alternatives from this scrap are being explored.

In Europe, bumpers from scrapped cars are already being recycled. Volkswagen blends reclaimed polypropylene with virgin materials and uses them in bumpers for new cars. BMW reclaims PBT/polycarbonate bumper material from scrapped cars and uses the recovered material in new auto parts. Through a pilot recycling plant, in partnership with a cement manufacturer and a scrap metal processor, Peugeot S.A. hopes to produce a granulate cement-making compound from 7000 scrapped cars over 2 years.

5.4 Germany's Green Television and DFE

As stated earlier, Germany leads the field as far as environmental awareness and action and, more importantly, environmental legislation geared toward the recycling/reuse of consumer products. German lawmakers introduced legislation requiring equipment makers to take back used televisions (TVs) and other electronics items for recycling. And even though "the government cannot tell industry how to make its equipment," DFE/DFR is a logical option for German manufacturers facing a real problem.

To this end, the German Ministry for Research and Technology and five European set makers have entered into a project to produce a fully recyclable TV; a prototype was introduced in 1994. The companies involved are Grundig AG, Loewe, RFT (all German), Nokia (Finland), and Thomson-Brandt (France). The motivation behind this novel partnership stems from the large number of TV sets (four million) that Germans discard annually. Naturally, the key areas under study are disassembly, recycling, and reuse of components. Another feature of this alliance is the division into different areas of DFE research. For example, Loewe is examining DFD aspects; Nokia, the recycling of picture tubes; Thomson, the construction angle; RFT, the recycling of coils and other assemblies; and Grundig, the selection of appropriate materials and production processes.

Referring back to the earlier guidelines, it is clear that one path to the solution of far-reaching and complex environmental problems is the idea of pooling resources from different areas of expertise. Teamwork will reap rich dividends, and the beauty of it all is that everyone benefits. The international flavor highlights an important point: United we stand, divided we fall.

5.5 Grundig AG's Environment Initiative

DFE takes firm commitment and perseverance. Take the overall manufacturing philosophy of Grundig AG, a $2.8 billion consumer electronics giant, which recently launched an "environment initiative" program. The goals of this program run the gamut from selecting suitable recycled materials to environmentally sound production and packaging. Upper management stands firmly behind these goals; a specialized team examines total environmental compliance in the production of all items and reports directly to the company's leaders.

Grundig consciously undertook a program to eradicate environmentally unsound chemicals from its production processes. Solvent-free water-based cleaning materials are used almost exclusively, and organic solvents and CFCs are no longer in use. All subassemblies and materials containing fire retardants are free of polymerized diphenylethers to prevent emission of cancer-causing dioxins and fumes. In addition, Grundig uses materials free of formaldehyde, asbestos, cadmium, and mercury. At Grundig, highly efficient robots and machines using water lacquers perform lacquering.

Finally, Grundig recycles electronic subassemblies such as TV tuners and remote control units in its own module repair center, which is the largest of its kind in Europe. After repair, the parts are ready for reuse.

The Grundig experience shows that an overall "green" vision must pervade all aspects of a company's operations for DFE to thrive. Unhindered by its large size, Grundig sets a DFE example for the world to follow.

5.6   DFE in the Consumer Appliance Industry

Appliance makers are probably on the cutting edge of DFE practice today; the U-Kettle is a good example of DFE design. Further, since the consumer appliance market is so aggressive, manufacturers must keep pace with new materials and find that happy medium between cost and performance to stay competitive. The same environmental pressures exist for appliance producers, and since plastics have a wide use in this industry, material selection is impacted not only by cost and/or performance factors but DFD, DFR, and DFE trends as well. In addition, CFC elimination and power consumption reduction dictated by current and forthcoming laws provide motivation to research and develop product designs and materials that are environmentally sound.

Refrigerator and freezer manufacturers face the problem of finding substitutes for CFCs, whose use in the United States is almost eliminated. Previously, CFCs were used as efficient refrigerators and insulators, but then it was discovered how their release could destroy the earth's protective ozone layer. The challenge, then, is to come up with an environmentally friendly substitute, which has the same properties. And the search continues.

For refrigeration, hydrofluorocarbon 134a has no effect on the ozone layer and could replace the soon-out-of-favor CFC 12, but it exhibits incompatibilities with some refrigerator parts. By the same token, CFC 11, used as insulation in refrigerator walls, has no real green substitutes that possess the same efficiency and noncorrosive properties. Designers have a choice: For better insulation, make refrigerator walls thicker. Unfortunately, given the limited kitchen space of most homes, larger appliances are not likely to be very popular with the consumer. But the search continues.

Also ongoing in the refrigerator arena is the use of engineering thermoplastics rather than CFCs as insulators. Designed by GE Plastics, one DFE design uses molded-in logos for material identification, a minimum of different materials, and compatible materials for combined recycling. DFMA features include parts consolidation and snap-fits. Once again, we witness the successful marriage of DFD, DFMA, DFE, and DFR in a single product.

As a final example of this marriage, consider Cleanworks, a compact modular dispensing system for industrial cleaning fluids. Designed for Scott/Sani-Fresh International, Inc. by Bally Design, Inc., it combines DFD, material reduction, reclamation, and reuse of bottles and product disposal. The system consists of six recyclable, blow-molded, and polyethylene bottles. Each is capped with an injection-molded polystyrene Venturi pump assembly that delivers a precisely mixed cleaning fluid from concentrate. The pump aspirates the concentrate and mixes it with water flowing through the dispenser. Once the module is empty, Scott picks it up and replaces it. The pump assembly is removed from the bottle, ground up, and recycled. The existing bottle is refilled and then capped with a new pump assembly.

During design, environmental concerns were at the forefront. Both the designer and manufacturer were concerned with the consequences of discarded refill bottles. The manufacturer (Scott) took it upon itself to retrieve the bottles and reuse them whenever possible. When that was not possible, Scott made sure a recycling mechanism was in place. As seen in other examples, the recycling factor drove the choice of materials in the original design. Also, Scott had to set up the collection and recycling infrastructure for the DFE objectives of the design to be satisfied.

Besides DFE concerns, engineers are sometimes faced with several other conflicting criteria. Take the case of microwave ovens. With power ratings for these units on the rise, manufacturers must turn to new high-performance engineering plastics to do be compatible in the units. Sharp Manufacturing Company of America boosted power in its Carousel 11 models from 800 to 900 watts. The impact on the design was that the control panel and door opener button now had to meet a higher UL relative temperature index. Moreover, the panel had to survive a drop impact test after withstanding a temperature of 204°F in a chamber. Sharp researched several materials before settling on a flame-retardant PC/ABS (acrylonitrile butadiene styrene) blend from Miles, Inc. At a 0.125-inch thickness, this material met all UL specifications. The designers also found high-impact resistance, dimensional stability, and resistance to ultraviolet and fluorescent light in this material. Aesthetic considerations were also satisfied as the material provided excellent colorability through color match with other components. DFMA principles were put to good use in this design; the control panel had clips molded to its backside, which are put through slots, slid down, and locked into place. The panel is fastened to the oven with just one screw. Sharp's design philosophy illustrates the necessity to conduct painstaking research on different materials, but it also shows how challenging the design task really is: Conflicting and diverse constraints must be dealt with simultaneously. Fortunately, the designers have tools such as DFMA, DFE, and DFR at their disposal.

5.7 DFE in Packaging Design

Our final DFE application will highlight the efforts of the packaging industry. As in other design and manufacturing operations, the design of packaging is driven by cost and/or performance. The added environmental constraint means that, for the most part, packaging must be reusable or recyclable and still maintain function. Surveys have shown that most consumers are very concerned about a package's disposability and these concerns influence their buying decisions. Along with recyclability, buyers want convenience, freshness, and tamper evidence in their packaging; designers of packaging materials are sometimes hard pressed to strike a balance between those requirements.

Most of the design emphasis for environmentally sound packaging is currently in the rigid packaging area. The most obvious solution appears to be that classic DFE premise: Use less material. But as discussed earlier, source reduction alone will not solve all our problems. As designers battle with the answers, they seek answers to some of the following questions: Should some resins and additives be avoided because, more and more, municipal waste is being incinerated? Should packaging consist only of single layers (as opposed to multilayers) to facilitate recycling? Will there be enough postconsumer waste collection and separation to justify designs using multilayers (for nonfood packaging) with recycled layers? The designers hope a public consensus on the matter of solid waste disposal will provide answers and help them shape their designs accordingly.

Current green practices in package design include the use of fewer materials; recyclable materials where possible; elimination of heavy-metal dyes, pigments, and inks; and the elimination of multimaterial lamination from packaging. Package designers are moving toward lightweight designs; structural layers get thinner as materials improve. More precise manufacturing process control reduces the need for "over-engineered barrier and adhesive layers." Designers continue the development of thinner, lighter, and stronger containers; as an incentive, Du Pont now hands out packaging awards for environmentally sound packaging.

Procter & Gamble (P&G) has explored source reduction in Germany where it markets Lenor fabric softener as a concentrate. Consumers get a refillable bottle into which they empty a small pouch of concentrate and enough water to produce 4 liters of solution. P&G claims this effort represents a 90% source reduction when compared with throwaway detergent bottles. But interestingly, P&G credits a large part of the program's success to the green mentality of the German consumer.

In the recycling arena, P&G once again has made significant inroads. It uses postconsumer scrap in the manufacture of its detergent bottles; P&G uses in excess of 100 million pounds of HDPE a year for its bottles. On a limited basis, it bottles Spic and Span, a detergent, in bottles made from 100% PET beverage bottle scrap. Other efforts include coextruded bottles from 20–40% reclaimed HDPE milk bottles. In the future, P&G hopes to double the recycle content of HDPE detergent bottles, even though the recycle content of a detergent bottle will never be 100% because of its contents. Detergents require at least a skin layer of the special grades of stress-crack resistant HDPE that now make up 100% of most detergent bottles. So the use of recycled milk bottles in detergent bottles will, for the near future, require multilayer bottles.

6 SUMMARY

This chapter started with a gloomy forecast for the future of our planet. True, the statistics about environmental damage and neglect can be mind numbing, but a look upon the ingenious efforts of the designers and manufacturers using DFE indicates a sense of greater optimism. Hope does spring eternal from the human breast; in the case of DFE today, enough constructive work is being accomplished in the arena of environmental protection that there is indeed the feeling that human creativity will endure once more.

But this is no time to rest on one's laurels, for much work must yet be done. The fact remains: The green designs are here to stay, but so are the DFE challenges. Engineers must try to come up with a green angle on every project, although it is difficult to make much progress. They could design the perfect product for disassembly, but newsprint, aluminum, and plastic PET (polyethylene terephthalate) are the only recycling channels currently open in the United States. Channels are not available to recycle most of the materials—the engineering plastics like ABS and polyurethane that lend themselves to more precision molding—that we work with. As designers, this is our biggest problem.

Recognition and definition of the problem derives half the solution; with slow and steady steps, we must capture the remainder. We have the tools at our disposal: DFE, DFR, DFMA, DFD, concurrent engineering. With careful manufacturing and materials research, top management commitment, increased consumer awareness and support, production and design alliances, improved collection and recycling infrastructures, and above all dedication, patience, and creativity, the "industrial ecosystem" will be well within our reach.

BIBLIOGRAPHY

1. Weissman, S. H. and Sekutowski, J. C. Environmentally Conscious Manufacturing: A Technology for the Nineties. AT&T Techn J 70:23–30, 1991.
2. Gaucheron, T., Sheng, P., and Zussman, E. Hierarchical Disassembly Planning for Complex Systems. Ann CIRP, 1997.
3. Pahng, F., Senin, N., and Wallace, D. Modeling and Evaluation of Product Design Problems in a Distributed Design Environment. Proceedings of the ASME DT Conferences, 97-DETC/DFM-4356, Sacramento, California, 1997.
4. Gosch, J. Leading-Edge Designers Design for Recycling. Machine Design 63:12–14, 1991.
5. Dvorak, P. Putting the Brakes on Throwaway Designs. Machine Design 65:46–8+, 1993.
6. Schofield, J. A. Programmable System Controls PC's Power Consumption. Design News 49:149–151, 1994.
7. Murray, C. Snap-in PC Parts Aid Recycling. Design News 49:45–51, 1993.
8. Machlis, S. A Peek at the Peripheral Grab Bag. Design News 49,:50–52, 1993.
9. Ishii, K. Incorporating End-of-Life Strategy in Product Definition. Invited paper, EcoDesign'99, First International Symposium on Environmentally Conscious Design and Inverse Manufacturing, February 1999, Tokyo, Japan.
10. Masui, K., Mizuhara, K., Ishii, K., and Rose, C. Development of Products Embedded Disassembly Process Based on End-of-Life Strategies. EcoDesign'99, First International Symposium on Environmentally Conscious Design and Inverse Manufacturing, February 1999, Tokyo, Japan.
11. Rose, C. M., and Ishii, K. Product End-of-Life Strategy Categorization Design Tool. J Electron Manufact, 1999.
12. Sheng, P., Carey, V., Bauer, D., Thurwachter, S., Creyts, J., and Bennett, D. Environmental Planning for Machining Operations and System. Proc. 1998, NSF Design and Manufacturing Grantees Conference, January 1998, Monterrey, Mexico.
13. Sheng, P., Bennet, D., and Thurwatcher, S. Environmental-Based Systems Planning for Machining. Trans CIRP, 1998.
14. Wallace, D., Abrahamson, S., and Borland, N. Design Process Elicitation Through the Evaluation of Integrated Model Structures. Proceedings of the ASME DT Conferences, DETC/DFM-8780, Las Vegas, Nevada, 1999.
15. Sousa, I., Wallace, D., Borland, N., and Deniz, J. A Learning Surrogate LCA Model for Integrated Product Design. Life Cycle Networks Proceedings of the 6th International Seminar on Life Cycle Engineering, CIRP, June, Kingston, 1999.

5

Environmental Laws in the United States

Mahendra S. Hundal
University of Vermont, Burlington, Vermont

1 INTRODUCTION

Environmental laws are enacted with the ultimate aims of protection of human health; protection of animal and plant life; protection of natural systems in general, thus promoting biodiversity and slowing species loss; conservation of natural resources (e.g., energy resources); and reducing the need for landfills. As can be seen after some reflection, these aims are interrelated to various degrees [1].

According to the Science Advisory Board of the U.S. Environmental Protection Agency (EPA), reduction in biodiversity and species loss pose risk to human health. The means by which these aims are to be achieved are primarily the reduction or elimination of pollution of air, water, and land and by conservation of materials and energy. Laws and regulations are a major tool in protecting the environment [2]. While all states and indeed many local governments have put into place environmental laws and regulations, what follows here deals with the laws at the federal level.

U.S. laws are enacted by the Congress through the legislative process. An act thus established describes a policy or program and may designate an agency or department of the government to develop details of the program.

The agency or department implements the law by issuing regulations, by publishing statements of policy, and by enforcement.

Environmental laws vary from country to country. These laws, as stated above, aim to address the problems of waste disposal, risks to human health and to ecosystems, global warming, and ozone depletion. In general, the public's environmental awareness, which eventually translates into laws, is proportional to a given country's national wealth. Thus, one sees the greatest progress in this direction in North America, Western Europe, and Japan. It is also true that under the present conditions of economies based on consumption, the amount of waste produced in a country is dependent on the size of its economy.

In Europe, as one might expect, environmental laws also vary by country. With the emergence of the European Union there is an increasing commonality among them. As shown by Dammert et al. (see Chapter 6), European laws tend to focus on environmental attributes of products. European countries also lead in developing labeling systems for products (ecolabeling). There are trends toward uniform product standards that tend, however, to set "minimum" standards. "Greener" (read: "richer") countries are able to exceed these standards, while poorer countries have difficulties meeting them.

In the United States the major laws at the national level are the Clean Water Act, the Clean Air Act, the Resource Conservation and Recovery Act (RCRA), and the Pollution Prevention Act (PPA). In addition, a number of Executive Orders can significantly affect the scope of the laws, for example, the 1991 Federal Recycling and Procurement policy, which mandates the use of recycled material in federal purchases. The RCRA regulates only the hazardous industrial waste; the management of nonhazardous solid waste is left to the individual states. Table 1 lists the major federal environmental laws, along with their chief attributes, and the years of their enactment [1, 3–6]. Until the passage of PPA, U.S. policies in general focused on regulating industrial wastestreams. The effect of laws and regulations on economy and innovation with a company are discussed later in this chapter.

2   THE MAJOR U.S. ENVIRONMENTAL LAWS

In the following, some of the major environmental laws are discussed briefly. The descriptions of laws administered by the EPA have been adapted [3]. The discussion here is brief. For details, see references 3–7.

2.1   National Environmental Policy Act

The National Environmental Policy Act (1970) was enacted to declare a national policy which will encourage productive and enjoyable harmony

TABLE 1   Environmental Laws in the United States

Law	Highlights	Year
Rivers and Harbors Act	Control construction, allow navigation	1899
Atomic Energy Act	Civilian use of atomic energy Mining, production, power, medical uses	1954
Solid Waste Disposal Act	First federal initiative on waste disposal	1965
Motor Vehicle Control Program	Control of CO, NO_X	1967
Resource Recovery Act	Waste reduction and source recovery Forerunner of RCRA of 1976	1970
National Environmental Policy Act	Created Environmental Protection Agency Environmental impact of federal projects Environmental Impact Statement	1970
Clean Air Act	Control of total suspended particulates, SO_2, CO, NO_2, O_3, Pb (airborne)	1970
Clean Water Act	Chemical, physical, biological Restoration to "fishable and swimmable" condition	1972
Marine Protection, Research and Sanctuaries Act	Ocean dumping of wastes Permit required for transportation	1972
Federal Insecticide, Fungicide and Rodenticide Act	Storage and disposal of pesticides Protection of groundwater Banning of pesticides	1972
Resource Conservation and Recovery Act (RCRA)	Generation, transport, storage of hazardous wastes Emphasizes conservation and recycling	1976
Hazardous and Solid Waste Amendments (to RCRA)	Protection of groundwater Landfills, undergrounds tanks	1984
Pollution Prevention Act	Source reduction Facilitate: government, businesses Methods of measurement Training programs Review regulations Federal procurement practices	1990

between humans and their environment, to promote efforts which will prevent or eliminate damage to the environment and biosphere and stimulate the health and welfare of humans, and to enrich the understanding of the important ecological systems and natural resources. The Council on Environmental Quality was established by this act. All agencies of the federal government are required to file an Environmental Impact Statement as part of their major Federal actions significantly affecting the environment. Courts have in certain instances extended this law to include state, local, and private projects.

2.2 Clean Air Act

The Clean Air Act (1970) is the comprehensive law that regulates air emissions from area, stationary, and mobile sources. This law authorizes the EPA to establish National Ambient Air Quality Standards (NAAQS) to protect public health and the environment. The goal of the Act was to set and achieve NAAQS in every state by 1975. The setting of maximum pollutant standards was coupled with directing the states to develop state implementation plans applicable to appropriate industrial sources in the state. The Act was amended in 1977 to set new goals (dates) for achieving attainment of NAAQS since many areas of the country had failed to meet the deadlines. The 1990 amendments to the Clean Air Act in large part were intended to meet unaddressed or insufficiently addressed problems such as acid rain, ground-level ozone, stratospheric ozone depletion, and air toxics.

2.3 Clean Water Act

The Clean Water Act (CWA) is a 1977 amendment to the Federal Water Pollution Control Act of 1972 for regulating discharges of pollutants to waters of the United States. The law gave EPA the authority to set effluent standards on an industry basis (technology based) and continued the requirements to set water quality standards for all contaminants in surface waters. The CWA makes it unlawful for any person to discharge any pollutant from a point source into navigable waters unless a permit (NPDES) is obtained under the Act. The 1977 amendments focused on toxic pollutants. In 1987, the CWA was reauthorized and again focused on toxic substances, authorized citizen suit provisions, and funded sewage treatment plants under the Construction Grants Program. The CWA has provisions for the delegation by EPA of many permitting, administrative, and enforcement aspects of the law to state governments. In states with the authority to implement CWA programs, EPA still retains oversight responsibilities.

2.4   CERCLA, or the "Superfund"

The Comprehensive Environmental Response, Compensation, and Liability Act (CERCLA), commonly known as Superfund, was enacted by Congress in 1980. This law created a tax on the chemical and petroleum industries and provided broad federal authority to respond directly to releases or threatened releases of hazardous substances that may endanger public health or the environment. Over 5 years $1.6 billion was collected, and the tax went to a trust fund for cleaning up abandoned or uncontrolled hazardous waste sites. CERCLA established prohibitions and requirements concerning closed and abandoned hazardous waste sites, provided for liability of persons responsible for releases of hazardous waste at these sites, and established a trust fund to provide for cleanup when no responsible party could be identified.

The law authorizes two kinds of response actions: short-term removals where actions may be taken to address releases or threatened releases requiring prompt response, and long-term remedial response actions that permanently and significantly reduce the dangers associated with releases or threats of releases of hazardous substances that are serious but not immediately life threatening. These actions can be conducted only at sites listed on EPA's National Priorities List (NPL). CERCLA also enabled the revision of the National Contingency Plan (NCP). The NCP provided the guidelines and procedures needed to respond to releases and threatened releases of hazardous substances, pollutants, or contaminants. The NCP also established the NPL. CERCLA was amended by the Superfund Amendments and Reauthorization Act (SARA) in 1986. The Superfund has been criticized as a law with the least benefit-to-cost ratios.

2.5   Superfund Amendments and Reauthorization Act

SARA amended the CERCLA in 1986. SARA reflected EPA's experience in administering the complex Superfund program during its first 6 years and made several important changes and additions to the program. SARA stressed the importance of permanent remedies and innovative treatment technologies in cleaning up hazardous waste sites, required Superfund actions to consider the standards and requirements found in other state and federal environmental laws and regulations, provided new enforcement authorities and settlement tools, increased state involvement in every phase of the Superfund program, increased the focus on human health problems posed by hazardous waste sites, encouraged greater citizen participation in making decisions on how sites should be cleaned up, and increased the size of the trust fund to $8.5 billion. SARA also required EPA to revise the

Hazard Ranking System to ensure that it accurately assessed the relative degree of risk to human health and the environment posed by uncontrolled hazardous waste sites that may be placed on the NPL.

2.6   Resource Conservation and Recovery Act

RCRA (1976) gave EPA the authority to control hazardous waste from the "cradle to grave." This includes the generation, transportation, treatment, storage, and disposal of hazardous waste. RCRA also set forth a framework for the management of nonhazardous wastes. Until that time only the disposal of hazardous wastes into the air and waters was regulated. Thus, land disposal and deep-well injection had become economically feasible approaches.

The 1986 amendments to RCRA enabled EPA to address environmental problems that could result from underground tanks storing petroleum and other hazardous substances. RCRA focuses only on active and future facilities and does not address abandoned or historical sites.

Hazardous and Solid Waste Amendments are the 1984 amendments to RCRA that required phasing out land disposal of hazardous waste. Some of the other mandates of this strict law include increased enforcement authority for EPA, more stringent hazardous waste management standards, and a comprehensive underground storage tank program. Whenever practical, conservation and recycling of wastes is emphasized by RCRA.

2.7   Toxic Substances Control Act

The Toxic Substances Control Act (TSCA) of 1976 was enacted by Congress to give EPA the ability to track the 75,000 industrial chemicals currently produced or imported into the United States. EPA repeatedly screens these chemicals and can require reporting or testing of those that may pose an environmental or human-health hazard. EPA can ban the manufacture and import of those chemicals that pose an unreasonable risk. Also, EPA can track the thousands of new chemicals that industry develops each year with either unknown or dangerous characteristics. EPA then can control these chemicals as necessary to protect human health and the environment. Polychlorinated biphenyls have been banned, as far as manufacture, processing, and distribution of products is concerned. TSCA supplements other federal statutes, including the Clean Air Act and the Toxic Release Inventory under Emergency Planning and Community Right-to-Know Act [3].

2.8   Pollution Prevention Act

PPA of 1990 focused industry, government, and public attention on reducing the amount of pollution through cost-effective changes in production, operation, and raw materials use. Opportunities for source reduction are often not realized because of existing regulations and the industrial resources required for compliance, focus on treatment, and disposal. Source reduction is fundamentally different and more desirable than waste management or pollution control. Pollution prevention also includes other practices that increase efficiency in the use of energy, water, or other natural resources and protects our resource base through conservation. Practices include recycling, source reduction, and sustainable agriculture. It is the most recently enacted federal environmental law.

University of Michigan's Center for Sustainable Systems provides information on PPA [4]. Following the passage of the PPA, EPA developed a formal definition of pollution prevention and a strategy for making pollution prevention a central guiding mission. Congress established a national policy as follows:

- Pollution should be prevented or reduced at the source whenever feasible;
- Pollution that cannot be prevented should be recycled in an environmentally safe manner whenever feasible;
- Pollution that cannot be prevented or recycled should be treated in an environmentally safe manner whenever feasible;
- Disposal or other release into the environment should be used only as a last resort and should be conducted in an environmentally safe manner.

Thus, prevention is given the highest priority. According to the EPA's official definition, pollution prevention means "source reduction" as defined in the PPA, but also includes "other practices that reduce or eliminate the creation of pollutants through (1) increased efficiency in the use of raw materials, energy, water, or other resources, or (2) protection of natural resources by conservation." Source reduction is defined under the Act as any practice that reduces the amount of any hazardous substance, pollutant, or contaminant entering any wastestream or otherwise released into the environment (including fugitive emissions) before recycling, treatment, or disposal and reduces the hazards to public health and the environment associated with the release of such substances, pollutants, or contaminants. Source reduction includes "equipment or technology modifications, process or procedure modifications reformulation or redesign of products, substitution of raw materials, and improvements in housekeeping, maintenance training, or inventory control" [4].

3 EFFECTIVENESS OF AND NEED FOR ENVIRONMENTAL LAWS

There is a continuing debate regarding the effectiveness and indeed the need for environmental laws. Firestone and Reed [7] state that:

At least in the United States, significant progress has been made on environmental matters. Also, the vast majority of Americans are today increasingly sensitive to the existence of environmental issues and now believe that those issues are of major personal concern to them. This awareness is bound to lead to further improvement of environmental conditions in the United States.

Also, if our environmental problems have been solved or are now being fully addressed by mechanisms other than governmentally imposed controls, then those controls should not continue. It is clear however, that, while government controls have been quite successful, our environmental problems are not even close to being completely solved.

Among the successes, and still remaining challenges, they cite the following [7]:

Lowering of toxic materials in the environment. Lead emissions have decreased, largely due to phaseout of leaded gasoline, ban of products which use lead, such as inks, dyes or adhesives, and increased recycling of the metal.

Concentration levels of toxic substances such as DDT and PCB have decreased due to the ban on their manufacture and use.

Exposure to asbestos has decreased.

Air and water quality have improved, though significant problems remain. This includes reduction in emissions of particulate matter, SO_2 and CO, despite the increase in vehicular traffic. Number of areas in the United States meeting EPA standards for air quality have increased, although over 100 cities still remain in non-compliance.

Sources of acid rain are sulphur and nitrogen oxides. While sulphur oxides have decreased 15% over 10 years, they remain significant. Nitrogen oxides have shown no reduction.

4 LAWS, ECONOMICS, AND PRODUCT/PROCESS INNOVATION

It appears to be a common belief in the society, and particularly in the industrial sector, that reducing pollution, making products and processes more environmentally friendly, increases costs. Studies have shown, as quoted by Porter and van der Linde [2], that this is not always the case. A company or industry that incorporates the environmental requirements in its product innovation process stands to gain and become an industry leader. The authors give many examples where an industry has made use of such an approach. One is that of the flower industry in the Netherlands, as shown in the following:

- Industry: Dutch flower growers.
- Problem: Contamination of soil and groundwater due to intensive use of fertilizers, pesticides, and herbicides. Limited availability of land.
- Solution: Flowers are now grown in greenhouses in water and rock wool, using closed-loop systems. The growing fluid is recirculated.
- Benefits/costs: Amount of fertilizers and pesticides required has been reduced. Quality has improved because of controlled growing conditions. Costs have been reduced by improved handling possibilities. Greenhouses are integrated with auction houses, where bidding takes place, thus improving global competitiveness.

Another example [2] is in the printed circuit board industry, 33 major process changes were initiated, of which 13 came from pollution control personnel. Of the latter, 12 changes resulted in cost reduction, 8 changes resulted in quality improvements, and 5 changes resulted in extension of production capabilities.

4.1 Waste Prevention

A study of waste prevention in chemical plants, motivated by waste disposal costs and environmental regulation, found that innovations resulted in increased resource productivity. Of the 181 waste prevention activities, only one resulted in cost increase. Of the 70 activities which led to changes in product yield, 68 were increases in the yield. Average increase was 7%. Of the 48 initiatives for which capital cost information was available, one-fourth required no capital investment. Of the 38 initiatives for which payback data were available, two-thirds had a payback period of 6 months or less. Finally, for each dollar spent on source reduction, an annual savings of $3.49 was reported.

Companies must look at pollution as a waste of resources. Within the company, packaging represents material waste and poor process controls. Often a company does not even know how many wastestreams there are. Outside the company, packaging has to be discarded by people down the supply chain: wholesalers, distributors, and customers. This represents wasted effort and energy, thus adding to the costs. Likewise, customers should not have to put up with polluting products.

4.2 Role of Innovation

A manufacturer has four inputs, or resources, to put into the enterprise: *raw material, energy, labor,* and *capital.* In the past a company with access to cheap inputs had a marketplace advantage. With the globalization of the economy, no single company, or indeed a country, has a lock on cheap resources. A company from a high-cost country can build a plant in a low-cost country. Rather, today, it is how a company uses its resources that make it competitive. Porter and van der Linde [2] call this "resource productivity."

A competitive product is one that can be produced cheaper than similar products or one that provides more value, for which customers are willing to pay more. Competitiveness and environmental friendliness are not mutually exclusive. Through innovation, by making better use of resources, products can be made more competitive and environmentally friendly.

As environmental awareness has grown worldwide, customers are more willing to pay higher prices for "green" products. Companies which have innovated to be the first in the marketplace with new or improved products enjoy advantages by being the leaders. By early introduction a product has a marketplace advantage by gaining early customers who lock on to it, develop loyalty, and are less likely to switch [8]. Germany adopted stricter environmental standards, including laws on take-back and recycling, earlier than other countries. Thus, German companies developed expertise in introducing products that require less packaging, are of lower cost, and cause less environmental burden.

If environmental regulations are seen as promoting end-of-the-pipe solutions and add-on pollution control systems, the products and processes will get more expensive and less competitive. Such has been the case in the United States, where regulations have promoted the use of "best available technology" to reduce pollution. If, on the other hand, regulations permit innovation and flexibility in finding solutions, then the companies can improve on the products and processes to meet the emission and other requirements. As an example, the Scandinavian pulp and paper industry developed a chlorine-free bleaching process, at the same time carrying out other process improvements. They made significant in-roads into the international market and, for some time, were able to charge higher prices for chlorine-free paper.

Just as Deming, Juran, Taguchi, and others proved 20 or more years ago that costs can be lowered while improving quality [9], today's companies must think of innovation as a means to reduce environmental burdens, as an additional outcome. In the past, higher quality was achieved by using more expensive processes and equipment and by rejecting output which fell below acceptable standards. Thus, if the product and processes were assumed to be "fixed," higher quality meant more rejects and hence higher costs. Today we recognize that quality can be designed into a product and that processes can be optimized for higher quality and lower cost.

The inputs (raw material, energy, labor, and capital) must be used more efficiently. The processes should eliminate unneeded activities and avoid generating waste products. Use of hazardous materials must be eliminated. While these points seem obvious, companies often do not take advantage of available knowledge. An example is the Green Lights program sponsored by the EPA, which promotes energy savings by improved lighting, with a 2-year payback period. Many companies have not taken advantage of it, because they are not aware of it or there is resistance to change.

4.3 A New Way of Looking at Environmental Requirements

Porter and van der Linde [2] make the point about the new way of looking at environmental aspects in design: "Environmental inputs must be embedded in the overall process of improving productivity and competitiveness. The resource productivity model, rather than the pollution control model, must govern decision making."

Innovative ways of utilizing the previously wasted byproducts yield generous dividends:

- A nylon manufacturer in France invested in recovering a byproduct called diacid, which was earlier incinerated. The investment of $12.5 million produced an annual revenue of $3.5 million.
- Dow Chemical in California using hydrochloric gas and caustic soda was required to close evaporation ponds containing waste water. By redesigning its process, the company reduced both the hydrochloric acid waste and the caustic waste. The company also used some of the wastestream as input for other parts of the plant. A cost of $0.25M provided annual savings of $2.4M.

- 3M produces adhesives in batches. Each batch is then added to a storage tank. A bad batch would spoil the whole contents of the tank, with resulting loss of product and time and increased hazardous waste. The company developed new techniques for determining the batch quality faster. The reduction in hazardous waste disposal alone saved it $200,000 annually.
- During the initial start-up of a chemical production or after an interruption, scrap material is produced. DuPont improved the monitoring equipment, reducing interruptions, waste generation, and increasing the plant uptime.
- Regulatory pressures led Ciba-Geigy in its dye producing plant to make changes that reduced pollution and also increased process yields by 40% and annual cost savings of $0.74M.

That regulations can lead to lower product costs and higher productivity has been proven by a number of cases:

- Upon being required to make easier to recycle products, Hitachi redesigned products for easier disassembly, at the same time reducing the parts count (by 16% in a washing machine and by 30% in a vacuum cleaner). The lower parts count also led to lower assembly time and overall lower costs.
- A 1991 regulation required drastic reduction in benzene emissions during coal tar distillation. It was generally opposed by the industry, since it required costly containment. Aristech Chemical Corporation chose to find solutions to the problem instead. The company found that removing benzene at an earlier processing step did away with the need for containment. Instead of a cost increase, the company saved $3.3M.

Porter and van der Linde [2] cited other industries that used innovation to achieve environmental benefits while at the same time off-setting any associated costs. The paper and pulp industry was required to stop chlorine bleaching, since it released dioxins. It introduced better cleansing processes and substituted O_2, O_3, or H_2O_2 for bleaching and changed to closed-loop processes. By making better use of energy from byproducts, it lowered operating costs by 25%. The companies could charge premium prices for chlorine-free paper. The paint and coatings industry had to remove volatile organic compounds (VOCs) in solvents used. It applied innovations such as low-solvent paints and water-based paints, with better application techniques and use of heat and radiation-cured coatings. The industry was able to charge premium prices for solvent-free paints, offer improved coating quality, and have safer working conditions. Materials savings reduced costs. The electronics industry also needed to remove VOCs in cleaning agents. The industry innovated by introducing closed-loop processes, no-clean soldering, and non-VOC cleaning agents. Companies achieved improved product quality and reduction in processing cost when cleaning could be eliminated. Pay-back periods of 1 year were often achieved.

Refrigerator manufacturers were required to remove CFCs used as refrigerants and reduce energy usage in operation. A propane–butane mix was substituted for the CFCs. Better insulation and gaskets and improved compressors were introduced. The companies showed higher energy efficiency and were able to charge premium prices for "greener" refrigerators. It has been documented that the German branch of the Greenpeace organization was instrumental in promoting this change in the refrigerant. Dry-cell manufacturers were faced with having to remove toxic metals such as cadmium, cobalt, lead, lithium, mercury, nickel, and zinc to reduce toxic wastes going to landfills and into the atmosphere if the cells were incinerated. They innovated by developing nickel–hydride and lithium rechargeable cells. The new products show higher efficiencies and higher energy capacities at competitive costs. The printing ink industry was also faced with the problem of VOCs in inks. They developed water-based and soy inks that have been demonstrated to have better efficiency and printability and brighter colors.

5 EFFECTS OF THE LAWS ON PRODUCT DESIGN AND MANUFACTURING

What should a design and manufacturing organization keep in mind regarding the environmental laws and regulations? From the viewpoint of toxic and hazardous substances, do not use hazardous substances in the product. The product should not cause hazardous discharges of any kind during use or after disposal. The manufacture of the product should not cause hazardous discharges. The materials used in the product should not have caused hazardous discharges during their extraction or preparation for manufacture.

From the viewpoint of source reduction, measure and account for all the wastes and emissions in the manufacturing process, the distribution, and in product use. Redesign the processes and product to eliminate the wastes and emissions. Find ways to use the wastes.

If a company exports its products, its products must conform to the laws of the importing country. It is necessary for U.S. manufacturers to be cognizant of all the pertinent environmental laws.

REFERENCES

1. R. E. Cattanch, J. M. Holdreith, D. P. Reinke and L. K. Sibik. Handbook of Environmentally Conscious Manufacturing. Chicago: Irwin, 1995.
2. M. E. Porter and C. van der Linde. Green and competitive: ending the stalemate. Harvard Business Review, pp 120–134, 1995.
3. U.S. EPA's website http:/www.epa.gov/ has information on environmental laws administered by the agency.
4. University of Michigan's Center for Sustainable Systems (formerly National Pollution Prevention Center, NPPC). http://www.umich.edu/~nppcpub/index.html
5. C. A. Wentz. Hazardous Waste Management. New York: McGraw-Hill, 1989.
6. G. A. Keoleian and D Menerey. Life Cycle Design Guidance Manual. Washington, DC: U.S.EPA. EPA/600/R-92/226, 1993.
7. D. B. Firestone and F. C. Reed. Environmental Law for Non-Lawyers. 2nd ed. So. Royalton, VT: SoRo Press, 1993.
8. P. G. Smith and D. G. Reinertsen. Developing Products in Half the Time. 2nd ed. New York: Van Nostrand Reinhold, 1998.
9. M. S. Hundal. Systematic Mechanical Designing: A Cost And Management Perspective. New York: ASME Press, 1997.

6

Legislation and Market-Driven Requirements
European Examples

**Taina Dammert, Eero Vaajoensuu,
Markku Kuuva, and Mauri Airila**
Helsinki University of Technology, Helsinki, Finland

1 INTRODUCTION

The need for a European Community (EC) environmental policy was first announced in 1972. Following this declaration the First Community Action Programme on the Environment was approved in 1973. Until that time the actions were reactive to existing problems. Legislative interventions like standards, controls, and prohibitions were favored. Even if the legislative approach still dominates, the trend has since the 1980s, been toward a more comprehensive environmental policy [1]. Precautionary principles, preventive actions, and "polluter pays" principles are the present objectives of the Community policy. In waste management this means that waste generation should be reduced by improving product design. Reuse and materials recycling are favored, and more responsibility is assigned to producers.

The environmental legislation in the European Union (EU) member countries is based on the environmental regulations and directives given by the Community. The regulations are to be implemented immediately in the whole Community. The implementation of the directives in the member states allows flexibility, but the given obligations and objectives must be met [2]. However, the implementation in practice in member states appears to be problematic.

Along with the legislative development, the voluntary and market-driven developments in environmental issues have become more important. The application process of the ISO 14001 environmental management system or Ecomanagement and Audit Scheme (EMAS) are often the starting points of better environmental performance of an organization or an industrial company.

Concerning products, the market-driven environmental approach is more diverse and the environmental labels are a good example of these. There are international, national, and organizational environmental labels, which have usually similarities within the product group concerning some basic criteria like use of certain substances during manufacturing or recycling aspects. On the other hand, depending on the market situation, a manufacturer shall consider which are the labels that give better market value when awarded.

Environmental declarations are a relatively new form of environmental information given by the manufacturers. Declarations provide materials and manufacturing process information of the product and are usually targeted to institutional buyers but in some cases also to the consumers.

2 EC LEGISLATION

Environmental legislation in the EC is based on treaties between the member states. The Treaty of Amsterdam of October 1997 specifies the principle of sustainable development as one of the aims of the EC. The treaty also specifies environmental policy and the objectives of the EC on the environment.

The objectives of the EC policy on the environment are formulated in the Treaty of Amsterdam, Article 174, as follows [3]:

- Preserving, protecting, and improving the quality of the environment;
- Protecting human health;
- Prudent and rational utilization of natural resources;
- Promoting measures at international level to deal with regional or worldwide environmental problems.

It is also stated that EC policy on the environment should aim at a high level of protection, taking into account the diversity of situations in the various regions of the EC. It shall be based on the precautionary principle and on the principles that preventive action should be taken, that environmental damage should as a priority be rectified at source, and that the polluter should pay [3].

Within the European Commission, Directorate General XI is responsible for Community policies for the environment, nuclear safety, and civil protection. Its actions are carried out within the strategy defined in 1992 by the EC Fifth Programme of Policy and Action in Relation to the Environment and Sustainable Development "Towards Sustainability" (1992–2000). The objective of the fifth program of action in relation to the environment is to transform patterns of growth in the Community in such a way as to promote sustainable development. The principles of the environmental policy are [4]

- The adoption of global proactive approach;
- The will to change current practices;
- Encouraging changes in social behavior—public authorities, citizens, consumers, and enterprises;
- Establishing the concept of shared responsibility;
- Using new environmental instruments.

The instruments to realize these principles are [4]

- Horizontal measures—improving information and environmental statistics; promoting scientific research and technological development; and improving sectoral and spatial planning, public information, and professional training.
- Financial support mechanisms—the LIFE program, the Structural Funds, the Cohesion Fund, European Investment Bank loans.
- Financial instruments—incentives for producers and consumers to protect the environment.
- Regulatory instruments.

The financial instruments can be listed in four groups [1]:

- Charges and levies, e.g., charges on emissions;
- Fiscal incentives, e.g., taxes on energy;
- State aids, e.g., subsidies, soft loans, or tax breaks offered by public authorities to encourage investments in pollution controlling technology or the use of "green" products, etc.;
- Environmental auditing—EMAS, which was adopted in 1993 is a mechanism by which companies may commit themselves to establish an environmental policy and management system.

Regulatory instruments are based on the treaties the Council of the European Union. The Commission gives the other ordinances, regulations, and directives. The regulations will be implemented immediately in the whole Community, and the implementation of directives is carried through national legislation. The implementation of the directives in the member states has some flexibility, but the given objectives and obligations must be met [2].

In Table 1 there are examples of regulatory instruments of the EC, which include principles of design for environment of machines and electrical equipment.

Community policy on waste management involves three strategies [14]: eliminating waste at source by improving product design, encouraging the recycling and reuse of waste; and reducing pollution caused by waste incineration. The Community's approach has been to assign more responsibility to the producer. The first objectives are given in the Waste Directive, the Directive on Packaging and Packaging Waste, and the Directive on Disposal of Discarded Batteries and Accumulators. The most recent approaches are the proposals for the Directive on End-of-Life Vehicles and on the Waste Electrical and Electronic Equipment. These are steps toward producer responsibility. These two proposals have caused much debate in recent years because they include tight targets, for example, on

TABLE 1 Examples of the Directives and Directive Proposals Concerning Design for Environment for Machines and Electrical Equipment

- Directive on pollutants from diesel engines in vehicles and the amendments [5]
- Proposal for a directive on end-of-life vehicles (9 July 1997)
- Directive on batteries and accumulators containing certain dangerous substances and amendments [6]
- Draft proposal for a directive on waste from electrical and electronic equipment (5 July 1999)
- Regulation on substances that deplete the ozone layer [7]
- Proposal for a council regulation on substances that deplete the ozone layer [8]
- Directive on packaging and packaging waste [9]
- Proposal for a directive on marking of packaging and on the establishment of a conformity assessment procedure for packaging [10]
- Decision of establishing the identification system for packaging materials [11]
- Directive on the limitation of emissions of volatile organic compounds due to the use of organic solvents in certain activities and installations [12]
- Directive on waste [13]

hazardous substances reduction and on treatment and recovery of end-of-life products. That is why the proposals may have amendments before the final approval.

2.1 End of Life: Vehicles

According to the amended proposal from June 1999 for a council directive on the end of life of vehicles [15] the measure aims at the prevention of waste from vehicles; the reuse, recycling, and other forms of recovery of vehicles and their components so as to reduce the disposal of waste; and the improvement in the environmental performance of the salvage operators. The directive proposal concerns prevention of the use of hazardous substances in vehicles. It also concerns coding standards/dismantling information, collection, treatment, reuse, and recovery of end-of-life vehicles. The directive proposal includes the following prevention measures for member states:

- Encourage vehicle manufacturers and material and equipment manufacturers to control and reduce the use of hazardous substances (under the directive 67/548/EEC) in vehicles;
- Promote the design for dismantling, reuse, and recovery—in particular the recycling of end-of-life vehicles, their components, and materials;
- Encourage manufacturers to integrate an increasing quantity of recycled material in vehicles to develop the markets for recycled materials;
- Prevent lead, mercury, cadmium, and hexavalent chromium contained in vehicles which are being shredded in vehicle shredders and disposed of as landfill or incinerated or coincinerated in waste. Lead used as solder in electronic circuit boards shall be exempted from this.

Regarding collection of vehicles at end of life, member states shall take the necessary measures to ensure that economic operators (producers of materials and of vehicles, distributors, dismantlers, shredders, recoverers, and recyclers) set up systems for the collection of all end of life vehicles; ensure that vehicles at end of life are transferred to authorized treatment facilities; and set up system according to which a certificate of destruction is a condition for deregistration of the vehicle.

Treatment, reuse, and recovery shall include the following actions:

- The removal of all fluids, tires, batteries, air-conditioning systems, air bags, catalysators, and other hazardous components and materials before further treatment.

- No later than January 1, 2005 the reuse and recovery of vehicles shall be a minimum 85% by weight per vehicle and the reuse and recycling of vehicles shall be to a minimum of 80% by weight per vehicle.
- No later than January 1, 2015 the reuse and recovery of vehicles shall be a minimum 95% and the reuse and recycling of vehicles shall be to a minimum of 85% by weight per vehicle.

According to the directive proposal, the necessary measures shall be taken to use common component and material coding standards to facilitate the identification of materials and components for reuse and recovery.

2.2 End of Life: Electronics

The objectives of the draft proposal from July 1999 of the Directive on End-of-Life Electronic Equipment [16] are prevention of waste from end-of-life electrical and electronic equipment; reuse, recycling, and other forms of recovery of such waste; and minimizing the risks and impacts to the environment associated with the treatment and disposal of end of life electrical and electronic equipment.

The draft proposal shall cover 11 categories of electrical and electronic equipment from large household appliances to electrical and electronic tools. According to the proposal, member states shall

- Promote the design and production of electrical and electronic equipment for repair, upgradability, reuse, dismantling, and recycling, and particularly to increase the use of recyclable materials.
- Ensure that ISO 1043-1, ISO 1043-2, and ISO 11469 on the generic identification and marking of plastic products is applied to plastic parts weighing more than 50 grams.
- Ensure that the use of lead, mercury, cadmium, hexavalent chromium, and halogenated flame retardants such as polybrominated biphenyls (PBB) and polybrominated diphenyl ethers (PBDE) are phased out by January 1, 2004. Exempted from this provision are some mercury-containing lamps and laboratory equipment, lead as radiation protection, lead in glass of cathode ray tubes (CRTs), light bulbs, fluorescent tubes, in some alloying elements, and in electronic ceramic parts and the use of cadmium and hexavalent chromium in some specific applications.
- Ensure that collection systems are set up for products at end of life from last holders and distributors.

- Ensure that the costs for collection, treatment, recovery, and environmentally sound disposal of waste electronics are borne by producers.
- Ensure that no later than January 1, 2004 the following reuse and recycling targets (% by weight) are attained by producers: large household appliances, 90%; small household appliances, 70%; radio, television, electroacoustics, 70%; gas discharge lamps, 90%; toys, 70%; electrical and electronic tools, 70%; equipment containing cathode ray tube, 70%.

It has been proposed that the pretreatment of end-of-life electr(on)ics shall include removal of all fluids and selective treatment of the following materials and components: lead (except lead in CRTs), mercury, hexavalent chromium, cadmium, polychlorinated biphenyls, halogenated flame retardants, radioactive substances, asbestos, and beryllium (> 2% alloys).

3 FINLAND

In Finland environmental legislation conforms to the EC legislation. Table 2 shows examples of the environmental legislation concerning environmental issues to be taken into account in electronics products design.

4 SWEDEN

Concerning the design for the environment (DFE) of electronics and electrical products, many ongoing reduction activities in Sweden aim at reducing the use of lead, brominated flame retardants, chlorinated paraffins, and nonylphenoletoxylates. The environmental policy, which was already adopted by the Swedish government in 1990, set goals for reduction of the following compounds: organic tin compounds, phthalates, brominated flame-retardants, and chlorinated paraffins. The National Chemicals Inspectorate has studied brominated flame retardants, and in 1995 the work resulted in a suggestion to phase out PBB and PBDE compounds. Also, flame-retardant tetrabromobisphenol-A (TBBP-A) has shown environmentally hazardous properties, and the basis for a risk assessment is under research. The use of chlorinated paraffins has been reduced mainly through activities undertaken by producers, importers, and users. The phase-out goal of short-chained highly chlorinated paraffins has been set by the year 2000 [18].

Table 3 shows examples of Swedish environmental legislation that contain aspects of DFE of electrical products. In recent years there has also been considerable discussion in Sweden on the producer responsibility

TABLE 2 Examples of the Finnish Legislation Concerning Environmental Issues and DFE Aspects To Be Considered in Electronics Design

Legislation	Objectives	Producer's responsibilities
Council of State Decision (105/1995) on batteries and accumulators containing certain dangerous substances and amendments	The decision refers to batteries and accumulators containing more than • 25 mg mercury in a battery cell except alkali–manganese batteries • Cadmium 0.025% per weight • Lead 0.4% per weight. The decision also refers to alkali–manganese batteries containing mercury > 0.025% per weight and batteries and accumulators containing mercury > 0.0005% per weight except button cells containing mercury max 2% per weight.	Markings, information, easy removal of spent batteries and accumulators.
Decision (262/1998) on ozone-depleting substances		The use of the following compounds will be banned: • CFCs • 1,1,1-Trichlorine ethane • HCFCs: for insulation materials production from 1.1.2000, for foamed plastics from 1.1.1999, for cooling

Council of State Decision (962/1997) on packaging and packaging waste	By the middle of 2001 82% of all packaging used will have to be reused, recycled, or recovered as energy. The targets of recovery and recycling as material are set for fiber-based, glass, metal, and plastic packaging.	According to the decree, packers and importers of ready packed products are obliged to
		• Prevent, reuse, and recover of packaging waste
		• Be responsible of the recovery costs (related to the amount and quality of the packaging sold).
		The packer shall ensure that the sum of concentration levels of lead, cadmium, mercury, and hexavalent chromium present in packaging or packaging components shall not exceed the following (by weight) 600 ppm after July 1, 1998; 250 ppm after July 1, 1999; 100 ppm after July 1, 2001.
Waste Act (1072/93) and Decree (1390/93), including amendments	The objective is to support sustainable development by saving natural resources and preventing harmfulness of waste to health and the environment.	Prevention of waste and reduction in the amount and its harmfulness.

CFCs, chlorofluorocarbons; HCFCs, hydrochlorofluorocarbons.
Source: Ref. 17.

TABLE 3 Swedish Environmental Legislations That Contain Aspects of DFE of Electrical Products

Legislation	Objectives	Producer's responsibility
Battery Ordinance (1997/645)	The ordinance refers to batteries and accumulators harmful to environment containing more than 0.0005% per weight of mercury; 0.025% per weight of cadmium; 0.4% per weight of lead.	Marketing of environmentally harmful alkali manganese batteries (except button cells containing >2% per weight of mercury) is prohibited. Environmentally harmful batteries and accumulators may be delivered only if they are labeled. In addition to this the producer is committed to make an announcement to Environmental Protection Agency of such batteries and accumulators and to pay the recycling fee.
The ordinance on chemicals and genetically modified organisms (1998/941). The ordinance on prohibition of treatment, import, and export of specific chemical substances (1998/944).	The ordinance 1998:944 gives upper limits for lead, cadmium, mercury, and hexavalent chromium in packaging. The upper limits are >600 µg/g, >250 µg/g after June 30, 1999, and > 100 µg/g after June 30, 2001.	The appendix of the ordinance (1998/941) contains a list of chemicals for which the producer or importer has to make a declaration. According to the ordinance 1998/944 it is forbidden • To use cadmium-containing substances in surface finishing, stabilizer, and color pigment; • To offer for sale or transfer to consumers for their private use chemical products which wholly or in part consists of methylene chloride, trichloro-ethylene or tetrachloro ethylene. • To professionally manufacture or sell certain mercury-containing thermometers, switches, measuring instruments, etc.

End-of-life management of packaging (1997/185)	The ordinance provides objectives for reuse and recycling of packaging.	Packaging shall be designed, manufactured, and marketed for reuse or materials recycling. The release of hazardous substances in landfilling or incineration of packaging waste shall be minimized by design solutions.
The PCB Ordinance (1985/837)	The ordinance is to control treatment and discard of PCBs and products contaminated with PCB.	The ordinance prohibits the production, resale, and use of PCBs and PCTs.
The ordinance on the ozone layer-depleting substances (1995/636)	The ordinance concerns, with certain exceptions, prohibition of CFCs, HCFCs (the use of HCFCs is prohibited in refilling of existing facilities for refrigeration, air conditioning, and heat pumps after December 31, 2001), halons, carbon tetrachloride, 1,1,1-trichloroethane, and HBFCs.	The ordinance includes, with some exceptions, prohibition against manufacture, use, transfer, and resale of the ozone layer–depleting substances.
Waste Ordinance (1998/902) and amendments	According to the ordinance §25 it will be prohibited to dump, incinerate, or granulate waste from end-of-life electronics by other than certified recycling companies. The Swedish Environmental Protection Agency will give the certificates. The enforcement of the §25 has not yet been decided.	

CFCs, chlorofluorocarbons; HBFCs, hydrobromofluorocarbons; HCFCs, hydrochlorofluorocarbons; PCBs, polychlorinated biphenyls; PCTs, polychlorinated terphenyls.
Source: Ref. 19.

of end-of-life electronics. There is a draft proposal for the take-back ordinance for electronics, and the prohibition to dump, incinerate, or granulate waste from end-of-life electronics by other than certified recycling companies is already set in §25 of the waste ordinance. The enforcement of §25 has not yet been decided (situation in December 1999). The Environmental Protection Agency will give regulations to specify the approved disassembly, separation, and disposal methods of end-of-life electronics.

According to the draft proposal for the take-back ordinance for electronics, the product groups of end-of-life electronics included in producer responsibility are [20]

- Home appliances and electric tools—according to the waste ordinance the treatment of freezers and refrigerators will be part of municipal waste management;
- IT and office equipment;
- Telecommunication equipment;
- Cameras, photo equipment;
- Television sets, audio and video equipment;
- Clocks and watches;
- Electronic toys;
- Lighting equipment;
- Medical equipment systems;
- Laboratory equipment.

End users of end-of-life electronic and electrical equipment can return their old equipment free of charge to retailers and producers upon purchase of new equivalent equipment. The producer will be responsible for environmentally friendly treatment of the returned products. Producers will be required to provide detailed information on product content to dismantlers, particularly on hazardous substances; inform about the take back obligation; and give Swedish Environmental Protection Agency the facts that are needed to ensure the compliance of the ordinance.

5   GERMANY

In Germany voluntary agreement between industry and government and the producer responsibility principles has been applied by the Packaging Ordinance (since 1991 and with amendment 1998), the Closed Substance Cycle and Waste Management Act (1994), End-of-Life Vehicles Decree (1998), and Batteries Ordinance (1998). Recently under preparation and discussion has been the ordinance on end-of-life electronics.

With the Dual System, Germany organizes the implementation of the packaging ordinance. The packaging materials concerned are glass, paper, tinplate, aluminium, composites, and plastics. The Dual System license fees are calculated by the "polluter pays" principle [21]. The battery ordinance calls for a common return system for all batteries. For car batteries the respective manufacturers have responsibility for their products [22]. The take-back degree for cars calls for industry to take back, at no cost to the last owner, end-of-life vehicles no more than 12 years old. The last owner of a car is obliged to take it to a licensed dismantling and recovery/recycling facility [23].

The proposed end-of-life electronics ordinance will cover all electrical and electronic products. According to the proposal, manufacturers are obliged to take back and recover the same quantity of used products as they have sold in the same year, to monitor the material flow, and to report to the waste management authorities. Manufacturers can build and run collective systems for the logistic and sorting of the products [24].

Table 4 gives the basic objectives and the producer's responsibilities of environmental legislation concerning machines and electronics.

6   ENVIRONMENTAL MANAGEMENT: BS7750, EMAS, AND ISO 14000

The first specified European environmental management standard was the British Standard BS7750, a specification for an environmental management system. The system was used to describe the company's environmental management system, evaluate its performance, and to define policy, practices, objectives, and targets for continuous improvement. The standard was first published in June 1992 and was subsequently reviewed and revised to a new January 1994 issue [27]. In 1997 the BS7750 was displaced by the International Standard ISO 14000 series.

EMAS is based on the Council Regulation [28]. Companies in the industrial sector are allowed voluntarily to participate in a Community ecomanagement and audit scheme. There was about 2500 registered sites in EU Member States and Norway in September 1999 [29].

The objectives of EMAS are as follows:

• Voluntary participation by companies performing industrial activities. The scheme is established for the evaluation and improvement of the environmental performance of industrial activities and the provision of the relevant information to the public.
• The objective of the scheme is to promote continuous improvements in the environmental performance of industrial activities by the establishment and implementation of environmental policies, programs, and management systems by companies, in relation to

TABLE 4 Examples of the Basic Objectives and Producer's Responsibilities of German Environmental Legislation Concerning Machines and Electronics

Legislation	Objectives	Producer's responsibility
Batteries Ordinance	To prohibit sales of batteries and accumulators containing hazardous substances; To organize a collection system for end-of-life batteries and accumulators; To lengthen the service life of batteries and accumulators. The ordinance refers to batteries and accumulators harmful to environment • Containing more than 25 mg mercury in a battery cell except alkali-manganese batteries; • Alkali-manganese batteries containing more than 0.025% per weight mercury; • Containing cadmium 0.025% per weight; • Containing lead 0.4% per weight	A recycling fee of 15 DM is defined for car batteries and is included in the price of a new product. The sales of batteries and accumulators containing hazardous substances are permitted only if the take-back and recovery of these is arranged. The spent batteries for appliances are to be collected free of charge by retailers, public waste disposal services, and scrap merchants.
Chlorofluorocarbons and Halons Prohibition Ordinance	Ban of certain ozone-depleting substances	
Chemicals Prohibition Ordinance	Includes among other things general ban of certain halogenated dioxins and furans	

Packaging Ordinance	By the end of 30 June 2001 65% per weight of packaging waste will be recovered and 45% per weight will be recycled as materials.	The ordinance gives recycling targets of 60–75% by weight for glass, tinplate, aluminum, paper, cardboard, and composite materials.
The Closed Substance Cycle and Waste Management Act		The waste law enlarges the manufacturer's responsibility so that it covers product's entire life cycle, including manufacture, distribution, use, and disposal. Priority is given to waste avoidance. The production process wastes should primarily be recovered in materials recycling, and disposal is permitted only if reuse and recycling is not economically feasible.
End-of-life vehicles		The car industry is obliged to take back, without any costs to the last owner, the end-of-life vehicles no more than 12 years old and registered after the agreement takes effect. An aim is also to reduce the share of nonrecoverable parts in cars to a maximum of 15% by 2002 and 5% by 2015.
Proposal: ordinance concerning the disposal of electrical and electronic equipment	Includes such products like information, office and communication technology devices, consumer electronic devices, household devices, small electric devices.	Manufacturers are obliged to take back the same quantity of used products, as they have put into circulation in the same year. Household near collection stays in the responsibility of communities. Manufacturers can build and run collective systems for the logistic and sorting of the products. Manufacturers are also obliged to monitor the material flow and to report to the waste management authorities.

Sources: Refs. 21–26.

their sites; the systematic, objective, and periodic evaluation of the performance of such elements; and the provision of information of environmental performance to the public.

- The scheme shall be without prejudice to existing Community or national laws or technical standards regarding environmental controls.

The European Commission has adopted in 1998 a proposal for a Regulation, which updates the EMAS. According to the new regulation, all organizations will be able to participate in this scheme. Organizations are allowed to use ISO 14001 as a building block in the implementation of EMAS [30].

According to the proposal, Member States shall promote organizations' participation in EMAS and shall, in particular, consider the need to ensure the participation of small and medium-sized enterprises (SMEs) by

- Facilitating the access to information and to support funds establishing or promoting technical assistance measures, especially in conjunction with initiatives from appropriate professional or local points of contact (e.g., local authorities, chamber of commerce, trade association);
- To promote participation of SMEs concentrated in well-defined geographical areas, local authorities, in participation with industrial associations, chambers of commerce, and interested parties, may provide assistance in the identification of significant environmental impacts associated with that area.

SMEs may then use these in defining their environmental program and setting the objectives and targets of their EMAS management system. Member States shall also consider how registration to EMAS may be used in the implementation and control of environmental legislation to avoid unnecessary duplication by organizations and authorities.

ISO 14000 is a series of international voluntary environmental management standards. These standards address the following aspects of environmental management [31]:

- Environmental management systems;
- Environmental auditing and related investigations;
- Environmental labels and declarations;
- Environmental performance evaluation;
- Life cycle assessment;
- Terms and definitions.

An environmental management system is a tool that enables an organization of any size or type to control the impact of its activities, products, or services on the natural environment. The ISO 14001 standard "Environmental management systems—Specification with guidance for use" is the standard that specifies the requirements of an environmental management system. The key elements of certification according to ISO 14001 are as follows [31]:

- Environmental policy—the environmental policy of the organization and specified objectives, targets, and environmental programs.
- Planning—the environmental aspects like processes, products, services, and goods of the organization shall be analyzed.
- Implementation and operation—the control and improvements of operational activities from an environmental perspective.
- Checking and corrective actions—monitoring, measurement, and recording the environmental impacts.
- Management review—the suitability and adequacy of the environmental management system shall be reviewed by the management of the organization.
- Continual improvement—environmental management system is a cyclical process of planning, implementation, checking, reviewing, and continual improvement.

7 ENVIRONMENTAL LABELING

Environmental labels may be classified into several groups: international (e.g., EU ecolabel, the Swan in Nordic countries), national (e.g., the Blue Angel in Germany), organizational, or labels awarded by different environmental organizations. In this chapter the focus is on examples of environmental labels that have criteria for electrical and electronics products and have an official status.

7.1 The Swan (Nordic Countries)

The Nordic (Denmark, Finland, Iceland, Norway, and Sweden) environmental label, the Swan (Figure 1), was created in 1989 to increase objective information on the environmental load of products. The Nordic environmental labeling program's Environmental Labeling Board has been established to decide the criteria for issuing the Swan Label. The members of the board represent consumer and environmental authorities, trade and industry, and organizations on consumer and environmental protection. The criteria will be valid for a specific time limit (e.g., 3–4 years) and only 5–40% of the products in the product group will achieve the criteria. The

FIGURE 1   The Swan (Nordic countries).

criteria have been approved for over 50 product groups and 750 labels have been awarded as of December 1999 [32].

An ecolabeling organization is set up in each country to manage the ecolabeling. For a product manufactured in Denmark, Finland, Iceland, Norway, or Sweden, the application for the ecolabel may be submitted for use in its own country. For products manufactured outside the participating countries, the ecolabel may be submitted in any of the Nordic countries. The manufacturer, importer, or reseller may submit the application. A charge to the right to use the label (0.4%) is based on the estimated annual turnover of the ecolabeled product. The holder of registration shall pay the relevant fee in the country in which the registration certificate has been issued [33].

There are environmental label criteria for such product groups of electrical and electronic equipment as primary and rechargeable batteries; copiers, washing machines, light sources, washing machines, refrigerators, freezers, printers, telefaxes, and personal computers. The criteria are under development for audio-visual products.

7.2   EU Ecolabel

The European ecolabel scheme was created in 1992 [34] and is valid in the EU, Iceland, Liechtenstein, and Norway (Figure 2). The first product groups were established in 1993. The relevant ecological issues and the corresponding criteria, in all product groups, have been identified on the basis of studies of the environmental aspects related to the entire life cycle of these products. They address energy consumption, water pollution, air pollution, waste production, sustainable forestry management, and in some cases noise or soil pollution. Representatives of industry, commerce, environmental, and consumer organizations and trade unions are taking part on the criteria setting. The views of producers outside the EU are also taken

FIGURE 2   EU ecolabel.

into account. Once most EU Member States and the European Commission adopt the criteria, they are valid for a period of 3 years. After this period, the criteria are revised and may be tightened up, depending on the market and advances in technology. A manufacturer or an importer may make an application to the National Competent Body of the EU ecolabel, and once approved and awarded the ecolabel, the company pays an annual fee for the use of the label. It is fixed at 0.15% of the annual sales volume of the product. The ecolabel is then valid until the criteria expire [35].

The EU ecolabel includes criteria for 15 product groups (the situation as of late 1999), for example [36], single-ended light bulbs; double-ended light bulbs, washing machines, refrigerators, dishwashers, and personal computers. Criteria for eight more product groups are under development, for example, portable computers and batteries for consumer goods. Concerning electric products, the ecolabel has been awarded for two manufacturers of washing machines and one manufacturer of refrigerators.

7.3   The Blue Angel

An example of the national environmental labels in Europe is the Blue Angel (Figure 3). The Blue Angel was first introduced in the Federal Republic of Germany in 1977. The environmental label is to inform the consumer about the existence of particularly environmentally compatible product groups or quality categories.

An award may be given for products that [37] compared with other products fulfilling the same function and considered in their entirety, taking into account all aspects of environmental protection (including the use of raw materials), are as a whole characterized by a particularly high degree of

FIGURE 3   The Blue Angel

environmental soundness without thereby significantly reducing their fitness for use and impairing their safety.

The basic structure for awarding the environmental label is worked out by the Environmental Label Jury, German Institute for Quality Assurance, and Labeling and German Federal Environmental Agency. Both German and foreign manufacturers can apply for the environmental label. The contract of the label is limited to a maximum of 4 years to guarantee that the product keeps up with latest technology. The applicant pays a one-time fee of DM 300 for the use of the environmental label and after signing the contract a graded annual contribution that depends on the annual turnover of the product [37].

In late 1999 there were basic criteria for the award of the environmental label for about 90 product groups, and about 4500 products are labeled with the Blue Angel. The criteria have been developed for electrical and electronic equipment like copiers, refrigerators, freezers, computers, printers, and television sets.

7.4   TCO

TCO (the Swedish Confederation of Professional Employees) has worked for technical development concerning ergonomics, emissions, energy consumption, and ecology issues of information technology products like personal computers. These requirements are included in TCO labeling schemes TCO'92, TCO'95, and TCO'99 (revision of TCO'95). TCO requirements have been developed in cooperation with The Swedish Society for Nature Conservation, The National Board for Industrial, and Technical Development (NUTEK) and a testing and certification organization SEMKO AB.

TCO'92 includes requirements for reduced electric and magnetic fields, energy efficiency, and improved electrical safety of displays. TCO'95 requirements are for the complete personal computer (visual display unit, system unit, and keyboard). The requirements concern ergonomics, electromagnetic emissions and safety, energy consumption, and environmental aspects. Preparing the environmental requirements, the criteria of the other environmental labels have been taken into account. TCO'99 is a revision of TCO'95. TCO'99 includes additional restrictions and tighter demands compared with the TCO'95.

TCO'95 will be open for applications up to December 31, 1999 and after that it will be replaced by TCO'99. No time limit has been set for TCO'92 to expire. In November 1999 TCO introduced new quality and environmental labeling schemes also for printers, faxes, and copiers. The scheme is based on the experiences of TCO labeling schemes for the computer branch [38].

Table 5 contains a summary of the main requirements with respect to ergonomics, emissions, energy consumption, and ecological aspects of TCO labels for computers.

8 ENVIRONMENTAL DECLARATIONS

The purpose of environmental declarations is to provide materials and manufacturing process information of the product and effects in the environment for consumers or institutional buyers. Environmental declarations may come in different forms depending on the target group. The benefits of these declarations are therefore that modifications may be carried out by the producer and according to the target group.

SITO (the Swedish IT Companies' Organization) developed an eco-declaration for such product groups as telefaxes, copiers, personal computers, and printers. These criteria are based on legislative requirements, international standards and the German environmental label the Blue Angel.

Another example of guides for environmental declarations is the ECMA (Standardizing Information and Communication Systems, Geneva, Switzerland) Technical Report TR/70: Product-related environmental attributes. It includes templates of ecodeclarations for communication technology and consumer electronics products like personal computers, printers, copiers, television sets, and mobile phones. It identifies and describes environmental attributes related to the use and end of life information [42] (Table 6).

9 OTHER STANDARDS AND GUIDES FOR DFE

Some additional standards, guides, and recommendations are presented in Table 7. Perhaps the most common standard for recycling is the ISO 11469

TABLE 5 Summary of the Main Requirements on Ergonomics, Emissions, Energy Consumption, and Ecological Aspects of TCO Labels for Computers

Requirements	TCO'92 Displays	TCO'95 Personal computers (visual display unit, system unit, and keyboard)	TCO'99 Visual display units, system units, and keyboards
Ergonomics		Usability characteristics	Ergonomic aspects included: strain injuries, visual ergonomics, and acoustic noise.
Emissions, safety	• Electromagnetic fields • European fire and electrical safety requirements	• Emission characteristics (e.g., electric and magnetic fields) • Electrical safety	• Emissions characteristics (e.g., electric and magnetic fields) • Electrical safety
Energy	• Automatic power-down function • Information of the energy consumption	Energy-saving function	• Energy-saving function • Energy declaration
Ecology		• No form of CFCs or HCFCs • The surface layer of the CRT shall not contain cadmium • Electronic components in the CRTs shall not contain mercury; batteries shall not contain a total of more than 25 mg/kg (ppm) mercury or cadmium	• Environmental policy, ecodocument, and environmental management system • The manufacturer must meet all environmental requirements of the district and the country • The CRT shall not contain cadmium • No use of ozone-depleting substances or chlorinated solvents

- Plastic components > 25 g shall not contain flame retardants that contain organically bound chlorine or bromine
- The shock-absorbing material in the packaging shall not be manufactured with the aid of CFCs
- All plastic components weighing more than 25 g shall be labeled in accordance with DIN 54 840 or ISO 11469

Preparation for recycling:

- The variety of plastics used is kept to a minimum and that possibly environmentally harmful parts are clearly marked
- On purchasing new equipment the customer shall be given the chance to return, without cost, an equal number of obsolete equipment

- Declaration of the total amount of mercury in background lighting system of flat panel displays
- The electronic components and batteries shall not contain any mercury or cadmium
- Plastic components weighing more than 25 g shall not contain flame retardants that contain organically bound chloride or bromide

Preparation for recycling:

- Plastic components > 25 g shall be labeled in accordance with ISO 11469
- All plastic components > 100 g shall be made from the same type of plastic material
- Limitations for painting, lacquering, and varnishing of plastic components
- No metallic paints, lacquers, or varnishes on plastic components

(continued)

TABLE 5 (contd.)

Requirements	TCO'92 Displays	TCO'95 Personal computers (visual display unit, system unit, and keyboard)	TCO'99 Visual display units, system units, and keyboards
		• The seller shall guarantee that the returned equipment will be processed by a professional recycling company	• The plastic housings shall have neither internal nor external metallization • The applicant company shall have an agreement with at least one professional electronics recycling • Use of recycled monitor glass: an applicant company should use no less than 2% recycled monitor glass in new products

CFCs, chlorofluorocarbons; HCFCS, hydrochlorofluorocarbons; CRT, cathode ray tube.
Sources: Refs. 21–26.

TABLE 6 Main Environmental Attributes of SITO and ECMA TR/70 Environmental Declarations

Attribute	SITO ecodeclaration for personal computers	ECMA ecodeclaration for communication technology and consumer electronics products
Company information	Environment and quality management	
Product description		Type, brand name, model, seller, weight, etc., extendibility, service warranty
Ergonomics	Monitor characteristics and visual ergonomics	
Design	Modularity, exchange of modules without the use of special tools, upgradability	
Energy consumption	Requirements of Energy Star and Swedish NUTEK	Operational/standby/sleep/off in watts
Emissions and safety	Electromagnetic fields, electromagnetic compatibility, noise characteristics, electrical safety	Electromagnetic and radio frequency emissions, acoustical noise, chemical emissions

(continued)

TABLE 6 (contd.)

Attribute	SITO ecodeclaration for personal computers	ECMA ecodeclaration for communication technology and consumer electronics products
Materials, components	The following substances shall not be present: asbestos, CFCs, HCFCs, PCBs, mercury (relays), cadmium (CRT, plastic parts > 25 g), lead (plastic parts > 25 g), brominated flame retardants (PBBs, PBDEs), flame retardants based on chloroparaffins, carcinogenic, or suspected carcinogenic flame retardants. No use of hazardous batteries. The manual is printed on nonchlorine bleached paper.	The following substances shall not be present: asbestos, cadmium (in plastic materials, packaging and inks), cadmium in CRTs, CFCs and/or HCFCs, limitations for chloroparaffins, lead limitations, mercury, PCBs, or PCTs, polybrominated biphenyls limitations. Statement that mechanical plastic parts > 25 g are marked according to ISO 11469. Declaration of batteries and packaging.
End-of-life management	Plastic parts > 25 g coded in accordance with ISO 11469, battery handling information, ecological recycling (information of recycling system, design for recycling), packaging (ecological, collection system, plastic packaging markings according to DIN 6120).	Data relevant to end-of-life management: ease of disassembly, parts that need special handling, reusable parts, markings of plastics, take-back information.

CFCs, chlorofluorocarbons; CRT, cathode ray tube; HCFCs, hydrochlorofluorocarbons; PBBs, polybrominated biphenyls; PBDE, polybrominated diphenyl ethers; PCBs, polychlorinated biphenyls; PCTs, polychlorinated terphenyls; NUTEK, Swedish National Board for Industrial and Technical Development
Sources: Refs. 42 and 43.

TABLE 7 Principles of DFE in Standards and Guides for Electrotechnical Products and Plastics Materials

Standard, guide	Main principles
IEC (International Electrotechnical Commission) Guide 109: Environmental aspects—Inclusion in electrotechnical product standards (first edition 1995-08)	Primarily the purpose of the guide was to familiarize technical committees with environmental considerations. On the other hand it includes information on environmental impact assessment (EIA) principles and guidance on DFE principles for the electrotechnical industry, which may be used as background information.
The Association of Engineers, Germany: VDI 2243 Designing technical products for ease of recycling; fundamentals, and rules for design (Konstruieren recyclinggerechter Produkte)	VDI 2243 includes information on general design for recycling principles for technical products.
IEC 1429: Marking of secondary cells and batteries with the international recycling symbol ISO 7000-1135 (first edition 1995-12)	The standard applies to lead–acid batteries (Pb) and nickel–cadmium batteries (Ni–Cd). In all cases cells have to be marked individually with the exception of those constituting a battery or a subassembly that cannot be dismantled.
ISO 11469: Plastics—Generic identification and marking of plastic products	The standard specifies a system for the uniform marking of products of plastic materials. The symbols given in ISO 1043-1 and ISO 1043-2 shall be used for this standard.

(*continued*)

TABLE 7 (contd.)

Standard, guide	Main principles
3B/244/CDV, IEC: Committee draft for vote (CDV) IEC:62079: Preparation of instructions	The standard is intended to become part of IEC 61082 Preparation of documents used in electrotechnology. The standard (draft) includes a collection of requirements and methodological rules to be followed when creating instructions for users of products. The environmental aspects included are as follows: • Preparing the product for use: advice about preparations for waste removal/disposal destruction: instructions should cover information as far as it is relevant on destroying of the product. • And/or waste materials with due regard to safety and environmental considerations. • Recycling: where specific procedures are necessary for safe disassembly of the product, recycling or disposal of waste materials, this shall be specified. • Disposal: Instructions shall convey important messages to the user about aspects related to the waste disposal and environmental considerations.

(generic identification and marking of plastic products) used for marking plastics products and packaging. The Association of Engineers in Germany (VDI) published a guide of design for recycling of technical products in 1993. Also, the International Electrotechnical Commission published guides of environmental aspects to be considered in electrotechnical products design.

REFERENCES AND OTHER INFORMATION SOURCES

1. Brouwer O, Comtois Y, van Empel M, Kirkpatrick D, Larouche P. Environment and Europe: European Union environment law and policy and its impact on industry. Stibbe Simont Monahan Duhot, Deventer, 1994, 221 p.
2. Kvist T, Hakkarainen, E. Ympäristönsuojelulainsäädännön perusteet [in Finnish]. University of Turku, The Centre for Extension Studies, Turku, Finland, 1998. Publications A 64. 192 p.
3. Treaty establishing the European Community, http://europa.eu.int/eur-lex/en/treaties/index.html, 1999-12-15.
4. Fifth European Community environment programme: towards sustainability, http://europe.eu.int/scadplus/leg/en/lvb/l28062.htm, 1999-10-20.
5. Commission Directive 97/20/EC of 18 April 1997 adapting to technical progress Council Directive 72/306/EEC on the approximation of the laws of the Member States relating to the measures to be taken against the emission of pollutants from diesel engines for use in vehicles, http://europa.eu.int/eur-lex/en/lif/dat/1997/en_397L0020.html, 1999-11-11.
6. Council Directive 91/157/EEC on batteries and accumulators containing certain dangerous substances, http://europa.eu.int/eur-lex/en/lif/dat/en 391L0157.html, 1999-11-11.
7. Council Regulation No. 3093/94 on substances that deplete the ozone layer, http://europa.eu.int/eur-lex/en/lif/dat/1994/en_394R3093.html, 1999-11-11.
8. Commisson proposal COM(1998) 398 final—98/0228 (SYN). Proposal for a Council Regulation (EC) on substances that deplete the ozone layer, http://europa.eu.int/eur-lex/en/com/dat/1998/en_598PC0398.html, 1999-11-11.
9. European Parliament and Council Directive 94/62/EC on packaging and packaging waste (20.12.1994), http://europa.eu.int/eur-lex/en/lif/dat/1994/en 394L0062.html, 1999-11-11.
10. Proposal for a European Parliament and council Directive on marking of packaging and on the establishment of a conformaity assessment procedure for packaging (COM(96) 191 final—96/0123(COD) 25 November 1996, http://europa.eu.int/eur-lex/en/com/dat/1996/en_596PC0191.html, 1999-11-11.
11. Commission Decision of 28 January 1997 establishing the identification system for packaging materials pursuant to European Parliament and Council Directive 94/62/EC on packaging and packaging waste (97/129/EC), http://europa.eu.int/eur-lex/en/lif/dat/1997/en_397D0129.html, 1999-11-11.
12. Council Directive 1999/13/EC (11.3.1999) on the limitation of emissions of volatile organic compounds due to the use of organic solvents in certain

activities and installations, http://europa.eu.int/eur-lex/en/lif/dat/1999/en_399L 0013.html, 1999-11-11.

13. Council Directive 91/156/EEC of 18 March 1991 amending Directive 75/442/ EEC on waste(15.7.1975), http://europa.eu.int/eur-lex/en/lif/dat/1991/en_391L 0156.html, 1999-11-11.

14. Environment—Current situation and outlook, http://europa.eu.int/scadplus/ leg/en/lvb/l28066.htm, 1999-11-8.

15. Proposal for a Council Directive on end of life vehicles COM(97) 358 final—97/0194 (SYN), http://europa.eu.int/eur-lex/en/com/dat/1997/en_597 PC0358.html, 1999-11-11.

16. Draft proposal of 1999-7-5 for a European Parliament and Council Directive on Waste Electrical and Electronic Equipment amending Directive 76/769/ EEC, http://www.icer.org.uk/, 1999-11-11.

17. Finnish legislation, http://finlex.edita.fi/index.html, 1999-10-13.

18. Planned and proposed legislation—Sweden, http://extra.ivf.se/halfree/legislation_swe.htm, 1999-10-13.

19. Swedish legislation, http://www.riksdagen.se/debatt/sfsr/index.asp, 1999-12-17.

20. Electric and electronic waste—coming regulation in Sweden, http://www.environ.se/www-eng/kortinfo/electr.htm, 1999-12-17.

21. The amendment to the Packaging Ordinance—and resulting changes for the Mass Flow Verification, http://www.gruener-punkt.de/e/content/daten/novelle.htm, 1999-12-20.

22. Battery recycling, http://www.umweltbundesamt.de/uba-info-daten-e/daten-e/ batterieverwertung.htm, 1999-12-20.

23. M.A.R.I.—Info, http://www.mari-germany.de/html/seite1a.htm, 1999-11-12.

24. CYCLE newsletter, http://www.fvit-eurobit.de/PAGES/AGS/CYCLE/newsletter/News02_07-07-99.htm, 1999-11-12.

25. Ordinance concerning the disposal of electrical and electronic equipment, http://www.fvit-eurobit.de/pages/ags/agcycle4_engl.htm, 2000-1-5.

26. Umwelt-online, http://www.umwelt-online.de/recht/abfall/krwabfg/krw_ges. htm, 2000-1-5.

27. British Standard 7750, http://www.quality.co.uk/bs7750.htm#Comparison, 1999-11-23.

28. The Council Regulation (EEC) No. 1836/93 of 29 June 1993 allowing voluntary participation by companies in the industrial sector in a Community eco-management and audit scheme, http://europa.eu.int/eur-lex/en/lif/dat/1993/ en_393R1836.html, 1999-12-7.

29. EMAS, http://europa.eu.int/comm/environment/emas/sitesummarystatistics_1. htm, 1999-12-7.

30. Commission Proposal—COM (1999) 313 final, Amended proposal for a European Parliament and Council Regulation (EC) allowing voluntary participation by organisations in a Community eco-management and audit scheme (EMAS), http://europa.eu.int/eur-lex/en/com/dat/1999/en_599PC0313.html, 2000-1-21.

31. ISO/TC207, ISO 14000, http://www.tc207.org/faqs/index.html, 1999-11-24.

32. Pohjoismainen ympäristömerkki, http://www.sfs.fi/ymparist/index.html, 2000-1-24.

33. Nordic Ecolabelling, the Swan, http://www.svanen.nu/nordic/HowToApply.htm, 1999-12-9.

34. Council Regulation (EEC) No. 880/92 of 23 March 1992 on a Community eco-label award scheme, http://europa.eu.int/eur-lex/en/lif/dat/1992/en_392R0880.html, 1999-12-9.

35. European eco-label, http://europa.eu.int/comm/environment/ecolabel/triptyque.htm, 1999-12-9.

36. The European Union ecolabel, http://europa.eu.int/comm/environment/ecolabel/index.htm, 1999-12-9.

37. Environmental Label: the Blue Angel, http://www.blauer-engel.de/Englisch/index.htm, 1999-12-9.

38. TCO Environmental Labelling, http://www.tco.se/datamil/datami_nt.htm,1999-12-10.

39. TCO'99 Certification. Ecology for displays, system units and keyboards. Report No. 5. TCO, The Swedish Confederation of Professional Employees, Stockholm, Sweden, 1998, 40 p.

40. Environmental labelling of display units—TCO 1992. The Swedish Confederation of Professional Employees, Stockholm, Sweden, 1993, 4 p.

41. TCO'95 Certification. Requirements for environmental labelling of personal computers. Report No. 1, 3rd ed. TCO, The Swedish Confederation of Professional Employees, Stockholm, Sweden, 1996, 62 p.

42. ECMA Technical Reports. ECMA TR/70. Product-related environmental attributes, 2nd ed. http://www.ecma.ch, 2000-1-24.

43. Eco-declaration—Personal computers. Swedish IT-companies'Organisation, Stockholm, Sweden, 1997, 4 p.

7

Ecodesign with Focus on Product Structures

Conrad Luttropp
KTH Royal Institute of Technology, Stockholm, Sweden

1 INTRODUCTION

Sustainability as defined by the Brundtland commission depends on many factors connected to product development. Materials choice, resource allocation, upgrading, and recycling have strong links to the product structure. Product structures, as means to achieve sustainable products, are the core ingredients in this chapter. Materials are to some extent a limited resource, and metals as an example have been recycled for a very long time since recycling is often cheaper than mining and refining. However, benefits from materials circulation is not obvious, since humankind has been very inventive in substituting different materials. Resources like energy, fresh water, and air are limited and industrial activities, and the products themselves have an impact on these resources that cannot be neglected. The biological circulation ("ecocycling") in nature, where water, coal, and oxygen circulate, is the basis for life on earth, and it is also an ideal model for the industrial ecology. Upgrading or recycling depends on many factors and is not always the best thing to do. What is the right thing to do under certain conditions can be totally wrong under other circumstances, and suboptimizations must be avoided.

Before upgrading or recycling is possible, products must be separated, sorted, and transformed into flows of materials and components. Components must be good enough for reconditioning or instant reuse and materials must be clean/pure enough, or possible to upgrade, for manufacturing of new products. Sorting is only possible if the parts and the different pieces of materials are possible to identify. With no identification, even energy recovery or deposit may be a problem. If the materials hold harmful or polluted components they must be properly treated, in both the incineration process and the deposit. Some basic statements can be formulated:

- Sorting is a key function in reuse, upgrading, or recycling activities.
- Sorting is not possible without separation except in single material products.
- Sorting is not possible without identification.
- If adapting to separation and sorting, this must be balanced against ecology, economy, and technology.

2 PRODUCT DEVELOPMENT

The most common way of describing the product development process is by seeing it as a chain of tasks with milestones and decisions at points along the way. This traditional linear way of looking at the design process is still relevant, especially when the product itself is at focus. Without going into details, the design process always starts with a conceptual/product-planning phase with three typical steps. The first step is an analytical phase where the problem is to be understood. What does the customer actually want and what are unnecessary functions, what price is the customer willing to pay for these functions, what basic principles will be used, and so on. In this step the base for future work is established. In the second step concepts are generated in a creative process, hard to describe and difficult to guide. Step three evaluates the concepts very thoroughly and chooses which concepts are useful.

Typically these three steps are repeated in an iterative way several times before a plan for the next phase, product design, can be decided on. In the product design phase the concepts from the first phase are developed with drafting, dimensioning, making prototypes, and further market analysis. In this later phase of the product design process, a number of methods are available and computer aid has taken over what was previously manual work.

Focused on the designer, the conceptual design work is like a patchwork where understanding the problem, analysis, concept generation and evaluation of possible design solutions are blended in a randomized sequence (Figure 1). The hatched areas inside the wide arrow illustrate this "designers patchwork." At the end of concept design and before product design there is a very important milestone. At this point, substantial information concerning the product is present but there are still possibilities to make major changes. This is illustrated in Figure 1 by the two graphs crossing one another. This situation may be looked upon as an "intellectual" break-even point in the product development process. This is the perfect point for strategic decisions concerning green structures of the new product. The exact location of this break-even point is impossible to establish since it is more of an interval than a strict point. It is anyway one of the best points at which to establish not only functional parameters connected to the use and work of the product but also end-of-life parameters and functions. In this way environmental demands will be balanced in the design core against other functional and economical requirements.

2.1 Ecoindustrial Methods

A variety of methods has been established to help in product development, among which are quality function development, fail mode effect analyses, and others. Life cycle assessment (LCA) is probably at present the best-known tool for green design activities.

The LCA method calculates the environmental impact of a product from "cradle to grave." The origin of raw materials used, manufacturing process, usage, and end-of-life treatment are all phases that should be covered in a full LCA. This makes the procedure complicated and time consuming. A basic problem is also that LCA needs a lot of data of which little is known in early product development. The necessary inventory needs a rather complete design, which of course makes it difficult to make major changes, since everything is already finalized.

One other major problem with LCA methods is that they do not involve the preferences of the customer. If several concepts are environmentally evaluated in an LCA process, the benefit to the customer from the different versions of the product must be the same; otherwise, the comparison is invalid. In practice it is almost impossible to present, say, three solutions to a design problem with radically different "LCA values" and still maintain the same performance to the customer. For example, a hood to a car may be evaluated in an LCA process and the alternatives may be steel, aluminum, thermoplastics, or glass-reinforced thermosetting plastic. These alternatives are not equivalent to the customer: A hood is not just a hood. Even the sound of it when you knock on the hood is a part of the impression.

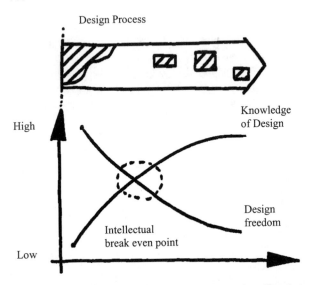

FIGURE 1 The thick arrow over the graphs illustrates all the information needed in a product development process over time. When a product design is finalized, the information is complete and the arrow is fully hatched. The graph from lower left to upper right illustrates the growth of knowledge over time. The graph from upper left to lower right shows, at the same time, how design freedom decreases with time. However, design knowledge is not established in a fully linear way with all principles established before all details. A materials choice and details concerning bearings might be settled before some of the working principles. Information is often gathered in a patched and randomized way. Pieces of information concerning basic working principles, customer needs, prototypes, manufacturing concepts, market constraints, and preliminary materials selection are assembled in an irregular way, called "designer's patchwork." This is illustrated by hatched "islands" in the wide arrow. The amount of information is set by the vertical scale and from which part of the design process is illustrated by the hatched areas inside the wide arrow. In the end of concept design and before product design there is a very important milestone. At this point, substantial information concerning the product is present but there are still possibilities to make major changes, the two graphs are crossing one another. This situation may be looked upon as an "intellectual" break-even point in the product development process; the possibilities of making major environmental improvements on the product are maximum at this point.

It is possible that customers may accept a slightly lower performance from a product if the environmental impact can be radically lowered. But, and this is important, then the benefit to the customer must be measured, evaluated, documented, and compared with the environmental impact. It is of course not a sustainable strategy to produce products with excellent environmental performance but with a functionality that does not apply to the customers.

2.2 Collaboration Between Designers and Management

Management and design must cooperate about green design since designers cannot make additional design efforts toward greener products if this is not allowed by the management, due to possible additional costs that these design efforts might raise. However, designers have functions as a main focus and management has profit or "red" figures in the accounts as main responsibilities. Therefore, it should to be possible to discuss the structure of the product and other green features early in product development, in a simplified way without loosing accuracy but without details. Semantically, the difference may be little, but precision and accuracy are not the same. An example:

> A person is standing in a telephone booth at Kings Road no. 123; he is ordering a cab to Kings Road no. 42. This order is precise but wrong. The taxi driver will stop at no. 42 and wait for the customer. If the customer orders the cab to Kings Road, this is a true order with low precision; there is a chance that the taxi driver will scan Kings Road for the customer. Therefore, Kings Road is an accurate order with low precision. The result for the taxi driver and the customer will probably be better. In other words, it is better to be approximately right than exactly wrong.

3 PRODUCT PLANNING

One way of making essential contributions to ecoperformance is to make a conceptual green structure of the forthcoming product at the intellectual break-even point, as defined earlier. A compromise helps not only nature but also the survival of the companies that make the products. Economical, technical, and environmental considerations merge into a conceptual green structure with high flexibility. The proposed procedure is divided into five main parts, steps A–E:

> Step A: An analysis of the market and benchmarking of products similar to the forthcoming new product.

Step B: Collect information concerning the new product through a questionnaire. Consider also main environmental targets for the new design. This can be done together in a meeting where management and the actual design group together provide the answers.

Step C: Modularize the new product according to green targets like materials, upgrading, and recycling. A worksheet is available for this. This document, "Module List" (Appendix 1), will contain facts on all levels of precision and conceptual assessments.

Step D: Make a "connection map" with the ecomodules as objects and the joints and connections as couplings. This map is just a concept and may be carried out in several versions depending on different design solutions. This is described below.

Step E: Put conceptual demands on the modules and their connections and joints. Use the module list and the connection map for this.

The motto for this procedure is it is better to be approximately right than exactly wrong.

4 STRUCTURE OF GREEN PRODUCTS

Green features, like reuse, upgrading, and recycling, rely on a good product structure. This can only be obtained if taken into account during the early design phases. The ecostructure of a product consists of pieces of homogenous materials, useful subassemblies, parts for reconditioning, upgrading, energy recovery, or deposit, called "ecomodules." The product then can be regarded as a set of ecomodules where each object or module consists of a homogenous piece of material, a useful subassembly, and so forth. Each of these objects has one or several surfaces where the object is connected or joined to other objects. These surfaces are called "separating surfaces" and indicate the border between ecomodules. If it is possible to identify the ecomodule and possible to use, it is surrounded by a "sorting border." The separating surfaces are held together by joints that are typically either separable (e.g., bolted, screw) or permanent (e.g., soldered, welded). Since joints play a crucial role in assembly and disassembly, they are characterized in this discussion by "load cases" as follows:

- Assembly load case: A force or torque is required for assembly.
- Resting (steady-state) load case: During use the joint is in a steady-state condition or "resting." At some point in time a force or torque would be needed for disassembly. The separation may be a destructive separation at a breaking point (Figure 2).

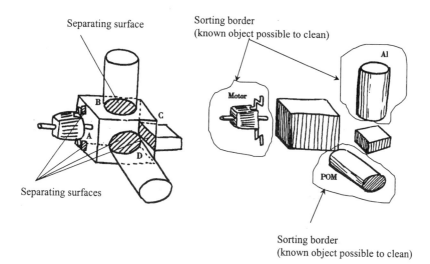

Separating surface

Sorting border
(known object possible to clean)

Al

Motor

POM

Sorting border
(known object possible to clean)

Separating surfaces

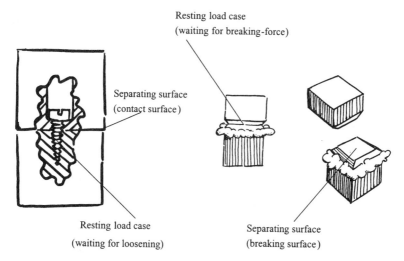

Resting load case
(waiting for breaking-force)

Separating surface
(contact surface)

Resting load case
(waiting for loosening)

Separating surface
(breaking surface)

FIGURE 2 A product can be regarded as a set of ecomodules where each module consists of a homogenous piece of material or a useful subassembly. Each of these objects has one or more surfaces where the object is connected or joined to other objects. These surfaces are called "separating surfaces" and indicate the border between ecomodules. If the ecomodule is possible to identify and possible to use, it is surrounded by a "sorting border." The separating surfaces are held together by "resting" or steady-state load cases, which are "sleeping" until the disassembly moment. Sometimes, as in a screw connection, the steady-state load case is just anticlockwise torque and the corresponding assembly joint load case is clockwise torque. In other cases it can be destructive separation at a breaking point. Separating surfaces can arise at several different joints such as screw joints, snapfits, and glue joints but also through a drop in strength somewhere inside the part. All joints that can be separated during disassembly are "resting load cases" since these joints are released by applying a force of some kind in a new way that is not present during service life of the product.

If we accept these ecostructural functions, features, and principles, some statements can be made.

I.   The ecostructure of a product can be regarded as a set of sorting borders, separating surfaces, and steady-state load cases.

II. Sorting borders enclose something that can be identified and if necessary upgraded, a subassembly, a labeled piece of material, etc.

III. Sorting borders must be congruent with a separating surface in all multimaterial products; otherwise, the border will not appear

IV. Reconditioning, upgrading, or recycling of a product starts with destructive or nondestructive disassembly to a certain degree; separation through steady-state load cases.

V.   In complex products sorting must be performed after disassembly and separation: sorting between useful parts, useful materials, and materials for energy recovery or salvage. To be useful they must be pure enough or possible to upgrade. Sorting borders must surround parts and fractions.

Separation and sorting are key features in green design:

Sorting border
- Encloses material or component.
- Can be identified and cleaned.

Separation surface
- Could be anywhere.
- Should be at sorting borders.

Steady-state load case
- Is activated during disassembly.
- Is giving a predefined separation surface at sorting border.

As an example, the PC monitor in Figure 3 has a very simple modular structure. There are just two steady-state load cases. The first one is four snapfits in combination and the second one is a screw joint, a hose hinge, that holds the electronics around the neck of the tube. This product then can be regarded as a set of modules all surrounded by sorting borders. The modules are printed circuit board; front and back in plastic; four rubber bushings; CRC tube; and tube neck electronics. Each of these objects has one or several separating surfaces where the actual object is connected or joined to other objects. That these objects each had a function before the disassembly event is not of interest, except for those that can be reused in their former function. Other parts are just pieces of material or amounts of energy.

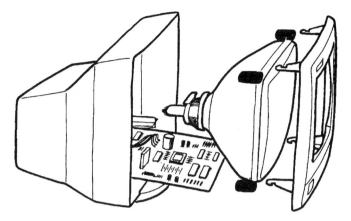

FIGURE 3 Exploded view of a simple disassembly case. The monitor has a very simple modular structure. There are just two steady-state load cases. The first one is four snapfits in combination and the second one is a screw joint, a hose hinge, that holds the electronics round the neck of the tube.

5 DISASSEMBLY ACTIONS COMBINED WITH PRODUCT STRUCTURE

Many products have a similar disassembly structure, which means that overall layout does not differ very much. Products, like wrenches and plastic watering cans, consist of just one material each and the main task for the scrappers is to identify and sort the materials in fractions. Other products, like computers and toasters, are built with separate components on a carrier with a cover. In this case the task will be more complex to identify and separate objects that are valuable, candidates for upgrading, and/or environmentally dangerous. Five different structures, important from a green design viewpoint, can be identified (Figure 4).

The PC monitor mentioned earlier has a main structure that can be visualized as a "hamburger" (Figure 4), with two halves, the front and the back, held together with four snapfits and between them, when the snap fits are released, the rest of the parts are separated. Other PC monitors or, for example, toasters or computers are made with a central carrier where everything is mounted and finally there is a cover, for electric security or just the looks of it. These products can be visualized as "dressed" designs (Figure 4).

Some products contain fluids or are meant to be waterproof and for that reason mostly have to be disassembled in a destructive way. Electric toothbrushes, ammunition, and gasoline tanks are examples of these "shell" designs (Figure 4). Many products contain just one material, like a watering can or a wrench, and from an ecostructural point of view these are all like a

Figure 4   Five different disassembly structures are shown. Top, from left to right: Shell, hamburger, twin. Bottom, from left to right: Rod and dressed designs.

"rod." Still other products have several easy identifiable fractions like a combination of rod of different materials like a "twin" design (Figure 4). More complicated products are usually combinations of these five concepts and often have subassemblies in several steps that have a structural variety of these five basic structures.

5.1   Hamburger Design

The first step is one nondestructive joint load case, and the second step is sorting. Typical products are mobile telephones, electric drills, toy cars, and remote controls (e.g., home electronics). These kinds of products have two halves, often plastic, that lock in and hold the interior, like transmission, motor, cables, and supporting parts. The first level of disassembly will then be to release the joint load case that keeps the two halves together. This joint load case requires good layout because the halves might be suitable for consumer disassembly. The hamburger concept is often used due to assembly matters, especially in products manufactured in large batches. It is also possible that the two casing halves are the most valuable recycling parts of the design. If this is the case, the joint between the two covering halves should be perfect, but the interior could be made with less accuracy. If, on the other hand, the interior contains something valuable, these objects should be easy to identify and separate. When the product is opened, cables, motor, and printed circuit board (PCB) can be directed to a reuse and/or fragmentation plus metal recovery facility and consumer plastics might go for energy recovery. If a product contains a battery, this must be easy to identify since this might be a problem for end-user waste handling.

5.2 Shell Design

The first step is one destructive joint load case, and the second step is sorting. Typical products are flashlights, ammunition, electrical toothbrushes, fuel tanks, gearboxes, and hydraulic jacks. The casing of these kinds of products has a closed shell structure with a smaller entrance into the shell than the main overall product dimension. The interior of complicated products from this family must enter through a small entrance. These kinds of products often have to be parted in a destructive way because the assembling is often not reversible. The interior can consist of transmission, motor, cables, supporting parts, and PCB. In-molded designs and hamburger designs, where the two hamburger halves are glued together, also belong to this group. These products, as a first step, have to be disassembled destructively in order to open the shell. The second task will then be to take care of the interior in a correct way. The main disassembling task connected to these product families is often to take care of dangerous waste materials inside the casing. If the product is a tank, the main task is to take care of what is in the tank like fuel, lubrication oil, and fluids from the cooling system. Electronic equipment often carries a lithium battery on the PCB, which should be taken care of.

5.3 Rod Design

The first step is sorting. Typical products are screws, screwdrivers, pliers, pipes for water, and television antennas. This product family has the characteristics of one or several pieces of the same homogenous material, and this "material body" is the main interest when it comes to recycling. These kinds of products often have minor attachments, such as plastic handles, that can be cleaned/burned away or a medium enclosed that does not demand special treatment. Historically, large pieces of homogenous materials have always been recycled, and these kinds of products just need to be labeled or possible to identify to make sorting possible.

5.4 Twin Design

The first step is separating, and the second step is sorting. Typical products are water taps, water closets, pieces of furniture with tubular legs, works of a clock, jewelry, and car wheels. In this case there is more than one important sorting object on the first sorting level, and the joint load case for this first level should be designed with great care. Steel, wood, aluminum, and brass could be combined in considerable pieces. The motive for the initial separating of the twin design is often that the two fractions involved need totally different treatment in upgrading or that both fractions are valuable, like a golden ring with a stone or a wheel with a steel hub and rubber tire. In these cases the sorting border between the two fractions should be good for economical reasons.

5.5 Dressed Design

The first step is one nondestructive joint load case, and the second step is a variety of joint load cases and sortings. Typical products are toasters, computers, audio equipment, and cars. These products are characterized by a carrier on which nearly all components are mounted and around this there is a cover/housing just as a protection and mostly with no other function. The first level of disassembly is always to remove the casing, and therefore this operation should be given a good nondestructive joint load case. The sorting border for the casing is most of the time quite natural. The carrier with the mounted parts contains a variety of different joint load cases, both destructive and nondestructive. This kind of design gives the opportunity to pick valuable components and leave components or subassemblies of minor interest. Parts to be taken care of that are mounted on the carrier should have good sorting border layout. They should be easy to identify and separate. This is a traditional layout for products with a lot of manual assembly work, and this type of design is common when there are large empty spaces inside the product. When designers try to make a certain product smaller, dressed designs can be transformed into hamburger designs.

In disassembly the first actions will be the most important ones, and the five product categories mentioned above can be organized in a table (Table 1) according to what the first disassembly action will be. The main objective with this approach is to get a better understanding of recycling layouts and conceptual disassembly principles. Each type of design has its own characteristics, but inside each family the recycling strategies can be quite similar. The focus will be on the two first levels in the scrapping procedure.

These concepts of recycling structures will guide designers to incorporate recycling in early design phases. Even if many products do not fit

TABLE 1 Properties Concerning Disassembly

	Hamburger	Shell	Rod	Twin	Dressed
First step	Sep-ND	Sep-D	Sort	Sep-ND/D	Sep-ND/D
Second step	Sort	Sort		Sort	Sep-Sort
Sorting_objects	(S_o) > 2	(S_o) > 2	(S_o)=1	(S_o)=2	(S_o) > 2

Separating can be destructive (D) or nondestructive (ND), and the number of sorting object (S_o) vary from family to family. This table of the five product structures shows how the activities differ on first and second levels.

perfectly into one of the product families, a conceptual recycling structure like this will be helpful in early design activities. The benefits from recycling will to a large extent lead to low costs on disassembling and sorting, and the first scrapping activities must be considered as the most important. In this model with five product "families," each family has a pronounced difference when it comes to the first or the second events in disassembly.

6 CONCEPTUAL EVALUATION OF DESIGN SOLUTIONS

Accepting the structure of products, as described earlier, requires a system for evaluation of different design solutions and for decision making in collaboration between designers and the management. Indices are used to put marks on different design solutions, and for practical reasons the scale is mostly simplified to an interval 0–1. The numerical values have quite some uncertainty as they are based on subjective estimations, and if insecure numbers are added, the result has a growing inaccuracy. The indices are therefore considered as accurate but imprecise.

There are four joint load case indices and four border indices. The joint load case indices give information on how a specific joint load case is organized, and the border indices give information on specific sorting objects that are surrounded by sorting borders. The four joint load case indices are as follows:

- Joint load case information $Li = [0;1]$: $0 =$ easy to understand, $1 =$ almost impossible to understand the joint load case.
- Joint load case equipment $Lq = [0;1]$: $0 =$ no tools, $1 =$ if special tools are needed to release the joint load case.
- Joint load case force $Lq = [0;1]$: how much force is needed to release the joint load case? $0 =$ just fingers, $1 =$ help from power tools.
- Joint load case time $Lq = [0;1]$: how much time is needed to release the joint load case?

In Figure 5 a simple "connection map" is presented with two Lego pieces put together in three different ways. The two balloons represent the two objects that are involved in this example. The upper left joint load case is very well understandable (Li = 0), no equipment is needed (Lq = 0), no extra force (Lf = 0), and very little time (Lt = 0). The other two joint load cases are more complicated, which can be seen also on the joint load case indices. A full description of the indices is presented in Appendix 2.

Through this index system it is possible for designers and management to make decisions together concerning the disassembling functionality of for example, a new mobile phone. If the hamburger concept is to be preferred

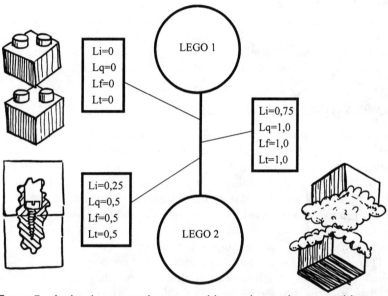

FIGURE 5 A simple connection map with two Lego pieces as objects, represented as two balloons. The straight line represents the presence of a steady-state load case, and in the squares the corresponding joint load case indices are shown. Upper left: A very simple joint load case. Simple to understand Li = 0, no equipment needed Lq = 0, very little force needed Lf = 0, and fast action Lt = 0. Lower left: A screw joint; a little bit more complicated. A practical person will understand Li = 0.25; a screwdriver is needed, Lq = 0.5; some force is needed, Lf = 0.5; and it will be a little bit more time consuming, Lt = 0.5. Right: A glue joint; much more complicated than the others. This is not easy to understand Li = 0.75; a variety of tools are needed, Lq = 1.0; more than normal manual power is needed, Lf = 1.0; and it will take some time, Lt = 1.0. A more detailed presentation of this index system is present in Appendix 2.

and the initial joint load case to part the two halves is to be made by the consumers, the opening of the two halves must have low joint load case indices. This will be an accurate but imprecise functional requirement to designers that the opening joint load case must be designed to be easy to comprehend, so that the average consumer will understand how to do it.

The inexperienced consumer needs much "easier" joint load cases, with lower joint load case indices than the experienced professional. A steady-state load case will look quite different for a professional than to a consumer. Sometimes it is even preferred that the consumer does not disassemble or even tries to do it because of safety.

There are also four sorting border indices, giving scores on sorting objects:

- Sorting border information $Bi = [0;1]L$: good understanding what is in the sorting border $= 0$; bad understanding $= 1$.
- Sorting border economy $Be = [0;1]$: valuable object $= 0$; objects that may cause extra costs $= 1$.
- Sorting border destiny $Bd = [0;1]$: possible to reuse in same position a second time $= 0$; deposit $= 1$.
- Separating surface efficiency $Bs = [0;1]$: surface follows the sorting border perfectly $= 0$; the surface does not follow the sorting border at all $= 1$.

7 GREEN CONCEPTUAL DESIGN IN FIVE STEPS

To achieve sustainable products, the economical, technical, and environmental properties must merge into a compromise, which helps not only nature but also the survival of the companies that make the products. Sustainable products could be looked on as a modular structure of subassemblies and pieces of materials, for instance, an ecostructure of sorting objects. The necessary separating surfaces and steady-state load cases are all supposed to be "sleeping" until the end of the first service life.

Modules consisting of materials for energy recovery can be, and for economical reasons should be, assembled and joined with minimum consideration for separating surfaces and steady-state load cases. There is no need to put design efforts into components that ecomanagement has forecast as suitable only for energy recovery.

Modules defined by a "valuable" sorting border should be connected to the rest of the design with great care because this is a module that very well might be reused, upgraded, or recycled at the end of the first service life. Let us suppose this valuable module is an electrical motor intended for reuse. In this case there are, of course, a lot of borders and surfaces inside this motor that could be designed "quick and dirty," since the motor will be reused on a higher sorting level. In upgrading strategies there should be good sorting borders around modules, which might be replaced during service life.

In line with the ideas of conceptual green design as presented earlier in this text, a procedure is developed. The procedure must be carried out in different ways depending on the circumstances, the structure of the company, type of product, and so on. The following description may be helpful in establishing a product specific conceptual ecostructure.

The proposed plan is divided into five main parts. The first step (A) is to conduct a benchmarking of the forthcoming product. The second step (B) is to collect information through a questionnaire and to plan the conceptual properties of the product. This can be done together in a meeting with management and the design group. The third step (C) is to document and organize known facts that are already decided on together with knowledge and assessments still not fully established. This document, the module list, will contain facts on different levels of precision and conceptual assessments (Appendix 1).

As fourth and fifth steps (D & E), a connection map of the possible sorting objects is established together with indices on steady-state load cases and sorting borders. The "map" is just a concept and may be carried out in several versions depending on different design solutions.

7.1 Step A. Examine the Market: Collect a Set of Products Similar to the New Product for Benchmarking

Disassemble present and relevant other products and use the "life cycle aspects" worksheet as an aid (Appendix 3). This worksheet is developed to be a checklist for possible aspects on the typical five life cycle phases that are standard in an LCA procedure. The best way to use it is to consider each life phase after another and for each phase work down the aspects. The aim of this is to document hangups, to get an overview, and to see to it that all aspects are covered. LCAs of similar products can also be a good information base for benchmarking and a good source for information on different positions in the life cycle aspect worksheet.

7.2 Step B. Product Planning: A Questionnaire to be Completed in a Meeting with the Design Group Together with Economical and Environmental Management

Question 1: What kind of product is it? A hamburger, a shell, a rod, a twin, or a dressed design? If the product is supposed to contain fluids or materials in the form of gas, there must be at least one subassembly with a shell layout. When the product is supposed to be water or gas resistant, the whole product is likely to be a shell design.

Products containing electronics or electric motors are mostly hamburger or dressed designs. When the product due to cosmetic, handling, or layout reasons must be larger than the interior, this actually calls for the dressed design concept. If the product "needs" all the space inside the cover for subassemblies and parts, the hamburger design is advantageous.

Design combinations are also possible. For example, a product containing several parts, including the cover, all suitable for incineration and also a PCB suitable for metal recovery and perhaps an Ni-Cd battery that must be taken care of. One way of optimizing this design from an end-of-life perspective would be to screw or glue the incineration candidates to the cover in what might be called a partly dressed concept. The cover in a hamburger concept could lock in the PCB and the Ni-Cd battery. This way there will be a concept with three sorting objects (four if the cover fully divides in two parts) all surrounded by sorting borders. In this case even a shell design is possible as long as the product clearly indicates where to break the cover to get the PCB and the Ni-Cd battery out without damage.

Question 2: Are there any sorting objects in the design with a positive economic value to the company? An electric motor that can be reused, a housing of engineering polymers, etc.? A conceptual bill of materials can be of help when answering this question, especially when the forthcoming product might be a rod or a twin design. In these cases it will, of course, be advantageous to minimize the number of materials used.

Dressed or hamburger designs often hold PCBs, electric motors, wiring, or heavy metal batteries, and each of them should be thought of as sorting objects. This is the time also to start a list of expected materials.

Question 3: What must be taken care of due to environmental impacts? Rechargeable batteries, toxic fluids, etc.? A list of sorting objects that will be connected with special waste handling or environmental taxes will be proper to establish at this point. This list may also contain all the objects for which the company, in the future, might be forced to take responsibility.

Toxic fluids are often present in products in a partly shell concept. Rechargeable batteries are furnished by suppliers and may therefore often be attached to the working parts through a hamburger concept.

Question 4: What will be the most likely scrapping scenario for this product? Consumer disassembly and then energy recovery? Professional disassembly and sorting? Retake? Try to consider the two or three first levels of separating and sorting. The following combinations will give some alternatives and will be very important for decisions during the product design phase. Disassembly and sorting will be carried out by consumer (initial sorting), maintenance company, scrapper or original manufacturer (take-back). What is the possible destiny of the sorting objects if the sorting border index (Bd) concerning destiny is used? This classification should later be transformed into indices and functional requirements suitable for designers to handle when evaluating different demands on steady-state load cases, separating surfaces, and sorting borders derived from manufacturing, market, and so on.

Question 5: Would it be possible to change some of the parts and in this way give the product a longer or shorter life and this way achieve a better ecoperformance? Estimate the product's length of life as a whole and of all the major subassemblies. Try also to estimate if any parts could gain from a shorter or a longer length of life. Will there be any service relevance? The information collected this way should be a database for future work and a main input to steps C–E.

7.3 Step C. Establish a Module List

The next step is to map the forthcoming design through its sorting objects by establishing a module list (Appendix 1). The worksheet is meant to be a core documentation of the conceptual balance of economic, environmental, and technical properties of a new product where a lot of design freedom still exists. The module list in order from left to right is as follows:

- Module no.: A simple counter.
- Type: Connects to the bottom, type description of possible modules (a–d). Is the proposed module also a natural module? Is it the material fraction in this module that is interesting? Does the module carry a positive-to-negative value? Is it a candidate for exchange, upgrading, etc.?
- Name: Try to find a name that characterizes the module.
- Material–manufacture–transport–use–end-of-life: These are the five phases in life cycle thinking, and this is the place for information concerning a specific module that is especially pertinent to a certain life phase of the product. This part of the sheet corresponds with the life cycle aspect worksheet.
- Index: This is the place for sorting border indices. Here, it is possible to communicate goals concerning the new product. How easy shall it be to understand the sorting border and what to do about it during end of life scenarios? How is the economic situation and are there any taxes connected to a specific module? What is the preferred destiny of the module? What demands are put on the potential separating surfaces?

- Notes: Free space for notes and comments.

Make imaginary modules of the expected product from a sustainable viewpoint. Fill out the form but leave the positions that are to uncertain or present alternatives or an interval. Set goals by giving each module Bi, Be, Bd, and Bs indices. In Table 2 as a demonstration example, a module list is made for the PC monitor from Figures 3 and 6, with the main sorting objects of interest and a list of the relevant sorting objects M1-Mx.

7.4 Step D. Establish a Connection Map of Sorting Borders

The next thing to do is to represent relevant sorting objects, M1-Mx, in a flowchart where the modules are for example circles and the connections are straight lines. If there is a joint load case between the modules use a straight line and use a dotted line if there is just contact.

A typical hamburger, as the PC monitor presented above, will have one initial joint load case that opens the design and then the rest will be free and ready for sorting. This kind of design will typically have one straight line representing the opening of the cover and several dotted lines representing the separating surfaces that are locked in by the cover. A connection map of the PC monitor is presented in Figure 7.

7.5 Step E. Put Relevant Indices on Steady-State Load Cases

The fifth step in this procedure would be to define indices for connections between different sorting objects and consider how and by whom the joint load cases should be released and in what way the sorting is likely to take place. The indices will be different depending on who is supposed to disassemble the product and what is to be done with the remains of the product. Sometimes the end user (consumer) would be the best one to perform a certain separation (e.g., removal of Ni-Cd batteries), which means that the joint load case information index in this case should be low. The rest of the product may have "cheaper" indices suitable for normal waste handling or take-back. Even the consumer who will not participate in take-back may have a "battery conscience." As an example, in Table 3, this has been done with the PC monitor from Figure 3. If the module list and the connection map are combined, we get Figure 7.

8 CLOSING REMARKS

Products with a well-designed set of sorting borders, separating surfaces, and steady-state load cases will have improved capability to close the envir-

TABLE 2 Module List for a PC Monitor

			Index								
M no.	Type	Name	M..	M..	T....	U....	E....	Bi	Be	Bd	Bs	Notes
M1	b	Housing										Flame retardants
M2	b	Front										Flame retardants
M3	a,b	PCB										
M4	a	Tube										
M5	b	Tube electricity|									
M6	b	Bushing										

This table is an example covering a present PC monitor, but the idea with this sheet is mainly to gather crucial facts about new products in the conceptual phase.

TABLE 3 Module List with Indices for a PC Monitor

M no.	Type	Name	M..	M..	T....	U....	E....	Index Bi	Be	Bd	Bs	Notes
M1	b	Housing						0.25	1.0	0.75	0	Flame retardants
M2	b	Front						0.25	1.0	0.75	0	Flame retardants
M3	a,b	PCB						0	1.0	0.5	0	
M4	a	Tube						0	0.5/1.0	0.5/1.0	0	
M5	b	Tube electricity						0	0.5	0.5	0	
M6	b	Bushing						0	0.5	0.75	0	

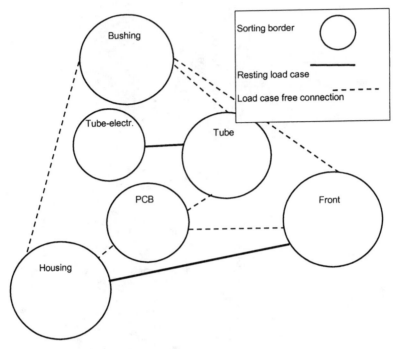

FIGURE 6   A connection map of the PC monitor from Figure 3. Solid lines stand for a steady-state load case and dotted lines indicate contact between these two objects.

onmental loops. If scrapping is made cheaper and sorting made possible, this will affect economy in cases where products have to be taken care of by the companies, from "cradle back to cradle." This can be achieved in the product development process after the concept design phase but ahead of more specific designing and drafting tasks. This will also make essential contributions to ecoperformance.

A structural viewpoint like this is a very complex issue and a lot of decisions taken by designers and management in collaboration must be governed by society, market conditions, and so on. These decisions must be transformed into instructions, functional requirements, and constraints given to designers. Environmental requirements have to exist together with the overall economic reality of the company since it still has to exist in a competitive business environment. The proposed procedure is intended to help management and designers when deciding on the appropriate recycling procedures for a new product. The main idea in this approach is the possibility to choose a basic recycling layout before the embodiment design is started.

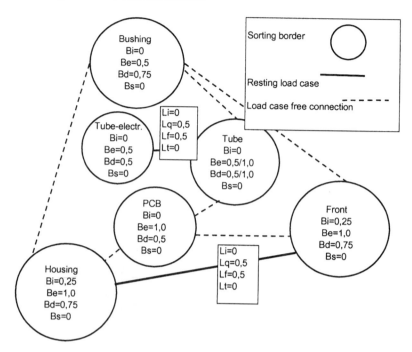

FIGURE 7 A complete connection map with both steady-state load case indices and sorting border indices noted. List with indices for the PC monitor in Figure 3.

BIBLIOGRAPHY

Lagerstedt J, Luttropp C. Functional priorities and customer preferences in the context of life cycle design. CIRP 6th international seminar on Life Cycle Engineering, Kingston, Canada, 1999.

Luttropp C. Design for disassembly-environmentally adapted product development based on prepared disassembly and sorting. Doctoral thesis, KTH Machine Design, Stockholm, Sweden, 1997.

Luttropp C. Eco-design in early product development. World Conference R99, Geneva, Switzerland, 1999.

Luttropp C, Züst R. Eco-effective products from a holistic view. CIRP 5th International Seminar on Life cycle Engineering, KTH, Stockholm, 1998.

Züst R. Sustainable products and processes. Proceedings of CIRP 3rd International Seminar on Life Cycle Engineering, ISBN3-85743-985-8, Zürich, Switzerland, 1996.

APPENDIX 1 Module List

M no.	Type	Name	Material	Manufacture	Transport	Use	End of life	Sorting border index				Note
								Bi	Be	Bd	Bs	
1												
2												
3												
4												
5												
6												
7												
8												
9												
10												

a) Natural module PCB, component bearing, motor, battery, LCD, frame, cover

b) Material fraction plastic, steel, copper, aluminum, incineration

c) Value dangerous, expensive, recycling demands

d) Time service, lifetime upgrading

APPENDIX 2 MATCHING INDICES

The matching indices all have ordinal scales. There is no absolute zero and there is no exact ratio between the numbers. The mark 1.0 is more than but not necessarily twice as much as 0.5. The indices Lt (time) and Be (economy) could have had ratio scales, but it is believed that it is better to standardize the scales to achieve accuracy more than precision.

Three main groups can be observed as operators:

A. The consumer or *end-of-life user*, who is not experienced in separating and sorting.
B. The *professional* who is experienced in disassembly and sorting of similar products but not necessarily a particular version of the present product.
C. The *company* that manufactured the product or an authorized retailer who has special knowledge and tools for a specific product.

Matching Indices on Steady-State Load Cases and Sorting Boarders

To evaluate steady-state load cases, four indices are proposed; each is supposed to give marks on one specific joint load case. The first index, the joint load case information index, depends a lot on the operator's competence and also on available information on the joint load case. The other three indices are more neutral but still related to the operator's abilities. The border indices say something about the interior expressed in sorting borders, sorting objects, and connected separating surfaces.

Joint Load Case Information Li = [0;1]: Ordinal Scale

Li = 0 . . . Operators A, B, and C: No extra information is needed/ joint load case self explanatory. Everyone understands the steady-state load case.

Li = 0.25 . . . Operator A: The steady-state load case will need a good label to be understood. Operators B and C: These groups will understand without any extra information due to their disassembly competence.

Li = 0.5 . . . Operator A: Will not understand the steady-state load case. One sample joint load case has to be released first. A label is not enough. Operator B: This group will understand the steady-state load case if it is provided with a good label/instruction. Operator C: This joint load case is possible to understand for operator group C even without a label due to special knowledge and experience.

Li = 0.75 . . . Operator A: Will not understand the steady-state load case and will probably not understand how a sample joint load case should be carried out. Operator B: Substantial information must be added or one sample joint load case has to be released first before this group can operate the steady-state load case. Operator C: This group will understand the steady-state load case if it is provided with a good label/instruction.

Li = 1 . . . Operators A and B: Will not understand the steady-state load case and will probably not understand how a sample joint load case should be carried out. Operator C: Substantial information must be added or one sample joint load case has to be released first. Special equipment may be necessary to develop to release the steady-state load case in a proper way.

This information/understanding index is independent of the time needed to understand how the joint load case should be released, since the time to understand how a certain joint load case should be released will differ a lot depending on experience, education, and so on. A consumer may give it the time needed, but a professional will probably start working on a perhaps destructive disassembly after 10 seconds even if he/she does not understand how to make an appropriate disassembly.

Some joint load cases are very simple and self-explanatory while others need considerable effort. A modularized product, where one module should be replaced very often, must have low indices on steady-state load cases and sorting borders, for example, the dustbin in a vacuum cleaner or the filter in a coffee machine.

Joint Load Case Equipment and Tools Lq = [0;1]: Ordinal Scale

Lq = 0 . . . No tools are needed, which may be suitable to group A, but it can be a poor index to groups B and C, for work injury reasons.

Lq = 0.5 . . . Simple tools (e.g., screwdriver, hammer) are sufficient. May suit ABC depending on the situation.

Lq = 1 . . . Special tools or a variety of tools are needed. This is suitable for B and especially C. It is also to be recommended if a large number of similar products are to be separated due to work injury reasons.

Simple tools are not a common understanding, but on this index my definition is that simple tools are a hammer, a knife, and a screwdriver with a 6 × 1-mm blade, which every consumer (operator group A) has at home. For the experienced scrapper (operator group B) simple tools could also mean that one and just one of his/hers standard tools is needed.

During the disassembly of a truck cab, in a thesis project, students wanted an access index to give marks on how easy it was to reach the actual joint load case. This might be interesting if the indices only were meant to be evaluation marks on "disassemblability." However, in this system the main goal for the indices is to be a "specification link" in conceptual design phases, which means that the accessibility is, more or less, built into the Lq and Lt indices. For example, a need for special tools, Lq = 1.0, can be derived from an unusual screw dimension and a common screw dimension but put in a place where it is hard to reach.

Joint Load Case Force Lf = [0;1]: Ordinal Scale

Lf = 0 . . . No extra force is necessary; two fingers are enough. This index points out joint load cases where several sorting objects are released by just one joint load case, for example, when a hamburger design is opened and the interior can be sorted without any extra joint load cases.

Lf = 0.5 . . . Normal manual power. Suitable to operator groups A and also BC depending on the number of similar movements (work injury considerations).

Lf = 0.5 . . . Extra power is needed such as two hand action with full power or extra power through a power tool or machine. Sometimes this is the most effective way with one destructive joint load case that will release all components. Suitable for BC.

There is no common understanding what will be normal manual power. In this index, normal manual power is what can be expected from a person who is *not* trained in practical matters. For example, the battery hatch of ordinary consumer goods should have Lf = 0. Otherwise, the product will be unusable to many people and there will be a constant risk that the product will be damaged during change of batteries.

Joint Load Case Time Lt = [0;1]: Ordinal Scale

Lt = 0 . . . The time to release the joint load case is less than 10 seconds.

Lt = 0.5 . . . The time to release the joint load case is 10–30 seconds.

Lt = 1 . . . The time to release the joint load case is more than 30 seconds.

The scale for this is classified as ordinal scale, but time itself is an absolute scale with a zero that can be expressed as nothing and an equally distinct scale with one as the smallest number. However, since the main objective for this system of indices is to be a conceptual specification language between designers and management, I have chosen to reduce absolute time to an index with an ordinal scale.

This index can be accepted with a high value to group A, but for economical reasons this index must be low to groups B and C. It may be possible to engage the consumer in disassembling his or her scrapped products, even if it takes some time, and thus doing something for the environment. To a disassembly company time is money, and this may give Lt to be the most crucial index for professional disassembly activities. It is quite possible that $Li = 1$, $Lq = 1$, $Lf = 1$, and $Lt = 0$ imply a joint load case perfect for take-back of a large number of similar products.

Understanding how to sort a specific sorting object is measured in the Bi index, and the additional economy is measured in the Be index. The Bd index shows the destiny of the sorting object. The Bs index gives marks on how well a separating surface follows the actual sorting border.

Sorting Border Information Bi = [0;1]: Ordinal Scale

$Bi = 0$ . . . Operators A, B, and C: No extra information is needed/border self-explanatory. Everyone understands the sorting border.

$Bi = 0.25$ . . . Operator A: The sorting border needs a good label to be understood. Operators B and C: These groups will understand without any extra information because of their disassembly competence.

$Bi = 0.5$ . . . Operator A: Will not understand the sorting border. A label is not enough. Operator B: This group will understand the sorting border if it is provided with a good label. Operator C: This sorting border is possible to understand to operators in group C even without a label due to special knowledge and experience.

$Bi = 0.75$ . . . Operator A: Will not understand the sorting border. Operator B: Substantial information must be added or an analysis has to be made. For example, the parts have no labels. Operator C: This group will understand the sorting border provided with a company-specific label.

$Bi = 1$ . . . Operators A and B: Will not understand the sorting border. Operator C: Substantial information must be added or an analysis has to be made. For example, the parts have no labels.

Sorting Border Economy Be = [0;1]: Ordinal Scale

$Be = 0$ . . . This sorting border contains something of value, even if this value may not be economically realistic to recover. This could be a valuable metal or a subassembly that can be used as spare part on the market.

Be = 0.5 . . . Economy of this sorting border is very much dependent on the situation.

Be = 1 . . . This sorting border is connected to special costs related to for example, toxic materials etc., that is outside of ordinary waste.

This is a very complex index, closely related to LCA. It is also related to market prices on recycled materials and to the costs for deposit. Statements on recycling economy are outside this work, but the index is very important and should always be considered. What is economically right to do in a large city may not be the right thing to do in sparsely populated areas.

Sorting Border Destiny Bd = [0;1]: Ordinal Scale

Bd = 0 . . . Reuse without any change of shape.

Bd = 0.25 . . . Reuse for another purpose such as old tires used as dock fenders.

Bd = 0.5 . . . Recycling when materials are reused as raw materials in new shape.

Bd = 0.75 . . . Energy recovery.

Bd = 1.0 . . . The sorting object goes to deposit.

This index is supposed to point out the destiny for each sorting object and of course Bd 0–0.5 need good sorting borders and also low Bi indices, if the destiny shall be realistic. It can be doubted however, if this index carries a scale at all, but if Bd = 0 is regarded as the shortest environmental loop Bd = 0.5 as a longer circle and Bd = 1.0 as a loss from the environmental loop, it might be possible to use a scale on the Bd index. This index can also serve as an example where it is convenient with more than three grades.

It must be stated at this point that these destiny preferences are not always the same, that major cities may have quite other preferences than sparsely populated areas. However, even if the indices can be used, Bd = 1 may sometimes be better than lower sorting border indices.

Separating Surface Efficiency Bs = [0;1]: Ordinal Scale

Bs = 0 . . . The separating surface follows the sorting border perfectly.

Bs = 0.5 . . . The separating surface does not follow the sorting border, but a good separating surface can be obtained through extra treatments.

Bs = 1 . . . The separating surface does not follow a sorting border.

To achieve sorting borders around homogenous pieces of materials or subassemblies, at the disassembly, the separating surfaces must follow the sorting borders as stated in section 6. This can be looked upon as a separating efficiency and points directly toward the joining operations in manufacturing and separating and sorting steps in disassembly.

This efficiency will directly influence the cost of the sorting border. If a product is planned for take-back, every divergence from a perfect separating surface will give rise to extra costs and therefore the index, in this case, should be close to Bs = 0. If the material is supposed to be recycled, the separating surface must not be perfect. A sorting border will arise anyway as long as it is possible to clean or recover the sorting object.

At this moment it can be observed that Bs = 0 points to a separating surface that once has been an assembly surface. However, all assembly surfaces shall and will not be separating surfaces.

APPENDIX 3 Life Cycle Aspects

	Material	Manufacturing	Transport	Use	End of life
Function					
Aesthetics					
Hazard					
Energy					
Life time					
Structure					
Mixture					

8

Decision Support for Planning Ecoeffective Product Systems

Rainer Züst[*]

Leiter KTC, Winterthur, Switzerland

1 INTRODUCTION

1.1 Resource Use

Our modern society uses enormous amounts of natural resources through various activities. Large natural areas throughout the world have already been exploited. For example, the nearly 7 billion inhabitants of the Earth used about 150,000 terawatt-hours of energy in the end of the 1990s (1 terawatt-hour $= 10^9$ kWh). This corresponds to an average use of energy of $\sim$20,000 kWh per inhabitant per year, or 2 kWh per inhabitant per hour. In Switzerland, the average use of energy per capita is at 5 kWh per hour, or at about 6.5 kWh, if the total energy consumption in the production phase is also considered (Imboden 1999). In North America, the energy use is more than 10 kWh per inhabitant per hour, whereas developing and threshold countries use, on average, only fractions of a kilowatt-hour per inhabitant per hour. This uneven distribution results in an energy use of nearly 75% of the total use in industrialized countries, although only a small part of the world's population lives here.

---

[*] *Current affiliation*: Alliance for Global Sustainability, ETH-Zentrum, Zurich, Switzerland.

Materials are also mined in large amounts and industrially processed into construction materials. In 1972, the "Club of Rome" first published an estimate of the availability of economically extractable metal reserves. Important metals such as tin, mercury, zinc, lead, copper, and wolfram would have been available only for a few decades according to the rate of use of that time. Later analyses show that the temporal availability spaces are a little larger. This is due to more efficient mining and dressing technologies and to material substitutions.

Practice shows an increased demand of energy and raw materials. Simultaneously, energy and, mostly, raw material prices are falling. Explanations for this are the lower costs per produced unit due to new technologies and larger amounts of production. Government supply and disposal guarantees often prevent effective and efficient energy and material management.

1.2 Relationship to Environmental Management

In the early 1990s, the International Organization of Standardization (ISO) started with the standardization work in the field of "environmental management." In working out this standard, the industry has expressed its willingness to assume self-responsibility in the field of environmental management. Companies of all kinds are to be supported in their efforts to determine, increase, and display their own environmental performance.

An environmental management system (EMS) according to ISO 14001 is part of the interdisciplinary management system containing the organizational structure, planning activities, responsibilities, methods, actions, processes, and resources for development, implementation, fulfillment, evaluation, and sustaining of environmental policy (CEN 1996). An EMS must guarantee that those activities, products, and services that can be influenced by the organization are continuously improved with regard to environmental impacts.

The basic model of the ISO standard 14001 (CEN 1996) is shown in Figure 1. Here, five steps are characteristic. They or parts of them are cyclically repeated in a certain rhythm.

The focus of EMS is not on the standard ISO 14001 but on the impact. An EMS is a means to an end. Keeping legal compliance and consideration of interested parties are required. Both aims orient toward the stakeholder requirements (Figure 1). EMS based only on stakeholder requirements are reactive systems. Sensible company-specific solution strategies, however, cannot be developed in this way (Züst 1998). Therefore, the standard ISO 14001 goes beyond these minimal requirements. Additionally, a company is expected to improve the environmental performance of its own activities,

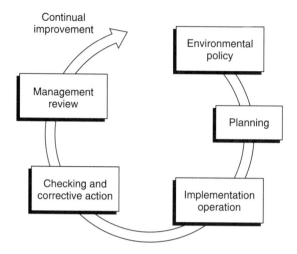

FIGURE 1 Basic structure of the ISO standard 14001. (From CEN 1996.)

products, and services continuously according to its specific potential. The assessment and improvement of the company's own specific environmental situation is important.

The improvement of the environmental performance represents a central target of standard ISO 14001. Therefore, further standard parts have been or are still being elaborated for the assessment and increase of the environmental performance in ISO/TC 207 environmental management. Especially, the standard ISO 14040 (CEN 1998) is of interest as far as environmental product design is concerned.

1.3 Meaning of the Early Planning Phase

Usually, the use of resources results in added costs. Seen from this view, wastage in production or high use of energy in product usage are not optimally managed resources and thus are a financial loss. Therefore, environmental management should mainly be a proactive resource management (Züst 1999). Thus, unnecessary use of resources can deliberately be avoided. Costs and environmental impacts of products are consequences of planning and decision processes. Opitz (1970) already examined the relation between "planned" and actual costs 30 years ago. His observations were limited to the development and production phase of technical products. The research showed that 70–80% of the planning and production costs are already set in the development phase. In the early 1990s, a case study also examined the relation between planned and actual costs. All life phases were examined, for instance, development and production and usage and disposal (Figure 2). The study examined warrantee costs in the usage phase and the taking back of products with a professional disposal in the disposal phase.

Ninety to 95% of the company costs arising during all life phases were already set in the development phase (Züst and Wagner 1992). It is similar to the environmental impacts. These too are largely determined in the development phase. Thus, modifications within the development phase represent the largest impact concerning environmental performance. These planned environmental impacts occur later in production, usage, and disposal. The development process of products is of the greatest importance with regard to the total environmental performance. If the development process is divided into individual subphases, as is typical in the field of engineering, there are important decisions to be made at the beginning of the development phase with rather little and also uncertain and abstract information (Figure 3).

In the later course of development, the information basis improves. Additionally, the description of the product to be developed becomes clearer. Possibilities of making changes decrease because of the previously made decisions. Disadvantageous solution principles can be adapted only in a limited measure. Therefore, it is important for the individual company to see the possible ecological meaning of each change already in the early planning phase.

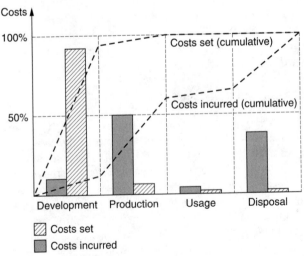

Figure 2 Relationship between set and incurred costs. (From Züst and Wagner 1992.)

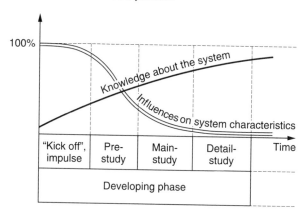

FIGURE 3   Influence on the system characteristics on the one hand and course of the system knowledge on the other hand in a qualitative presentation. (From Daenzer 1976; Züst 1997.)

1.4   Central Questions

The systematic and methodically correct handling of uncertain, incomplete, and/or abstract information in the early planning phase has an important meaning. At present, corresponding methods and procedures are missing in this field. Therefore, the central questions are as follows:

- How can energy and material flows caused by products and services in production, usage, and disposal be systematically registered, visualized in a receiver-friendly way, and implemented into the planning and decision process? (→ presentation of environmental information)
- Which methodical components can effectively support the planning and decision process in the early product development phase with regard to improved performance? (→ processing of environmental information)

2   PRODUCT LIFE CYCLE MANAGEMENT

The individual project parts are based on common model representations. These concern the product life cycle management. In the following, some aspects of product life cycle management are described.

2.1 Product Life Phases

Products pass through various life phases or transformation processes (Figure 4). In the first step, products are developed. These consist of market and customer research, development of concepts, and determining the detailed product design. After the development phase, construction plans, such as work piece sketches, part lists, and operation plans, are developed. Then, the products are manufactured. This life phase is characterized by material and energy supply, production of semifinished goods, component parts, subassemblies and assembly to final products, and necessary transport to the customer. Additionally, the corresponding operational means, such as tool machines, assembling machines, and transportation systems, are required.

The usage of the products follows after the production. Often there is an owner change. Thus, the original producer has only limited influence on the usage. Further, during usage often service and repair cycles occur that prolong the lifetime of the product or improve the state of operation.

If the product is no longer used, it is put out of operation. The obsolete product can be disposed off in various ways or can be integrated in the recycling processes. Thus, products pass through phases of development, production, usage, and disposal. To achieve good solutions in general, the individual life phases must be optimized as a whole. Here "life cycle engineering" is mentioned (e.g., see Alting and Legarth 1995) or product life cycle management. Further, all necessary auxiliary processes, which are necessary for the proper operation of products, must be considered. An isolated product turns into a product system, which must be evaluated and optimized (also see Caduff and Züst 1996; CEN 1996). This chapter therefore emphasizes complete product systems.

2.2 Special Product Life Cycles

Generally, products do not pass the previously described life phases in a linear sequential order. Various cycles within or between the life phases can occur. These cycles can be planned (e.g., service interval) or occur coincidentally (e.g., repair). Four product life cycles of physical products are described as an example:

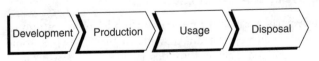

FIGURE 4 Product life phases. (From Züst and Wagner 1992.)

1. *Repair and service cycle.* In the usage phase the product is checked several times. If necessary, worn or defective parts and components are replaced, so the product can further fulfill its intended functions. A typical example is that of a car.

2. *Recycling loops.* In the production, usage, and disposal phases, production waste, defective parts, subassemblies, and the product itself can be materially recycled or used in a different way. VDI document 2243 describes these loops in detail (VDI 1993). In practice, various recycling loops can be carried our successfully. Usually, it is the question of the ecological and economic meaning of this material reintegration that remains unanswered.

3. *Reverse manufacturing system.* After usage, the product is generally disassembled into components, checked and tested, possibly upgraded or repaired, and sent on for further usage. A typical example is the special camera where film and case are built as a unit and are offered to the customer as such. Here the customer buys the opportunity (or the service) to take pictures.

4. *Upgrading.* Existing products are upgraded in a service and repair process and the functionality is increased. This function enhancement is easier to achieve with modular products. Typical examples are in the area of computer technology and copying machines.

At least a qualitative and a quantitative estimation are necessary to determine weaknesses and improvement potentials in a concrete application. In the following the possibility of a theoretical evaluation is shown. It consists of several steps. Then, the example of a clothes washing machine shows the rather simple construction of a first life cycle model and its interpretation with regard to ecological weaknesses.

2.3 Environmental Performance Evaluation in Five Steps

The environmental assessment represents the basis for the continual improvement of the environmental performance of an organization. Therefore, the standard ISO 14001 requires the development and introduction of corresponding procedures. The activities, products, and services that the company can influence and design must be analyzed with regard to possible environmental effects and must be assessed with regard to the environmental relevance. Planning activities—as shown before—have considerable environmental relevance because the environmental impacts occurring later are largely determined during planning. This relevance must be taken into consideration in finding environmental weaknesses of an organization. Therefore, the following series of steps is recommended for an environmental assessment (Züst 1997, 1998):

1. First, those activities, products, and services must be listed that are decided on by the organization or which are decisively influenced (Figure 5). In the development phase of new products and processes, the environmental impacts occurring later are determined, as far as feasible. If, for example, an organization develops products, it has its share of responsibility for the environmental impacts in production, use, and waste disposal of this product. This relation must be seen and shown in the first step.

2. Second, their environmental effects must be shown. Every activity, product, and service or parts of them can have environmental effects. First, these effects must be shown qualitatively.

3. In a further step, the attempt must be made to quantify these relations between "company and product" and the ecological environment. This step can be methodically supported by means of an input-output analysis. It is important that no partially or fully aggregated data are used. So, the energy and material flows must be depicted in elementary flows (see also ISO 14040 ff. CEN 1998). If the data situation is uncertain or incomplete, this must be taken into consideration in the following assessment (Figure 5).

Activities, products & services ...influenced or designed by the enterprise	QUALITATIVE					QUANTITATIVE		
	Relation to the environment					Input/output		
	Yes	No		Elementary flows	Environ-mental problems		Elementary flows	Assessment
....		×						
Cleaning process	×		Emission	VOC	Ozone depl., global warming	5 Mio. workpieces	1 kg VOC	High signif.
					....	....	....	
Transport	×		Resources	......	....	25000 ℓ gasoline		
			Emission	NO$_x$	Summer smog		y NO$_x$	Signif.

FIGURE 5 Matrix as a possible form to depict the results. Definition of the activities, products, and services that the organization can influence, showing the environmental relation first in a qualitative and second in a quantitative way and assessment with regard to the environmental relevance. (From Züst 1998.)

4. The fourth and main step contains the assessment and interpretation concerning environmental relevance. The question must be answered which activities, products, and services or parts of them have or could have an important environmental impact. Its local, regional, and global relationships and temporal relationships are important. In this planning step, it is especially important to consider all environmental impacts of activities, products, and services, even if the data situation is uncertain or incomplete. The results have to be interpreted critically. The aim must therefore be "better approximately correct, instead of precisely wrong" (Chapter 7). In this planning step, ecobalancing methods, such as ecopoints (UBP) according to BUWAL (1996) or ecoindicator according to Goedkopp et al. (1995), could possibly also be applied. Here, the corresponding interpretation of all evaluation steps is vital to avoid misinterpretations.

5. Finally, a list with those activities, products, and services or parts of them can be made that have or could have a considerable environmental impact. Here, the priorities are important and not an absolute value. In this connection, it is important to mention that the standard ISO 14001 does not require life cycle assessment in its strictest sense. However, the principles described in ISO 14040 ff. must be implemented.

In practice these steps are carried out iteratively. After a first overview, the required detailing grade can be approached successively. But simplified or adapted sequences also make sense. The next section discusses a simplified application.

2.4 Discussion of an Approximate Life Cycle Model

At the beginning of a development project, the question always arises as to how possible weaknesses can be shown as fast as possible with as little effort as possible. The following example of a clothes washing machine shows how with little effort an initial life cycle model can be developed and improved with regard to product improvement potentials (Züst 1999).

1. In a first step it is determined what should be modeled and examined with regard to possible environmental impacts. With a clothes washing machine, the function "washing clothes" is considered, for instance, the impacts of a certain number of washing procedures. One washing procedure describes the washing of 5 kg of clothes with a standard washing process at 60°C. The standard series ISO 14001 "Environmental Management Systems" speaks here of the so-called function-oriented measure basis or functional unit (see ISO 14041 Life Cycle Assessment). Thus, the definition of a measure basis should be clarified at the beginning of the enquiry.

2. Next, the main impacts in relation to this function are shown. For this, subdivision into life phases and the depiction of the corresponding

input and output factors makes sense. First, a clothes washing machine is developed. This can be either a new design or a further development. As a result, construction plans will be available. Then a number of washing machines are produced. For this, energy and materials for the preproduction and for the production itself must be provided. The usage by the customer follows as the third life phase. Here, repair and service cycles are also possible. At the end of the product life (e.g., after a number of wash cycles), the washing machine must be disposed off or parts of it must be brought to a further usage cycle. Now the individual input and output factors can be determined for the individual life phases—at first in a mere qualitative form. During usage, a washing machine needs water, soap, and energy for the function of washing clothes. During washing, waste heat, noise, and vibrations emanate. As an output, wastewater exits, besides the clean wet wash. The result is a qualitative life cycle model.

3. Then the individual quantities can be measured. With 10,000 washing procedures at an average water use of 50 L per washing procedure, 500 m^3 water is used. This means costs of $1000 or more for the fresh water and for the treatment of the wastewater. For the aspect "energy," 10,000 kWh electricity or nearly $1000 are used if 1 kWh is needed per wash. Here it becomes clear that the energy use is clearly higher in usage than in the total production phase (including material supply). For the rough model, the individual input-output factors are not depicted in the form of elementary flows.

4. As a further step, the quantitative model can be interpreted with regard to weaknesses and improvement potentials. The washing machine example shows that important input and output flows can arise in the usage phase. Thus, an important improvement potential could be the improvement of the washing process.

5. As a first working hypothesis there is a possibility that the function "washing" could represent an important environmental aspect. In a further procedure, the rough model could be refined. The example clearly shows that it makes more sense to view the whole life cycle in an overview instead of deepened examination in one life phase (e.g., in production).

2.5 Developing Common Model Representations

An organization can easily determine the use in energy and material at the site. One reason for this is that the individual input and output quantities are measurable by classical instruments and processes of enterprise economics. The products and services can, however, cause multiple environmental impacts over all the life phases, not just during manufacturing. The main difficulty consists in developing sensible interdisciplinary model representations within planning and decision processes. This specific system understanding represents the necessary basis for early recognition of possible environmental impacts of the products and services.

The model-development process should also help to understand the actual customer requirements and to develop customer-specific solutions. Asking about the basic customer requirements helps to question existing solution approaches and mainly to show new activity options. If current solution approaches are only "face lifted," often there is only little or even negative economic benefit.

Ecology-oriented analyses over all life phases as described below can provide useful hints in the planning phase as to how the products and services are to be checked and improved. The question whether an organization should continue to sell only physical products or only its services should also be discussed in this context. It may well be that in the future customers might prefer to ask for services rather than buying and owning products. These new services will surely be arranged differently. Thanks to modular construction, these products could be adapted to changing customer requirements during usage.

The following description relates to model construction or to product life cycle management.

3 SEARCH HEURISTIC FOR ECOLOGICAL IMPROVEMENT POTENTIALS

3.1 Introduction

This part discusses and evaluates cases of industrial and commercial products from the areas of machine construction and electrical technology. The aim is to support the analysis and search process in the early phase of product development through a search heuristic.

The search heuristic presented here builds primarily on the perception that there are both active and passive and movable and immobile objects. From this perception, various work hypotheses can be derived—both with regard to the situation analysis and to the solution search. The heuristic can be refined and made more concrete through further parameters.

3.1.1 Context

An organization has a range of activities with regard to designing its products and services that go beyond the legal frameworks, such as laws, ordinances, and standards. This entrepreneurial scope should be regarded in the following.

As mentioned earlier, in the early planning phase important decisions are made with regard to the quality of the solution. Therefore, methods and procedures supporting an actual "business innovation" in the early planning phase are vital. Obviously, the ecological improvement potential increases with the expansion of innovation perspectives beyond mere process and product innovations (e.g., see Minsch 1996; Schneidewind 1994).

While process innovations can still content themselves with technical solutions (possibly even in the sense of end-of-pipe solutions), the ecological design scope increases considerably with the product innovations. The product life phases are optimized in the meaning of life cycle engineering. Additional ecological improvement potentials arise when function and need innovations are also considered. Here, there is a chance to satisfy given needs through totally new products, ideas, and "benefit instead of selling goods." With need-oriented innovations, the given needs should also themselves be questioned. This also shows that the solutions increasingly go beyond company limits by approaching the ecologically interesting innovation perspectives. Finally, they even represent social projects (Minsch 1996).

This part considers the processing of methodical bases, which support the product, function, and need innovation in the early product-development phase.

3.1.2 Objectives

In the early development phase, the use of checklists and guidelines is not or is only partially possible (Luttropp and Züst 1998). This is due in particular to incomplete and/or inexact existing information at this stage. Further, calculations, as for example within life cycle assessments or a simulation, are only possible if there are many assumptions made. These assumptions, however, lead again to great uncertainties with regard to the evaluation and interpretation of the results. Therefore, the question arises to what extent these solution approaches make sense for the present problem task.

Therefore, this contribution discusses case examples of industrial and commercial products of the areas of machine construction and electrical technology and evaluates them with regard to a corresponding heuristic. The aim is to develop a search heuristic supporting the analysis and search process within an innovation plan in the early product development phase.

3.1.3 Aspects

Within the heuristic, primarily the ecology-oriented aspects are considered. Therefore, the energy and material flows caused by the company's products and services are supplemented by corresponding environmental requirements (Caduff and Züst 1996). The input stream is subdivided into materials, energy, and other resources. While materials and forms of energy go through defined functions, other resources are drawn directly from the ecosphere. Output, in addition to the product itself, involves undesirable flows, which may be categorized in terms of byproducts and emissions. The term "emissions" signifies several output flows delivered directly into the ecosphere. Products and byproducts undergo existing functions (Figure 6).

Many input and output streams may be specified in an in-depth manner. In this way, products, byproducts, and emissions reemerge as materials, energy, or resources. By joining the corresponding elementary functions one after the other, a total life cycle can be depicted (Caduff 1998; Caduff and Züst 1996).

Here it is important that the direct environmental relationship is visible in form of environmental impacts. This corresponds to the direct resource extraction and to emissions. Accounting for and depicting the environmental impacts form the basis for a systematic environmental evaluation.

3.2 Examples of Environmental Product Evaluations

Within a search project about environmental information systems, the so-called important environmental aspects according to ISO 14001 (CEN 1996) had to be determined with representative product groups. Since it is the industrial partners who determine the functions and the individual characterizations of the products, the whole life cycle, from material and energy supply up to and including disposal, was considered. The product entry and evaluation was performed as described previously. For the quantitative evaluation both existing ecoinventories (e.g., BUWAL 1996) and evaluation methods (e.g., Ahbe et al. 1990; Goedkoop 1995) were used. The intermediate results and the individual evaluation steps were continuously interpreted and evaluated, as described in ISO 14040 (CEN 1998).

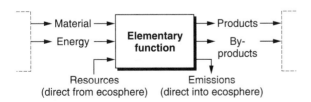

FIGURE 6 Input and output flows. (From Caduff and Züst 1996.)

3.2.1 Example 1: Products from the Sanitary Field

In this example, products from the sanitary field were examined. These are products for supply and disposal and active products such as flushing systems or the urinal electronics to control the flushing. (For the distinction between "active" and "passive" products, see De Winter and Kals 1994.) The toilet fixtures were also part of this analysis (Figure 7).

The results show clear differences. It can be seen that the total environmental impact during production phase, when compared to the whole life cycle, is exceedingly small. There are enormous differences, however, in the usage phase. Passive products cause no or only irrelevant environmental impacts during their usage. Active products, however, such as flushing systems, cause 95% and more of the entire environmental impacts during the usage phase. The production and disposal phases are irrelevant here.

The results presented here must be applied for effective environmental management. For product developers, work becomes easier with such information. So, with the flushing systems the flushing function is the one to be improved. Here, the designer does not need environmental knowledge in its narrow sense; he or she can focus on the flushing function. Further, the customer benefit is easy to communicate. In the future, he or she can reduce

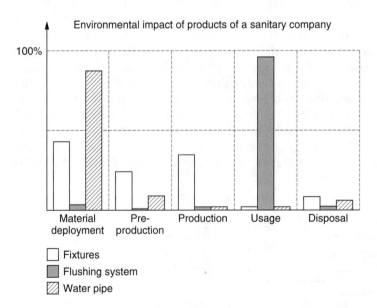

FIGURE 7 Relative environmental impacts of various sanitary products. (From Frei and Züst 1998; Gerber 1997.)

the use of water and thus also costs.

3.2.2   Example 2: Various Cable Types

The next example describes the relative environmental impacts of various cable types with various ways of usage. On the one hand, these are stationary performance and signal cables. On the other hand, cables that are moved in the usage phase, as in cars, railway wagons, or planes, are examined. The three examples are shown in Figure 8 in an overview.

Here, we have a similar picture as before. Passive products, such as signal cables, cause important environmental impacts by their material intensity (choice of material and processing process). The loss performance caused by the ohmic resistance leads to low ecological impacts during usage. With energy cables this is just the opposite. The loss performance during usage is rather high. The only possibility to reduce the transfer losses is an increase of the cross-section and thus an increase in material intensity. Further, if the cables are moved, the moved mass plays an important role. In this case the material intensity must be reduced despite simultaneous increase of the loss performance.

Both examples show different situations for similar products or for products of the same branch. The solution strategies must be adapted cor-

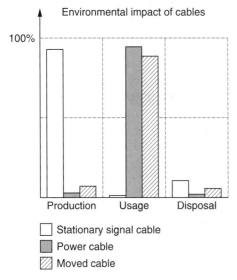

FIGURE 8   Relative environmental impacts of various cable types. (From Frei 1999; Seipelt 1998.)

respondingly. The example of the stationary energy cables also shows that a material-intensity reduction for cables, as demanded in many publications, is counterproductive. This would even worsen the environmental performance of the energy transfer. Here one should consider to what extent the cable cross-section could be increased.

3.2.3 Conclusions

To conclude, six statements are made that can be deduced from the basics and examples presented above. The following hints should help to recognize environmental weaknesses more effectively in practical application (see also Züst 1998):

1. The planning processes have a high environmental relevance. There, the environmental impacts occurring later in various places are largely determined. It is irrelevant whether physical products or services are designed. With both, the planning-decision consequences on the later life phases must be considered.
2. Analyses, which consider only the production, do not normally make sense because critical problem and solution fields may go unrecognized.
3. Environment-critical materials can cause significant environmental impacts already in small amounts. Input-output analyses regarding only the large material flows are not suitable in this case.
4. With so-called passive products, for instance, products causing no further environmental impact during use, the material intensity often represents an important environmental aspect. If the company has extensive production facilities, its production processes are also relevant. Here, environmental improvements can be achieved by a modified product design (choice of material, good use of material, design, etc.) or by other production sequences and processes (less rejects and less waste products caused by processes) (Caduff 1998; Mannhart 1996).
5. With active products, for instance, products causing significant environmental impacts during use, the way functions are realized often plays an important role. Cars and electronic devices but also cables (energy transfer) and flushing systems, as shown above, represent active products.
6. With mobile objects (e.g., use of vehicles), light methods of construction also should be considered. The way and manner of material deployment and waste disposal is often subordinated

with regard to a holistic improvement of environmental performance.

3.3 Search Heuristic for Ecological Improvement Potentials

For ecology-oriented product design there is always a need for a long product life. Depending on the way and usage of the product, this characteristic can even be counterproductive. Therefore, in the following the question is examined how the demands with regard to function, light construction, and recyclability may be balanced against each other.

From the environmental perspective, two central features based on the previous considerations are method of use, whether the objects can have an "active" or "passive" behavior, and place of use, whether during usage the objects remain at the same place, for instance, "stationary" or they are regularly moved and are thus mobile.

3.3.1 Active Product, Locally Immobile

Active products are characterized by causing further environmental impacts during the usage phase through the usage of the intended functions and services. Here, the desired function can only be achieved if additional processes can be introduced. Examples of active and immobile products are current meters in a building, TV sets in a room, refrigerators in a kitchen combination, flushing system in a toilet, and stationary energy cables (as described earlier).

For active and locally immobile objects, the following solution strategies emerge:

- *Improved function and lower losses.* This demand primarily concerns the product functionality.
- *Strategies for material reintegration.* Immobile objects can have a heavier construction. With a material-intensive construction the material choice must be made with regard to the material provision, processing, and assembly/disassembly/material recycling.
- *Life duration.* If no important product function-related technical innovations are to be expected, a long life should possibly be achieved. If important technical changes are to be expected, the demand for a long life duration must be reconsidered.

3.3.2 Active Product, Frequently Moved

Active objects can also be regularly moved. Examples of active and moved objects are cooling aggregations in a vehicle, railway (transportation of people and goods), or TV sets in a plane.

For active and locally moved objects, the following solution strategies are developed, contrary to immobile / active objects:

- *Improved function and lower losses.* This demand primarily concerns the product functionality.
- *Light construction/miniaturization.* Reduced mass enables easier acceleration and slowing down.
- *Life duration.* Demand for long life duration is rather secondary. Here, a (too) long life duration can prevent the introduction of new product technologies and the demand for reduced mass.
- *Strategies for material reintegration.* Demand for recycling ability is rather secondary because thus an improvement in the function principle and a light construction is hindered.

3.3.3 Passive Product, Locally Immobile

In the usage phase, passive products cause no further environmental impacts by using the provided functions and services. Examples for passive and immobile products are furniture and fixtures in a building. Handyman and household devices can be nearly passive products because of their infrequent usage, although they do cause further environmental impacts during usage.

For passive and close to passive and locally immobile objects, the following solution strategies result:

- *Long life duration.* As no further environmental impacts result through usage and as the object is not regularly moved, the life duration must clearly be optimized by reconsidering production and disposal.
- *Strategies for material reintegration.* Immobile objects can have a heavier construction. For a material-intensive design, the material choice must be made with regard to processing and assembly/disassembly/material recycling.
- *Light construction and miniaturization.* These are rather meaningless because they shorten the life duration and reduce the environmental performance.

3.3.4 Passive Product, Frequently Moved

Passive objects can also be moved frequently. Examples for passive and movable objects are seats and desks in a vehicle (plane, car, railway) and reusable package and transportation materials.

For passive and locally moved objects, the following solution strategies result:

- *Light construction/miniaturization.* Reduced mass enables easier acceleration and slowing down.
- *Life duration.* Demand for long life duration is rather secondary. Here, a (too) long life duration can counter the demand for reduced mass.
- *Strategies for material reintegration.* Demand for recyclability is rather secondary because then a light construction is difficult to achieve.

3.4 Discussion

The examples given above have clearly shown the large differences within the same trade and within the same products with regard to the whole life cycle. At the beginning of the planning, effective strategies are required both in the analysis phase and within a first solution search for ecoeffective product development. One possibility consists of the application of the previously described search heuristic. First, structuring and estimations can be made at an early planning stage. Thus, sensible work hypotheses can be derived and the development process can be arranged into sensible steps.

The search heuristic can be supplemented further (Frei 1999); Frei et al. 1996). Consider the following:

- *Kind of materials or substances.* Materials have an influence on the environmental performance of an object based on the supply, processing, and recycling processes.
- *Mixture grade.* Usually, materials are available in a "mixed" form. Therefore, it must be examined how the various materials are mixed or alloyed with each other.
- *Geographical distribution.* The individual objects can have a local, regional, or global distribution. Thus, various distribution service and/or recycling concepts emerge. This again influences the environmental performance or services of a product.
- *Ownership.* Products often change the owner after the production phase. Therefore manipulation and changes made by the new owner are possible. Thus, taking-back concepts or "up-grading" possibilities are only limited.

At present, an attempt is made to apply the previously presented search heuristic in development projects. The search heuristic should be further specifically supplemented and evaluated with regard to the effectiveness.

REFERENCES

Ahbe S, Braunschweig A, Müller-Wenk R. Methodik für Ökobilanzen auf Basis der ökologischer Optimierung (Schriftenreihe Umwelt Nr. 133). Bern: Bundesamt für Umwelt, Wald und Landschaft (BUWAL), 1990.

Alting L, Legarth J. Life Cycle Engineering and Design. In: Annals of the CIRP Vol. 44/2/1995. Bern: Hallwag Verlag, S. 569–580.

BUWAL–Bundesamt für Umwelt, Wald und Landschaft. Ökoinventare für Verpackungen. Band I und II, Schriftenreihe Nr. 250. Bern: BUWAL, 1996.

Caduff G. Umweltoreintierte Leistungsbeurteilung. Wiesbaden: Gabler Verlag, 1998.

Caduff G, Züst R. Increasing Environmental Performance via Integrated Enterprise Modelling. In: ECO-Performance—3rd International Seminar on Life Cycle Engineering CIRP, edited by Züst R, et al. Zürich: IO Verlag, 1996, pp. 39–46.

CEN-Zentralsekretariat (Hrsg.). EN ISO 14001 Umweltmanagementsysteme— Spezifikation mit Anleitung zur Anwendung. Brüssel: Europäisches Komitee für Normung, 1996.

CEN-Zentralsekretariat (Hrsg.). EN ISO 14040 Environmental Management Systems—Principles and Framework. Brüssel: Europäisches Komitee für Normung, 1998.

Daenzer W. (Hrsg.). Systems Engineering. 1. Auflage, Zürich: IO Verlag, 1976.

De Winter A D, Kals J A G. A methodic approach to the environmental effects of manufacturing. In: Feldmann K. (Hrsg.). CIRP 2nd (International Seminar on Life Cycle Engineering. Bamberg: Meisenbach Verlag, 1994, pp. 287–301.

Frei M. Öko-effektive Produktentwicklung. Wiesbaden: Gabler Verlag, 1999, especially p. 101 ff.

Frei M, Caduff G, Züst R. Eco Effectiveness—Systematic Inclusion of Ecological Aspects in Product Development. In: NordDesign '96, Helsinki, 1996, pp. 130–140.

Frei M, Züst R. Die öko-effektive Produktentwicklung—Die Schnittstelle zwischen Umweltmanagement und Design. In: Markt- und Kostenvorteile durch Entwicklung umweltverträglicher Produkte (VDI-Berichte, Nr. 1400). Düsseldorf: VDI Verlag, 1998, pp. 51–71.

Gerber R. Bestimmen der bedeutenden Umweltaspekte der Produkte der Firma Geberit. Unveröffentlichte Diplomarbeit. Zürich: Betriebswissenschaftliches Institut (BWI) der ETH Zürich, 1997.

Goedkoop M. The Eco-Indicator 95—Weighting Method for Environmental Effects that Damage Ecosystems or Human Health on a European Scale. Amersfoort (NL): Preconsultants, 1995.

Goedkoop, Demmers, Collingnin. The Eco-Indicators—Manual for Designers. Internet http:/www.pre.nl., 1995.

Imboden D. Wahrnehmungslücken und Handlungsbedarf. In: Züst R, Schlatter A (Hrsg.). Umweltmanagementsysteme in der öffentlichen Verwaltung. Zürich, Verlag Eco-Performance, 1999, p. 173.

Luttropp C, Züst R. Eco-effective products: Holistic viewpoint. Keynote Paper, 5th International Seminar on Life Cycle Engineering, Proceedings Life Cycle Design '98, KTH Stockholm, 1998, pp. 45–55.

Mannhart M. Umweltmanagement für KMU. Unveröffentlichte Projektarbeit am Betriebswissenschaftlichen Institut der ETH Zürich. Zürich, 1996.

Minsch J, et al. Mut zum ökologischen Umbau. Basel, Boston, Berlin, 1996, especially p. 65 ff.

Opitz H. Moderne Produktionstechnik, Stand und Tendenzen. 3. Aufl., Essen: Verlag W. Girardet, 1970, p. 525.

Schneidewind U. Mit COSY—Company-oriented Sustainability—Unternehmen zur Nachhaltigkeit führen. Diskussionspapier Nr. 15. St. Gallen: IWÖ-HSG, 1994.

Seipelt D. Bestimmen der bedeutenden Umweltaspekte von Energie- und Signalkabel der Firma Huber + Suhner. Unveröffentlichte Diplomarbeit. Zürich: Bebtriebswissenschaftliches Instiut (BWI) der ETH Zürich, 1998.

VDI-Richtlinie 2243. Konstruieren recyclinggerechter technischer Produkte: Grundlagen und Gestaltungsregeln. Düsseldorf: VDI Verlag, 1993.

Züst R. Einstieg ins Systems Engineering—Systematisch denken, handeln und umsetzen. Zürich: IO Verlag, 1997a.

Züst R. Umweltmanagement—Beurteilen als zentrale Tätigkeit. In: Umwelt Focus 1/9. Zürich: SeculMedia, 1997b, pp. 19–21.

Züst R. Öko-Performance. In: Umwelt Focus 4, Zürich/Forch: SecuMedia, 1998a, pp. 29–31.

Züst R. Principles of Environmental Assessment in the Context of Design Process. Keynote Paper at the 5th International Design Conference—Design 98, edited by D Marjanovic, WDK Zürich, 1998b, pp. 53–63.

Züst R. Life-Cycle-Thinking—oder: Wie erkenne ich rasch und einfach Verbesserungspotentiale? Manage Qual 7 + 8:63, 1999a.

Züst R. Umweltmanagement ist Ressourcenmanagement. Manage Qual 4:63, 1999b.

Züst R, Wagner R. Approach to the Identification and Quantification of Environmental Effects During Product Life. Ann CIRP 41:473–476, 1992.

9

The Ecodesign Checklist Method
Design Assessment and Improvements

Wolfgang Wimmer
Vienna University of Technology, Vienna, Austria

1 INTRODUCTION

The transformation of more natural resources into products with shorter service lives and then into waste is widely regarded as a normal regular way for companies to do business and for consumers to fulfill their needs. The ensuing problems, such as mountains of waste, widespread environmental damage, and social problems resulting from increasing consumption of resources, are mainly attributable to the Western industrialized countries. The Brundtland definition of sustainable development—development that meets the needs of the present without compromising the ability of future generations to meet their own needs—seems far removed from reality. Wackernagel and Rees [1] assessed this in terms of the "ecological footprint" and found that worldwide resource consumption dramatically exceeded ecological capacity—a clear warning to revise our attitude toward the environment. A factor of 10 for the reduction of resource consumption has been internationally agreed as necessary to achieve sustainability.

In addition, growing environmental awareness means that an increasing number of consumers are demanding ecologically sound products and services. Ecodesign is a strategy enabling product development engineers and designers to integrate environmental aspects into their daily work so as to improve the ecological soundness of their products [2,3]. At first glance, this issue appears to be just another specification for which engineers are expected to comply. But in detail, it becomes increasingly complex, developing into a discussion of social values. As important and interesting as such a discussion may be, the environment cannot be treated in this way in the interests of better product design. The other extreme is to narrow this fairly broad topic down to one of appropriate choice and designation of materials. However, neither approach will result in any significant changes or improvements.

2   A DEFINITION OF ECODESIGN

Looking at nature, we find a complex system that is not necessarily efficient, as sometimes resources are used in a very wasteful way. But at the same time, nature is very effective, which has made this "company" sustainable over quite a long period of time. This shows us that "technology" and "organization" are both important elements of any sustainable system. If we are looking for ways to integrate environmental issues into (product) development, we can identify three dimensions: technology, organization, and culture.

When we refer to environmental technology, we are generally talking about a flue gas purification plant or a waste incineration facility, for example. And yet we do not associate the management or avoidance of waste with environmental technology. However, based on the theory of appropriate technology [4], in principle we can identify three subsystems for the classification of a technology:

- A technical and economic system, the technology. Products would be manufactured without waste, effluents, or emissions and process materials would pass through a cycle. The use of renewable raw materials (materials and energy) would outweigh that of fossil materials. Consequently, the products manufactured could simply be repaired, upgraded, dismantled, and recycled largely without structural destruction (reuse and continued use). Materials would be retained in a cycle of matter, and unusable materials could be simply integrated in biological cycles. Suitable information systems accompanying each product would make it possible to ensure efficient utilization of manufactured products throughout their entire lifetime (use, degree of utilization, purpose, nature of reuse or continued use, etc.).

- A material system consisting of a structure of institutions, organization, and infrastructure. Effective organization is an important aspect in dealing with the available resources, since it may lead to a separation of economic growth and resource consumption. Thus, in new utilization scenarios, products can be used and reused and collected after use in decentralized structures so as to be offered for reuse. Here the problem solutions previously proffered in the form of products could be replaced by professional service providers. These would have to be locally organized so as to take optimal advantage of prevailing regional conditions (resources) to meet local needs.
- A nonmaterial system of standards, values, and concepts. Sufficiency represents a new approach in which sustainability becomes established as a separate concept, being regarded as a central element of individual responsibility. The products and services needed for everyday life are required and supplied in the form of solutions adaptable over time and customizable to individual requirements. In many instances, consumers would become users as a result of using, rather than possessing, products. However, they would receive equal consideration in the design of new products.

Thus, technology always consists of an interplay of all three subsystems. Hence, whenever we need to reflect upon technology, we must always examine technology, organization, and development. In the case of product development, the question of technology involves questions as to the actual project, the technical object (e.g., reparability). The question of organization relates to the processes, technology management (e.g., energy-saving measures), whereas the question of development probes the underlying standards and values (e.g., environmental awareness).

The following definition is derived from the above considerations so as to obtain an understanding of ecodesign:

Ecodesign is a process with the purpose of arranging technology and organization in such a way that, through intelligent use of all available resources with minimum environmental impact, the greatest possible benefit is obtained for all the individuals involved, and consumer satisfaction is also guaranteed.

In this context, resources refer to the flow of both energy and materials and to human resources and consequently the available abilities and knowledge. However, it also embraces the available time and motivation.

The following rough action guidelines can therefore be established on the basis of this definition of ecodesign:

- Products and services should be oriented toward customers' needs and benefits (creation of service systems, product development for new need fulfilment concepts).
- Production of products and goods by means of appropriate technologies and materials (avoiding production of waste and harmful substances, reduction of energy and water consumption, use of renewable resources).
- Design of products for intelligent use of resources and establishment of resource management (extension of product service life, introducing circulation of products, components and materials, avoidance of toxicity, etc.).

3 ENVIRONMENTAL DESIGN ASSESSMENT

In a survey of Austrian small and medium-sized companies certified in accordance with ISO 14000 and EMAS, it was found that virtually all these companies had used checklists and simple ABC evaluation schemes or value analysis for environmental assessment. Software tools were used significantly less often.

One reason may have been the time available and another the differing quality of the results obtained. On the one hand, many small companies are unable to carry out a time-consuming environment audit for reasons of cost. On the other, the expected results cannot often be utilized directly in product development and require special interpretation. This frequently overtaxes even the most enthusiastic companies.

Thus, the purpose of presenting the ecodesign checklist method (ECM) is, on the one hand, to show that it is a familiar easy-to-handle tool (checklists). On the other, the evaluation of products is based on a general design evaluation. In other words, this is an approach that attempts to include all the design elements involved in ecodesign and to evaluate the degree to which they are implemented.

The product developers themselves should be the ones to carry out this environmental audit. On the one hand, they know their products best, and on the other hand, they can directly implement the potential for improvement revealed by such an audit.

The environmental audit proposed below therefore assumes that general rules of environmentally sound design (such as the possibility of dismantling the product) can be laid down for ecodesign and checked for compliance (e.g., removability of fastening elements) in the environmental audit. Obviously, the extent to which such rules and specifications are always applicable will vary, depending on the individual product under investigation.

4 ECODESIGN CHECKLIST METHOD (ECM)

Starting from this approach to the environmental evaluation of products, the ECM has been developed as an easy-to-apply guideline for engineers and designers [5,6]. This tool provides a methodical instrument to assist decision making in the design of products with respect to the new theme of environmental issues.

ECM identifies measures to be taken by an engineer in product development and incorporates a win-win-win strategy benefiting the company, the user, and the environment.

All aspects relevant to reducing the impact of a product on the environment or the use of resources (such as durability, reparability, reusability, and recyclability, etc.) are considered and incorporated in a methodology for studying the product at an early stage in its development [7–11]. ECM allows opportunities for reducing environmental impact and resource consumption to be identified. Taking the product cycle and the product environment into account, all the relevant requirements for environmentally sound product development/ecodesign are derived from the ECM targets for product improvement. On the basis of the formulated targets and requirements, analytical procedures are established using checklists.

4.1 Modules

The ECM analysis consists of three modules: part analysis, function analysis, and product analysis (Figure 1). Part analysis assesses the degree of compliance of each part or subassembly with the stated requirements. To pinpoint any weaknesses, in each case parts are rated according to the criteria material, manufacturing, lifetime, functionality, maintenance, repair, disassembly, and recycling. Function analysis involves the compilation of a functional hierarchy for the product. The functions are allocated to individual components in accordance with the value analysis, enabling them to be evaluated with regard to the product function with the highest theoretical environmental impact. This provides a further reference point for improvements to the product. Product analysis involves an assessment of the degree to which the product design complies with the requirements. This is in terms of environmental deficiencies with regard to the criteria use, functionality, consumption, emissions, and distribution.

4.1.1 Part Analysis

The first step in the part analysis is to take the product apart into its major components or subassemblies and determine its parts as construed

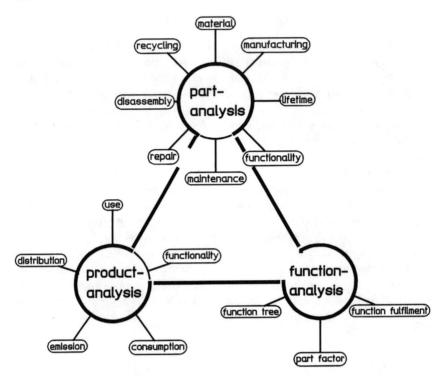

FIGURE 1   Modules of the ecodesign checklist method.

by the method. There is no need to dismantle the product totally. The best approach is to go through the parts list to define the part structure for this first step in the analysis. The checklists are then applied to every defined part to determine whether the ecodesign requirements are met. Going through the checklists for all eight criteria ensures that relevant requirements are not overlooked and that all aspects receive the required focus.

After going through the checklists, an evaluation step follows that ends up in a part profile showing the environmental performance (expressed as environmental deficit) of a single part or the whole product, as shown in Figure 2. This clearly shows the criteria for which weak points could be identified, allowing an engineer in product development to come up with initial improvements at part level by simply going back and finding out why the parts did not meet the ecodesign requirements for specific criteria. Furthermore, a so-called part factor is calculated by summing up the assessment ratings for each criterion.

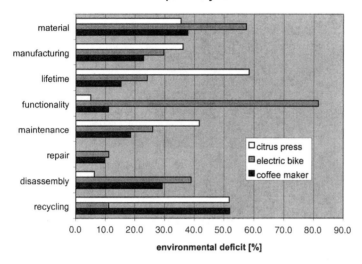

FIGURE 2   Results of ECM part analysis.

4.1.2   Function Analysis

In the function analysis module, any functions causing high environmental impact are determined as another starting point for redesigning the product. A value analysis approach is used, and either an existing function structure can be used or a new one created (Figure 3). The function table results in a list of all the functions of the product (Figure 4). It also associates them with parts via fulfillment of function. This represents the function structure of the product to be analyzed. In the next step, the part factor for each part is entered, resulting in a function assessment showing which product functions fall most short of the ecodesign requirements.

The advantage of this module is the possibility of easily carrying out the same analysis using a comparison with part costs. Potential improvements with respect to both environment and reduced manufacturing costs can therefore be pinpointed. Typically, redesign at this stage would involve integration, omission, and creation of functions—yet another important starting point for product improvement.

4.1.3   Product Analysis

As in the part analysis, the whole product is investigated in this module, based on an in-depth knowledge of the functionality and—more importantly still—the performance of the product in the utilization phase. To

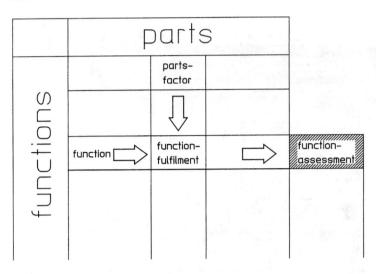

FIGURE 3   Function assessment.

achieve this, the requirements for the five criteria (Figure 1) are applied
again using checklists.

This level of examination addresses the whole product concept: How
do the parts work together? The result is a product profile as shown in
Figure 5. Environmentally related improvement options are pinpointed on

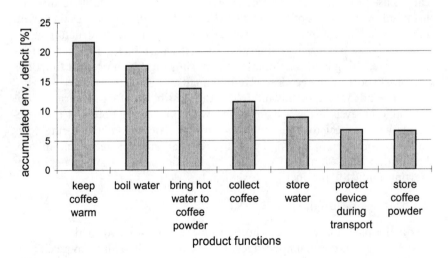

FIGURE 4   ECM results at function level (coffee maker).

ECM - product analysis

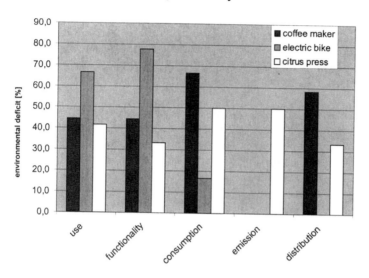

FIGURE 5   Results of ECM product analysis.

the basis of the weak points identified. Reengineering of the product concept can therefore be initiated.

4.2   Product Comparison

In most cases it is important to compare products with respect to their environmental performance. This is another field where ECM can be used. Comparisons can be easily carried out using the results of a full product evaluation. For each criterion, an average value is calculated for all product parts and shown at both part and product level.

In a product comparison, this approach to ecodesign delivers differentiated results for the relevant criteria instead of being limited to a single value as in many other methods of assessment. With this advantage, the strengths and weaknesses of a product concept can be identified using a comparison.

ECM was applied to two very different citrus presses [12]. Model A is a direct flow model that can be characterized as a durable (lifetime, 200 hours), high-quality, expensive product. By contrast, model B is a container model, with a short lifetime (80 hours), a cheap low-quality product. Both are ideal candidates for the performance of a product comparison to verify the ECM developed.

The citrus press (model A) comes out well in the ECM comparison at product level for the criteria use, functionality, and emission. Model B is better with respect to the criteria consumption and distribution. At a parts level, model A was found to be much better with respect to the criteria lifetime and maintenance and nearly equal in functionality and recycling. At this level, model B is ahead in material, manufacturing, repair, and disassembling.

Considering the possible ways of improving the products, it is evident that the model A concept could be redesigned with ease. This would therefore be the solution for the design of a new model. Some of the redesign tasks identified were material reduction and function integration in the housing, increasing recyclability of housing and gearbox, recycling of production waste, improving disassembly time, and reuse of the motor.

Methods of environmental audit based mainly on the evaluation of materials are often unable to provide concrete results that can be implemented in practice [13]. As a result, the product developer then faces only a choice of different materials to improve the environmental performance of the product. However, the results of an ECM product comparison go significantly further, indicating a wide variety of possibilities and methods for possible product improvement.

Reducing ecodesign just to material or consumption considerations at the utilization phase misses opportunities to improve a product environmentally. The approach to ecodesign needs to be set up on a wider footing, based on more requirements, to probe all the relevant possibilities for reducing the environmental impact of a product. The ECM is one way of doing this.

4.3 Product Redesign Tasks

For a final definition of redesign tasks, one has to go through the assessment sheets to look at the requirements not fulfilled. Environmentally related improvement tasks for the product analyzed are obtained by translating deficiencies into actions to be taken. This shows where the whole product concept should be improved; where to create, omit, or integrate functions; and how to redesign parts and components.

Some of the redesign tasks identified in the product examples—electric bike, citrus press, and coffee maker [14]—are shown in Table 1. It is only a small step from the ECM results to detailed redesign tasks. At this point the search for sound environmental solutions to the weak points identified can be ideally supported by any product innovation method (e.g., TRIZ, brainstorming, etc.).

5 ECODESIGN PILOT

The next step in developing ECM further was the ecodesign product investigation, learning, and optimization tool (PILOT) based on the experience gained from company workshops held by the Austrian Ecodesign Information Point (http://www.ecodesign.at). These showed that the implementation of ecodesign in companies can be supported by the use of specific tools that

- Present information on ecodesign and stimulate additional interest in environmentally related product innovation;
- Encourage an increasing awareness for sustainable product development;
- Point out possibilities for the implementation of ecodesign in specific products;
- Provide working documents for independent use within the companies.

These workshops were the starting point for the development of the PILOT on CD-ROM for interactive learning about ecodesign and its application to products.

The target groups for PILOT may be defined as follows:

- Executives from industry and business, particularly in areas close to production, such as technology and development, and also marketing.
- Employees in product development, environmental experts, and designers.
- Employees involved in the implementation and realization of environmental management systems under EMAS regulations or ISO 14000.
- Employees in public institutions involved in green procurement.

5.1 Content of Ecodesign

The whole content of the PILOT is built up in form of a database containing ecodesign requirements. Each requirement is formulated in to-do instructions (e.g., ensure separability of materials, avoid hazardous substances, reduce consumption, etc.). All these represent possible directives to be acted on to improve the environmental performance of a product. Since this is quite a large source of information, it naturally also contains contradictions. That means not all ecodesign requirements listed can be fulfilled at the same time by a specific product, but all may be relevant for fulfillment.

TABLE 1 Identified Redesign Tasks

Criteria	Redesign tasks generated
Electric bike	
Use	• Enable individual adjustment of whole seat position • Integrate battery charger into housing for higher flexibility of use • Change housing for multifunctional use (expandable storage space for baggage transport)
Material	• Reduce weight of frame by reducing thickness of frame tubes and by reducing the sheet steel corner plates • Leave out dynamo energy supply for lighting, use system battery instead • Reduce weight of housing of drive unit (lightweight design) • Reduce cables required by integrating starter switch and recharge plug into housing
Functionality	• Improve damping system of seat • Capsule (water- and dirtproof) electric components and motor by closing drive unit housing
Citrus press	
Consumption	• Improve cleanability of cone sieve, housing, and container (reduce amount of water and washing-up-liquid required) • Leave out function "collecting juice" and change to direct flow principle
Lifetime	• Ensure higher lifetime of motor • Avoid easily scratched surfaces
Recycling	• Avoid recycling incompatible materials (e.g., SAN - PP) • Ensure separability of materials • Enable easy disassembling of motor (for reuse) • Apply identification of materials
Coffee maker	
Consumption	• Change principle for water heating from flow-type to immersion heater • Avoid energy consumption for function "keeping coffee warm" by using insulated pot • Implement permanent filter instead of disposable paper filters
Material	• Use PP recycled material for nonvisible parts
Recycling	• Avoid material mixes (PP, PVC, glass) and ensure separability • Ensure easy discharge of electrical components (for reuse or recycling)

To carry out actual product improvements, one has to find now the way through the interdependence and pick out the most relevant ecodesign instructions. To do this efficiently, every instruction contains the following six information parts:

- Part 1. Basic idea: a short description of the idea behind the particular instruction, explaining the background and providing the relation between design instruction and the environmental consequences.
- Part 2. Example: a picture or drawing, either in a small detail or in a whole product example, demonstrating a possible realization of the specific design instruction,
- Part 3. Interdependence: links from one instruction to another related instructions showing their interaction,
- Part 4. General question: addressing aspects one has to think about to do an ecodesign assessment based on the given instructions. The questions highlighting the surrounding of the specific instruction and help to identify ways to think about before doing a detailed product design assessment.
- Part 5. Assessment question: a single sentence to do the design assessment. It refers to the specific instruction.
- Part 6. Additional information: gives hints where support for the decision making process in designing and in assessing the product is externally available and can be found (Internet).

5.2 Information Access

As pointed out earlier, the ecodesign content is a collection of to-do instructions, a big nonstructured information source. There is a need for specific access to these instructions and a demand for a procedure how to access this information. Adjusted to the different users, the PILOT is designed for product developers, designers, engineers, and consultants and the purpose of use (learning about ecodesign, assessing product designs, etc.). The following four main gates to access were established:

- The product life cycle (PLC) is probably the most natural access since it reflects the actual product life and groups all the ecodesign instructions along the product life cycle, as shown in Figure 6.
- The product development process (PDP): The PDP access addresses mainly engineers and designers in product development. It groups all the ecodesign instructions along the product development process. For example, a designer in the conceptual design phase finds the ecodesign requirements to meet in this phase. In this way it

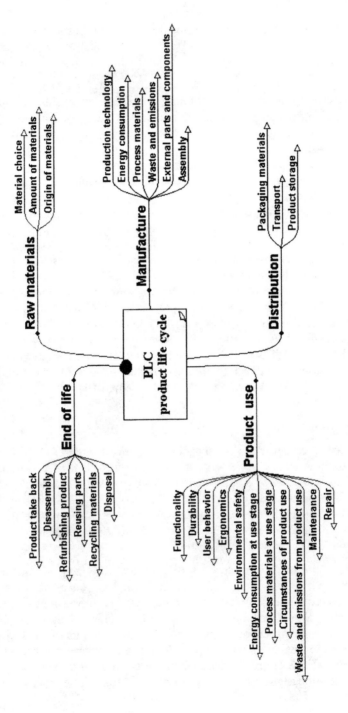

FIGURE 6 Product life cycle.

helps not to overlook important aspects. It makes clear when certain requirements are addressed along the PDP and when information support is necessary.

- The product development strategies (PDS): The so-called strategies for product development serve as access to carry out a design assessment that can be done with the PILOT. There will be a strategy-selection matrix available guiding to the relevant strategies for a product to be improved. Depending on the type of product (e.g., active, passive, etc.), respectively, the environmentally most relevant product life cycle, improvement strategies are suggested by the PILOT.
- The Index: As an additional feature a smart index enables one to search for all ecodesign instructions linked to a specific topic. This thematic information access is dedicated mainly to the user interested in learning about different topics or having a problem with a product and looking up some help for quick trouble shooting.

5.3 Navigation System

The intention of using the information with its different possibilities to access is generally spoken either as "learning" or "assessing." According to these two main purposes, a navigation system with special features was established to guide one in the learning or assessment part.

The learning section starts here with a tutorial showing the key aspects. It leads then to simple product examples (packaging) everyone is familiar with and guides one deeper and deeper into the topic of ecodesign. The product examples are used to touch on environmental issues and help to intensify a broad understanding of what is ecodesign about. Furthermore, it allows one to jump thematically into the information source and find out more aspects and finally, by using the interactions between the instructions, to learn what is all covered. This leads to a complex picture of connected environmental aspects. The implemented navigation system ensures that the user always knows where he or she came from and how to go back to a point he or she already has been before. In that way it steers one through and manages the process of learning.

The assessing section enables one to make a design assessment of a product by checking where requirements are met. The process of assessing consists in selecting ecodesign instructions and checking their fulfillment. Since the PILOT should be open to nearly "every" product, the selection is done more generally via the PDS explained earlier. Once the individual strategy(ies) is(are) found, one can check the instructions and identify weak points as starting points for product improvements.

The selection of the strategies itself comes out of a process of reflecting the whole product life cycle in terms of environmental main issues (material, energy consumption, hazards, etc.). This rough quantitative approach brings up the ecodesign instructions to carry out. By this the product performance can be assessed and possible improvements can be worked out following the instructions.

5.4 PILOT's Output

At all different stages—PLC, PDP, PDS, Index, learning and assessing module—the PILOT allows at any time the generation and printing of working documents in the form of checklists to help users get along in realizing ecodesign. Therefore, a variety of these checklists can be useful:

- Checklists for designers to help them considering environmental issues for clarifying the task;
- Checklists for production managers to clean up processes;
- Checklists for product developers to improve an existing product according to one or more environmental targets (strategies);
- Checklists for "normal users" or "environmentalists" to learn what is connected with specific topics, like "material," "disassembling," and so on.

These checklists can be assembled individually based on the information structure explained earlier. All checklists can contain any of the six information parts from there. Examples are the following or any other useful combination:

- Ecodesign instruction and an example;
- Ecodesign assessment question and the basic idea.

6 CONCLUSION

With the ECM, a tool is available that addresses engineers and designers in the product development process. The ECM records and evaluates all environmentally significant product characteristics. It purposefully points out redesign tasks to increase the environmental performance of a product. Based on a holistic view of the product, the results of the investigations are showing specific design tasks to environmentally reengineer an analyzed product. The ECM is applicable to a wide range of products. This approach is easy to learn and accept and to be integrable into the daily design work. Bringing together a systematic analysis and an investigation of a product in three analysis levels (part, function, and product level), the method shows clearly where the weak points of a product are. It also shows how to realize reuse, recycling of parts, where to integrate, omit or create functions, and where to reduce consumption or increase efficiency and usability of the whole product.

The method enables an engineer or designer in product development to directly derive environmental improvements concerning the analyzed product. Therefore, the ECM is one way to learn and make use of the various aspects and possibilities of ecodesign in the product creation process.

The ecodesign PILOT is an interactive tool for addressing environmental issues in product development and has been developed to persuade the target group to promote the implementation of ecodesign projects in their respective companies. To achieve this goal, this method is both communicating ecodesign and providing for in-depth information. An interesting part of this tool is the connection to worldwide available environmentally relevant information via the URL http://www.ecodesign.at. Furthermore, the PILOT is showing the potential for innovation in product design and development and simultaneously indicating related possibilities to approach new business areas. This is done by presenting comprehensive possibilities to implement ecodesign by introducing successful product examples.

On the basis of a broad understanding of sustainable product development, the PILOT offers the opportunity to improve company-specific products via an established design assessment procedure.

REFERENCES

1. Wackernagel M, Rees W. Unser ökologische Fußabdruck. Basel, Boston, Berlin: Birkhäuser, 1997.
2. VDI–Berichte 1400. Markt- und Kostenvorteile durch Entwicklung umweltverträglicher Produkte. Düsseldorf: VDI Verlag, 1998.
3. Züst R. Einstieg ins Systems Engineering: Systematisch denken, handeln und umsetzen. Zürich: Verlag Industrielle Organisation, 1997.
4. Riedijk W. Appropriate Technology in Industrialized Countries. Delft: Delft University Press, 1989.
5. Wimmer W. Design for Environment—ECODESIGN: A Tool for Product Improvement Based on Function Assessment. Proceedings of the 5th International Design Conference, Design 98, Dubrovnik, 1998.
6. Wimmer W. The ECODESIGN Checklist Method: A Redesign Tool for Environmental Product Improvements. 1st International Symposium on

Environmentally Conscious Design and Inverse Manufacturing. Ecodesign'99, Tokyo, 1999.

7. Birkhofer H. Umweltgerechte Produktentwicklung: Ein Leitfaden für Entwicklung und Konstruktion. Berlin, Wien, Zürich, 2000.

8. Brezet H, van Hemel C. ECODESIGN: a Promising Approach to Sustainable Production and Consumption. United Nations Publication, 1997.

9. Brinkmann, Ehrenstein, Steinhilper. Umwelt- und recyclinggerechte Produktentwicklung. Augsburg: Weka-Fachverlag, 1995.

10. Goedkoop M. The Eco-Indicator 95 Final Report. NOH report 9523, Amerfoort, 1995.

11. VDI–Richtinie VDI 2243. Konstruieren recyclinggerechter technishcher Produkte, Grundlagen und Gestaltungs-regeln. Düsseldorf: VDI Verlag, 1993.

12. Wimmer W. Environmental Improvements of a Citrus Press using the ECODESIGN Checklist Method. Interntional Conference on Engineering Design ICED 99, August 24–26, Munich, 1999.

13. Ernzer M, Wimmer W. From Environmental Assessment Results to DFE Product Changes—an Evaluation of Quantitative and Qualitative Methods. Proceedings of the Engineering Design Conference 2000, Brunel, U.K., June 2000.

14. Wimmer W. ECODESIGN Experiences from Case Studies—How to Consider Environmental Issues in Product Development? Proceedings of the 6th International Design Conference, Design May 2000, Dubrovnik, 2000.

10

Holistic Design for Environment and Market
Methodology and Computer Support

Herbert Birkhofer and Chris Grüner
Darmstadt University of Technology, Darmstadt, Germany

1 INTRODUCTION

Environmentally friendly products have been required by public demand for many years. On the other hand, we can recognize an enormous stream of new product releases on markets worldwide. This high amount of new product releases causes doubts about the ecological impacts of these products. Especially in countries with high industrial and economic standards, we observe extreme forms of customer markets forcing companies to focus on customer wishes and to settle even in very small market segments.

Therefore, it seems simple to ask for green products optimized only in terms of ecological impacts. The problem is that only few people are willing to reduce their lifestyle to an ecologically oriented one in regard to a sustainable dealing with nature [25]. The real challenge for designers in leading industrial countries is to fulfill the requirements and needs from customers and at the same time to reduce the impact on environment as much as possible. This chapter gives an overview about present and future possibilities to achieve this goal.

2 FRAMEWORK FOR DESIGN FOR ENVIRONMENT AND MARKET

When a company wants to sell a product, the company must find a target market for this product (Figure 1). The product is then developed by the design department of the company and produced. During production, the environment is always affected by the extraction of raw materials, the emissions from production, and so on. When customers buy the product they will use it, and this usage also results in a pollution of the environment (e.g., if the product runs on electricity). At the end of the product's lifetime, the user disposes of it and it pollutes the environment again. Viewed from this perspective, it may be best if a product does not have any market success because then the pollution from operation is eliminated.

On the other hand, a product serves a certain need of the customer. This need served by the product is also called the *service* of the product. Viewed in a simplified way, customers do not care, which particular product serves their needs. So if one product does not suit customers' needs, they just buy another product and the product which does not serve the purpose is produced in addition to the fitting product. This means that

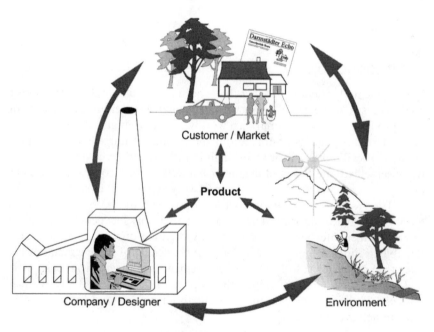

FIGURE 1   Marketability and environmental friendliness of products.

there is an additional but useless product on the market that has been produced and has to be discarded without ever have been of value. An environmentally friendly product must always fit the needs of a customer (or a market) in addition to being environmentally friendly. Because of this, designers always have to look at the needs of the customers in the target markets if they want to minimize the ecological impact of a product. In other words, the designer has to create a product with a service that meets the customer's needs.

3   PRODUCTS AND ENVIRONMENT

Products are used by customers especially for easing life and improving living conditions (Figure 2). These products are always closely connected to the processes in their life cycle. For example, to ride a bike for leisure, this bike has to be produced and discarded after its use phase. The processes are in turn responsible for environmental burden in a product's life. It is often said that products cause all kinds of environmental damage. If one takes a closer look at a product, however, it is the processes in a product's life that are responsible for this damage. A vacuum cleaner by itself does not cause any environmental impact. Harmful, though, are the energy consumption and the dust emission while cleaning a carpet with it.

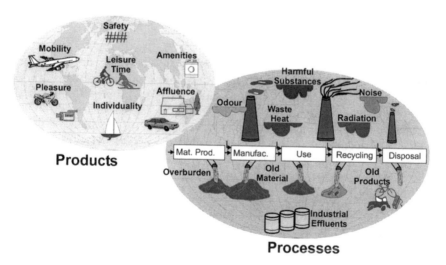

FIGURE 2   Customer utilities and environmental loads.

3.1 Product Life Cycle

A product's life consists of different life phases beginning with the extraction of raw materials and ending with the disposal of the product (Figure 3). In all life phases, processes transform energy or material from one state into another. A lawnmower, for example, needs sheet steel or plastic granulate for manufacturing a cooling device or a housing. After the usage of the lawnmower, the parts are disassembled and partly recycled or incinerated. A life cycle can therefore also be regarded as a model for materialization and dematerialization of a product.

3.2 Process Model

Within each life phase of a product, hundreds and often thousands of processes happen. A process can be modeled using a black box representation with inputs and outputs, an operation, and several operators that cause the transformation from input to output (Figure 4).

Typical operators are machines with tools, a human operator, and a process management whose data could be stored either in the mind or in hard- or software. This universally applicable process model is a prerequisite for a holistic modeling of all product life phases. It can be used for describing different processes like digging for oil, bending a steel plate, cutting a lawn, disassembling an AC motor, or burning shredded disposals. This process model is the basis for performing an inventory analysis, which in turn is the basis for a life cycle assessment (LCA) [7,11].

3.3 Ecological Impacts

As shown above, all concepts to minimize the environmental damage caused by the product must take its life cycle into account [18,28]. This is easy to demand but extremely difficult to achieve. Keeping in mind the hundreds and thousands of processes in a products life cycle, designers should be aware of all environmental impacts and should try to minimize them. Besides the problems of the huge number of processes and the uncertainty of environmental data, we have to accept the following:

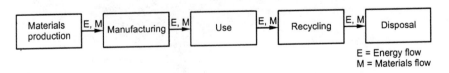

E = Energy flow
M = Materials flow

FIGURE 3 Life phases of a product.

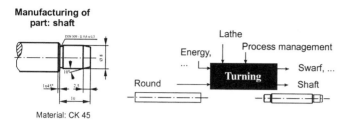

Manufacturing of
part: shaft

Material: CK 45

FIGURE 4   Process and process model for manufacturing a shaft.

- There is no ecological zero-impact product, and an absolute minimum of the environmental burdens caused by a product cannot be defined. This means that there is no reference point for the ecological quality of a product.
- Therefore, the ecological quality of a product can only be judged in relation to another one.
- Designers can only try to reduce environmental damages.
- It is especially important to reduce the ecological damages in the total life cycle.

This means that "green" products might cause quite severe environmental damages in one process if they are able to compensate for this additional damage with reductions of the environmental burdens caused in the rest of the product's life cycle. In design for environment (DFE), designers always have to focus on the minimization of the overall impact on the environment during the whole life cycle. Figure 5 shows the 3L-Lupo

FIGURE 5   A green product—the Volkswagen 3L-Lupo.

from Volkswagen as an example. It is mainly made from light-weight materials like aluminum. These materials require a high amount of energy in production and therefore cause higher environmental burdens during production. But, because of its low weight (among other facts), this car uses 50% less energy in operation than a VW Golf, which easily compensates the higher environmental burdens in production (scenario: product lifetime 10 years, total distance 150,000 km).

4 PRODUCTS AND MARKET

To survive in a market, every product has to meet the customer expectations of its target market. The realization of marketable and environmentally friendly products often leads to target conflicts, especially if the production costs will rise as a result of the minimization of product-related environmental impacts. In the end, the market acceptance always determines a product's success.

The marketability of any product largely depends on the consumer's buying decision. The purchase decision of a customer is determined by a large number of aspects [30]. Especially important is the consumer's perception of the product's utility. From the consumer's point of view, a product delivers a certain utility. This utility can be differentiated in the basic functional utility and the additional utility that can consist of a social and/or an individual utility (Figure 6).

The basic functional utility of a car, for example, is to transport people and things (service of the product). An additional individual utility may be safety, whereas the brand may deliver a social utility. However, a product's utility is always opposed by its price. Thus, the marketability of a product is given if the perceived difference of the product's utility and its price meets the demanded expectations in such a way that the product is purchased and a significant market share is reached.

5 CONFLICTS BETWEEN MARKETABILITY AND ENVIRONMENTAL FRIENDLINESS

5.1 Market Limitations for Environmentally Friendly Products

The purchase of environmentally friendly products has been subject to a number of investigations in recent years. In Europe, for example, a strongly visible sensitization for environmental problems, an increasing environmental knowledge, and an increased willingness to act "pro-nature" developed

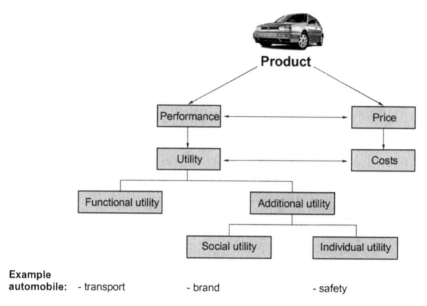

**Example
automobile:** - transport   - brand   - safety

FIGURE 6   Utility dimensions of products.

[27,48]. However, there is still a large gap between the attitude and the actual behavior. Many reasons for the lack of market acceptance of environmentally friendly products have been identified [6]. Besides the general availability of a product, often the asymmetric distribution of product information between the consumer and the producer are mentioned. This results in doubts about the authenticity of the environmental quality of products. Other reasons include disadvantages in the product's efficiency and aesthetic deficiencies.

However, the most important reason is conflicts in the buying motives if the price of the environmentally friendly product exceeds that of an ordinary substitute. This situation is especially common when the additional utility is of a social kind that cannot be individually internalized. It is therefore obvious that requirements of the market establish priorities in designing products as customers pay and companies want to profit. Compared with price increases, the protection of the environment usually always takes second place unless

- There is a market for ecologically optimized products and customers are willing to pay additional costs (e.g., green products distributed by environmental organizations such as Greenpeace or World Wildlife Fund);

- Ecological optimization is economically neutral for customers and seems to be a desirable feature (e.g., the use of ecologically friendly insulating materials in cars);
- Ecological and economical goals correspond (e.g., reducing waste in punching processes reduces the consumption of raw material and at the same time saves money for the manufacturer because he can buy less sheet metal);
- Ecological optimization is required by law (e.g., the prohibition of asbestos in the early 1980s in Germany);
- Companies adopt a green image as a marketing strategy in itself to convince customers of their ecological awareness (e.g., "green" clothing companies).

At present, strong market requirements for all products competing in a market can be observed. An environmentally friendly product only has a chance in these mass markets if it can compete with other less environmentally friendly products. This situation usually leads to a limitation on environmental measures taken in product design. As trivial as it sounds, green products must be sold; otherwise, they do not save the environment at all.

5.2 Possibilities to Deal with the Market Limitations

From the mentioned market limitations, a number of requirements for the design of environmentally friendly products result. To facilitate the information transfer between the producer and the consumer of a product, the design team should offer information about the product life cycle. This information can then be used in marketing. This attempt will be more effective if well-known environmental certificates are used. The related requirements should be taken into account early in the design process (e.g., defined as requirements in the stage "clarification of the task"). Another possibility is to realize easily recognizable product alterations. This is normally the case if an additional individual utility is realized or if widespread public concerns are addressed. An example would be to reduce the water consumption of a washing machine (cost savings out of the water reduction as an additional utility) or to minimize the packing effort (popular in Germany because of legislative measures).

Efficiency disadvantages should be avoided in any case. Environmentally friendly products should deliver at least the same functional utility to the user as their substitutes. For example, a lower water consumption of a washing machine should not result in laundry that is less clean.

Aesthetic deficiencies should also be avoided. This is of special importance if direct competition with ordinary products is attempted. However, in the case of specific (eco-) product differentiation to address environmentally concerned customers, this may not be advisable. In such cases, it might be better to clearly signalize the improved environmental quality of the product (e.g., gray recycling paper).

Generally, environmentally friendly products should be offered at the same price as their substitutes. This is probably the most important requirement of the market. If this is not possible, either a repositioning in a new market segment, possibly accompanied by the formation of a special green brand, or the realization of changes in the area of the individual utility should be considered (Figure 7).

6 PRODUCT DESIGN: MODELING A PRODUCT AND DEFINING ITS PROPERTIES

Product design plays a major role in the value chain of companies. Because other departments, such as marketing, sales, manufacturing, assembly, and purchasing, are closely linked to the design department, product development should be integrated in the whole value chain of a company. When analyzing DFE in terms of knowledge and methodology, two different but strongly interrelated process chains have to be considered: the product life cycle chain and the design process chain.

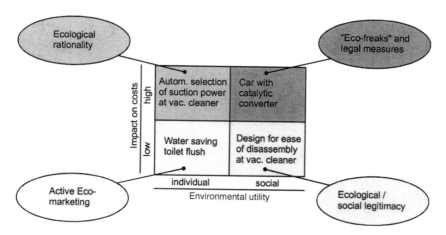

FIGURE 7 Impacts of environmental improvements on the product's utility and costs.

6.1 Product Life Cycle Chain

The life cycle of a product consists of all energy and material flows in a product's life. As described in Chapter 4, the use phase plays a major role in regard to the requirements of customers. Products are bought by customers due to their utilities. All other life phases only help to fulfill this main task. In addition, the use phase is the only phase where the product is used as an operator, while in other life phases processes are transforming the product by means of operators like injection molding machines or recycling plants.

6.2 Design Process Chain

Product design can be seen as a chain or network of processes dealing with different types of product models. Usually, product development starts with a product idea or proposal coming from a customer, from marketing, or from other company departments. By shaping this product idea with processes like abstraction, concretization, variation, combination, and generation, designers develop the final product model with a detailed product description in CAD models and parts lists.

Product planning and design processes can therefore be seen as a modeling of a product from a very abstract to a concrete level using information carriers such as sketches, technical drawings, or CAD models. This design process can be divided into several phases (Figure 8).

The first and most important phase is the clarification of the task and the defining of a requirement list. After that, the product concept with its functions and working principles is developed, which is especially important for the design of very innovative products. In the phase of embodiment design, the geometry and materials are defined, and in the phase of detailed

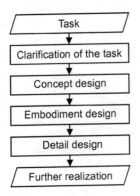

FIGURE 8 Phases in product design. (From Ref. 50.)

design, all the information for manufacturing, assembly, and further life cycle phases are elaborated. The design model in Figure 8 after Guideline VDI 2221 of the German Association of Engineers is only a very abstract one, which might be comparable with the life cycle model. Design in reality is a highly iterative process chain where one process might be left out while another must be performed multiple times. Modern product development uses design methods [12,35,50] and computer support like CAD, CAM tools, or simulation and calculation tools.

6.3 Relation Between Product Design and Product Life Cycle

The result of a product development is the product documentation. The product is then manufactured according to the product documentation. In this context, the product documentation can be seen as the link between product development and a product's life cycle (Figure 9).

During a product's design properties (e.g., geometric data of parts, material, or the topology of parts in a product unit) are defined. These product properties are defined directly in drawings, CAD models, or parts lists. They correspond with indirectly defined product properties like its functionality, quality, costs, or appearance. These are of special interest for customers and are directly related to the product's marketability. The properties also cause processes in the whole life cycle, which in turn influence the environment by harmful impacts. Because of this, design work

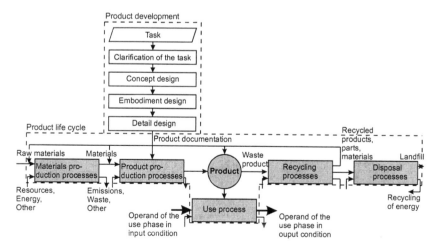

FIGURE 9   Product life cycle and product design.

can be considered as the definition of topology, geometry, and material of parts and units that best fulfills the requirements of the market, the company, and the environment. Good designers should therefore know the relations between a product design and the life cycle of the product. Due to the very special nature of design work, it is evident that many contradictions within the huge amount of design knowledge occur. Design work is therefore a good example for an extremely complex optimization process.

7 STRATEGIES AND RULES FOR DFE

When developing environmentally friendly products, two types of design support are important. The support of the assessment of environmental impacts caused by a product (analysis support) helps the designer to find weak points in existing designs. On the other hand, support in the creation of solutions helps to find new or more environmentally friendly products (synthesis support).

7.1 Analysis Support by Assessment of Environmental Impacts During Design

To assess the environmental impacts caused by a product, LCA studies are currently carried out. Figure 10 shows the generic steps of an LCA. The general framework of an LCA is already standardized (e.g., in the ISO standard 14040 [26]), which is not true for the exact procedures,

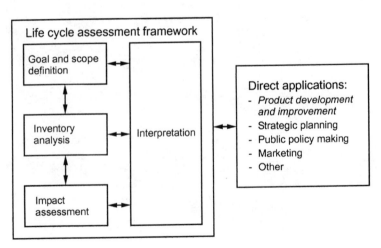

FIGURE 10 Generic phases of an LCA (after ISO 14040).

especially in the phases impact assessment and interpretation [17,24,42,43,45,46]. The step interpretation varies very much, depending on the application of the LCA. When performing an LCA, the definitions from the goal and scope definition step guides the complete LCA process [22,23,32,52].

Current practice in DFE is that an LCA of an existing product is carried out using one of the different LCA methods before the design of the new product begins [10,54,55]. With the knowledge of the weak points of the existing product, a product design process is started with the goal to eliminate these weak points. After product design, another LCA has to be carried out to verify that the new product really is environmentally friendlier than the previous one.

A major drawback of LCAs is that they are very time consuming [16]. Because of this, it is impossible to perform an LCA study for each of the continuous decisions made during the design process. Even abridged LCA concepts like the Eco-Indicator 95 method [19] can only be used in design at a few milestones. Currently, a detailed and continuous ecological evaluation during the design process is more fiction than reality.

7.2 Synthesis Support by Creation of Environmentally Friendly Solutions

The targets to reduce the environmental burdens of a product can range from incremental improvements in the embodiment design of the product to a radical change of the service provided by a product. To achieve these targets, many methods and instruments are available for ecological improvement. Some common methods that are already available today are mentioned in the following section.

7.2.1 Harmful Materials and Critical Processes Lists

Ecologically conscious companies usually have set up lists for harmful materials and critical processes (Table 1, e.g., as proposed in [9]). Designers generally must avoid the substances within the list. They may only use a substance if they can prove that other materials or processes would cause unreasonably high costs.

7.2.2 Ecologically Oriented Design Rules

Because of the problems when performing an ecological assessment in design, designers today usually rely on rules and strategies for DFE alone instead of combining these rules and strategies with an ecological assessment (sometimes combined with examples for DFE, e.g., [29]). Design rules link

TABLE 1 Harmful Materials List (Excerpt)

| Substance | Harmful for health/environment | | Laws (German) |
	Classification	Remark	
Metals			
Arsenic alloys	Forbidden	High toxicity of arsenic, carcinogenic	GefahrstoffV Anhang IV
Lead	Avoid if possible	Toxicity of lead and its compounds, carcinogenic	GefahrstoffV Anhang V
Plastics			
Polyvinyl chloride (PVC)	Avoid if possible	Toxic decomposition products when overheated	
Plastics containing brominated flame retardants	Forbidden	Formation of dioxins and furans when heated or incinerated	Dioxinverordnung

features of the product model together with required ecological properties of the product [31]. Designers get a concrete hint at what they can do to achieve a better ecological quality of a product. An environmentally oriented rule is often dedicated to a special processes in certain phases of the product life cycle (Figure 11). For example, the rule in Figure 11 should suggest the selection of process chain 1 because these manufacturing processes need less cutting volume which results in a low energy consumption for the manufacturing of the part.

The high number of processes in a product's life require a high amount of rules in development. Within this high amount of rules, target conflicts occur (e.g., the development of light-weight products out of compound materials vs. the use of recyclable metals). Without any support, designers have to decide on their own which rule should be used and which should not. This results in a limitation in the application because designers have to evaluate the ecological impact of different often contradicting rules. The problem of contradicting rules will be reduced when a joined ecological, technical, and economical assessment can be performed [20].

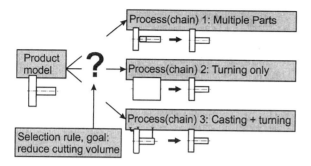

FIGURE 11   A typical design rule for ecological improvement.

7.2.3   Strategies for the Design of Green and Profitable Products

In DFE, every product must be optimized in a different way to address the individual weak points of the product and to find the best ecological product design. For example, when designing a vacuum cleaner, the most important goal should be to reduce the energy consumption of the product during operation since more than 80% of the total environmental burden in its life are caused during the use phase.

These different design goals can be addressed by so-called strategies [21,49]. In our approach, strategies are defined as an *overall plan to achieve a given (eco-) design goal*. Strategies help to find the correct direction for DFE actions and to realize the blueprint of a product's design. On the other hand, strategies for DFE do not consider market aspects while all strategies affect economical properties of a product.

In the following section, some examples for DFE strategies are presented. In a first approach, the relations between marketability and DFE strategies can be visualized using a matrix (Table 2). Since the effects of the strategies largely differ depending on the individual situation of a product, a household vacuum cleaner is used as an example.

Use Environmentally Friendly Materials.   The use of materials that cause less environmental damage are a popular measure when performing DFE. Since most of the products today are cost optimized, a change in the materials composition of a product potentially increases the costs of that product. The selection of materials often also affects the visual appearance of a product. This has to be considered when developing the visual design of a product. For example, the bodies of vacuum cleaners today are usually manufactured using ABS plastics because the customer demands a glossy bright surface with the current smooth design. If the design could be chan-

TABLE 2 Relations Between DFE Strategies and the Marketability of Products

Strategy	Market requirement				
	Avoid extra costs	Inform customer	Additional utility	Avoid efficiency disadvantages	Avoid aesthetic deficiencies
Clean materials	–				(–)
Clean production	–				–
Eco-Packaging	(+)	+			
Use efficiency	(–)		+	+	
Avoid false handling			+	+	
Extend lifetime	–	(+)	+		–
Optimize recycling	–	+	+		–

ged to a matte "decent" appearance, PP plastics, which cause significantly less environmental damage, could be used for these parts.

Realization of Environmentally Friendly Production. The designer of a product strongly influences the production processes by the material and the geometry defined. Like materials, environmentally friendly production processes tend to cause cost increases because of the extra expenses for recycling loops or filtration equipment. In addition, the production is often affected by site-specific legislation (so-called push-oriented) measures, because this part of the product life cycle (at least partly) takes place within the company itself.

Ecopackaging. An environmentally friendly packaging system often results in less costs, because less and less costly packaging material is used (e.g., folded cardboard boxes instead of styrofoam). Environmentally friendly packaging also causes high public interest because of legislation measures (e.g., German packaging law).

Increased Efficiency During Use. The efficient use of products often corresponds with cost savings in use because the product needs less energy and/or working materials (e.g., water-saving washing machine). An example for energy losses during operation is presented in Figure 12.

However, these efficiency measures often cause the product to cost more. This higher price is often accepted by the customer because of the

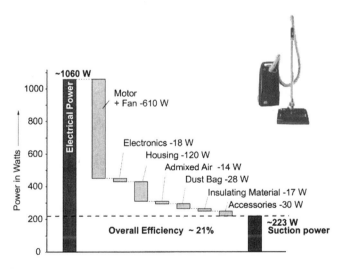

FIGURE 12   Energy losses when operating a vacuum cleaner.

additional utility of cost savings during use. In some markets, such as the vacuum cleaner market, the performance of the product is judged by its power consumption. (The higher power consumption suggests a higher cleaning performance.) A change of this market situation could be caused by using the suction or cleaning power as a performance criterion. In this scenario, more efficient vacuum cleaners are preferred.

Avoid False Product Handling.   False product handling can lead to a higher wear of the product, decreasing the available total operation time. This means that the user has to replace products more often. If false product handling results in a waste of energy or working materials, this additional effort has to be purchased by the user, but there is no increase in the utility of the product.

Since both types of erroneous behavior cause higher costs, they should be avoided by design measures such as a self-explaining product design or signs with operation notices. The drawback of these measures is that the costs saved during product operation do not cause a reduction in the price. This means that expensive measures to prevent false product handling have to be explained to the customer. Only if the customer can take the reduced operation costs into account during the purchasing decision is there an increased chance of market success of the product.

Extension of Product Lifetime.   A product with a longer lifetime has to be purchased less often and therefore tends to cause fewer costs. However, a longer operation time can be achieved by designing the product more robustly, which means that the product will be more expensive. Since the product has to compete with cheaper products with a shorter lifetime, the consumer must be informed about the additional operation time compared with the extra expenses for the product.

Optimization of Recycling.   Recycling can be performed by reusing the entire product or parts or by reusing the materials of the product. Many push-oriented measures are performed in this area, such as the German take-back legislation of cars.

7.3   Profit and Limitations of Rules and Strategies for DFE

Rules and strategies provide a good possibility for a first consideration of environmental aspects in product development. However, an intelligent presentation of the rules is needed to reduce the conflicts within the presented rules. This presentation should also be combined with a technical, economical, and ecological assessment. As shown before, the combined product and process view on the life cycle of a product is needed to achieve this. The following section shows an approach of how the life cycle of a product can already be considered in product development.

8 HOLISTIC DESIGN FOR ENVIRONMENT AND MARKET: THE INTEGRATED PRODUCT AND PROCESS DESIGN

Because of the aforementioned problems and difficulties, a holistic and preventative design of environmentally friendly products should be the most effective means for the reduction of environmental impacts (Figure 13). During product design, a designer must foresee the processes in a product's life to assess the environmental impacts caused by the product. On the other hand, the designer must define the product in such a way that the overall environmental impacts caused by the processes are reduced.

Today, in the beginning of a DFE project, a reference product is usually analyzed. If a limited number of weak points cannot be identified during the analysis, the designer is forced to carry out a certain number of LCAs until he or she is sure that the new design is environmentally friendlier than the previous product version (Figure 14). This results in a wide control loop of the design process, because the designer must always define the complete embodiment design and to model the life cycle of the product to carry out the LCA.

In the concept of the so-called integrated product and process design (IPPD) based on the framework of the standard VDI 2221 of the German Association of Engineers (also see section 6.2) [50], the life cycle of a product

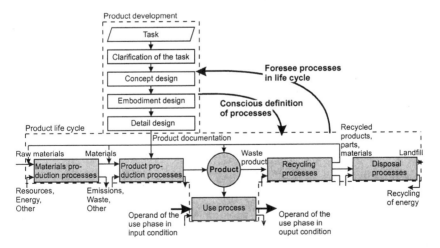

FIGURE 13   Product design and a product's life cycle.

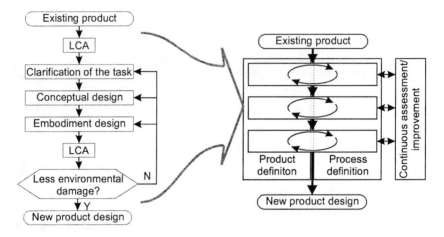

FIGURE 14   LCA studies versus the IPPD.

is defined simultaneously to the product itself [40]. When designing a product, many processes in a product's life are at least predefined. Besides this,
product-related information, as well as general and company-specific information, is captured and used for the completion of the life cycle. The IPPD
takes advantage of this by capturing all the information about the life cycle,
so the designer only has to add a relatively small amount of information to
assess a product design. With the IPPD, the designer models the life cycle of
the product concurrently to the design of the product [39]. This means that
he or she should be able to measure the environmental damage caused by a
product while designing it and to minimize the damage interactively. In this
way, a continuous assessment and improvement of his or her design can be
achieved (Figure 14).

9   COMPUTER SUPPORT FOR THE IPPD

9.1   Integrated Design Environment

To provide the type of support given by the IPPD, a computer support
system is currently being developed at the Darmstadt University of
Technology (Figure 15). The system is based on an information model [1]
and consists of the following parts:

- A CAD system allows the designer to model a product.
- A so-called life cycle modeler combines the different data to support the definition of the life cycle of a product. At the same time,

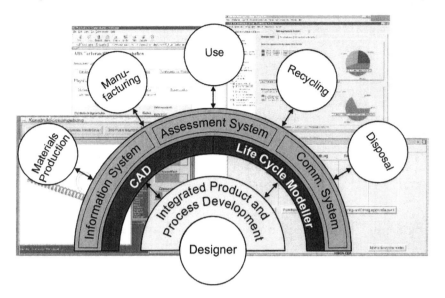

FIGURE 15   Computer support system for the IPPD.

it provides a user interface for the designer where he or she is able to look at the life cycle processes and to manage different scenarios for a product's life cycle (e.g., recycling vs. incineration scenario).

- An assessment system then evaluates the product and its life cycle [4]. The system gives an overall score for the environmental consciousness of the design. The system also tracks the environmental weak points of the design.
- An information system provides the designer with product design related environmental knowledge such as rules or descriptions of methods for DFE.

9.2   Working with the Computer-Assisted Integrated Product and Process Design

Due to the holistic nature of environmental aspects, information and decision-making processes have to be carried out by all responsible departments. A computer support system should therefore be a company-integrated system. The following sections suggest what decisions have to be made within each of the life cycle phases and how life cycle relevant data can be collected.

9.2.1 Raw Materials Extraction

The processes of raw materials extraction and their properties themselves cannot be directly influenced by the designer. The designer can only select different types of process chains by choosing different preproducts. The choice of a certain preproduct or material includes a lot of uncertainties, because the type of preproducts used in manufacturing also depends on decisions in the stockholding or purchasing department. Even the selection of a certain material includes a lot of uncertainties, because the designer is unable to choose the country where the material is coming from. For example, in Germany steel from Korea will be less environmentally friendly than steel from Germany due to the less restrictive laws in Korea and the necessary transport. A lot of the above problems can be solved by a company-specific set of preferences in the system. These preferences can come from different departments (e.g., the purchasing department could determine the steel supplier).

In a first prototype of the support system, the selection of preproducts is divided into two areas: The preproducts and their referring material are put into the system once and are stored in a database. During the design process, the designer simply has to choose a preproduct out of the database and specify the characteristic property of the preproduct (e.g., when the designer selects "round steel" as a preproduct, he or she must specify the length of the preproduct). The system is then able to create a preproduct as one "starting point" for the processes in the product's life cycle. Examples for materials selection concepts can be found elsewhere [3,33].

9.2.2 Manufacturing

In the life cycle phase of manufacturing, the designer has a great influence on production planning and therefore also on the definition of individual processes. The designer defines the type of manufacturing process (e.g., turning) and the basic properties of the process can often be extracted from CAD data (e.g., cutting volume). The problem in this life cycle phase is that the designer is not able to define all the properties and details of the processes because of the great amount of data needed to specify (tool up the machine, clamp part, etc.). Support in this phase of the product life cycle is usually concentrated on providing DFM rules and checklists for the capabilities of certain manufacturing processes (e.g., as presented in [8,15,47].

Rather than defining the processes necessary to manufacture a part in detail, the designer should only have to select the type of the manufacturing process and the system automatically selects the preferred machinery for the selected process (e.g., the selection of a lathe depends on the largest diameter of the part). It is essential that the machine is defined, because the environmental damage caused by manufacturing processes depends heavily on the manufacturing machine itself [37]. The manufacturing machines and their properties and capabilities can be predefined in the system (e.g., by manufacturing experts). They are defined once and stored in a database.

9.2.3 Use

The use phase of a product is already defined in the early stages of the design process (clarification of the task or concept design). The processes can only be influenced in detail during embodiment design. In addition, the use phase of a product usually is defined in cooperation with other departments, such as marketing or sales. However, the designer still defines many "detail processes" in the use phase and therefore decides how well the product performs in operation (e.g., handling processes, ergonomic properties, or maintenance processes) [13,14].

It is impossible to store all the potential process of the use phase in the support system, because the use phase differs too much depending on the individual product. This means that the designer often has to model the use process and its properties manually. Nevertheless, it might be possible to standardize single basic blocks of the use phase that are common for a certain type of product and to integrate these in the support system (e.g., changing of 1 L of oil).

9.2.4 Recycling and Disposal

The processes in the recycling and disposal phase of the product life are usually defined by the company's environmental policy and by legislation and the particular design of the product (so-called end-of-life scenario). The product designer cannot define the end-of-life scenario directly, although the product design has a great influence on the environmental friendliness of the processes. But the designer can try to analyze how the design will perform in the different end-of-life scenarios (e.g., in recycling processes by using disassembly graphs or computer-based support systems for disassembly and recycling, as presented in [41]).

Because the designer usually has a limited influence on the process selection in the recycling/disposal phase, the support for the synthesis phases of design are limited to methods and instruments that determine how a product can be designed for optimal recycling/disposal. Great efforts have already been made for design support in the recycling phase [5,36,53,56].

10 DISCUSSION AND OUTLOOK

The framework of the integrated product and process development and the design environment system presented offers a good possibility for the design of environmentally friendly products. First implementations and easy-to-use methods are currently developed, and results show that is seems possible to use the concept for an efficient support for DFE. For the future we see several problems that have to be solved to support the designer:

- In early design phases, life cycle processes are not known. It is therefore necessary to use experiences from other design projects for a design support. Without these experiences it is hardly possible to predict environmental impacts that are linked to requirements or concepts. One potential approach is to use screening LCAs such as CED (Cumulative Energy Demand), KEA, MIPS, or EcoIndicator (with the problems mentioned in the evaluation of LCA methods) to predict environmental impacts based on the experience of the designer [19,38,51]. One example for this is to predict the materials composition of a new product from similar products already designed (especially useful for redesign projects).

- The problem of the lack of available data concerning the inventory, classification/characterization of impacts, and their evaluation can only be diminished when the impacts of the preproducts are assessed by the producer of these preproducts. This effort can only be delivered with a consensus about rules for LCA. This implies governmental regulations and definitions about environmental goals.

- Another current problem is that more expensive environmentally friendly products are not demanded by the markets if the environmental friendliness is just an additional feature. In this case, the environmental friendliness does not result in an additional buying reason, and therefore the market usually does not allow a higher price. A promising approach seems to keep the customer considering the environmental properties of a product alone and to enable a designer to foresee the consequences of DFE activities on the marketability of the product.

Nevertheless, there are still enough possibilities to develop environmentally sound products even if not all the problems stated above are solved. The next steps in the development of the support system presented in this chapter are as follows:

- The modeling of an example with respect to all levels of concretization of a product and the related processes from requirements and concepts to detailed embodiment layout;

- The collection of LCA-related data concerning the most important processes ("key issues" [34]) in the life cycle of the example;
- The support of the processes selection and the synthesis of product properties;
- Design of a so-called life-cycle-modeler that supports the designer in completing the life cycle information of a product. This step is currently done manually.

The long-term goal is the development of a computer-based system based on an integrated product and process model [2] and easy-to-use methods and instruments for DFE together with strategies and measures to introduce DFE in a companies' organization. With this "collection" of DFE support methods and instruments, it seems to be possible to find a fitting DFE support for every company.

ACKNOWLEDGMENT

The research presented is part of the collaborative research center SFB 392 "Design for Environment—Methods, Working Aids and Instruments" funded by Deutsche Forschungsgemeinschaft (DFG).

REFERENCES

1. Anderl, R.; Daum, B.; John, H.; Pütter, C. (1999) Information Modelling Using Product Life Cycle Views. In: Mills; J.J.; Kimura, F. (Ed.): Information Infrastructure Systems for Manufacturing—IFIP TC5 WG5.3/5.7 Third International Working Conference on the Design of Information Infrastructure Systems for Manufacturing. Kluwer Academic, Chicago, pp. 152–162.
2. Anderl, R.; Daum, B.; Weißmantel, H.; Wolf, B. (1999) Cooperation During the Design Process—A Computer Based Method for All Experts in the Entire Life Cycle. In: Proceedings of the 6th International Seminar on Life Cycle Engineering, Kinston, Canada, June, pp. 295–303.
3. Ashby, M.F. (1989) Materials Selection in Mechanical Design. Materials Sci Technol 5:517–525.
4. Atik, A.; Schulz, H.; Pant, R.; Jager, J. (1999) A Decision Supporting Tool for Environmentally Conscious Product Design. Proc. of the 1st Int. Symposium on Environmentally Conscious Design and Inverse Manufacturing, IEEE, Tokyo, 1999, pp. 718–722.
5. Baier, C.; Weißmantel, H.; Kaase, W.; Thomas, A. (1997) Benefit Function for the Calculation of the Best Recycling Option of Products and Parts. Proceedings of the 4th International Seminar on Life Cycle Engineering, Chapman & Hall, New York, pp. 276–288.

6. Bänsch, A. (1990) Marketingfolgerungen aus Gr nden für den Nichtkauf umwelt-freundlicher Konsumgüter. In: GfK Jahrb. der Absatz- und Verbrauchsforschung Nr. 4, pp. 360–379.
7. Boustead, I. (1995) Life Cycle Assessment: An Overview. Energy World: the Magazine of the Institute of Energy, 230:7–11.
8. Bralla, J.G. (1986) Handbook of Product Design for Manufacturing. New York: McGraw-Hill.
9. Brezet, H.; van Hemel, C. (1997) Ecodesign—a Promising Approach to Sustainable Production and Consumption. United Nations Environment Programme (UNEP).
10. Bundesamt für Umwelt, Wald und Landschaft (BUWAL). (1991) Methodik für ökobilanzen auf der Basis Ökologischer Optimierung. Bern: BUWAL 132.
11. Bundesamt für Umweltschutz (BUS). (1984) Ökobilanzen von Packstoffen. Schriftenreihe Umweltschutz Nr. 24, Bern.
12. Cross, N. (1996) Engineering Design Methods: Strategies for Product Design. Chichester: John Wiley & Sons.
13. Dannheim, F.; Schott, H.; Birkhofer, H. (1997) The Significance of the Product's Usage Phase for Design for Environment. In: Riitahuhta, A. Proceedings of the International Conference On Engineering Design ICED 97, Tampere, Finland, pp. 641–646.
14. Dannheim, F.; Birkhofer, H. (1998) Human Factors in Design for Environment. Proceedings of the 5th International Seminar on Life Cycle Engineering, Stockholm, September 1998, pp. 13–24.
15. DeGarmo, E.P.; Black, J.T.; Kohser, R.A. (1988) Materials and Processes in Manufacturing. New York: Macmillan Publishing Company.
16. De Winter, A.; Kals, J.A.G. (1996) On Manufacturing Processes and the Environment, Proceedings of the 3rd International Seminar on Life Cycle Engineering, Zürich, Switzerland, pp. 177–186.
17. Environmental Protection Agency (EPA). (1995) Life-Cycle Impact Assessment: A Conceptual Framework, Key Issues, and Summary of Existing Methods. Report No. EPA-452/R-95-002.
18. Fiskel, J. (1996) Design for Environment—Creating Eco-Efficient Product and Processes. New York: McGraw-Hill.
19. Goedekoop, H. (1995) The Eco-Indicator Final Report. NOH Report 9523. PRé consultants, Amerfoort, Netherlands.
20. Grüner, C.; Dannheim, F.; Birkhofer, H. (1998) Use of Environmental Knowledge in DFE. In: Luttrop, C.; Persson, J. Proceedings of the 5th International Seminar on Life Cycle Engineering, Stockholm, pp. 121–130.
21. Grüner, C.; Birkhofer, H. (1999) Decision Support for Selecting Design Strategies in DFE. In: Lindemann, U.; Birkhofer, H.; Meerkamm, H.; Vajna, S. (Hrsg.). Proceedings of the 12th International Conference on Engineering Design, ICED 99, München, 24–26 August 1999, pp. 1089–1092.
22. Guinée, J. B., de Haas, U. (1993) Qualitative Life Cycle Assessment for Products. 1: Goal Definition and Inventory. J Cleaner Product 1:81–91.

23. Heijungs, R.; Guinée, J. B., Huppes, G. (1992) Environmental Life-Cycle Assessment of Products: Backgrounds. Leiden: CML.

24. Hunt, R.; Sellers, J. D.; Franklin, W. E. (1992) Resource and Environmental Profile Analysis: A Life-Cycle Environmental Assessment for Products and Procedures. Environ Impact Assess Rev 12:245–269.

25. Hutchinson, C. (1992) Corporate Strategy and the Environment. Long Range Planning 25:9–12.

26. ISO 14040. (1997) Environmental Management—Life Cycle Assessment. Principles and Framework. International Organisation for Standardisation, Geneve, Switzerland.

27. Johnson, M. (1995) Europeans Agree: Green is Good. Manage Rev November 1995, pp. 9–12.

28. Keldmann, T. (1995) The Environmental Part of the Product Concept. Proceedings of the International Conference on Engineering Design ICED'95, Praha, HEURISTA, Zürich, Switzerland, pp. 1048–1053.

29. Klose, J.; Heinevetter, G.; Schön, T. (1997) COMMET—Design of environment-friendly products by using examples. In: Krause, F. (Hrsg). Proceedings of the IFIP WG5.3 4th International Seminar on Life-Cycle Engineering—Life Cycle Networks, Berlin (Germany). London: Chapman & Hall, pp. 142–151.

30. Kotler, Ph.(1996) Principles of Marketing. London: Prentice-Hall.

31. Legarth, J.; Alting, L.; Erichsen, H.; Gregersen, J.; Jorgensen, J. (1994) Design of Environmental Guidelines for Electronic Appliances. Proceedings of the IEEE International Symposium on Electronic and the Environment, San Francisco.

32. Müller-Wenck, R. (1994) The Ecoscarcity Method as a Valuation Instrument within the SETAC-framework. SETAC: Integrating Impact Assessment into LCA. SETAC-Europe, October 1994, pp. 115–120.

33. Navinchandra, D. (1991) Design for Environmentability. Design Theory Methodol 31:119–125.

34. Nordic Council of Ministers (NORDIC). (1992) Product Life-Cycle Assessments—Principles and Methodology. Århus.

35. Pahl, G.; Beitz, W. (1996) Engineering Design—A Systematic Approach. London: Springer.

36. Port, O. (1996) "Green" Product Design. Business Week, June 10, p. 109.

37. Schiefer, E. (1998) Methodik zur ökologischen Bilanzierung von Bauteilen. In: Kolloquium zur Entwicklung umweltgerechter Produkte, TU Darmstadt, 3.-4. November 1998, pp. 66–70.

38. Schmidt-Bleek, E. (1993) Wieviel Umwelt braucht der Mensch? MIPS—das Maß für ökologisches Wirtschaften, Berlin.

39. Schott, H.; Grüner, C.; Büttner, K.; Dannheim, F.; Birkhofer, H. (1997) Design for Environment—Computer Based Product and Process Development. In: Krause, F. (Hrsg). Proceedings of the IFIP WG5.3 4th Int. Seminar on Life-Cycle Engineering—Life Cycle Networks, Berlin (Germany). London: Chapman & Hall, pp. 174–188.

40. Schott, H.; Grüner, C.; Birkhofer, H. (1997): Sustainable Product Development—a Challenge for Design Science. In: Riithahuhta, A. (Hrsg). Proceedings of the ICED'97, Tampere (Finland), Schriftenreihe WDK 25.

41. Seliger, G.; Müller, K.; Perlewitz, H. (1997) More Use with Fewer Resources— a Contribution Towards Sustainable Development. In: Krause, F. (Hrsg). Proceedings of the IFIP WG5.3 4th Int. Seminar on Life-Cycle Engineering—Life Cycle Networks, Berlin (Germany). London: Chapman & Hall, pp. 3–16.

42. Society of Environmental Toxicology and Chemistry (SETAC). (1991) A Technical Framework for Life-Cycle Assessments. August 18–23, Smuglers Notch, Vermont. SETAC, Penascola, Florida.

43. Society of Environmental Toxicology and Chemistry (SETAC). (1993) Guidelines for Life-Cycle Assessment: A Code of Practise, Sesimbra.

44. Society of Environmental Toxicology and Chemistry (SETAC). (1994) Integrating Impact Assessment into LCA, October 1994.

45. Society of Environmental Toxicology and Chemistry (SETAC). (1992) A Conceptual Framework for Life-cycle Assessment, February 1992, Penascola 1993.

46. Steen, B.; Ryding, S. O. (1994) Valuation of Environmental Impacts within the EPS-System. SETAC: Integrating Impact Assessment into LCA, SETAC-Europe, October 1994, pp. 155–160.

47. Todd, R. H.; Allen, D. K.; Alting, L. (1994) Manufacturing Processes Reference Guide. New York: Industrial Press.

48. Umweltbundesamt (Ed) (1996) Umweltbewußtsein in Deutschland Bonn: Bundesministerium für Umwelt, Naturschutz und Reaktorsicherheit.

49. Van Hemel, C. (1998) EcoDesign Empirically Explored. Design for Sustainability. Research Programme Publication No. 1, TU Delft.

50. VDI-Richtlinie 2221. (1993) Methodik zum Entwickeln und Konstruieren technischer Systeme und Produkte, VDI, Düsseldorf.

51. VDI-Richtlinie 4600 Entwurf. (1995) Kumulierter Energieaufwand - Begriffe, Definitionen, Berechnungsmethoden, VDI-Richtlinie 4600 (Entwurf), VDI, Düsseldorf.

52. Vignon, B. W.; Tolle, D. A. (1994) Life-Cycle Assessment. Boca Raton, FL: CRC Press Lewis Publishers.

53. Wallace, D. R.; Suh, N. P. (1993) Information-Based Design for Environmental Problem Solving. Ann CIRP 42:175–180.

54. Wenzel, H.; Hauschild, M.; Alting, L. (1997) Environmental Assessment of Products. London: Chapman & Hall.

55. Wenzel, H.; Hauschild, M.; Jørgensen, J.; Alting, L. (1994) Environmental Tools in Product Design. Proceedings of the IEEE International Symposium on Electronic and the Environment, San Francisco, pp. 100–105.

56. Wolf, B.; Wansel, A.; Boestfleisch, I.; Weißmantel, H.; Schmoeckel, D. (1999) A New Approach Considering Recycling in Steel Product LCA. Proceedings of the 6th International Seminar on Life Cycle Engineering, Kinston, Canada, June, pp. 342–349.

11

Development of an Integrated Design Environment for Ecological Product Assessment and Optimization

Reiner Anderl, Bernd Daum,
Harald John, and Christian Pütter
Darmstadt University of Technology, Darmstadt, Germany

1 INTRODUCTION

A product causes environmental burdens throughout its entire life cycle—from raw material extraction and processing through production and use to recycling and disposal. The great variety of processes accompanying its life cycle requires an effective and efficient computer support to assess and optimize the product while considering ecological aspects.

Current software systems supporting the ecological evaluation and optimization of products are based on the methodology of life cycle assessment (LCA), whose principles and framework are standardized in CEN ISO 14040 [1]. Despite the relatively advanced standardization of this method, LCAs appear to be problematic in practical application. This becomes obvious while assessing complex technical products, where extensive data acquisition is required—a time-consuming and cost-intensive endeavor. In addition, a great number of specifications must be made that can have a strong influence on the result of an LCA, for instance specification of system boundaries, allocation procedures, and so on. Studies show that most of the life cycle analysis conducted are of retrospective nature [2], meaning that existing products are examined for weak points or are compared with competitive products. Prospective studies (i.e., analyses of products planned) are made in exceptional cases only.

The LCA software tools increasingly feature improved graphical user interfaces and partially offer the selection of different evaluation models [3]. From the design engineer's point of view, however, these software systems are stand-alone solutions with two substantial deficits:

1. Evaluation cannot be made on current product developments since there are no interfaces to the design environment, especially to the computer-aided design (CAD) system. Therefore, geometry data or physical properties cannot be accessed from the product model. As a result, it is not possible to analyze alternative design variants based on shape variations for their ecological effects during the development process.

2. Performing an assessment is very time consuming since a great variety of information has to be entered manually. This information on inputs and outputs of process steps is partially unknown to a product developer. Direct access to information resources such as enterprise resource planning or environmental management information systems is not facilitated.

A simple black box integration of LCA software into the design engineering workplace is possible using a product data management (PDM) system. PDM systems offer management of product structures and configurations and support workflow and cooperation during product development. Similar to CAD and other CAx applications used to simulate product behavior or production processes, LCA data could also be managed within PDM and linked with the appropriate LCA software tool. However, this solution does not realize any data exchange and sharing with CAx applications.

The standard for data exchange in product development is ISO 10303 (STEP) [4]. Various application protocols offer data format specifications for different branches enabling data exchange and sharing between different CAx applications. Typically, these specifications do not contain any ecological product data, as CAx applications themselves do not handle such information.

Data exchange between different LCA software systems is realized by the SPINE and Society of the Promotion of Life Cycle Assessment Development (SPOLD) data formats. SPOLD and various suppliers of LCA software systems [5] defined a communication format to enable the consistent presentation of LCAs accompanied by additional information such as data quality or data source references. The SPINE database format

[6] resulted from the work of Centre for Environmental Assessment of Product and Material Systems. An ISO standardization of a common LCA data documentation considering both formats based on STEP commenced in 1998.

All these initiatives do not solve the problem of exchanging or sharing data between LCA and CAx applications. As a result, the realization of an integrated design environment needs research and software development efforts.

2 MODEL-BASED APPROACH

To develop an integrated design environment for ecological product assessment and optimization, a product data model must be integrated with process models representing the knowledge about environmental impacts in product life cycle phases. The resulting information model should be independent from a particular product. It should also include a life cycle inventory to reduce the efforts needed for LCA.

2.1 Model Architecture

The information model can be realized in a three-layer architecture, as shown in Figure 1. First, the core model layer contains product data from all product development phases. It defines, for example, product structure, geometry, and other information being independent from a particular life cycle phase. Second, the process model layer is used to represent knowledge on processes in a particular life cycle phase such as production, use, recycling, and disposal. Inclusion of new processes is possible. Finally, the inventory data layer contains information about environmental impacts. Information of this kind can be derived from product and process data represented in core and process model layers.

To distribute the modeling task among several experts, the information model can be divided into partial models. Each partial model uses the relevant part of the core model, adds process data for a life cycle phase or special activities, and derives the life cycle inventory data.

2.2 Modeling Technique

The model development should include the following major steps (Figure 2):

1. As environmental impacts result from processes in product life cycle phases, activity modeling can be used to analyze them, thereby determining all relevant input and output streams. In contrast to the activity modeling performed in LCA tools, the model is product independent and thus does

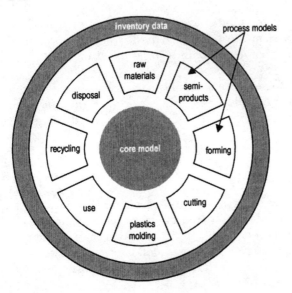

FIGURE 1   Information model architecture. See text for explanation.

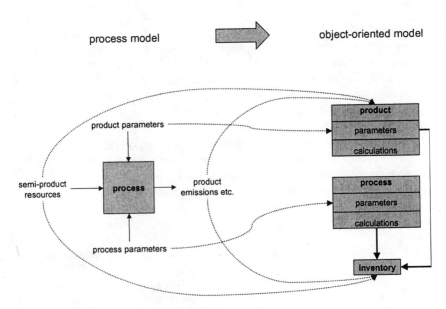

FIGURE 2   Modeling technique. See text for explanation.

not contain any values. An example for a suitable modeling language is IDEF0 [7].

2. To integrate this informal activity models with a product data model representing geometry, materials, product structure, and other product properties, a transformation from the process-oriented view in a data-oriented view is required. Inputs and outputs have to be quantified using functional relations or data from LCAs already performed. This requires the facility to model functional and rule-based dependencies. Object-oriented modeling languages, such as the Unified Modelling Language [8], offer these constructs. Suitable modeling tools support the transformation of models into an implementation usable for a database schema. With this, time and resources during implementation tasks can be saved using rapid prototyping of an appropriate database system.

2.3 Modeling Task

Usually, information modeling is carried out by information modeling experts. Knowledge, which has to be represented in an information model, comes from possibly various domain experts, who are interviewed by a modeling expert to acquire the knowledge relevant to the application context. Afterward, the resulting formal information model or a prototype of a software system using the information model is reviewed by the domain experts. Similar techniques are used in many different areas such as knowledge engineering, business process reengineering, object-oriented software development, and even in the area of product data technology.

In contrast to such modeling tasks, the information modeling for environmentally sound design should be done by domain experts themselves. In general, this approach has advantages but may also cause problems. The main benefits are as follows:

- There is no loss or misinterpretation of knowledge. While interviewing the domain experts, the modeling expert has to understand and therefore interpret the information from the domain experts. Usually, this is a source of error because the domain experts' terminology differs from the modeling expert's.
- The modeling of complex process chains or application areas where the participation of several domain experts is required can be done in parallel. The domain experts model particular parts of the information model, representing information from their individual domain. Therefore, the information modeling process can be accelerated.

- While domain experts can be efficiently taught suitable modeling methods, modeling experts can hardly be expected to learn all of the domain experts' different knowledge to the required detail (refs. 9 and 10 come to similar conclusions).

The main problem is to ensure that the information model developed in cooperation with the domain experts is consistent and free of redundancy. In general, domain experts do not have much experience in information modeling. Therefore, the modeling technique must be as simple as possible to meet the requirements of these users. In addition, all languages involved in the modeling process should be completely graphical to avoid learning syntactical details of a lexical modeling language. This would simplify the modeling process for users who are not familiar with modeling or implementation. Finally, partial models have to be integrated by a modeling expert to form a coherent information model.

3 IMPLEMENTATION

The implementation of the information model requires additional considerations concerning information sources, software components, suitable user interfaces, and system architecture.

3.1 Information Sources

The sources of information that supply the data for instantiating the information model need to be specified. This includes departments involved in the product development process and software systems used by them. For example, data relating to production steps have to be supplied by other departments of the company and by subcontractors. Information sources can be determined by performing a process analysis, for example, using the Structured Analysis and Design Technique [11]. Their degree of integration and the importance for the process of development differ (Figure 3) and depend on the stage of the development.

In practice, not all information required to derive the life cycle inventory can be covered in existing software systems typically used for product development. Additional queries or facilities need to be integrated into these systems using programming interfaces. Furthermore, the entire product life cycle cannot be specified once for any product. As a result, the life cycle is best modeled in parallel to the product modeling [12]. Similarly to the detailing of the product model, the processes are more and more detailed during the development process as it proceeds.

Functional area	Importance for the development of environmentally sound products	
	low	high
Management		X
Controlling	X	
Marketing		X
Development		X
Design		X
Simulation	X	
Operations scheduling		X
Montage	X	
Customer service		X
Recycling		X
Legal department		

FIGURE 3 Participation on the development of environmentally sound products.

3.2 System Components

The required system components not only result from the information sources determined but also from the process to be supported. For designing environmentally sound products, both product and life cycle model must be assessed and optimized in iterations (Figure 4).

Parallel product and process development requires the use of one common logical database. This database does not only guarantee the consistence of the data during multiuser operation, it also allows parallel access to the same data. The information model can be implemented by means of an object-oriented database system.

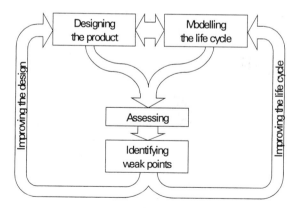

FIGURE 4 Process of designing environmentally sound products.

The CAD system is the central tool of a design environment supplying product data directly to the database. This includes, for example, the assembly structure, the volume of each component, and the characteristic dimensions of form features. Functions to link process related information, such as surface treatment to the geometry inside CAD, might be added to avoid manually entering geometry values during process definition.

A life cycle modeler is needed to describe and compare various life cycle scenarios. This life cycle modeler includes components for each life cycle phase to add product-specific process information during product development and enterprise-specific information during configuration of the system.

Based on the information provided by CAD and the life cycle modeler, an evaluation system for environmental and also technical and economical properties of products is able to access the life cycle inventory data automatically calculated within the database system based on the information model. It performs an impact assessment that is presented to product developers in a suitable manner for optimizing the product's environmental properties [13]. To this end, the evaluation system allows the designer to trace harmful impacts to their causes. Thus, the designer can determine the processes or rather the product components that contain the biggest potential of improvement.

The need to consider all phases of a product life cycle while designing calls for a close cooperation of experts for particular life cycle phases. This can be achieved by means of computer-supported cooperative work such as conference systems. To store and retrieve design rules, best practices, and recommendations, a hypertext information system is useful.

3.3 System Architecture

New systems such as life cycle modeler or evaluation system are necessary for the design of environmentally sound products but increase the complexity for the user. A computer-based design environment should reduce this additional complexity. It should also provide specialized interfaces to designers and experts from all life cycle phases.

A suitable system architecture is the Common Object Request Broker Architecture (CORBA) specified by the Object Management Group [14] (Figure 5). The architecture shown in the figures enables internal communication between various systems and provides basic services necessary for cooperative working in a distributed heterogeneous environment. This approach keeps the entire system flexible—efforts for integrating new applications from other product development phases (e.g., job planning) are minor. The integration of individual systems on data level is based on the information model transformed into the interface definition language.

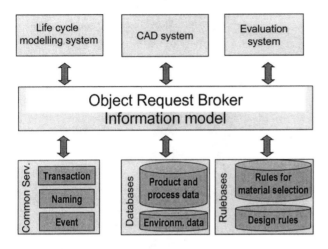

FIGURE 5 Architecture of a design system environment.

4 PROTOTYPE OF AN INTEGRATED DESIGN ENVIRONMENT

In the research project "Development of Environmentally Sound Products"* at Darmstadt University of Technology, a CORBA-based prototype of an integrated design environment for product assessment and optimization (Figure 6) has been developed according to the methods and ideas proposed.

For the purpose of creating the underlying information model, a modeling tool was implemented integrating both process and object-oriented modeling languages. The tool incorporates the generation of a database schema specialized for the object-oriented database system O2 and interfaces for the object request broker by IONA Technologies. Both evaluation system and life cycle modeler have been developed using the object-oriented programming language JAVA. As an example for a commercial CAD system, Pro/Engineer from the company PTC has been extended and integrated. A web-based information system provides information about how to design for environment throughout the development process.

The main activities applying this prototype are as follows.

1. The product shape is being modeled within the CAD system as usual. This includes modeling of geometry and selection of materials. In addition, processes are linked to shape elements using new function calls inserted into the standard CAD menu. Processes from product life cycle

---

*Granted by the German research association DFG as SFB 392.

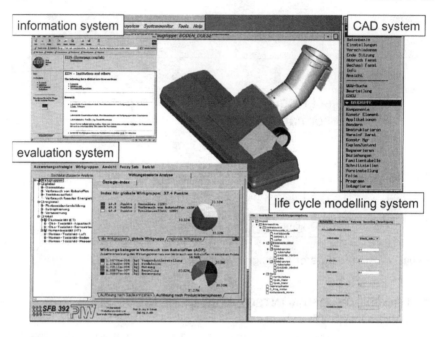

FIGURE 6   Screenshot of a prototype.

independent from geometry are selected and combined to a life cycle scenario within the life cycle modeling system.

2. Product data and related process data relevant for the assessment
are transferred to the database system and synchronized during product
design activities. When starting product assessment from within the CAD
system, formulas specified in the information model are evaluated to generate data for the life cycle inventory. This process does not require any user
interaction.

3. The evaluation system accesses the life cycle inventory data from the
database, calculates the life cycle inventory for the entire product, and performs a life cycle impact assessment according to the assessment method
implemented. A selection between various assessment methods could also be
realized based on this concept. Finally, an ecological index is derived to
facilitate an easy comparison of alternative product solutions.

4. Within the evaluation system, results of the assessment are presented as trees enabling an individual exploring and charts showing very
fast the core results. They can be traced back to the responsible processes in
product life cycle and the shape elements or other product characteristics
they are linked to.

5. After recognizing weak points or improvement potentials in product design, the designer can modify the product shape and processes being linked to or the entire life cycle scenario and compare results for original and changed product. Determining the effect of changes on results, an iterative optimization of ecological product performance can be achieved.

By calculating the life cycle inventory and impact assessment based on product data and a life cycle scenario, the efforts to evaluate product design can be reduced to an amount comparable with other kinds of product simulation. The close integration of all system components enables a fast optimization process. Altogether, the prototype shows that the model-based approach and the system architecture are suitable to realize an integrated design environment for ecological product assessment and optimization.

REFERENCES

1. CEN European Committee for Standardisation. ISO 14040: 1997–06: Environmental Management—Life Cycle Assessment—Principles and Framework. Brussels: CEN, 1997.
2. Grotz S, Scholl G. Application of LCA in German Industry Results of a Survey. Int J Life Cycle Assess 1:226–230, 1996.
3. Rice G. LCA Software Review—A Review of Commercial LCA Software, with Specific Emphasis on European Industrial Applications. Centre for Environmental Strategy, University of Surrey, July 1996.
4. ISO 10303-1.1994 Industrial Automation Systems and Integration—Product Data Representation and Exchange. Part 1. Overview and Fundamental Principles. ISO TC184/SC4, 1994.
5. Singhofen A. Introduction into a Common Format for Life-Cycle Inventory Data. Status Report, Brüssel, Belgien: SPOLD, January 1996.
6. Stehen B, Carlson R, Löfgren G. SPINE—A Relation Database Structure for Life Cycle Assessments. Centre for Environmental Assessment of Product and Material Systems, Chalmers University of Technology, Göteborg, Sweden, 1995.
7. Tipnis VA. Towards a Comprehensive Life Cycle Modeling for Innovative Strategy, Systems, Processes and Product/Services. PROLAMAT '95, Berlin, 1995.
8. Jacobson I, Booch G, Rumbough J. UML—Unified Modelling Language (Version 1.3). Rational Software, 1999.
9. Fowler M. Analysis Patterns—Reusable Object Models. Boston: Addison-Wesley, 1997.
10. Vesterager J. Product-Modelling and Concurrent Engineering. Proceedings of the European PDT Days, 1996.
11. Marca AD, McGowan CL. SADT: Structured Analysis and Design Techniques. New York: McGraw-Hill, 1988.

12. Schott H, Grüner C, Büttner K, Dannheim F, Birkhofer H. Design for Environment—Computer Based Product and Process Development. 4th International Seminar on Life Cycle Engineering, Berlin, 1997.

13. Atik A, Schulz H, Pant R, Jager J. Method and Computer Aided Software System for Ecological Evaluation of Products Accompanying to Development Process. C. Lutrop (ed.): Life Cycle Design '98. Proceedings of the 5th International CIRP Seminar on Life Cycle Engineering, Stockholm, 1998.

14. Object Management Group. The Common Object Request Broker: Architecture and Specification. Revision 3.0 (Draft), 1999.

12

Utilization Phase as a Critical Element in Ecological Design

Jürgen Sauer, Bettina S. Wiese, and Bruno Rüttinger
Darmstadt University of Technology, Darmstadt, Germany

1 CHALLENGES IN ECOLOGICAL PRODUCT DESIGN

Product design and development are critical activities for many organizations in ensuring their long-term survival. The requirements during this process are manifold and of great complexity (Pahl and Beitz 1996). Because considerable time and budget constraints are permanent companions, a primary concern is to minimize the duration of the design and development cycle, for example, by employing approaches such as simultaneous engineering.

Ecological product design is no different, as it needs to follow the same general principles of the process, such as not exceeding cost limits and meeting tight deadlines. However, the complexity of the design process is significantly enhanced, as there are a number of additional issues to be considered.

First, an approach based on life cycle assessment (LCA) needs to be used (Wenzel et al. 1997; Birkhofer and Grüner, see Chapter 10). This enables the designer to assess the environmental friendliness of a given product across its entire life cycle. The assessment also needs to include a comprehensive evaluation of the utilization phase as a crucial part of the life cycle. Furthermore, to obtain a precise assessment of the environmental impact of the utilization phase, all critical subphases of the utilization phase need to be examined.

Second, it is necessary to determine ecological product properties and subphases that are pertinent to a specific product. This cannot be achieved by carrying out a general examination of the utilization phase. Instead, the specific behaviors of users need to be analyzed and subsequently judged against ecological criteria (e.g., behavior that increases energy consumption or reduces longevity). A list containing such criteria has been proposed by Schmidt-Bleek and Tischner (1995). However, since the weight of each criterion varies between products, the assessment needs to be made on a case-by-case basis.

Third, a complex trade-off between economic and ecological requirements of a given product needs to be considered. This necessitates detailed market research to identify the ecological features considered relevant by the customer. It is important to note that these features may differ from those relevant to the environment. Without completion of this market-oriented step, one would take the risk of the product not breaking even.

A number of specific problems are associated with the utilization phase, mainly concerning LCA and the difficulty to strike a balance between commercial and ecological requirements. If one wishes to provide a precise assessment of the environmental impact of a product, LCA needs to take user behavior into consideration. A major problem is the unavailability of suitable data for the description of user behavior. Furthermore, an important aspect of the utilization phase refers to product purchase, where the conflict between economic and ecological criteria becomes apparent. For these reasons, it is not surprising that the utilization phase has been rather neglected in LCA.

In this chapter we provide an analysis of the problems associated with assessing user behavior during the utilization phase. This is followed by presentation of a number of approaches, methods, and instruments that can be used to deal with these problems.

2 LCA-BASED ENVIRONMENTAL IMPACT ASSESSMENT

An LCA-based analysis is indispensable to determine the extent to which user behavior has an influence on the environmental impact of a product. If there were little impact as a function of user behavior (e.g., as for passive products*), little attention would need to be paid to human factors issues in the utilization phase. In that sense, LCA has an important *filter function* in identifying the critical elements of the product (e.g., longevity, energy consumption).

If the utilization phase does play an important part in LCA, we need to discuss three issues that are central to carrying out a user-centred environmental impact assessment: multiple-phase model, environmental impact score, and empirical model of user behavior.

2.1 Multiple Phase Model of Product Life Cycle

Important elements of LCA are phase models that describe the various stages of a product's life cycle. The research literature provides several models of this kind, which differ slightly in emphasis and in the number of phases. A model by Wenzel et al. (1997) comprises five phases (extraction of raw material, production of materials, product manufacturing, use, and disposal). This model may be considered production centered because of its emphasis on industrial processes. Birkhofer and Schott (1996) have also suggested a model with a strong focus on industrial processes. It consists of six separate phases: product development, extraction of raw materials, production, distribution and transport, utilization, and recycling and disposal. A four-phase model by Schmidt-Bleek and Tischner (1995) places somewhat more emphasis on postutilization phases and may therefore be considered disposal centered. It proposes the following phases: production, utilization, recycling, and disposal. Recently, Dannheim (1999) suggested a five-phase model that gives more equal weight to the stages of the product life cycle. It consists of the following phases: extraction of raw materials, production, utilization, recycling, and disposal. However, it still may be considered slightly production centered, as the perspective of the manufacturing organization prevails.

While all models may have demonstrated their utility in their target application, their weakness from our point of view is that they have neglected the utilization phase. This may be due to two reasons. First, there are no "off-the-shelf" instruments available to determine the impact of user behavior. Therefore, assessments of this kind are costly to carry out and require some expertise. Second, organizations may have paid less attention to the utilization phase because it was considered the sole responsibility

---

*The distinction between active and passive products is important (see De Winter and Kals 1994). While active products have an environmental impact during employment by the customer (e.g., lawnmower, motorbike, coffee machine), passive products have very little impact or none at all (e.g., cutlery, carpet).

of the user while the focus was on minimizing environmental damage during the phases under direct control of the organization (e.g., manufacturing).

To redress the balance, we present a model that acknowledges the criticality of the utilization phase for ecological design. The model is displayed in Table 1. The product life cycle is divided into three main phases, with each phase comprising a number of subphases. In each main phase, a principal agent determines the environmental impact of that phase.

A phase model alone is insufficient, however, to assess the environmental impact of a product. While it provides a structure of the product's life cycle, methods and instruments are needed that allow an assessment of the environmental impact, separated by phases, across a product's lifespan. The phase-based approach permits the limitation of the assessment to a specific time period of the lifespan (e.g., manufacturing, usage by customer). Each phase is associated with particular general tasks (e.g., the utilization phase includes purchase, maintenance, repair, etc.), which all have a specific impact on the environment.

2.2 Environmental Impact Score

For a comparative assessment of product variants, the designer requires a quantifiable result of the assessment, for example, in the form of an environmental impact score (EISc). EISc as a quantitative measure indicates the

TABLE 1 Three-Phase Model of Product Life Cycle

Principal agent	Main phase of product life cycle	Subphases
Manufacturing organization	Development and manufacturing	• Market research • Design • Processing of raw materials • Manufacturing
User/customer	Utilization	• Purchase • Assembly • Operation • Disposal
Government institutions/waste management organization	Disposal	• Dismantling • Recycling • Dumping

relative degree of environmental damage caused by a product during its complete life cycle (Atik et al. 1998). It is based on the single-score principle, which aggregates all EIScs from different ecological aspects (e.g., toxicity, energy consumption) into a combined index.

The calculation of EISc is a complex and demanding exercise. This is because a comprehensive database needs to be created, comprising all processes (e.g., cutting, forming, and milling) and parameters (e.g., material and volume) that are part of the product's life cycle. Furthermore, it is difficult to compare different kinds of environmental impact, such as CO_2 released into the air and nitrate discharged into the water. A transparent procedure is therefore required for aggregating these into a single index.

Although an aggregated EISc permits the selection of the better product variant, on the basis of a compiled index the designer is generally unable to pinpoint product modifications, which lead to a substantial reduction of EISc. Therefore, the overall score must be dividable into its components, as the designer requires this information for completing effective modifications. Components may be subphases (e.g., repair) or ecological product properties. Ecological product properties may also be considered ecological criteria, such as those proposed by Schmidt-Bleek and Tischner (1995). For the utilization phase, they have suggested a total of 24 generic criteria, such as size and weight, energy consumption, reparability, multifunctionality, opportunity for multiple-user employment, and ease of maintenance. Both subphases and product properties may provide guidance to the designer for the best intervention to achieve ecological improvements.

2.3 Empirical Model of User Behavior

Within the context of the multiple phase model of the product life cycle, the impact of the utilization phase may have been underestimated. For many products, the largest ecological impact occurs during the utilization phase (see Wenzel et al. 1997). This is based on several LCA-based analyses of electric household appliances and gardening tools, which have demonstrated that about 75% of the energy consumption occurs in the utilization phase. The analyses were based on such different products like refrigerators (Mose et al. 1997), TV sets (Nedermark et al. 1997), and high-pressure cleaners (Sand et al. 1997). Similarly, an assessment of the potential for environmental damage using different criteria (e.g., global warming and toxicity) provided considerable percentage scores for the utilization phase. These were 33% for the refrigerator (here the disposal phase is dominant), 87% for the high-pressure cleaner, and even 99% for the television.

These analyses are based on a *normative* model of user behavior, which assumes that the user operates the system in a near-optimal manner. This is in contrast to an *empirical* model, which is concerned with actual user behavior. Actual user behavior is rarely optimal if judged against the designer's conception of user-product interaction. In fact, a considerable proportion of users is likely to show behavior patterns that are rather unexpected. Generally, user behavior is difficult to predict by the designer without using human factors assessment methods (see section 5).

As the analyses above have shown, the utilization phase was even predominant when a normative model of user behavior was applied. While the use of a normative model is acceptable to gain a rough estimate, an empirical model is required if a more precise estimate is needed. When applying an empirical model, the predominance of the utilization phase is very likely to increase.

3   EMPIRICAL ANALYSIS OF USER BEHAVIOR

In this section, we present a model that helps us understand user behavior by outlining the main elements in the interaction of the user with the system. We then discuss the implications of domestic use compared with use in a work context. This distinction is important as many products are employed in a domestic context.

3.1   User-Product System

The interaction between user, task, and product may be described by a model providing the theoretical basis for identifying suboptimal interactions between system elements (Sauer and Rüttinger 2000). The model of the user-product system assumes the occurrence of three types of mismatches as a result of suboptimal interactions (Figure 1):

1. Functional problems (task-product interaction): A task cannot be completed according to ecological principles because the product does not have the required function.
2. Efficiency problems (user-product interaction): The product is not sufficiently transparent for the user to carry out an ecological action.
3. Effectiveness problems (user-task interaction): The user sets himself/herself a nonecological task because of a lack of motivation or lack of knowledge to pursue ecological goals.

The model emphasizes the importance of examining the interaction of elements instead of merely investigating each element in isolation. This is because the question of whether a given product is environmentally friendly is strongly dependent on the kind of user and the kind of task pursued with the product. The model implies that poor ecological performance is not due

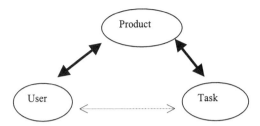

FIGURE 1   Model of the user-product system.

to a single element of the model but may be caused by the poor match between elements (e.g., efficiency problem).

A task analysis (Kirwan and Ainsworth 1992) is required to identify mismatches in the user-product interaction. The ensuing environmental impact of these mismatches is to be assessed, which allows determining whether intervention is effective at all. This is followed by a prioritization of areas, which allows the identification of the most promising areas for intervention.

If one seeks improvements of user-product-system performance, interventions can be carried out on all three elements. However, from the viewpoint of design, interventions focus primarily on product-task match and product-user match. This implies two principal goals of ecological design: ecological functionality and transparency for ecological use. Ecological functionality requires that within the functional limit of the product, any ecologically relevant task should be executable with the product (product-task match). Transparency for ecological use requires that the design of the product permits the user to understand easily how to execute this task (product-user match). The interaction between user and task is of lesser importance since the designer has little possibility of influencing factors such as insufficient motivation to fulfill ecological goals or lack of environmental knowledge.

3.2   Implications of Domestic Product Use

The user-product system may be looked at as a special case of a human-machine system.[†] This is a useful approach due to the rich knowledge base of ergonomic design for work environments. The domestic use of products is

---

[†] If we are concerned with a technical system for human use, we employ the term *product* when referring to domestic usage in the context of the user-product system and the term *machine* when referring to industrial use in the context of the human-machine system.

characterized by a number of differences compared with machine use in a work context. These differences have important implications for design and thus need to be discussed. Machine use is characterized by the following features:

1. There is some form of performance control while this is usually absent in domestic product employment.
2. Machines are generally used more frequently and for a longer duration.
3. Machine operators generally have a higher level of experience in operating the system.
4. Operators are generally better prepared for using the machine (e.g., by selection and training).

The goal of organizations is to obtain an optimal match between system, user, and task to improve overall human-machine-system performance. While there are clearly mismatches of various degrees in organizations that impair overall performance, the four principal strategies mentioned above may be used for corrective measures (Gebert and von Rosenstiel 1989):

1. Selection: By using various selection tools (tests, interviews, etc.), the organization is able to select a person with a suitable skills and ability profile for a particular job. Prior to that, the required profile should have been determined by a job analysis.
2. Training: The organization provides training to the job incumbent concerning the skills and abilities needed for showing adequate performance on the job.
3. Design of task: This refers to the (re)design of work to make tasks more manageable for human operators, for example, by dissecting them into smaller components and allocating parts thereof to a machine.
4. Design of technical system: The technical system (which may be a simple tool) used by the operator to fulfill work goals is redesigned to improve human-machine-system performance.

If the product is to be employed in a work context, any of the four strategies may be used. However, in the context of domestic use, the possibilities to shape user behavior are much more limited. First, it is not possible to select domestic users according to ecological criteria by restricting the sale of appliances to certain user groups. Second, the possibilities for training domestic users are very limited since pressure cannot be exerted on individuals to participate in a training program. Third, tasks of users are usually self-defined and there would be little possibility to modify the task if it was nonecological (e.g., to heat the kitchen with the oven). Thus, the fourth option, design of the technical system, is the primary mechanism for intervention to achieve behavioral modification. The designer needs to be aware that domestic users have generally a lower level of knowledge of how to use an appliance. In contrast to an industrial context, there is little peer control of which appliance is bought and how it is operated (i.e., fewer attempts are made to improve performance through exchange of information). Due to the less elaborated and less formalized decision-making processes in a domestic context, there is a higher risk of an appliance being bought that does not fulfil task requirements.

4 SYSTEM DESIGN FOR ECOLOGICAL USE

While many different agents impinge on the EISc of a product (e.g., user, production workers), the designer exerts the single biggest source of influence. The decisions of the designer determine to a very large extent the product properties, which impinge on the EISc in all three life phases. They determine materials (e.g., steel vs. PVC) and tools and processes used during manufacturing (e.g., milling vs. turning). Furthermore, they determine functional properties and product interface and herewith the interaction between user and product. Finally, they determine the recyclability of the product through the materials used and their compositions. Due to the complexity of the data and interactions across subphases, the demands on the designer are considerable. For example, changing the material of one component may have repercussions in all three life phases. Therefore, the designer requires support from an integrated product design environment. This computerized decision support system allows making predictions about the effects of design decisions across life phases. Anderl et al. (see Chapter 11) describe the development of such an integrated product design environment.

The process of system design for ecological use is not fundamentally different from general system design. A primary goal of system design has been the improvement of human-machine-system performance. This was measured by a range of performance parameters relevant to the system under investigation. If we add ecological performance to that bundle of measures, we can use the same steps of system design but need to be guided by the results of LCA to focus on the system elements with the greatest ecological impact. Based on an approach suggested by Czaja (1997), we consider the following steps important to meet the multiple demands of designing for ecological use: task allocation, task analysis, interface design, and design of support materials. All work-specific requirements (e.g., job design, design of work environment), included in Czaja's approach, do not need to be considered in the present case.

In the proposed four-step approach, communication plays a central role, in particular, for the design of interfaces and support materials. Therefore, due consideration should be given to aspects of communication theories because this may enhance performance of the human-machine system.

4.1 Task Allocation

This refers to the designer's decision to assign tasks or functions (thus also referred to as functional allocation) to the human or to the system or to allow for a dynamic allocation between them (Sharit 1997). Task allocation is a very complex issue when designing systems for highly complex work environments, such as aircraft and space vessels. Although in the case of domestic appliances complexity levels are generally much lower, sufficient care is still required to consider the design options and their respective influence on user behavior.

A commonly used form of allocating a function to the machine is by automating it. If human performance of this function has been unsatisfactory (and it is technically feasible), the designer has often chosen this option. The automatic devices designed to aid users in the operation of the appliance may range from simple decision support to full automation (Endsley and Kiris 1995). Compared with feedback devices, automatic devices are less dependent on the motivation of the user (e.g., if the user does not pursue ecological goals). With regard to the model of the user-product system, the implementation of automatic devices is a suitable means to deal with effectiveness and efficiency problems.

4.2 Task Analysis

An appropriate method of task analysis (Kirwan and Ainsworth 1992) can be employed to identify critical ecological activities of the user. With a task analysis method, the user-product interaction during the utilization phase can be dissected into smaller units. Based on the theory of action regulation, one can distinguish between several hierarchical layers of goal setting and task execution (Frese and Zapf 1994; Volpert 1982). The subphases presented in Table 1 are associated with goals set by the user at different levels of specificity. At the highest level of the model of action regulation, there is the principal goal of the individual (e.g., use of high pressure cleaner to clean garden wall), which may be divided into subservient goals (e.g., start operation of appliance). These subservient goals may be divided again into further subordinate goals (e.g., ensuring water supply, ensuring electricity supply).

At the lowest level of the model, visible sequences of action are described (e.g., connecting appliance to tap, plug appliance into socket). Visible actions are open to observation while the various levels concerned with goals are only accessible to methods relying on self-reports.

Task analysis allows the identification of those goals and actions that lead to environmental damage. After having identified critical behavior, an assessment of the environmental impact needs to be made, which allows a prioritization of areas most promising for intervention. The results of the task analysis will primarily feed into interface design.

4.3 Interface Design

The interface is the basis for interaction of the human with the machine and has therefore a strong influence on user behavior. Based on a number of fundamental design principles proposed by Norman (1988), we consider three concepts central to interface design: system feedback, affordances, and constraints. These concepts have inherent strengths and weaknesses and differ with regard to their main thrust for the improvement of user-product-system performance. Table 2 presents a summary of the qualitative evaluation of the concepts in the context of designing for ecological use, followed by a more detailed discussion of each concept.

TABLE 2   Principal Interface Design Measures to Influence User Behavior

	Primary thrust of intervention	Strength	Weakness
System feedback	Efficiency problem	• High levels of user control • Can improve system management skills	• Can be ignored by user • No prevention of intentional non-ecological behavior
Affordances	Efficiency problem	• High levels of user control • Implicit behavior guidance	• Limited to non-complex system features
Constraints	Effectiveness problem	• Prevents occurrence of nonecological behavior	• Reduced user acceptance due to restricted choice

4.3.1 System Feedback

The concept of system feedback refers here to the flow of information from the machine to the user. This may range from simple information sources such as the transparency of system elements (e.g., fuel gauge) to more sophisticated information displays (e.g., petrol consumption per km). Thus, system feedback subsumes related terms, such as visibility and transparency. Improving feedback facilities of a system is a primary measure in system design to improve human-system performance (Wickens 1992). It provides the user with information about the current state of the system. Poor performance is often due to the user not being able to determine the current state of the system or its environment (e.g., light indicating operating temperature of iron). Well-designed feedback facilities also affords learning opportunities to the user to improve system management strategies, as appropriate information is given to whether intended goal states have been attained. Enhanced feedback is, however, only beneficial if we are dealing with an efficiency problem (i.e., the user pursues ecologically relevant goals but does not succeed in achieving them). If there is an effectiveness problem (e.g., the user pursues no ecological goals), this measure is very limited in its power.

4.3.2 Affordances

The concept of affordance refers to perceived and actual properties of an object, which determine the way in which it may be used, that is, they give information about the operation of an object (Norman 1988). This is primarily done by labeling and design of controls. Within human factors research, the importance of adequate labeling and design of controls has been widely recognized. In particular, safety-critical systems (e.g., nuclear power plants, aircraft cockpit) have attracted a great deal of research to improve the design of controls so that the prevalence of unintended errors (e.g., slips) is minimized (Wiener and Nagel 1988; Woods et al. 1987). Household appliances (e.g., cooker) have also been examined in that context (Shinar & Acton 1978). However, there is little research on labeling and design of controls for ecological use. Nevertheless, based on some general principles, the designer needs to select labels for control settings where too high settings (i.e., those resulting in excessive energy consumption) produce negative connotations and appropriate settings generate positive connotations (e.g., "extremely dirty" vs. "slightly dirty" in case of a vacuum cleaner).

4.3.3 Constraints

Implementing constraints to limit the user's choice in interacting with a system is a very powerful measure for a designer to prevent undesired user behavior from occurring (Norman 1988). It allows the elimination of nonecological behavior, which is considered to have a particularly high negative environmental impact. While system feedback depends on the user's willingness to respond appropriately to the information provided, constraints do not require that the user pursue proenvironmental goals. Despite the high effectiveness of constraints in achieving behavioral modifications, the designer needs to be aware of possible risks associated with implementation. First, the product might not be bought because the user does not like the constraint. Second, the user does purchase the product but hampers with the product to remove the constraint. For example, this is often done with constraints implemented for safety reasons, as these frequently have the disadvantage of lowering work efficiency. Despite the overall effectiveness of constraints in modifying human behavior, remarkably little use has been made of this means in the field of ecological design.

4.4   Design of Support Materials and Services

This refers to the development of materials that facilitate the interaction of the user with the system, such as instruction manuals and training devices (Czaja 1997). This is, however, not limited to written materials but may also include different services that support the user. In the context of designing for domestic use, this is primarily the instruction manual and support services.

4.4.1   Instruction Manual

Even though instruction manuals appear only to be of moderate effectiveness as a means of influencing user behavior, it seems worthwhile to provide some information on ecological use of the product. The limited effectiveness is due to the fact that a considerable percentage of users do not read instructions. This is because users already feel familiar with the product (Gebert 1988; Sanders and McCormick 1992) or because manuals do not adhere to the basic principles of good design, discouraging the user from reading them thoroughly (e.g., easy language, avoidance of excess words, physical separation of points) (Bailey 1989; Oborne 1995). Despite their limitations, instructions are an important instrument of the designer to communicate with the user. Improving the quality of instructions (e.g., by following the principles of good design) will increase the likelihood of the instruction manual being read and understood. It may also be helpful to present advice about ecologically desirable user behavior under the angle of monetary savings (e.g., reduced energy consumption) rather than solely referring to environmental benefits.

4.4.2 Support Services

In the case of products for domestic use, the user may gain information about ecological product employment from two main sources: from the retailer during the sales process or from after-sales services (e.g., telephone helpline). At first sight, the involvement of sales staff appears useful because it allows for two-way communication and important points can be demonstrated to the buyer even if the buyer has not specifically asked for them. On the other hand, the success of this approach may depend too strongly on the willingness of the retail business to take this issue seriously and requires staff to be trained on environmental issues. At present, it appears that sales staff has very little knowledge of environmental issues (Dannheim 1999). The implementation of an after-sales services (e.g., telephone helpline) has the advantage of not having to rely on the retailer but is dependent on the willingness of the user to establish contact with the manufacturer. Furthermore, a telephone helpline may be quite costly to run. Overall, the effectiveness of support services for modifying human behavior is somewhat limited but may be useful under certain circumstances.

4.5 Guideline for Optimization of Machine-User Communication

After having presented a step-by-step guide to designing systems for ecological use, we present an approach that helps the designer to deal with information that needs to flow from the machine to the user.

The principles of communication between humans can also be applied to communication in the context of human-machine interaction. We may therefore make use of the basic communication formula of Laswell (1964): who says what, in which channel, to whom, and with what effect? In the context of our user-product system, the formula may be decomposed into the following elements: information content, organization of information, and presentation of information (Johannsen 1993).

4.5.1 Information Content

The first step for the designer is to determine the kind of information the user needs to achieve improvements in ecological performance. This involves the generation of a range of information sources that may influence user behavior toward a stronger ecological orientation. The data provided by the LCA-based analysis may be used to identify the critical areas. If energy consumption were the critical area, the important design issues would be different from a case where the excessive use of consumables (e.g., oil and tires) was the main problem. In the case of petrol consumption, the list of possible information to be presented may include present and historic data about energy consumption (e.g., mean consumption over previous 4 weeks), ecological performance variables (consumption related to performance), or appropriateness of strategies used (e.g., selected gears for chosen speed).

4.5.2 Organization of Information

A structured organization of the information presentation process ensures that information is provided at the required time and that subsystems interact such that information from different sources can be processed for presentation to the user (Johannsen 1993). In our case, it also refers to the question of which information carrier is to be used to provide information most effectively. Based on the previous analysis, we may distinguish between four information carriers: product, sales staff, instruction manual, and after-sales service. For example, the information that an appliance (e.g., lawn-mower) needs to be well maintained to operate efficiently (e.g., regular sharpening of blade) may be presented by different information carriers. The designer needs to decide which one is most effective for that purpose. However, the designer may use more than one carrier to produce some redundancy in the information transmission process. Table 3 summarizes the strengths and weaknesses of information carriers, providing some guidance to designers to make a good choice.

4.5.3 Presentation of Information

This refers to the best form of presentation of a piece of information, whether written in text form, graphical form, or in speech (Dillon and Pellegrino 1991; Johannsen 1993). The text or graphical information can be provided on hardcopy or display. The information may be static (e.g., color coding of optimal power setting) or dynamic (e.g., indication of current parameter level). The presentation of graphical information is often preferable to language-based information because it can usually be processed more quickly and does not require an understanding of the language. As a detailed discussion of the pros and cons of different display modes is beyond the scope of this chapter, the interested reader is referred to other sources (e.g., Sanders and McCormick 1992).

5 HUMAN FACTORS METHODS FOR ASSESSING USER BEHAVIOR

We have argued for the use of a user-centered approach to ecological design for a comprehensive LCA. As the prediction of user behavior is a central concern for designing products for ecological use, different human factors methods are presented for gaining a realistic assessment of the range of

TABLE 3 Strengths and Weaknesses of Different Information Carriers for Optimizing Machine-User Communication

Information carrier	Strength	Weakness	Estimated effectiveness of influencing behavior
Product	• Permanent accessibility of information • Close functional link between presented information and user behavior	• Limits for presenting complex information	High
Sales staff	• Providing information through demonstration • Conveying information prior to use	• Depending on motivation and expertise of sales staff	Low
Instruction manual	• Potentially accessible at any time • Support for identification of routine faults	• Often not read or misplaced	Moderate
After-sales service	• Support for identification of routine and nonroutine faults	• Depends on motivation of customer to initiate contact	Low

actual user behavior and its contribution to the overall EISc of the product. Once a task analysis has been completed (see section 4.2), its results help identify promising areas for human factors research.

5.1 Indirect Assessment Methods

5.1.1 Literature-Based Impact Assessment

The method of meta-analysis (e.g., Hedges and Olkin 1985) represents a scientifically sound approach, which may be used to determine how design features of interest affect relevant outcome variables. Up to the present date, meta-analysis has only rarely been used in the context of ecological behavior (e.g., Hines et al. 1986). Its utility is largely dependent on the quality of the research conducted ("garbage in–garbage out") and the degree of relevance to the present issues. A meta-analysis has the advantage of relying on a larger database than single studies and it requires less time for completion, as there is no need for data collection. However, in practice the analyst often encounters difficulties in that the reported data in research papers do not always provide the parameters needed for meta-analysis. Furthermore, one may be faced with the "apples and oranges" problem, that is, the meta-analysis looks at studies that are so different that they are not really comparable (Hunter and Schmidt 1990). Nevertheless, using knowledge gained from previous research may be a highly effective way of addressing current research questions, as it is a time-saving and economical exercise.

5.1.2 Human Factors Expert Assessment

If little research has been conducted in the area of interest or if very fast decisions need to be taken, one may use a panel of human factors experts to evaluate the impact of different design options. While this approach produces very fast results, the quality of the assessment strongly depends on the cumulative knowledge of the expert panel. It is advisable to use experts who have extensive experience of using human factors research methods. Sole reliance on experts without that experience might result in a too strong emphasis being placed on the designer's model of the product instead of user needs being considered. Otherwise, the goal of designing for ecological use may be jeopardized. Finally, one also needs to be aware of the cost implications if expertise is not available within the organization.

5.2 Direct Assessment Methods

Conducting empirical research is naturally preferable to nonempirical methods since it usually yields better results, though it is costly and time consuming and therefore may not always be an option for organizations. In human factors, one may distinguish between three research methodologies (Sanders and McCormick 1993): descriptive studies, experimental research, and evaluation research. Within each methodological approach, a number of methods are available for data collection: performance measures, physiological measures, observation, questionnaires, and interviewing. These are considered in turn after the principal differences between the methodological approaches have been outlined.

5.2.1   Empirical Research Methodologies

Descriptive Studies.   This kind of research is important in providing data about general population parameters, such as age, body size, ecological knowledge, and attitudes toward environmental issues. These data are the fundamental building blocks of the ecological design process. More descriptive studies take place in the field than in a laboratory. The drawback of descriptive studies is that they do not allow us to draw conclusions about cause-effect relationships. While descriptive studies can describe relationships between variables (e.g., personal norms and domestic energy saving behavior) (Black et al. 1985), the direction of the relationship cannot be determined (behavior influences personal norms or personal norms influence behavior).

Experimental Research.   In these studies, variables of interest are manipulated to examine the effects of certain design decisions on important measures, such as user behavior and performance. Experimental research may be conducted in the laboratory or in the field.

In laboratory-based studies, the critical elements of real operational environments are reproduced (e.g., by using a driving simulator), allowing the investigation of variables of interest under comparatively high levels of control. The laboratory setting may vary with regard to the fidelity of being able to model the operational environment (full-scale driving simulator vs. software running on PC). However, high fidelity is not always an advantage since the inherent complexity may distract from features of primary interest (Meister 1991). In the laboratory, one can examine levels of performance on tasks of different priorities gradients (primary and secondary tasks), system management strategies (control actions and information sampling behavior), subjective state measures (fatigue, effort expenditure, etc.), and psychophysiological parameters, such as heart rate variability (Wickens 1992). An example of how such a methodological approach may be used in research is provided in Hockey et al. (1998).

Similar methodological approaches may also be used in field studies, though the selection and manipulation of experimental variables is subject to more constraints. On the other hand, user behavior observed in field studies has generally a higher level of validity for real behavior than results obtained in the more artificial lab situation.

Evaluation Research.   While experimental work is concerned with generic research into the ecological behavior of humans to develop general guidelines, evaluation research is concerned with the evaluation of a particular system to increase its efficiency and effectiveness. Evaluation research may address two kinds of questions. First, it may determine whether a particular system satisfies a set of criteria. Second, it may carry out a comparative analysis of several systems to take an informed decision about selecting the best alternative.

An important form of evaluation research is *usability testing* (Nielsen 1997). The goal of this approach is to evaluate newly designed equipment with a group of users (i.e., individuals for whom it was actually developed). This permits the identification of design deficiencies before the product is released onto the market. The closer the match between participants in the usability testing sessions and the target population is (e.g., in terms of levels of expertise), the more valid the results will be. Originally developed in the context of human-computer interaction, principles of usability testing apply equally well to other products that involve some kind of user-product interaction.

Before a newly designed product is manufactured in large numbers, it is advisable to perform some human factors testing with the prototype to ensure than primary demands from the product specification list have been met. While it is easy to test whether "hard" criteria (e.g., technical specification criteria, such as engine performance) have been fulfilled, it requires more effort to evaluate the fulfillment of "soft" or behavior-related criteria (e.g., frequency of filter changes). It is acknowledged that human factors testing is costly and prolongs the product development process. Therefore, it should be limited to the verification of the most important design criteria. Human factors testing of the final product is necessary to reduce the probability of unwanted side effects.

5.2.2 Data Collection Methods

Performance Measures. This is a central aspect of system development, as the goal of many system design processes is to achieve improvements of user-product-system performance. To gain the most comprehensive picture possible of user-product interaction, it is advisable to collect a battery of performance variables to obtain performance data from critical performance aspects as well as subsidiary task elements. Performance measures may also include efficiency measures of task performance, which capture the strategies used to achieve certain performance levels. Analyzing efficiency measures may be critical in gaining an understanding of why there are differences in performance levels between users. In experimental settings one may assign priorities to different task elements, which allows a distinction between primary and secondary task performance (Proctor and Dutta 1995). Secondary tasks (i.e., low-priority tasks) may be used as a workload assessment technique to examine primary task performance, as secondary tasks are much more sensitive (and hence a better indicator) to variations in overall task load when comparing two technical systems (Wickens 1992).

Observation. The observation of people is a fundamental data collection method in human factors research. It is particularly relevant in field studies, as it allows examining user behavior in its full complexity in a natural environment. Depending on the circumstances, one may make use of different types of observation (Wilkinson 1995), such as casual observation (exploratory inspection of user behavior), formal observation (systematic data collection by using categories of behavior, etc.), and participant observation (the observer is part of the group). A video camera is often useful to make accurate recordings of user behavior, permitting a subsequent in-depth analysis. Furthermore, the analysis of videotaped observations combined with the interviewing technique leads to an enhancement of the methodological strength (see below).

Interviewing. The interviewing technique has the distinct advantage of being sufficiently flexible to address basically any research question of interest (Breakwell 1990). It enables the interviewer to respond to any issue raised by the respondent. An interview generally provides very rich data. However, this may be a disadvantage at times, too. If the data have to be processed in a very structured manner (e.g., content analysis), the time and resources needed may be considerable. Interviewing also requires adequate levels of skill, as untrained interviewers may produce too much bias in the data. Furthermore, they may have more difficulties in establishing sufficient rapport with the respondent, the basis for any successful interview. Interviewing is generally useful when dealing with behavior that is the result of a rational decision-making process and with low-frequency tasks (i.e., it would take a long time to observe these in natural settings). The use of the interviewing technique is more limited when referring to habitual behavior since users find it difficult to verbalize this (Rüttinger and Lasser 1998). In experimental settings, this limitation can be overcome by using a retrospective walk-through, that is, the user provides a verbal description of the behavior previously videotaped (Kirwan and Ainsworth 1992). This technique has proven to be very useful in the context of designing for ecological use (Rüttinger and Lasser 1998).

Questionnaires. In comparison with the interview, using a questionnaire is a more objective method of data collection. This is mainly due to higher standardization of the data collection process, in which "interviewer" effects are considerably reduced. It is a more economical instrument of data collection since it allows the use of larger samples. However, designing a good questionnaire is time consuming and requires considerable expertise.

Therefore, a number of important points need to be taken into account during the questionnaire design process to maximize quality parameters of the questionnaire study, such as reliability, validity, and return rate (Oppenheim 1992). The questionnaire format is suitable for capturing subjective responses of participants concerning their experience in using the product (e.g., comfort, anxiety, frustration, fatigue).

Physiological Measures. These allow the assessment of user workload by measuring various aspects of the activity of the peripheral and central nervous system. Physiological parameters provide an additional indicator of overall task demands, complementing self-report measures of stress. While physiological measures allow a continuous measurement of the parameter and do not interfere with the primary task, some of them may be physically obtrusive during data collection (Wickens 1992). Many different physiological measures have been used in research, such as heart rate variability, blood pressure, eye movements and pupillary response, skin conductance, breathing patterns, and hormones.

6 EVALUATION OF TRADE-OFF BETWEEN ECOLOGICAL AND COMMERCIAL REQUIREMENTS

This is a critical step in the design process, as it determines whether the product can be sold in sufficient numbers to yield a profit. There would be no benefits to the environment in developing a highly ecological product if it was not bought by customers to replace less ecological products.

Therefore, the designer needs to assess design options in terms of their impact on both criteria, the product's ecological value and its general attractiveness to the consumer. The designer may benefit from a model that describes the relationship between the two evaluation criteria. For this task, the designer requires the support from marketing to make these assessments.[‡] The task is to determine the level of congruence between ecological and marketing goals. As shown in Table 4, a distinction between five major outcomes of the evaluation may be made when comparing ecological value and product attractiveness. There are three types of congruence between ecological and marketing criteria. Congruence may be positive, negative, or neutral (i.e., no benefits and no damage), depending on whether it is desirable or not to implement the design element. However, any of these

---

[‡]While the designer benefits from the knowledge of the marketers about customer preferences, marketers are also likely to benefit from the collaboration since they may become more aware of the concept of ecological design. This may enable them promote those features better when devising marketing strategies.

TABLE 4 Result Matrix of Comprehensive Product Evaluation Combining Ecological and Commercial Criteria

Result of evaluation	Ecological value	Attractiveness to consumer	Examples of product features from model product "vacuum cleaner"
Congruent positive	+	+	Feedback device for state of filters and bag
Marketing dilemma	+	−	Low-energy motor
Congruent neutral	0	0	Form of filters and bags
Ecological dilemma	−	+	Booster
Congruent negative	−	−	Heavy weight

+, positive; 0, neutral; −, negative evaluation of criteria.

do not represent a problem for the designer because there is no conflict between ecological and commercial interests. More critical are cases where these interests are in conflict.

In the first case, the design feature is highly ecological but it is not appealing to consumers. This may be called a *marketing dilemma*. The low-energy motor for a vacuum cleaner is an example of this. While there is very little customer acceptance for this feature, there is little doubt about the ecological value of an engine of approximately 750 W, compared with current models of 1500 W or more (Dannheim 1999). Energy consumption would be reduced by half while the reduction in suction power would be negligible.

Conversely, in the second case a highly attractive product feature is ecologically undesirable. This may be termed an *ecological dilemma*. An example is the booster of a vacuum cleaner, which may be much appreciated by customers, though it leads to increases in energy consumption while its performance-enhancing effect is doubtful.

In the case of dilemmas, the *gain-volume trade-off* should be taken into account, too. This suggests that only moderate improvements (=low gain) of the ecological product value may still result in considerable benefits for the environment if the product enjoys large sales (=large volume). In contrast, a product that aims for substantial improvements (=high gain) of the ecological product value may be less beneficial if it sells only in small numbers. The product of the formula (ecological gain × sales volume) is therefore critical in assessing the overall ecological value of the product.

The designer also needs to be aware that many ecological features are so-called credence attributes. This goes back to a theoretical distinction between three types of product attributes: search, experience, and credence features (Darby and Karni 1973; Ford et al. 1988; Nelson 1970). Search attributes are those product features than can be inspected before purchase (e.g., automatic gear change). Experience attributes concern features that the consumer has already had experience of from previous use of a similar product (driver comfort). The last type, credence attributes, can neither be inspected nor can previous experience aid in assessing the feature (e.g., nontoxic materials). Since credence attributes are generally less prevailing in the purchase decision process and many ecological features belong to this category, the difficulties of promoting ecological features during the sales process become clear. One may strengthen the potential influence of credence qualities by using internationally recognised ecological quality certificates.

Research into consumer psychology of ecological design has shown that the relevance of ecological issues in consumer decision making may have been underestimated (Rüttinger 1999). While consumers did regard product properties such as longevity and ease of maintenance as highly important for their purchase decision, they subsumed these under properties of product quality rather than ecological properties. The studies suggested that consumers have a model of ecological product qualities that is different from the scientific model. Highly valued ecological product properties were typically associated with personal advantages of the consumer in terms of monetary benefits (e.g., longevity, availability of parts) and reduced health risks (e.g., use of nontoxic materials, no harmful emissions). Therefore, those ecological product properties considered important by the consumer were mainly related to the utilization phase.

7 CONCLUSION

In this chapter, we have argued for a stronger consideration of the utilization phase as a significant part of life cycle-based environmental impact assessments. This has been demonstrated in several LCAs in the field of household appliances. However, it is not sufficient to base an environmental impact assessment on nominal user behavior, as there is usually a considerable difference between nominal and actual behavior. Therefore, relying on a nominal model of user behavior is very likely to lead to an underestimate of the environmental impact of the utilization phase.

The use of an empirical model of user behavior allows us to make a more precise assessment of actual user-product interaction. For that purpose, various human factors research methodologies (e.g., experimental research) and data collection methods (e.g., observation) were presented. These may be used to explain and predict user behavior more accurately.

When evaluating design options with regard to their effect on ecological user-product-system performance, the assessment needs to consider two critical criteria that place ecological performance into a wider context. First, the expected gains from encouraging more ecological user behavior need to be judged against possible increases in other aspects of the environmental impact score. Second, the expected gains from encouraging more ecological user behavior need to be compared with the effects on the attractiveness of the product to the consumer.

The concepts, methods, and instruments outlined in this chapter provide the designer with effective tools for reducing the environmental impact of the user-product-system. The human factors approach represents an essential and promising complement to the advanced methods employed by engineers in carrying out LCA-based analyses.

REFERENCES

Atik, A., Schulz, H., Pant, R. and Jager, J. (1998). Method and computer aided software system for ecological evaluation of products accompanying to development process. Proceedings of the 5th CIRP seminar on Life Cycle Engineering, Stockholm, pp. 251–264.

Bailey, R. W. (1989). Human performance engineering. London: Prentice Hall.

Birkhofer, H. & Schott, H. (1996). Die Entwicklung umweltgerechter Produkte— eine Herausforderung für die Konstruktionswissenschaft. Konstruktion 48:386– 396.

Black, J.S., Stein, P.C. & Elworth, J.T. (1985). Personal and contextual influences on househould energy adaptations. J Appl Psychol 70:3–21.

Breakwell, G.M. (1990). Interviewing. London: Routledge.

Czaja, S.J. (1997). System design and evaluation. In G. Salvendy (Ed.), Handbook of human factors and ergonomics. New York: Wiley and Sons, pp. 17–40.

Dannheim, F. (1999). Die Entwicklung umweltgerechter Produkte im Spannungsfeld von Ökologie und Ökonomie. Düsseldorf: VDI.

Darby, M.R. & Karni, E. (1973). Free competition and the optimal amount of fraud. J Law Econ 16:67–88.

Dillon, R.F. & Pellegrino, J.W. (Eds.) (1991). Instruction: theoretical and applied perspectives. New York: Praeger.

Endsley, M.R. & Kiris, E.O. (1995). The out-of-the-loop performance problem and level of control in automation. Human Factors 37:81–394.

Ford, G.T., Smith, D.B. & Swasy, J.L. (1988). An empirical test of the search, experience and credence attributes framework. Adv Consumer Res 15:239–243.

Frese, M., & Zapf, D. (1994). Action as the core of work psychology: a German approach. In H.C. Triandis, M.D. Dunnette & L.M Hough (Eds.), Handbook of industrial and organizational psychology. Palo Alto, CA: Consulting Psychologists Press.

Gebert, D. (1988). Gebrauchsanweisungen als Marketinginstrument. Wiesbaden: Forbel.

Gebert, D. & von Rosenstiel, L. (1989). Organisationspsychologie. Stuttgart: Kohlhammer.

Hedges, L.V. & Olkin, I. (1985). Statistical methods for meta-analysis. New York: Academic Press.

Hines, J.M., Hungerford, H.R. & Tomera, A.N. (1986). Analysis and synthesis of research on responsible environmental behavior: a meta-analysis. J Environ Educ 18:1–8.

Hockey, G.R.J., Wastell, D. & Sauer, J. (1998). Effects of sleep deprivation and user interface on complex performance: a multilevel analysis of compensatory control. Human Factors 40:233–253.

Hunter, J.E. & Schmidt, F.L. (1990). Methods of meta-analysis. Newbury Park: Sage.

Johannsen, G. (1993). Mensch-Maschine-Systeme. Heidelberg: Springer.

Kirwan, B. & Ainsworth, L.K. (1992). A Guide to task analysis. London: Taylor & Francis.

Laswell, H.D. (1964).The structure and function of communication in society. In L. Bryson (Ed.), The communication of ideas. New York: Cooper Square.

Meister, D. (1991). Psychology of system design. Amsterdam: Elsevier.

Mose, A.-M., Wenzel, H. & Hauschild, M. (1997). Gram: refrigerators. In H. Wenzel, M. Hauschild & L. Alting (Eds.), Environmental assessment of products. Vol. 1. London: Chapman & Hall, pp. 319–368.

Nedermark, R., Wesnos, M. & Wenzel, H. (1997). Bang & Olufsen: Televisions. In H. Wenzel, M. Hauschild & L. Alting (Eds.), Environmental assessment of products.Vol. 1. London: Chapman & Hall, pp. 369–414.

Nelson, P. (1970). Information and consumer behavior. J Polit Econ 78:311–329.

Nielsen, J. (1997). Usability testing. In G. Salvendy (Ed.), Handbook of human factors and ergonomics. New York: Wiley and Sons, pp. 1543–1568.

Norman, D.A. (1988). The design of everyday things. New York: Currency and Doubleday.

Oborne, D.J. (1995). Ergonomics at work. New York: Wiley and Sons.

Oppenheim, A.N. (1992). Questionnaire design, interviewing and attitude measurement. London: Printer Publishers.

Pahl, G. & Beitz, W. (1996). Engineering design: a systematic approach. London: Springer.

Proctor, R.W. & Dutta, A. (1995). Skill acquisition and human performance. London: Sage.

Rüttinger, B. (1999). Umweltorientierte Produktbeurteilung und Kaufentscheidung. In G. Krampen, H. Zayer, W. Schönpflug & G. Richardt (Eds.), Beiträge zur angewandten Psychologie. Bonn: Deutscher Psychologen Verlag, pp. 241–243.

Rüttinger, B. & Lasser, M. (1998). Markt- und Nutzungsaspekte der Entwicklung umweltgerechter Produkte. In R.W. Scholz & A. Heitzer (Eds.), Erfolgskontrolle von Umweltmaßnahmen. Heidelberg: Springer.

Sand, A., Sørensen, A. & Caspersen, N. (1997). KEW Industri: high pressure cleaners. In H. Wenzel, M. Hauschild & L. Alting (Eds.), Environmental assessment of products. Vol. 1. London: Chapman & Hall, pp. 415–449.

Sanders, M.S. & McCormick, E.J. (1992). Human factors in engineering and design. New York: McGraw-Hill.

Sauer, J. & Rüttinger, B. (2000). A new framework for the design of ecological domestic appliances: design-centred product development. Proceedings of the 14th Triennial Congress of the International Ergonomics Association and the 44th Annual Meeting of the Human Factors and Ergonomics Society in San Diego, California, July 30–August 4, 2000. Santa Monica, CA: HFES.

Schmidt-Bleek, F. & Tischner, U. (1995). Produktentwicklung: Nutzen gestalten—Natur schonen. Wien: WIFI Österreich.

Sharit, J. (1997). Allocation of functions. In G. Salvendy (Ed.), Handbook of human factors and ergonomics. New York: Wiley and Sons, pp. 301–339.

Shinar, D. & Acton, M.B. (1978). Control-display relationships on the four burner range: population stereotypes versus standards. Human Factors 20:13–17.

Volpert, W. (1982). The model of the hierarchical-sequential organization of action. In W. Hacker, W. Volpert & M. Cranach (Eds.), Cognitive and motivational aspects of action. Berlin: Hüthig, pp. 266–277.

Wenzel, H., Hauschild, M. & Alting, L. (1997). Environmental assessment of products. Vol. 1. London: Chapman & Hall.

Wickens, C.D. (1992). Engineering psychology and human performance. New York: Harper-Collins.

Wiener, E.L. & Nagel, D.C. (Eds.) (1988). Human factors in aviation. London: Academic Press.

Wilkinson, J. (1995). Direct observation. In G.M. Breakwell, S. Hammond & C. Fife-Schaw (Eds.), Research methods in psychology. London: Sage.

Woods D.D., O'Brien, J.F. & Hanes, L.F. (1987). Human factors challenges in process control: the case of nuclear power plants. In G. Salvendy (Ed.), Handbook of human factors. New York: Wiley and Sons, pp. 1724–1770.

13

Modeling and Control for Environmentally Conscious Design and Manufacturing

**Panicos Nicolaou, Donna Mangun, and
Deborah L. Thurston**
University of Illinois at Urbana-Champaign, Urbana, Illinois

1 INTRODUCTION

Designers no longer throw their finished product design "over the wall" to let manufacturing engineers deal with quality control. With increasing demands for pollution prevention and impending product take-back legislation, neither can they throw products over the wall for others to deal with environmental impacts. Planning for the management of life cycle environmental impacts calls for ever more precise control of each life cycle stage, ranging from raw material selection through manufacturing, customer use, disassembly, reuse, and recycling. This requires design engineers to employ more sophisticated and complex analytic approaches to concurrent engineering which integrate cost, quality, and environmental impacts. This chapter first describes a general, domain-independent mathematical model for design and manufacturing process analysis. Two examples illustrate the model. The first focuses on product design and the second on manufacturing process design, although the model can be used to analyze both concurrently. The first example addresses the problem of product take-back requirements, proposing a multiple take-back and component reuse model for long-range product portfolio planning. The second example addresses machining processes and illustrates the effect of cost, quality, and environmental trade-offs in determining optimal dry and wet machining parameters.

2 DOMAIN-INDEPENDENT MULTICRITERIA OPTIMIZATION MODEL

The general form employed here is multicriteria optimization. Figure 1 shows a "house of quality" type matrix organization, illustrating the interplay between evaluation criteria listed on the vertical axis and engineering design decision variables along the horizontal axis.

Initially, the designer seeks to optimize several objectives x_i, such as minimize pollution, maximize quality, and minimize manufacturing cost. This is where analytic design expertise is employed to identify design specifications that optimize each objective, driving the iterative design process to define the Pareto optimal frontiers (Thurston 1991). However, the designer is often thwarted in the attempt to simultaneously optimize all objectives, since the decisions that improve one objective can worsen another. For example, substituting plastic materials for steel might improve cost but can worsen recyclability. When the Pareto optimal frontier is reached, all opportunities for improving one objective without worsening another have been exploited. Here we make an important semantic distinction; once we reach the Pareto optimal frontier, minimizing pollution is no longer an objective. Instead, maximizing some function of attributes x_i, which includes pollution, becomes the objective, as shown in Eq. 1.

$$\text{Maximize } f(x_1, x_2, \ldots, x_n) \tag{1}$$

Several methods for determining the form of the objective function exist, including weighted averages, the analytic hierarchy process (Saaty 1980), and multiattribute utility analysis (Keeney and Raiffa 1993). The reader is referred to (Thurston 1991) for a discussion of the advantages and disadvantages of each approach within the context of engineering design.

The relationships between the engineering design decision variable vector $\mathbf{y_j} = (y_1, \ldots, y_m)$ and the resulting effect on each of the attributes x_i can be expressed as constraint functions $\mathbf{g} = (g_1, \ldots, g_n)$ shown in Eqs. 2–4.

The table below ("House of Quality") maps Multiple Evaluation Criteria (rows) against Engineering Design Decision Variables (columns). Columns are grouped by life‑cycle stage.

Engineering Design Decision Variables →	Raw Materials / Material Selection	Raw Materials / Process Selection	Manufacturing	Assembly	Customer Use	Product Take back & Disassembly / Collection Method	Disassembly Method	Solvent type	Amount of water used	Reuse / Cleaning	Reuse / Assembly	Reuse / Customer Use	Reuse Take back & Disassembly / Collection Method	Disassembly Method	Energy use	Process used	Shredding & meltdown	Recycle Manufacturing / Material Selection	Process Selection	Recycle / Assembly	Recycle / Customer Use	Recycle Take back & Disassembly / Collection Method	Disassembly Method	Disposal / Landfill	Disposal / Incineration
Cost																									
Cost of new materials	-	X	-	-	-	-	-	-	-	-	-	-	-	-	-	-	-	-	-	-	-	-	-	-	-
Cost to (re)manufacture parts	-	X	X	-	-	-	X	-	-	-	-	-	X	-	X	X	X	-	-	-	X	-	-	-	-
Cost to assemble parts	-	X	-	X	-	-	-	-	-	X	-	-	-	-	X	-	X	-	-	-	-	-	-	-	-
Product take back costs	-	-	-	-	-	X	-	-	-	-	-	X	-	-	-	-	-	-	X	-	-	-	-	-	-
Disassembly costs	-	X	X	X	-	-	X	-	-	X	-	-	X	-	-	X	X	X	-	-	X	-	-	-	-
Cleaning costs	-	-	-	-	-	-	-	X	X	-	-	-	-	-	-	-	-	-	-	-	-	-	-	-	-
Shredding & Melt down costs	-	-	-	-	-	-	-	-	-	-	-	-	-	X	X	-	-	-	-	-	-	-	-	-	-
Disposal costs	-	X	-	-	-	-	-	X	-	-	-	-	-	-	-	X	-	-	-	-	-	X	X	-	-
Energy costs	-	X	X	X	-	-	X	-	X	X	-	-	X	X	X	X	X	-	X	-	X	-	-	-	-
Reliability																									
1st life cycle	-	X	-	-	X	-	-	-	-	-	-	-	-	-	-	-	-	-	-	-	-	-	-	-	-
2nd & Subsequent life cycles	-	X	X	-	-	-	-	X	-	-	X	-	-	-	X	X	X	-	X	-	-	-	-	-	-
Environmental Impact																									
Air Pollution	-	X	X	X	-	X	X	-	-	X	-	X	X	X	X	X	X	X	-	X	X	-	X	-	-
Wastewater	-	-	X	-	-	-	-	X	-	-	-	-	-	-	X	-	X	-	-	-	-	-	-	-	-
Solid waste	-	X	X	-	-	-	X	-	-	-	-	X	-	X	X	X	-	-	-	X	X	-	-	-	-

FIGURE 1   House of quality.

$$x_1 = g_1(y_1, \dots, y_m) \tag{2}$$

$$x_2 = g_2(y_1, \dots, y_m) \tag{3}$$

$$\vdots \qquad \vdots \qquad \vdots$$

$$x_n = g_n(y_1, \dots, y_m) \tag{4}$$

For a linear model, these constraint functions can be written as a matrix where a_{ij} are the coefficients of $g(y) = x$, and a_{i0} are constants as shown in Eq. 5.

$$
\begin{bmatrix} a_{10} \\ a_{20} \\ a_{30} \\ \vdots \\ a_{n0} \end{bmatrix} + \begin{bmatrix} a_{11} & a_{12} & a_{13} & a_{14} & \cdots & a_{1m} \\ a_{21} & a_{22} & a_{23} & a_{24} & \cdots & a_{2m} \\ a_{31} & a_{32} & a_{33} & a_{34} & \cdots & a_{3m} \\ \vdots & \vdots & \vdots & \vdots & \cdots & \vdots \\ a_{n1} & a_{n2} & a_{n3} & a_{n4} & \cdots & a_{nm} \end{bmatrix} \cdot \begin{bmatrix} y_1 \\ y_2 \\ y_3 \\ \vdots \\ y_n \end{bmatrix} = \begin{bmatrix} x_1 \\ x_2 \\ x_3 \\ \vdots \\ x_n \end{bmatrix} \tag{5}
$$

This general formulation provides a structure into which a great breadth of analytic tools can be placed. For example, in Carnahan and Thurston (1998) the results of a statistical manufacturing process control experiment were used to define a set of constraints to predict (and thus control) manufacturing cost, quality, and hazardous air pollution on a floor tile manufacturing line. The first example in this chapter illustrates how reliability analysis can similarly be employed to define constraints that predict (and thus control) product failure rates when components are reused from one product generation to the next. The second example integrates activity-based cost analysis and statistical quality control analysis.

Realistically, trade-offs are most often allowed or feasible only within a certain range, and the additional constraints shown in Eqs. 6 and 7 must also be determined, where x_{iw} is the worst tolerable and x_{ib} is the best achievable, defined as in Thurston (1991).

$$
x_i \geq x_{iw} \quad \text{for each } i = 1, \ldots, n \tag{6}
$$

$$
x_i \leq x_{ib} \quad \text{for each } i = 1, \ldots, n \tag{7}
$$

3 CLOSED-LOOP PRODUCT PORTFOLIO DESIGN

Legislation that holds manufacturers financially responsible for their products at the end of their useful lives is in various stages of being enacted in Europe, Japan, and the United States. As such legislation approaches, it is in the best interests of manufacturers to plan ahead for product take-back during the initial stages of product design. Thoughtful material selection, component specification, manufacturing, and assembly process design can increase the economic viability of disassembly and component reuse at the end of one product life cycle. This example provides an analytic framework for such an analysis.

3.1 Closed-Loop Life Cycle

A reduction of waste and raw materials is accomplished by creating a closed-loop product life cycle. This means that at the end of its service a product is taken back, is disassembled, and each component is reused, recycled, or disposed of. An example of a closed-loop life cycle is shown in Figure 2.

Reusing or recycling product components sometimes results in additional benefits such as increasing profits or improving competitiveness by appealing to "green" consumers. However, this is not always the case. Disassembly and reuse of components instead can increase costs and lower product reliability. Consequently, design engineers need a tool that will allow them to individually analyze each product component to determine whether it should be reused, recycled, or disposed of. In addition, long-range product design planning over several take-back periods must be performed to maximize the residual value of reused components.

3.2 Analyzing a Portfolio of Products

Traditionally, companies that have incorporated environmentally conscious design and manufacturing and product take-back planning into product design have tried to simultaneously meet the demands of all its customer groups with the development of a single product. However, for any given product there are multiple customer groups displaying different levels of environmental consciousness and therefore having different demands for the products they purchase. To ensure maximum customer satisfaction, it is necessary to consider each customer group's demands individually by adopting a portfolio approach to meet all these demands most efficiently. This approach allows a specific product within the portfolio to be tailored toward a specific customer group.

3.3 Long-Range Product Portfolio Model

A spreadsheet-based decision tool has been developed to aid design and manufacturing engineers in the development and long-range planning of a product portfolio (Mangun and Thurston 2000). Cost, quality, environmen-

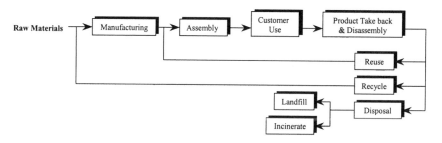

FIGURE 2 Closed-loop product life cycle.

tal impact, and customer satisfaction for multiple customer groups are considered in the multiobjective function shown in Eq. 8. Decision variables indicate when each product component should be reused, recycled, or disposed of throughout multiple life cycles.

$$\max U(\mathbf{X}) = \sum_{l=1}^{r} \sum_{p=1}^{z} U(X_p)_l \tag{8}$$

subject to

$$Y_{1,i} + Y_{2,i} + Y_{3,i} = 1 \quad \text{for each } i = (1, \ldots, k) \tag{9}$$

$$U(X_p)_l = \frac{1}{K} \left[\prod_{j=1}^{3} (Kk_j U(X_j) + 1) - 1 \right] \quad \text{for each } p = (1, \ldots, z) \tag{10}$$

$$X_{1,p,l} = \sum_{i=1}^{k} \left[Y_{1i} \left(C_{8i} + \sum_{n=1}^{5} C_{ni} \right) + Y_{2i} \left((2 - Z_i) \right.\right.$$
$$\left.\left. \sum_{n=3}^{6} C_{ni} \right) + Y_{3i} \left(C_{7i} + \sum_{n=2}^{5} C_{ni} \right) \right] \tag{11}$$

$$X_{2,p,l} = \prod_{i=1}^{k} \{ (Y_{1i} + y_{2i}) R_i(t) + Y_{2i} [Z_i R_i(t) + (1 - Z_i)$$
$$\left[1 - (1 - R_i(t))^2 \right]] \} \tag{12}$$

$$R_i(t) = (Y_{1i} + Y_{3i}) e^{-(t/\theta_i)^b} + Y_{2i} e^{-(u_i t/\theta_i)^b} \tag{13}$$

$$X_{3,p,l} = \sum_{i=1}^{k} \left[Y_{1i} \left(E_{6i} + \sum_{m=1}^{3} E_{mi} \right) + Y_{2i} \left((2 - Z_i) \sum_{m=2}^{4} E_{mi} \right) \right.$$
$$\left. + Y_{3i} \left(E_{5i} + \sum_{m=1}^{3} E_{mi} \right) \right] \tag{14}$$

$$E_{mi} = \frac{A_{\max} - A_{mi}}{A_{\max} - A_{\min}} + \frac{S_{\max} - S_{mi}}{S_{\max} - S_{\min}} + \frac{W_{\max} - W_{mi}}{W_{\max} - W_{\min}} \tag{15}$$

$$C_{p,\min} \le X_{1,p} \le C_{p,\max} \tag{16}$$

$$R_{p,\min} \le X_{2,p} \le R_{p,\max} \tag{17}$$

$$E_{p,\min} \le X_{3,p} \le E_{p,\max} \tag{18}$$

where:

$i = (1, \ldots, k)$ components for product p

$j = (1, \ldots, 3)$ attributes
$p = (1, \ldots, z)$ products in the portfolio, one for each customer group
$l = (1, \ldots, r)$ life cycles being modeled
$U(x_p)_l$ = multiattribute utility of product p with respect to attributes j
  for life cycle l
$U(X_j)$ = total utility of attribute j
$X_{1,p,l}$ = total cost in dollars of product p summed over all components
  i for life cycle l
$X_{2,p,l}$ = total reliability of product p for life cycle l measured as percent
  of components that fail before the expected lifetime
$X_{3,p,l}$ = environmental impact of product p for life cycle l summed
  over all components i as a normalized value
$Y = (Y_1, Y_2, Y_3)$ = decision variables
$Y_{1i} = 1$ if component i is new, 0 otherwise
$Y_{2i} = 1$ if component i is reused, 0 otherwise
$Y_{3i} = 1$ if component i is recycled, 0 otherwise
K = normalizing parameter
k_j = independent scaling factor for attribute j
$n = (1, \ldots, 8)$ categories of cost
t = operating time in hours = $x \cdot d \cdot h$
d = days a year product p is used
h = hours a day product p is used
x = length of time until product p is taken back
u_i = number of life cycles component i has been reused
$R_i(t)$ = reliability, or probability that component i will not fail by time
  t
θ_i = characteristic life of component i (hours)
b = slope of the Weibull distribution curve
$Z_i = 0$ if an additional component is added in parallel with a reused
  component, 1 otherwise
$m = (1, \ldots, 6)$ categories of environmental impact
E_i = overall environmental impact incurred for component i
S_{max} = maximum allowable level for solid waste
S_{min} = minimum achievable level for solid waste
S_i = measured solid waste level
A_{max} = maximum allowable conc. for air pollution
A_{min} = minimum achievable conc. for air pollution
A_i = measured air pollution concentration
W_{max} = maximum allowable conc. for wastewater
W_{min} = minimum achievable conc. for wastewater
W_i = measured wastewater concentration
$C_{p,min}$ = minimum cost for product p

$C_{p,\max}$ = maximum allowable cost for product p
$R_{p,\min}$ = minimum allowable reliability for product p
$R_{p,\max}$ = maximum reliability for product p
$E_{p,\min}$ = minimum environmental impact for product p
$E_{p,\max}$ = maximum allowable environmental impact for product p

3.3.1 Constraints

Cost, reliability, and environmental impact are related to the decision variables in Eqs. 11, 12, and 14, which serve as constraints on the objective function. An additional set of constraints in Eqs. 16, 17, and 18 account for customer satisfaction.

Cost X_1. Cost is defined to include all manufacturing costs. This can further be broken down into eight distinct categories:

- C_1 = Cost of new materials
- C_2 = Cost to manufacture or remanufacture parts
- C_3 = Cost to assemble
- C_4 = Product take-back costs
- C_5 = Disassembly costs
- C_6 = Cost to clean parts
- C_7 = Cost of shredding and melting down used parts
- C_8 = Cost of disposal

The total cost for one product p over a single lifetime l is shown in Eq. 11.

Reliability, X_2. Reliability is defined as the probability that a component will not fail within its expected lifetime. As shown in Eq. 19, reliability is a function of operating time, and therefore reliability of a component is decreasing over time (O'Connor 1985).

$$R_i(t) = e^{-(t/\theta_i)^b} \tag{19}$$

where

$R_i(t)$ = reliability, or probability that component i will not fail by time t
t = operating time (hours)
θ_i = characteristic life of component i (hours)
b = slope of the Weibull distribution curve

Typically, there are three stages for a component lifetime: early failures, chance failures, and wearout failures. This is illustrated in Figure 3,

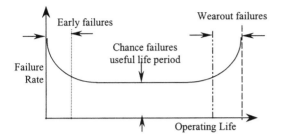

Figure 3 Component failure rate as a function of operating life.

which plots failure rate for a population of components against component lifetime, commonly known as a Weibull distribution (O'Connor 1985).

As can be seen in Figure 3, the failure rate is constant during most of the useful life of a component. Therefore, the slope of the Weibull distribution curve, b, is taken to be 1 during the useful life of the component. For the purpose of this example, manufacturing defects and early life failures are not included. In addition, wearout failures have not been modeled because the tolerable attribute range constraint (Eq. 17) is defined such that the wearout stage will not be reached.

Reliability of a system is dependent on whether the components in the system are connected in series or parallel, shown in Figures 4 and 5, respectively. The reliability of components connected in series is simply the product of each component's reliability. For the purposes of this model, it is assumed that all components are initially connected in series. The reliability of components connected in parallel is shown in Eq. 20.

$$R_{\text{sys}} = 1 - (1 - R_1(t))(1 - R_2(t)) \tag{20}$$

Introducing a redundant component in parallel with a reused component can increase the reliability of the system. For example, if a system initially consists of one component with 90% reliability, then the system reliability is 90%. If an additional component with the same reliability as the first component is added in parallel, using Eq. 20, the reliability becomes 99%.

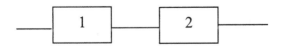

Figure 4 Components connected in series.

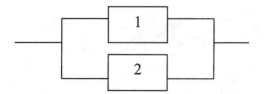

FIGURE 5   Components connected in parallel.

A decision variable, Z_i is included in the model to allow the decision maker to determine if an additional reused component should be added in parallel to increase system reliability. Therefore, the total reliability of product p at the end of the second and subsequent time periods is given by Eq. 12. Since an additional component will also alter the total cost and total environmental impact of the product, the decision variable, Z_i is also included in Eqs. 11 and 14 to account for this change.

The variables d and h are assumed to be equal among the customer groups. However, the product take-back time, x, is a decision variable determined by the model for each product p and is constrained to be an integer value greater than 0 years. As shown in Eq. 13, if a part is recycled or composed of new material, its reliability at the end of the second time period will be the same as its reliability at the end of the first time period. However, a reused component with minor repair or refurbishment will have a lower reliability at the end of the second time period.

Environmental Impact, X_3.   To measure the total environmental impact of a product or its parts, it is necessary to measure it over all media. The environmental impact attribute can be divided up into six categories that apply to solid waste, air pollution, and wastewater streams:

E_1 = environmental impact created from manufacturing and remanufacturing operations

E_2 = environmental impact created from assembly operations

E_3 = environmental impact created from disassembly operations, including disposal if applicable

E_4 = environmental impact created from cleaning operations for reused components

E_5 = environmental impact created from "melt down" operations for recycled components

E_6 = environmental impact created from disposal

The total environmental impact for product p, over two time periods, is described by Eq. 14. Since environmental impact is measured over different media, it is important to convert the values to a uniform scale, as shown in Eq. 15. Each category of pollution for each media is scaled from 0 to 1. Since there are six categories of environmental impact and three different media, each component has a total possible environmental impact score of 18. This is admittedly a gross oversimplification of the very complex problem of evaluating trade-offs between impacts on different environmental media. Weighting factors could be assigned for each media to improve on this method, as is commonly employed in commercially available life cycle analysis packages.

Incorporating Customer Preferences. Since customer response is a driving force behind product design, the decision tool also incorporates a method of including customer preferences explicitly within the design process. The model assumes that in a portfolio of products, each product will be geared toward one specific customer group. Equations 16, 18, and 19 allow each customer group to determine the minimum and maximum acceptable values for cost, reliability, and environmental impact, thereby including their preferences in the design process.

Each customer group's acceptable attribute range is used to assess the utility function for that attribute. The independent scaling factors, k_j, and the normalizing parameter, K, are determined and then the multiattribute utility function is found using Eq. 10. For a complete discussion of multiattribute utility theory, refer to Keeney and Raiffa (1993).

3.3.2 Example: Portfolio of Computer Products

The model is illustrated on a line of computers. The product line is made up of three separate models designed to appeal to the product's three distinct customer groups: the "environmentalists," who are willing to pay more for a product with the lowest environmental impact and perhaps lower reliability; the "family" group, who would like to do their part for the environment but do not want it to interfere with the cost or the reliability of the product; and the "executive" group, who demands the highest reliability in their products with the latest features and are willing to pay more for them but are not concerned about environmental impact.

For our purposes, the computer consists of six components: the monitor, steel components, plastic components, metal wire, printed circuit boards, and mixed metals. Each component is modeled for two time periods. The first time period includes the acquisition of raw material, manufacture, and assembly of the product with new material for all the components. The second time period begins when the product is taken back and disassembled and ends when the product has been fully reassembled. Cost data, characteristic life of each component, and environmental impact data are estimated.

To calculate the multiattribute utility function for each product, each customer group's attribute range is assessed. These values are shown in Table 1. Once the data are entered into a spreadsheet, Eqs. 8 through 20 are entered into a linear programming solver.

Table 2 shows the results of the traditional approach of designing one product from all new materials to meet the needs of all customers. This results in a product whose cost is too high for the family or environmental customer group and whose reliability is too low for the executive customer group to find acceptable.

Another option would be to design the product using all reused components. The resulting cost, reliability, and environmental impact values from designing a product with all reused components are shown in Table 3. Although the cost and the environmental impact of the computer shown in Table 3 are acceptable to the three customer groups, the reliability is too low for both the family group and the executive group. Introducing reused components in parallel with some of the existing components would increase the product reliability, however, that would raise the cost and environmental impact levels. In this situation, the model would be helpful in determining the trade-offs associated with introducing additional components into the system.

Another alternative to meet the demands of the three customer groups with the design of one product is to use all recycled components. These results are shown in Table 4. The computer shown in Table 4 would not satisfy any of the customer groups because its cost is too high. In addition, its high environmental impact would not satisfy the environmentalists and the low reliability would not satisfy the executives.

None of the products shown in Tables 2, 3, and 4 can meet the needs of all three customer groups simultaneously. However, by using the model presented in this chapter to design a portfolio of products, each customer

TABLE 1   Attribute Ranges

Customer group	Cost ($)	Reliability (%)	Environmental impact
Environmental group	850–1100	85–100	5.5–7.5
Family group	850–1000	90–100	5.5–9
Executive group	1000–1200	95–100	7–12

TABLE 2 Traditional Design: One All
New Product for All Customers

	All new components
No. of new	6
No. of reused	0
No. of recycled	0
No. of redundant	0
Take-back time	2 years
Cost	$1162.48
System reliability	93.2%
Env. impact	5.63
Total portfolio utility	**1.58**

group's needs will be met. The results of running the model are shown in Table 5. By employing the optimization model as shown in Table 5, the environmental customer group, who demands products with low environmental impact, will be satisfied by product 1; the family customer group, who wants a low-cost reliable product, will be satisfied by product 2; and the executive customer group, who demands high reliability products, will be satisfied by product 3. Additionally, the take-back time is adjusted for each product to better meet the needs of each customer group. The final row in Table 5 shows the resulting total utility of 2.74 realized by the three customer groups if that portfolio of products were designed. Comparing that utility with the utility in Tables 2, 3, and 4 (1.58, 1.71, and 0.49, respectively)

TABLE 3 Reuse: One All Reused
Components Product for All
Customers

	All reused components
No. of new	0
No. of reused	6
No. of recycled	0
No. of redundant	0
Take-back time	2 years
Cost	$852.82
System reliability	86.8%
Env. impact	6.28
Total portfolio utility	**1.71**

TABLE 4   Recycled Components:
One Product for All Customers

	All recycled components
No. of new	0
No. of reused	0
No. of recycled	6
No. of redundant	0
Take-back time	2 years
Cost	$1322.14
System reliability	93.2%
Env. impact	8.65
Total portfolio utility	**0.49**

TABLE 5   Combination of New, Reused, and Recycled Components for Portfolio of Three Products

	Product 1	Product 2	Product 3
Results of decision model	Environmental group	Family group	Executive group
No. of new components	2	1	4
No. of reused components	4	5	2
No. of recycled components	0	0	0
No. of redundant components	3	0	3
Take-back time (year)	1	1	1
Cost	$904.73	$841.20	$1143.01
System reliability	96%	93.9%	98.3%
Environmental impact	5.67	6.26	8.52
Total portfolio utility		**2.74**	

shows that implementing the model to design a portfolio of products improves the total utility over designing a single product to meet the needs of the three customer groups.

4   COST MINIMIZATION MODEL IN MACHINING PROCESSES

The machining process can play a significant role in determining product manufacturing cost, quality, and impact on the environment. The model developed here focuses first on machining decisions and their impact on cost. Additionally, it takes into consideration quality and environmental restrictions.

Initially, all decision variables that might impact product performance attributes are considered. The qualitative house of quality (HOQ) serves as a base for a mathematical model. The matrix of relationships between product attributes and engineering decision variables indicates many influencing factors. It is necessary to reduce the number of rows and columns so that only the most important elements remain. This has been done by a reduction method, based on four operations: combining, eliminating, separating, and redefining of the product performance attributes and the engineering decision variables (Locascio and Thurston 1998). The result of this process is the attribute set shown in the reduced HOQ, shown in Figure 6.

Product Performance Attributes Xi			General Production parameters				Feature parameters			Cutting parameters			Machining process							Lubrication parameters		
			Program life	Annual prod. volume	Material choice	Physical volume	Depth of hole	Diameter of hole	End mill diameter / Drill length	Cutting speed / spindle speed	Feed rate	Depth of cut	Tool nose radius	Cutting angles	Tool material	Square Footage	Number of stations	Machine selection	Load orientations	Type of cutting fluid	Cutting fluid concentration	Disposal method
Cost	X1																					
Variable Costs	X11	Casting unit costs	-	-	+	+	-	-	-	-	-	-	-	-	-	-	-	-	-	-	-	-
	X12	C-machining	-	-	+	-	+	+	-	-	+	+	-	+	-	+	+	+	+	-	-	-
	X13	C-coolant	-	-	-	-	+	+	-	+	+	-	-	-	-	-	+	-	-	+	+	+
	X14	C-tool	-	-	+	-	+	+	+	+	+	+	+	-	+	-	-	-	-	+	+	-
	X15	C-material	-	-	+	+	-	-	-	-	-	-	-	-	-	-	-	-	-	-	-	-
Fixed Costs	X16	Init. Inv. Equipment	+	+	-	-	-	-	-	-	-	-	-	+	+	+	+	+	-	-	+	+
	X17	C-setup	-	-	-	-	-	-	-	-	-	-	-	+	+	+	+	+	-	-	+	+
	X18	C-overhead	+	+	-	+	-	-	-	-	-	-	-	+	+	+	-	+	-	-	+	+
Quality	X2																					
Casting	X21	No cracks	-	-	+	+	-	-	-	-	-	-	-	-	-	-	-	-	-	-	-	-
Milled Surfaces	X25	Surface finish	-	-	+	-	-	+	+	+	+	+	+	+	-	-	+	-	-	+	+	-
	X26	Surface flatness	-	-	+	-	-	+	+	+	+	+	+	+	-	-	+	-	-	-	-	-
Environment	X3																					
	X31	FOG	-	-	-	+	-	-	+	+	+	-	-	+	-	+	+	-	+	+	+	
	X32	BOD	-	+	+	+	+	-	-	+	+	+	-	-	-	-	-	-	+	+	+	
	X33	Worker Risk	-	+	-	-	-	-	-	+	-	-	-	-	-	-	-	-	+	+	+	
	X34	Resource Depletion	+	+	-	+	-	-	-	+	+	+	+	+	-	-	-	-	+	+	+	
Weight	X4		-	-	+	+	-	-	-	-	-	-	-	-	-	-	-	-	-	-	-	-

+ Engineering parameters related to customer performance attributes
− Engineering parameters not related to customer performance attributes

FIGURE 6 The reduced house of quality.

4.1 Cost Model

The cost estimation model includes casting, machining, and cutting fluid costs. For casting and machining, cost is estimated through an activity-based approach, which distinguishes between unit, batch/product, and facility level activities. Different cost drivers serve as the basis for each activity (Cooper and Kaplan 1991). The structure of the cost estimation model presented here reflects attributes of each product type (Schreiner 1999). Cutting fluid cost estimation (Berry 2000) was mainly based on information taken from the U.S. Environmental Protection Agency (1983a, 1983b, 1995). Specifically the cost structure is as follows:

1. Casting costs
   - Unit level costs for casting (machining and material costs)
   - Product/batch level costs for casting (tooling costs)
   - Facility level costs for casting
2. Machining processes costs
   - Unit level costs for operations (machining and tooling costs)
   - Product/batch level costs (setup and investment costs for the equipment)
   - Facility level costs for machining
3. Cutting fluid costs
   - Fluid system costs
   - Recycling system costs
   - Industrial wastewater pretreatment costs
   - Disposal costs

4.2 Quality Model

Quality estimation models for machining processes can be developed based on multiple regression analyses of experimental data. Here we consider only milling operations. The reader is referred to Nicolaou and Thurston (2000) for a more detailed discussion of a drilling machining model.

The milling model (Urlichs 1999) was developed based on the end-milling simulation model (EMSIM), developed by the Machine Tool Agile Manufacturing Research Institute at the University of Illinois (website: mtamri.me.uiuc.edu). The model calculates minimum, average, and maximum surface error. Surface error is defined as the deviation of the finished machined surface from the desired surface in the Y direction (Figure 7) that would be produced with a completely rigid system of cutter, work piece, and fixture. The surface error is not necessarily uniform over the machined surface (i.e., in direction of axial depth of cut). Both the average and maximum surface error are estimated.

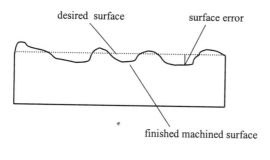

FIGURE 7   Milling surface errors.

The relevant decision variables are end-mill diameter, radial rake angle, feed per tooth, spindle speed, and axial depth of cut. Virtual experiments were designed for both aluminum and cast iron. In both cases, an uncoated carbide cutter was used. The variables and the range of values used for the two simulation experiments are shown in Table 6. The ranges are defined in accordance with guidelines provided by EMSIM and other machining handbooks (Walsh 1999).

Regression analysis revealed that the exponential model provides the most appropriate fit for both work piece materials and quality measures. The model is

$$c \prod_{j=1}^{5} e^{b_j x_j} = y_i \tag{21}$$

which was transformed to

$$\ln c + \sum_{j=1}^{5} b_j x_j = \ln y_i \tag{22}$$

where:

x_j = end-milling decision variables ($j = 1 \ldots 5$)
y_i = milling errors ($i = 1,2$)

The results of this regression are shown in Tables 7 and 8.

4.3   Environmental Model

The use of cutting fluids in machining operations generates large amounts of wastewater, which often requires costly pretreatment prior to disposal to a

TABLE 6 End-Milling Decision Variable Ranges for Aluminum and Cast Iron

	Low value	High value	Units
A 356 T51 cast aluminum			
End-mill diameter, D	0.5	1.8	Inches
Radial rake angle, alpha	5	30	Degrees
Feed per tooth, f	0.003	0.00769	Inches
Spindle speed, N	1550 (0.5), 420 (1.8)	2990 (0.5), 830 (1.8)	rpm
Axial depth of cut, d	0.04	0.3	Inches
Gray cast iron (150-220 BHN)			
End-mill diameter, D	0.5	1.8	Inches
Radial rake angle, alpha	5	30	Degrees
Feed per tooth, f	0.003	0.00769	Inches
Spindle speed, N	1950 (0.5), 530 (1.8)	3000 (0.5), 750 (1.8)	rpm
Axial depth of cut, d	0.04	0.3	Inches

TABLE 7   End-Milling Regression Results with Exponential Transformation for 32 Experiments with A 356 T51 Cast Aluminum

	Maximum surface error		Average surface error	
	Coefficient	Standard error	Coefficient	Standard error
Intercept	−6.09	0.86	−6.05	0.74
D	−4.91	0.37	−4.97	0.32
Alpha	−0.03	0.01	−0.03	0.01
f	87.29	55.56	94.66	48.05
N	−2E-4	2E-4	−2E-4	2E-4
d	13.59	1.00	11.00	0.87
Multiple R	0.9826		0.9862	

publicly owned treatment works (POTW). The impact on the environment is measured in terms of the biochemical oxygen demand (BOD) and fats, oils, and greases (FOG) content of spent cutting fluids. A method for estimating BOD and FOG content of various cutting fluids from the stoichiometric balances of cutting fluid components was developed by Skerlos et al. (1998). Typical formulations of three common cutting fluids are shown in Table 9 (Childers 1994).

After defining the stoichiometric equations for each component in the cutting fluid, the BOD and FOG content of spent cutting fluids can be calculated. If industrial pretreatment is carried out prior to discharge to a POTW, then BOD and FOG are reduced, as shown in Table 10 (U.S. Environmental Protection Agency 1983b).

TABLE 8   End-Milling Regression Results with Exponential Transformation for 32 Experiments with Gray Cast Iron

	Maximum surface error		Average surface error	
	Coefficient	Standard error	Coefficient	Standard error
Intercept	−9.04	1.06	−9.08	0.76
D	−3.40	0.47	−3.45	0.33
Alpha	−0.02	0.01	−0.02	7E-3
f	94.28	49.54	125.36	35.28
N	1E-3	3E-4	1E-3	2E-4
d	13.04	0.89	10.89	0.64
Multiple R	0.9872		0.9929	

TABLE 9 Typical Oil Fluid Formulations (Soluble, Semisynthetic, and Synethetic Oil)

Function	Component	Percent by weight
Soluble oil		
Oil	100/100 napthetic hydrotreated oil	68
Emulsifier	Petroleum sulfonate	17
EP lubricant	Phosphate ester	5
Boundary lubricant	PEG ester	5
Rust inhibitor	Alkanolamide	3
Biocide	Proprietary additions	2
Semisynthetic oil		
Emulsifier	Petroleum sulfonate	5
Emulsifier	Alkanolamide	15
Oil	100/100 napthetic oil	15
Corrosion inhibitor	Amine dicarboxylate	6
Coupler	Clycol ether	1.5
Biocide/Fungicide	Proprietary additions	2
Diluent	Water	55.5
Synthetic oil		
Diluent	Water	70
Rust inhibitor	Amine carboxylate	10
pH buffer and inhibitor	Triethanolamine	5
EP lubricant	Phosphate diester	4
Boundary lubricant	PEG ester	5
Boundary lubricant	Sulfated castor oil	4
Fungicide	Proprietary additions	2

4.4 Mathematical Model

The cost, quality, and environmental estimation models are integrated into an optimization model that minimizes the total casting, machining, and cutting fluid costs for an automotive steering knuckle, shown in Figure 8. The general form of the model, a classical nonlinear optimization problem with binary variables, is as follows :

$$\text{Objective function: } MIN \; C(Y_1, Y_2, \ldots, Y_{23}) \tag{23}$$

Subject to

$$Y_1 + Y_2 = 1 \tag{24}$$

TABLE 10 Median Removal Rates for Industrial Pretreatment

Pretreatment technologies	Median removal rates
Primary treatment	
Sedimentation	BOD, 50%
	FOG, 78%
Gravity oil separation	BOD, 0%
	FOG, 62%
Second treatment	
Dissolved air flotation	BOD, 61%
	FOG, 79%
Ultrafiltration	BOD, 82%
	FOG, 99%
Tertiary treatment	
Activated carbon adsorption	BOD, 49%
	FOG, 24%

$$Y_3 + Y_4 = 1 \tag{25}$$

$$Y_5 + Y_6 = 1 \tag{26}$$

$$Y_{7\min} \le Y_7 \le Y_{7\max} \tag{27}$$

$$Y_{8\min} \le Y_8 \le Y_{8\max} \tag{28}$$

$$Y_{9\min} \le Y_9 \le Y_{9\max} \tag{29}$$

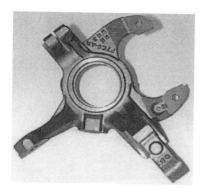

FIGURE 8 A steering knuckle.

$$Y_{10\min} \leq Y_{10} \leq Y_{10\max} \tag{30}$$

$$Y_{11\min} \leq Y_{11} \leq Y_{11\max} \tag{31}$$

$$Y_{12\min} \leq Y_{12} \leq Y_{12\max} \tag{32}$$

$$Y_{13} + Y_{14} = 1 \tag{33}$$

$$f_1(Y_1, Y_{13}) \quad \leq \text{max of maximum surface error} \tag{34}$$

$$f_2(Y_1, Y_{13}) \quad \leq \text{max of average surface error} \tag{35}$$

$$f_3(Y_2, Y_{13}) \quad \leq \text{max of maximum surface error} \tag{36}$$

$$f_4(Y_2, Y_{13}) \quad \leq \text{max of average surface error} \tag{37}$$

$$Y_{15} + Y_{16} + Y_{17} = 1 \tag{38}$$

$$Y_{18} + Y_{19} + Y_{20} = 1 \tag{39}$$

$$f_5(Y_{15}, Y_{16}, Y_{17}, Y_{21}, Y_{22}, Y_{23},) \leq \text{maximum limit allowed}$$
$$\text{for BOD per disposal} \tag{40}$$

$$f_6(Y_{15}, Y_{16}, Y_{17}, Y_{21}, Y_{22}, Y_{23},) \leq \text{maximum limit allowed}$$
$$\text{for FOG per disposal} \tag{41}$$

$$Y_1, Y_2, Y_3, Y_4, Y_5, Y_6, Y_{13}, Y_{14}, Y_{15}, Y_{16}, Y_{17}, Y_{18}, Y_{19}, Y_{20}, Y_{21},$$
$$Y_{22}, Y_{23} \in \{0 \text{ or } 1\}$$
$$\tag{42}$$

where:

$Y_1 = 1$ if aluminum, 0 otherwise
$Y_2 = 1$ if cast iron, 0 otherwise
$Y_3 = 1$ if squeeze casting process for aluminum, 0 otherwise
$Y_4 = 1$ if permanent mold casting for aluminum, 0 otherwise
$Y_5 = 1$ if permanent mold casting for cast iron, 0 otherwise
$Y_6 = 1$ if sand casting for cast iron, 0 otherwise
$Y_7 = $ end-milling diameter (mm)
$Y_8 = $ radial rake angle (deg)
$Y_9 = $ feed per tooth (mm)
$Y_{10} = $ spindle speed (rpm)
$Y_{11} = $ axial depth of cut (mm)

Y_{12} = diameter of the hole (mm)
Y_{13} = 1 if dry machining, 0 otherwise
Y_{14} = 1 if wet machining, 0 otherwise
Y_{15} = 1 if soluble oil, 0 otherwise
Y_{16} = 1 if synthetic oil, 0 otherwise
Y_{17} = 1 if semi synthetic oil, 0 otherwise
Y_{18} = 1 if recycling-sedimentation, 0 otherwise
Y_{19} = 1 if recycling-pasteurization-centrifugation, 0 otherwise
Y_{20} = 1 if recycling-gravity filtration, 0 otherwise
Y_{21} = 1 if primary treatment, 0 otherwise
Y_{22} = 1 if secondary treatment, 0 otherwise
Y_{23} = 1 if tertiary treatment, 0 otherwise

The parameters, inputs by the user in the model, are shown in Table 11.

TABLE 11 Parameters, Inputs by the User in the Model

Parameter	Value
Physical volume per part	2000 cm^3
Duration of production	5 years
Annual production volume	150,000 parts
Facility cost rate for casting	20 \$/ft^2
Area required for casting process	20,000 ft^2
Ratio of productive and nonproductive time	50%
Number of teeth per cut	4
Hourly machine rate	100 \$/hour
Number of features per category	2
Max of the maximum surface error	10 μm
Max of the average surface error	10 μm
Volume of the fluid system	7560 L
Oil concentration	2%
Number of disposals per year	4
Make-up concentration	240 L/week
Flow rate of sedimentation	5 L/sec
Size of the system for gravity filtration	200 m^2
Flow rate for secondary treatment	10 L/sec
Flow rate for tertiary treatment	3 L/sec
Number of workers for the fluid system	4
Wage of workers in the fluid system	15 \$/day
Maximum level for BOD	0.04 kg/L
Maximum level for FOG	0.0002 kg/L

Based on the above parameters, the optimal solution yields the values of the decision variables shown in Table 12. Cast iron was chosen as the work piece material, while sand casting was used as the casting process. Dry machining was superior to wet machining.

The optimal solution reflects a variety of user inputs. This is a great advantage, since it allows interaction between the user and the model. Computational results showed that if more stringent quality restrictions are applied, then wet machining is included in the optimal solution (Table 13, second column). Next, we increase the oil concentration from 2% to 5%, maintaining the same maximum limits allowed for BOD and FOG per disposal. As a result, secondary pretreatment is included in the optimal solution (Table 13, third column). This was expected, since the increase in the oil concentration results in higher values of BOD and FOG per disposal and additional pretreatment becomes necessary to satisfy the environmental constraints (Eqs. 40 and 41). Finally, a concurrent increase in the oil concentration to 10% and a decrease in the maximum limits allowed for BOD (0.03 kg/L) and FOG (0.00015 kg/L) yield an optimal solution (Table 13, fourth column) where all types of pretreatments are employed (primary, secondary, and tertiary) to maintain a feasible solution to the problem (satisfy Eqs. 40 and 41). Note that the use of secondary and tertiary pretreatment significantly increases the cost per component.

TABLE 12 Optimal Level of the Decision Variables for Cost Minimization

Decision variables	Optimal level of the decision variables
Work piece material	Cast iron
Casting process choice	Sand casting
End-milling diameter (mm)	30.24
Radial rake angle (deg)	30.00
Feed per tooth (mm)	0.08
Spindle speed (rpm)	530.00
Axial depth of cut (mm)	1.00
Diameter of the hole (mm)	7.45
Wet or dry machining	Dry machining
Choice of cutting fluid	None
Choice of recycling technology	None
Primary treatment	None
Secondary treatment	None
Tertiary treatment	None
Total cost per part ($/part)	**18.33**

TABLE 13 Optimal Level of the Decision Variables for Cost Minimization with Strict Quality Restrictions

Decision variables	Optimal level of the decision variables			
	Oil conc. = 2% BOD = 0.04 kg/L FOG = 0.0002 kg/L	Oil conc. = 5% BOD = 0.04 kg/L FOG = 0.0002 kg/L	Oil conc. = 10% BOD = 0.03 kg/L FOG = 0.00015 kg/L
Work piece material	Cast iron	Cast iron	Cast iron
Casting process choice	Sand casting	Sand casting	Sand casting
End-milling diameter (mm)	45.70	45.70	45.70
Radial rake angle (deg)	30.00	30.00	30.00
Feed per tooth (mm)	0.10	0.08	0.08
Spindle speed (rpm)	530.00	530.00	530.00
Axial depth of cut (mm)	1.00	1.00	1.00
Diameter of the hole (mm)	9.06	7.35	7.50
Wet or dry machining	Wet machining	Wet machining	Wet machining
Choice of cutting fluid	Synthetic oil	Synthetic oil	Synthetic oil
Choice of recycling technology	Centrifugation-pasteurization	Centrifugation-pasteurization	Centrifugation-pasteurization
Primary treatment	Yes	Yes	Yes
Secondary treatment	None	Yes	Yes
Tertiary treatment	None	None	Yes
Total cost per part ($/part)	**19.68**	**23.62**	**30.86**

5 SUMMARY

This chapter has described a domain-independent, constrained, nonlinear, multiattribute optimization problem formulation for environmentally conscious design and manufacturing. Two examples were presented. The first focused on design of a portfolio of products to facilitate product take-back, disassembly and component reuse over multiple life cycles. The second example focused on integrating statistical analysis of a manufacturing process (machining) into a general cost versus quality versus environment tradeoff problem. This general structure provides the design engineer with a robust decision-making framework into which a wide range of analyses can be placed, including activity-based cost estimation, statistical manufacturing process control, market-based portfolio analysis, and environmental impact assessment.

ACKNOWLEDGMENT

We are grateful for the support provided by NSF grant DMI 95-28629, the Manufacturing Research Center, and the NSF I/UCRC.

REFERENCES

Berry M. 2000. A machining and cutting fluid decision tool for cost, quality and environmental tradeoffs. Master's thesis, University of Illinois at Urbana-Champaign.

Carnahan JV, Thurston D. 1998. Tradeoff modeling for product and manufacturing process design for the environment. J Industrial Ecol 2:1.

Childers J C. 1994. The chemistry of metalworking fluids. Metalworking Fluids. New York: Marcel Dekker, pp. 165–189.

Cooper R, Kaplan RS. 1991. Profit priorities from activities based costing. Harvard Business Review, May–June, pp. 130–135.

Keeney R. Raiffa H. 1993. Decisions with multiple objectives. Cambridge: Cambridge University Press.

Locascio A, Thurston D. 1998. Transforming the House of Quality to a multiobjective optimization formulation. Structural Optimization. 16:2, pp. 136–146.

Mangun D, Thurston D. 2000. Product portfolio design for component reuse. Proceedings IEEE International Symposium on Electronics and the Environment.

Nicolaou P, Thurston D. 2000. Machining: quality, cost and environmental estimation and tradeoffs. Proceedings 2000 ASME Conference on Design for Manufacturing.

O'Connor P. 1985. Practical reliability engineering. Chichester, England: John Wiley and Sons.

Saaty TL. 1980 (revised 1988). The analytic hierarchy process. New York: McGraw-Hill.

Schreiner T., 1999. Cost estimation for tradeoff decisions in design for manufacturing. Thesis, University of Illinois at Urbana-Champaign.

Skerlos SJ, DeVor RE, Kapoor SG. 1998. Environmentally conscious disposal considerations in cutting fluid selection. IMECE Proc ASME.

Thurston D. 1991. A formal method for subjective design evaluation with multiple attributes. Res Eng Des 3:2.

Thurston D. 2001. Real and perceived limitations to decision based design with utility analysis. ASME J Mech Design, 123:2..

Urlichs B. 1999. A predictive quality model in a design for machining decision tool with consideration of uncertainty. Thesis, University of Illinois at Urbana-Champaign.

U.S. Environmental Protection Agency. 1983a. Treatability manual: cost estimating. Vol. IV. Washington DC: Office of Research and Development, Publication No. EPA-600/2-82-001d.

U.S. Environmental Protection Agency. 1983b. Treatability manual: technology for control/removal of pollutants. Vol. III. Washington DC: Office of Research and Development, Publication No. EPA-600/2-82-001c.

U.S. Environmental Protection Agency. 1995. Development document for the proposed effluent limitations guidelines and standards for the metal products and machining phase I point source category. Springfield, VA: U.S. Department of Commerce, Publication No. 821-R-95-021.

Walsh R. 1999. McGraw-Hill machining metalworking handbook. New York: McGraw-Hill.

14

Environmental Issues in Collaborative Design

Ram D. Sriram and Robert H. Allen
University of Maryland, College Park, Maryland

D. Navin Chandra*
Carnegie Mellon University and TimeØ Inc., Cambridge,
Massachusetts

> [S]ustainable development ensures that future generations have
> access to the social capital—human natural and physical capi-
> tal—to create a life at least equal to that of this generation.
> *From the 1987 report of the World Commission on Envi-
> ronment and Development (the Brundtland Commission)*

1 MOTIVATION

Proper diet, regular exercise, and relaxation techniques are proven to have a
beneficial effect on leading a healthy life (Sears 1995). Another aspect to
living a healthy life is living in a clean environment. When adverse conditions exist in the environment—in the form of air pollution, solid pollution
and liquid pollution—the impact on the quality of life can be substantial

---

* *Current affiliation*: NovaSpike Inc., Boston, Massachusetts.

(Naar 1990; Pal 1998; Wenzel et al. 1997). The greater the quantity of pollution, the greater the adverse effect on public health. And pollution quantities are staggering.

A principal measure of air pollution is the amount of carbon dioxide (CO_2) in the air. Until about 1900, the global concentration level of CO_2 hovered around 280 mg/L (or ppm) for nearly a millennium. There has been a drastic increase in the chemicals being released into the atmosphere in the last century. This has been due in large part to increased use of coal, fossil fuels, and natural gas—20×10^9 tons annually. This has pushed the global level of CO_2 from 280 to 360 mg/L (ppm), and much of this emission comes from the United States. Increased CO_2 in the atmosphere leads to global warming (as is demonstrated by scientific studies) with detrimental effects to our fragile planet (Naar 1990; Pal 1998). Other gases, such as chlorofluorocarbons emitted by a variety of household products, also add to heat entrapment and ozone depletion, with considerable health risks to the human race.

In the United States, the manufacturing, mining, and farming industries generates around 5×10^9 tons of solid waste, with homes and businesses contributing to another 230×10^6 tons of garbage every year (Naar 1990). The above waste contains hazardous chemicals, such as pesticides, nuclear waste, and toxic metals. Some of these cannot be reduced to harmless products in waste management units, such as incinerators. Unless disposed of properly, these eventually find their way into our water resources and air.

Liquid pollutants result as a byproduct of many industrial processes, such as mining and manufacturing, and from a variety of other sources, including our homes. The United States produces about 700,000 tons of toxic waste every day. These include benzyl chloride (a byproduct from drug and perfume manufacture), chlorine, and hydrogen cyanide. Production of these is far above the quantity deemed dangerous to humans. Added to these toxic chemicals, we also spray 12×10^6 tons of pesticide annually (Naar 1990).

There are a number of ways to reduce the amounts of pollution generated, many of which are interconnected: change in industry practice, economic incentives, technological development, shift in material use, change in consumer behavior, and regulation. As shown in Figure 1, for example, end products (such as automobiles) produce the most toxic chemical waste among different categories of producers. One reason for this is the automobile, which, over the course of traveling 200,000 km, will consume about 21,000 kg of hydrocarbons and emit 59,000 kg of carbon dioxide (Sullivan et al. 1998). Clearly, one way to limit pollution is to design more fuel-efficient

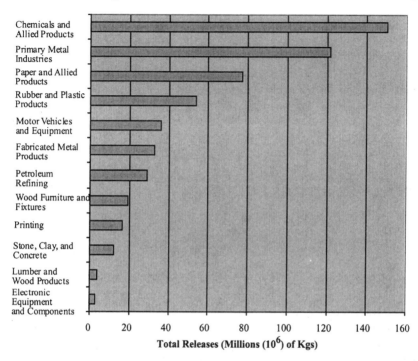

FIGURE 1 Annual releases of toxic chemicals by the U.S. manufacturing industry. (From Pal 1998.)

vehicles. Another way is to develop alternative less-polluting transportation systems.

Figure 1 also shows that the contribution from the manufacturing industry to toxic chemical releases is substantial. A large percentage of this comes from the chemicals and primary metal industries, which often provide the raw materials for consumer goods. In addition, of the 2.5×10^9 metric tons of materials consumed by the United States in 1990, only 10% were recycled (Pal 1998).

Fortunately, there is a wealth of opportunities for design teams and manufacturers to create products that are less disruptive to the environment. To do this effectively, however, design teams need to have knowledge about environmental issues and access to environmental databases from the early conceptual design phase through manufacturing and ultimate disposition. This requires expertise that would likely come from different members of a collaborative design team. The purpose of this chapter is to explore several strategies for incorporating environmental impacts into the product life cycle and to see how collaborative techniques can be used to ultimately achieve pollution reduction in product design, manufacture, and use.

This chapter is organized as follows. In the next section, we discuss factors that provide the impetus for environmentally conscious design and manufacturing. Then, we outline the various product development stages. We briefly discuss the design for recovery problem. This is followed by an architectural description of a collaborative design framework. Finally, we discuss one component of this framework—access to various heterogeneous databases containing environmental information.

2 DESIGNING GREENER PRODUCTS

Green engineering design (or green design) is an approach to product and process design that reduces environmental impact without compromising a product's quality or its commercial viability. The aim is to identify, develop, and exploit new technologies that can bolster productivity while minimizing impact to the environment. Over the past decade, considerable research and effort have been put into understanding issues such as waste management and materials recovery, as they relate to products *after* they enter the waste-stream (Sullivan et al. 1998). Attention is now being focused on the product design. The idea is to inject environmental considerations into the design process where the assessment of environmental impact is based on a life cycle view of the product. This includes the product's manufacturing process, distribution, use, and final disposal. Hence, green design takes a *proactive* approach to environmental problems rather than a *reactive* approach, which tries to fix problems only after they occur. The notion of designing products for environmental concerns was introduced a decade ago (Navin Chandra 1990) and has since been adopted by many researchers and practitioners as part of the concurrent engineering process (Caspersen and Sarenson 1998, Hersch 1998). This approach requires that design teams perform their duties not only for specification, production, and maintenance but for disassembly, reuse, repair, materials recycling, remanufacturing, and reassembly as well. This is a significant change from traditional design and manufacturing practices.

But why should designers and manufacturers change their processes and products to be green? In the U.S. economy, a company must be profitable to be viable. If green design increases design and production cost, this may well undermine the ability to be profitable. One way to overcome this hurdle is to create products with environmental differentiation. For example, textile manufacturers are willing to pay more for a certain type of dye that requires less salt for absorption into materials (Reinhardt 1999). The reason for this is that the dye ends up paying for itself by lowering processing costs that result from less salt consumed, reduced wastewater treatment, and improved quality control. Two factors critical in the success of this product are finding customers willing to pay more and communicating the products environmental benefits to the textile industry.

An indirect economic reason for doing so is the rising cost of waste disposal. In certain instances, it may be less costly to recycle than to pay for disposal (Chen et al. 1994). External factors as well encourage green design. Environmental legislation often mandates specific processes to protect the environment. Another factor, curiously enough, is customer demand. As consumers become more environmentally conscious, their inclination for products made in an environmentally conscious way increases. As a result, another factor is corporate image in the corresponding public perception. All these factors, primarily the economic ones, make it attractive for a company to invest in redesign, retooling, and the establishment of recycling facilities. Specific cost-reducing concepts that encourage green design include the following:

1. *Holism and simplification.* Taking a holistic view of a product, in terms of recycling and material compatibility, can lead to simplification. For example, in one of our disassembly experiments (Navin Chandra 1994; Wivell and Navin Chandra 1992), we found a product to have dozens of different types of plastics. Using reverse engineering, however, we found only six different plastic specifications were needed. In retrospect, this product could have been designed with fewer types of materials. Such a simplification reduces the number of vendors used, material inventories, the different types of joining methods used, and the number of assembly operations. These changes reduce cost *and* improve recyclability at the same time.

2. *Remanufacture and reuse.* A product can be designed for ease of remanufacturing and reuse. Starting with materials recycling, which is the simplest form of recovery, other options include repairing the item, its disassembly to recover separable materials, reuse of components, and remanufacture of parts. Automobile parts are classic examples of this process. Every time a part is reused, all the energy and emissions that were produced in its original manufacture and processing are partially salvaged.

3. *Materials mortgage.* If a customer leases a new product every few years and returns it to the manufacturer for recovery, then one can consider a long-term mortgage. For example, if we wanted to improve the vibration reliability of a computer, we could do this by introducing more gold on the connectors. Normally, this would undesirably increase the price of the computer. One the other hand, if we knew computers—hence the gold—would be returned, then the customer could mortgage the gold over a longer term. In this way, manufacturers can provide very high-quality products without

sticker shock. With recycling, the large initial investment is spread out over several product lifetimes (Navin Chandra 1994).

4. *Redefining markets.* By redefining its business model, a company can actually reduce overall costs by incorporating environmentally desirable activities (Reinhardt 1999). For example, up until about 1980, Xerox had market dominance in copier equipment and was lax about cost savings and machine disposal (a nontrivial problem). When competition threatened this dominance in the 1980s, Xerox responded by retaining disposal rights to the equipment they sold. They set up an infrastructure to disassemble, remanufacture, and incorporate new technology into existing machines; these were then resold at considerable profit. By 1995, Xerox estimates it was saving more than several hundred million dollars annually. More importantly, it redefined the market, forcing competitors such as Kodak, IBM, and Canon to follow suit.

The above strategies represent new ways in which products can be designed, marketed, and recovered. Some other forward-looking organizations such as Dupont and Intel have already adopted some of these new strategies (Resetar et al. 1999). With the right kind of approach and design tools, it is possible to make environmentally compatible products that are also commercially profitable.

3 PRODUCT DEVELOPMENT PROCESS

As shown in Figure 2, there are three fundamental phases in the life cycle of a product: engineer, use, and dispose/recycle. The engineering phase consists of several activities in design and manufacturing, which are described in detail by Barkmeyer (1995, 1996) and itemized below. The first four focus on design and the last four focus on manufacturing.

1. *Plan products.* Depending on (potential) market needs and customer requirements, develop the idea for a product and characterize it in terms of function, target price range, and relationship to existing products of the manufacturing firm. Other activities include defining cost constraints, performance constraints, and other marketability factors; performing market analysis, cost-benefit analysis; and developing product development and marketing plans.

2. *Generate product specifications.* From the conceptual product specification, formulate an engineering specification for the product. This involves mapping the customer requirements into engineering requirements and refining the engineering requirements in consideration of the relevant laws, regulations, product standards, and also of the existing patents in the same area. This process may involve determination of the relationship of the new product to the firm's library of existing product designs.

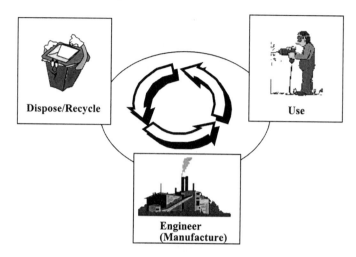

FIGURE 2 Fundamental phases of a product life cycle.

3. *Perform preliminary design.* Decompose the design problem into a set of component design problems and develop the specifications for each component problem. Define the integration of the components into a product in a set of interface specifications and a preliminary layout model. This process is iterative, as the early phases of the component design will generate new considerations and changes. Primary results are the product layout drawing and annotations and the component design specifications. The preliminary design activity involves generation of various alternatives and evaluation of these alternatives against criteria.

4. *Produce detailed designs.* For each subsystem (or component) that is not off the shelf (or identical to an existing in-house design) and for the component integration, produce all specifications needed to describe completely the subsystem for manufacture. This includes drawings and geometry, materials, finish requirements, fit requirements, and assembly drawings and tolerances.

5. *Engineer manufacture of product.* The process of making the product is defined. This includes determining the elementary stock materials and components to be acquired, the equipment, tooling and skills to be used, and the details of that usage. Details include the exact sequence of setups and operations to be performed and the complete instructions for each operation, whether by human or automated resources. For engineering purposes, every product is decomposed into a collection of component *parts*, each of which is either a fabricated (piece) part or an assembly, including embedded parts that can be produced by rapid prototyping.

Any part, however, may be subjected to inspection and finishing processes. The final product is itself a part—it may be a single fabricated part or a final assembly.

6. *Engineer production system.* New or modified production facilities for the manufacture of a particular collection of parts are designed. A "facility" may be a plant, a shop, a line, a manufacturing cell, or a group of manufacturing cells. This activity encompasses both design-from-the-walls of such a facility and reengineering of all or part of such a facility to improve the production of certain products. It includes identification of the parts, products, and processes for which the production system is to be tailored, identification of the equipment to be installed or replaced, (re)design of the floor layout, and development of an implementation plan for the (re)designed production system.

7. *Produce products.* The production facilities needed to produce the parts according to the specifications in the process plans are developed and maintained. This involves defining the production schedules and controlling the flow of materials into and out of the production facility, scheduling, controlling and executing the production processes themselves, providing and maintaining the production equipment and the human resources involved, and developing and tracking the tooling and materials.

8. *Manage engineering workflow.* This activity would involve the specification of engineering tasks, controls, reviews, and approvals. The sequence of these engineering activities and the required resulting information objects, and their due dates, if appropriate, are defined.

For complex design and manufacturing tasks, these activities typically involve thousands of personnel working in smaller collaborative teams toward a particular subgoal. The next section expands some of the activities and details how those activities relate to the environment.

4   PRODUCT LIFE CYCLE AND THE ENVIRONMENT

The three basic stages of the product life cycle shown in Figure 2 and can be elaborated as shown in Figure 3. Each aspect of this product life cycle— from mining the material, transporting material and goods, to disposing the product—has an impact on the environment. Identifying these impacts is important in developing green products. Our focus in this chapter is on assessing environmental impacts during the design stage. In particular, we explore pragmatic ways to deal with recovery issues (Section 6) and how to access heterogeneous databases that contain environmental data (Section 7).

Figure 4 depicts an input/output diagram identifying the basic activities that impact the environment. By minimizing waste, designers and man-

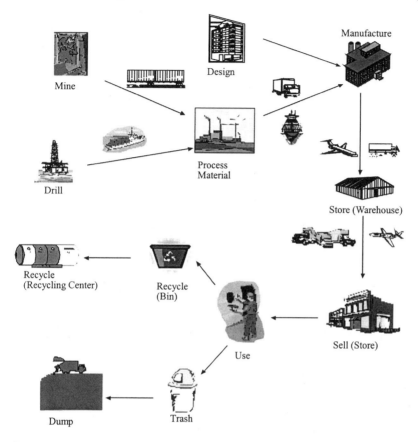

FIGURE 3   Product life cycle: from mining to reuse.

ufacturers achieve one green goal; even better, would be to have waste products be fed back into input.

Consider, for example, the automobile. A life cycle inventory study conducted by the United States Automotive Materials Partnership Life Cycle Assessment Special Topics Group (USAMP/LCA) identified over 30 materials utilized in a generic vehicle, which was based on the "generic" 1995 Intrepid/Lumina/Taurus cars (Sullivan et al. 1998). A few of these materials are shown in Figure 5, and a comprehensive list is provided in Table 1. The USAMP/LCA study identified reuse and materials recycling for five major stages of the product life cycle: raw materials acquisition and processing, parts and subassembly manufacturing, vehicle assembly, use, and disposal. An example of one process input/output diagram (for steel

TABLE 1 Materials Used in a Generic Vehicle

Material	Mass (kg)	Mass (%)	Material	Mass (kg)	Mass (%)
Plastics			**Metals (ferrous)**		
ABS-PC (acrylonitrile butadiene styrene-polycarbonate blend)	2.8	0.18	Ferrite (Fe)	1.5	0.10
Acetal	4.7	0.31	Cast iron (Fe)	132	8.59
Acrylic resin	2.5	0.16	Pig iron (Fe)	23	1.48
Acrylonitrile butadiene styrene (ABS)	9.7	0.64	Steel (cold rolled)	114	7.46
Acrylonitrile styrene acrylate (ASA)	0.18	0.012	Steel (EFA)	214	13.94
Epoxy resin	0.77	0.050	Steel (galvanized)	357	23.29
PA 6-PC (polyamide-polycarbonate blend)	0.45	0.030	Steel (hot rolled)	126	8.23
Phenolic resin	1.1	0.072	Steel (stainless)	19	1.23
Polyamide (PA 6)	1.7	0.11	Total metals (ferrous):	985	64
Polyamide (PA 66)	10	0.67	**Fluids**		
Polybutylene terephthalate (PBT)	0.37	0.024	Automatic transmission fluid	6.7	0.44
Polcarbonate (PC)	3.8	0.25	Engine oil (SAE 10w-30)	3.5	0.23
Polyester resin	11	0.75	Ethylene glycol	4.3	0.28
Polyethylene (PE)	6.2	0.40	Glycol-ether	1.1	0.069
Polyethylene terephthalate (PET)	2.2	0.14	Refrigerant (R 134a)	0.91	0.059
Polypropylene (PP)	25	1.6	Unleaded gasoline	48	3.1
Polypropylene (PP, foam)	1.7	0.11	Water	9.0	0.59
Polystyrene (PS)	0.0067	0.00044	Windshield cleaning additives	0.48	0.031
Polyurethane (PUR)	35	2.3	Total fluids:	74	4.8
Polyvinyl chloride (PVC)	20	1.3	**Other materials**		
PP-EPDM (polypropylene-ethylene propylene diene monomer blend)	0.10	0.0067	Ethylene propylene diene monomer (EPDM)	10	0.68
PPO-PC (polyphenylene oxide-polycarbonate blend)	0.025	0.0017	Adhesive	0.17	0.011
PPO-PS (polyphenylene oxide-polystyrene blend)	2.2	0.14	Asbestos	0.4	0.026
Thermoplaastic elastomeric olefin (TEO)	0.31	0.020	Bromine (Br)	0.23	0.015
Total plastics:	143	9.3	Carpeting	11	0.73
			Ceramic	0.25	0.016
			Charcoal	0.22	0.014

Metals (nonferrous)			Other materials (contd.)		
Aluminum oxide	0.27	0.018	Corderite	1.2	0.081
Aluminum (cast)	71	4.663	Desiccant	0.023	0.0015
Aluminum (extruded)	22	1.438	Fiberglass	3.8	0.25
Aluminum (rolled)	3.3	0.2	Glass	42	2.8
Brass	8.5	0.55	Graphite	0.092	0.0060
Chromium (Cr)	0.91	0.060	Paper	0.20	0.013
Copper (Cu)	18	1.1	Recycled textile fibers	12	0.78
Lead (Pb)	13	0.85	Rubber (except tire)	23	1.5
Platinum (Pt)	0.0015	0.00010	Rubber (extruded)	37	2.4
Rhodium (Rh)	2.9E-04	0.000019	Sulfuric acid (HSO4)	2.2	0.14
Silver (Ag)	0.0034	0.00022	Tire	45	3.0
Tin (Sn)	0.067	0.0044	Wood	2.3	0.15
Tungsten (W)	0.011	0.00073	Total other materials:	192	13
Zinc (Zn)	0.32	0.021			
Total metals (nonferrous):	138	9.0	**Total weight of generic vehicle:**	**1532**	**100**

Source: Sullivan et al. 1998.

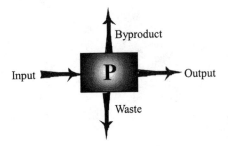

FIGURE 4   Process input/output diagram for products.

manufacture) is shown in Figure 6. By having many of these type diagrams for alternative materials, design teams can assess the environmental impact of each material and make it part of their design decisions, not unlike the way cost, performance, and safety considerations are normally considered.

Another example—polyvinyl chloride (PVC) manufacture—is shown in Figure 7. Note that the process input diagram includes numerical esti-

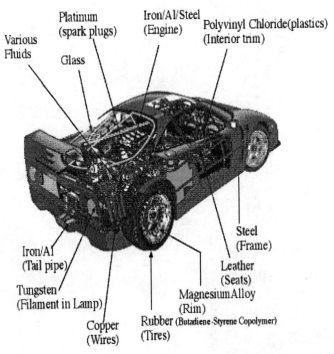

FIGURE 5   Some typical material components used in a car.

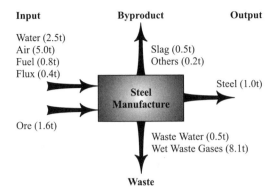

Figure 6 Input/output diagram for steel manufacture (t, metric ton).

mates per unit of the product produced (e.g., kilowatts of energy and grams of PVC). Figure 7 also shows the toxic PVC as a waste product. Most manufacturers claim that it is fed back into the input to minimize potential environmental impacts. It will be useful to generate such flows for all activities, from cradle to grave.

5 ISSUES IN DESIGN FOR RECOVERY

In the process of recovery, the optimal solutions represent a trade-off between cost, time, and environmental distress. One cannot expect that the best strategy for dealing with a discarded item is going to always involve

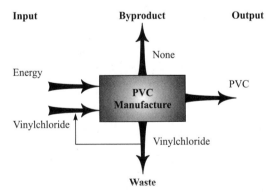

Figure 7 Input/output diagram for PVC manufacture (note that the above should also include various numerical quantities per unit of the product).

100% recycling. For example, it sometimes takes far more energy—and hence pollution—to recycle a product than it takes to make it in the first place. In such cases, landfilling may well be the most environmentally benign option and the less expensive one

In the process of recovering a product, some parts may be reused while others may be recycled and the rest may be incinerated or landfilled. Such a process might represent a balance that has to be struck between the amount of emissions, cost of recovery, energy usage, and the environmental impacts of landfilling. The engineering team must find a balance such that, to the extent possible, one objective should always be to design so that the land-filled volume is reduced. For example, in an automobile dashboard there are many plastic parts that are worth recycling, but often the cost of isolating the parts is more than the value of its materials. Consequently, these parts end up in shredder fluff. Through clever redesign, it is possible to make recycling more attractive by making recyclable components more accessible. Finding the balance point for a product and redesigning to move the balance point in a favorable direction is the aim of the green design process. We call this the recovery problem:

> For a given design (or product) to find a recovery plan that balances the amount of effort (e.g., energy) that is put into recovery and the amount of effort that is saved by reusing parts and materials. In this way, recovery is a leveraged process.

A recovery analysis can be used to determine the pragmatic recovery process for a given product or design. In recycling a photocopier, for example, one rather complicated component is the control panel, with its many small plastic parts. It takes more energy to recover these materials than it will take to make them from original sources. It is hence important to understand that the pragmatic aspect of recovery might call for some environmentally undesirable actions as part of the recovery process, actions such as landfilling or incineration. The aim is to detect break-even points and points of maximum payoff, where payoff can be measured in terms of any accumulative such as emissions, energy, money, or disposal volume.

Many aspects of recovery are considered during disassembly. Each step in disassembly involves a decision on the next action. At each decision point, the following issues need to be considered for each subassembly:

1. *Dismantle further?* At some stage in the disassembly process one reaches a point at which the cost of further disassembly might be more than the value of parts that will be disassembled.
2. *Send to shredder?* If a subassembly of compatible materials is reached, it might make more sense to send it to a shredder

than to disassemble it further. If a subassembly has a small number of parts that are of incompatible materials, the subassembly could still be shredded without further disassembly. This, of course, requires knowledge about acceptable levels of impurities.

3. *Sell?* A subassembly can be worth more than the sum of parts. In an automobile alternator, for example, the rotor can be sold for refurbishment as an assembly: The actual value of the materials in the rotor is substantially lower than the rotor itself.

4. *Remanufacture?* While remanufacturing operations, such as cleaning, plasma spraying, and machining, have environmental impacts, they are often less then the original materials processing and manufacturing processes that are used to make the part from scratch.

5. *Hazardous materials in subassembly?* If a subassembly contains some hazardous materials, it is imperative to perform the disassembly until the hazardous material is reached. Whether further disassembly should proceed depends on the expense of sending the leftover subassemblies to a hazardous landfill.

These ideas have been captured in an automated disassembly and recovery analysis tool called ReStar (Navin Chandra 1994). At every stage, the system evaluates whether it would be better to continue dismantling or whether the parts should be reused, sold, or sent to a shredder for material recycling. This process is recursively applied to each subassembly that is generated through the disassembly process.

Having established that environmental issues in design and manufacture are necessary to consider, we focus in the remaining two sections on how design teams can collaborate to achieve greener goals.

6 FRAMEWORK FOR COLLABORATIVE DESIGN

Recent trends in computing environments and engineering methodologies indicate that the future engineering infrastructure will be distributed and collaborative, where designers, process planners, manufacturers, clients, and other related domain personnel communicate and coordinate using a global weblike network (Nidamarthi et al. 2001). The designers may be using heterogeneous systems, data structures, or information models, whose form and content will probably not be the same across all disciplines. Hence, appropriate standard exchange mechanisms are needed for realizing the full potential of sharing information models. The various applications are coordinated by a work flow management system, which acts as a project manager. They are connected to one another by a design net, which provides the infrastructure for high bandwidth communications. These applications retrieve relevant design data and knowledge from distributed design repositories and the evolving design (or designs) is stored in a database. This database provides various snapshots of the evolving design, with design artifacts and associated design rationale stored at various levels of abstraction. Finally, design applications communicate with other manufacturing applications and databases through various nets, such as production, process planning, and user networks (Figure 8).

The information exchange between various applications does not occur at only one level. We envision the interoperability problems between heterogeneous engineering applications to occur at several levels:

- *Physical*: This level is concerned with the physical transmission medium, such as Ethernet, and fiber optics.
- *Object*: At this level, the engineering objects are transported using appropriate object transfer modes, such as CORBA (Common Object Request Broker Architecture), EJB (Enterprise Java Beans) or COM (Microsoft's Common Object Model).
- *Content*: This level deals with the communication of engineering artifacts and should include feature, constraint, geometry, material, process, and so on. Information at this level can be expressed in an appropriate modeling language, such as STEP's EXPRESS (http://www.nist.gov/sc4), KIF (Knowledge Interchange Format, http://logic.stanford.edu/kif/kif.html), or XML (Extensible Markup Language, http://www.w3.org/XML/).
- *Knowledge/design rationale*: This level deals with design rationale and design history issues, which provide additional information (including inference networks, plans, goals, justifications, etc.) about the engineering objects at the content level.
- *Communication*: This level provides additional detail to the content and knowledge/design rationale levels. Such details include the specification of engineering ontologies used, sender, recipient, and so on, as defined by the KQML standard (Knowledge Query Manipulation Language, http://www.cs.umbc.edu/kqml/).
- *Negotiation*: Any multiagent activity will involve negotiation activity. The protocols needed to conduct such negotiations will be defined at this level.

The design net provides the infrastructure for supporting the above communication levels. Next, we discuss issues involved in accessing heterogeneous environmental databases during various design stages. We focus on the content level protocol.

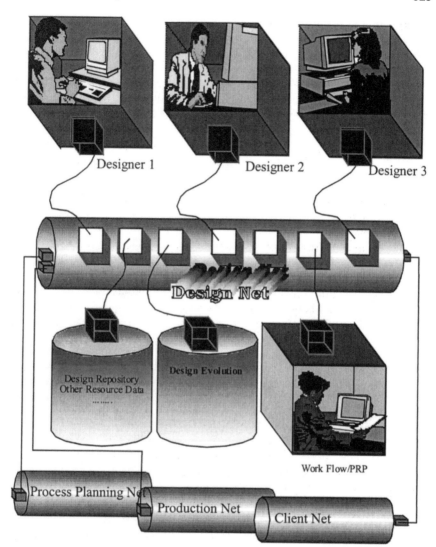

Designer 1   Designer 2   Designer 3

Design Net

Design Repository
Other Resource Data

Design Evolution

Work Flow/PRP

Process Planning Net

Production Net

Client Net

FIGURE 8   One framework for distributed design.

7 ENVIRONMENTAL DATABASE ACCESS

During the course of product design and manufacturing, engineering teams typically access several large databases, such as Numerica and Sigma-Aldrich's Hazardous and Regulatory Data Library. These databases typically contain physical and chemical characteristics, environmental impact data of chemicals, equipment information, and regulatory standards and guidelines (such as those addressing hazardous waste transport) (Graedel 1996; Liang and Garrett 2000). However, engineers encounter several problems accessing appropriate information. A few of these problems are enumerated below:

1. The data reside on heterogeneous databases, with different formats that require knowledge about multiple query languages, i.e., we have several independent databases, where each database has its own schema, is expressed in its own data model (relational, object-oriented, etc.), and is accessible through its own query language.
2. Query mechanisms provided with the databases are not in a format that is natural to the engineer.
3. An answer to a query may be spread across different databases, which may require generating multiple queries and their integration.
4. The engineer has to navigate through several pages before he or she gets the information required.
5. The database developers may have used a wrong data model, which results in inefficient retrieval strategies.

Here, we describe an architecture for an intelligent interface for integration of heterogeneous chemical and environmental databases and application programs. The architecture provides a virtual integrated database management system with a high-level data manipulation language (DML) for the engineering teams. The main issues that need to be addressed for integrating the heterogeneous databases are the following:

1. Resolving the incompatibilities between the different databases; these could include conflicting schema names and data types.
2. Resolving any inconsistencies between the databases; these could be inconsistencies in copies of the same information stored in different databases.
3. Transforming the query expressed in the high-level DML posed by the designer into a number of subqueries that need to be posed to various databases in their respective DMLs. The resulting responses to the subqueries are to be collected and an appropriate answer is to be filtered from them.

These concepts have been incorporated in a commercial software tool called Enchilada™, an information integration technology (TimeØ 2000) that is used to automatically access heterogeneous databases and web databases in real time and to integrate the results into a uniform data schema. The data can then be exported in XML, RDF (Resource Description Framework, http://www.w3.org/TR/rdf-schema/), or some other formats into a design application. Figure 9 depicts a schematic of Enchilada's function. The recipes serve as the basis of information integration that underlies our collaborative design framework and the user contexts place the heterogeneous information in a form that is most usable to the user.

7.1   Environmental Data Manager

Within the collaborative design framework, we include an *environmental data manager* (ENVDM), which provides an intelligent interface between various databases and application programs, as shown in Figure 10. Other data managers, such as Numerica and Dippr (http://www.tds-tds.com/), can also be attached to the design net as the need arises. (Pitts et al. 1998). In this example, we assume pharmaceutical design because its associated manufacturing process uses and produces chemical solvents. These have a profound impact on the environment. The interactions in this framework are as follows.

Any new design application that plugs into the design net has to provide interfaces to various design net services, which would involve

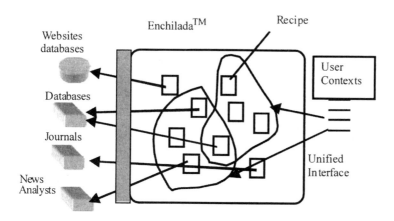

FIGURE 9   Enchilada accesses multiple databases via recipes "wrappers" that operate as independent information agents. The agents collect, normalize, and repurpose collaboratively the data into a user's context via a unified interface.

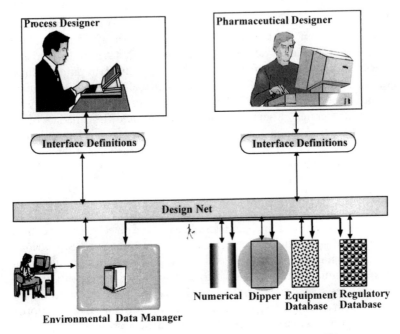

FIGURE 10 Schematic view of the intelligent database interface.

appropriate CORBA's interface definition language entities. In the current framework, each application should also map its domain terms into the terminology of ENVDM (as described below). By deploying intelligent agents on the net, ENVDM constantly monitors the environmental (and other associated) databases accessible on the design net. ENVDM's agents gather various metadata, such as local schemas of each database. The metadata is processed by ENVDM, as described below. When a designer queries the ENVDM, it accesses the appropriate databases and returns the needed information to the design team.

The ENVDM consists of the following modules.

1. Global data schema or an ontology, where an ontology is a collection of class objects with certain relationships between these objects.
2. Syntactic query translator, which transforms the queries expressed on the objects in ENVDM.
3. Query decomposer and access path selector, which translates a query over the internal model into a plan for processing the query. The plan consists of a set of queries each of which is

posed on exactly one database, a set of move operations to ship the results of these queries to each of the databases in an appropriate format, and a set of queries executed locally to integrate the results of the above queries.

4. Ontology builder, which generates the ontologies (this can be automated or user defined).

5. Persistent store, which stores the ontologies in a database. This was EXODUS—an object-oriented database management system developed at University of Wisconsin–Madison—in the initial prototype.

6. Knowledge-based inference engine, which is utilized by various modules. COSMOS, developed at MIT, was used in the initial prototype.

7. Query composer, which takes the results of a set of queries and presents them in an appropriate format to the designer.

8. Integration schema module, which keeps track of all metadata obtained from various databases and mapping between these database entities and the ENVDM ontology. For example, consider two databases each containing some physical properties of organic chemicals. One database might store the heat of fusion (at melting point) of benzene as 30.1 cal/g, whereas the other might record the same information as 2.370 kcal/mol. To integrate these two databases, we need information about mapping these two scales. Such information is stored in an integration database, whose schema is based on an object-oriented data model. The integration database might also contain information necessary for reconciling any inconsistencies between copies of the same data stored in different databases.

9. Intelligent agents, which roam the net and gather meta information about various databases. The meta information would include database type (relational, object-oriented, network), data/class dictionaries, access pathways and privileges, and other relevant information needed to retrieve data. It is assumed here that the databases have interfaces to the Internet. Huhns and Singh (1998) categorize agents (including many of the modules described in our work) into user, broker, ontology, mediation, and resource agents. Agents communicate using KQML, which is a knowledge-based query manipulation language developed with DARPA (Defense Advanced Research Projects Agency) sponsorship. At the content level, we believe that XML is the appropriate language for communicating domain data.

7.2 Example

The following example provides a flavor of querying on heterogeneous chemical and environmental databases. For simplicity, the transformations here are evident. However, in actual practice these transformations involve considerable processing. We assume one database to be in a relational format, which has the following relation:

	Substance	
Slot	Type	Constraint
Boiling point	Number	> 0
CAS registry	Number	Integer
Chemical formula	String	
Density	Number	> 0
Molecular weight	Number	> 0
Phase	Enum	One-of solid, liquid, gas
Toxicity	Number	> 0

The CAS Registry (Chemical Abstract Service Registry) allows a unique identifier for materials with multiple names. Benzene, for example, is also known as benzol, benzolene, and carbon oil. Let the other database be in an object-based format, with the following objects:

Environmental data
- Common name
- Half-life: (pointer to half-life)
- Fate-rate-constant: (pointer to fate-rate-constant)

Half-life
- Air
- Surface water
- Ground water
- Soil

Fate-rate-constant
- Volatization
- Photolysis
- Oxidation
- Hydrolysis
- Biodegradation
- Bioconcentration

Integration schema to link the synonyms of a chemical with its unique CAS registry number:
The above integration schema is attached to ENVDM.

CAS		
CAS registry number	Synonyms	
71432	Benzole, benzene, cyclohexatriene, ...	
100414	Ethylbenzene, phenylethane, ...	

Assume that the following query is posed to ENVDM: Find the half-life of phenylethane in surface water. The processing of this query involves the following two-stage process:

1. Generate the query: Find the CAS registry number of phenylethane. The response would be 100414.
2. Using the above response, generate the query: Find half-life frame for CAS number 100414 and find value of the surface-water slot. The response would be 5 hours.

With such a framework, environmental factors can be brought to bear on any design or manufacturing consideration and help play a pivotal role in the decision making of design and manufacturing teams.

8 SUMMARY

To create environmentally conscious products, product designers should consider environmental impacts of design when decision making during the product development process. Green engineering design is an approach to product and process design that achieves environmental consciousness without compromising a product's quality or its commercial viability. It takes a proactive approach to environmental problems rather than a reactive approach. The factors motivating green design were explored in this chapter.

The future engineering infrastructure will be distributed and collaborative, where designers, process planners, manufacturers, clients, and other related domain personnel communicate and coordinate using a global weblike network. We also presented an architecture for realizing effective collaborative green design, where designers access heterogeneous environmental-related databases. This access can be achieved through an object-oriented data management framework, which utilizes ontologies for semantic resolution.

DISCLAIMER

The bulk of the work reported here by the first author was conducted during his tenure at MIT. Commercial equipment and software, many of which are either registered or trademarked, are identified to adequately specify certain procedures. In no case does such identification imply recommendation or endorsement by the National Institute of Standards and Technology nor does it imply that the materials or equipment identified are necessarily the best available for the purpose.

REFERENCES

Barkmeyer EJ. Background study—requisity elements, rationale, and technology overview for the Systems Integration for Manufacturing Applications (SIMA) program. NISTIR 5662, Gaithersburg, MD, 1995.

Barkmeyer EJ. SIMA reference architecture: activity models. NISTIR 5939, Gaithersburg, MD, 1996.

Caspersen NI, Sorenson A. Improvements of products by means of lifecycle assessment: high pressure cleaners. J Cleaner Product 6:371–380, 1998.

Chen RW, Navin-Chandra D, Prinz FB. A cost-benefit analysis model of product design for recyclability and its application. IEEE Trans Comp Packag Manufact Technol A 17:502–7, 1994.

Graedel TE, Allenby BR. Design for Environment. Englewood, NJ: Prentice Hall, 1996.

Hersh MA. Survey of systems approaches to green design with illustrations from the computer industry. IEEE Trans Systems Man Cybernet C Appl Rev 28:528–540, 1998.

Huhns MN, Singh MP. All agents are not created equal. IEEE Internet Comput 2:94–96, 1998.

Liang V-C, Garrett J Jr. Java-based environmental regulation broker. J Comput Civil Eng 14:100–108, 2000.

Lippiatt B. Building for environmental and economic sustainablity (BEES). In CIB World Building Congress 1998: Construction and the Environment, Sweden, 1998 (http://www.bfrl.nist.gov/oae/publications/proceedings/bees.pdf).

Naar J. Design for a Livable Planet. New York: Harper and Row, 1990.

Navin Chandra D. Steps Toward Environmentally Conscious Engineering Design, A Case for Green Engineering. Technical Report, Robotics Institute, Carnegie Mellon University, CMU-RI-TR-90-34, 1990.

Navin Chandra D. Exploration and Innovation in Design: Towards a Computational Model, New York: Springer-Verlag, 1991.

Navin Chandra D. The recovery problem in product design. J Eng Design 5:67–87, 1994.

Nidamarthi S, Allen RH, Sriram RD. Observations from supplementing the traditional design process via internet-based collabortion tools. J Comput Integrated Manufact 14:95–107, 2001.

Pal U. Report on Green Manufacturing, Report submitted to the National Institute of Standards and Technology, copies can be obtained from Dr. Uday Pal, Department of Manufacturing Engineering, Boston University, MA, 1998.

Pitts G, Fowler J. Collaboration and knowledge sharing of environmental information: the EDEN Project. In Proceedings of the 1998 International Symposium on Electronics and the Environment. URL, 1998: http://www.mcc.com/projects/env/pubs.html.

Resetar SA, Lachman BE, Lempert RJ, Pinto MM. Technology forces at work. Rand Document No. MR-1068/1-OSTP, Arlington, VA, 1999.

Reinhardt FL. Down to Earth: Applying Business Principles to Environmental Management. Harvard Business School Press, 1999.

Sears B. The Zone: A Dietary Road Map. New York: HarperCollins, 1995.

Sullivan JL, Williams RL, Yeser S, Cobas-Flores E, Chubbs ST, Hentges SG, Pomper SD. Life Cycle Inventory of a Generic U.S. Family Sedan: Overview of Results USCAR AMP Project, SAE Paper #982160, 1998. http://www.sae.org.

TimeØ. Enchilada™ Information Integration Service, 2000. http://www.time0.com/Platform/Utility_services/index.html

Wenzel H, Hauschild M, Alting L. Environmental Assessment of Products: Methodology, Tools, and Case Studies in Product Development. New York: Chapman and Hall, 1997.

Wivell CD, Navin Chandra D. Disassembly Analysis Manual. EDRC-05-63-92. Technical Report, Green Engineering Project, Carnegie Mellon University, 1992.

Useful URLs:
http://greenmfg.me.berkeley.edu/green/Home/Index.html
http://www.ce.cmu.edu/GreenDesign/
http://www.uwindsor.ca/imse/people/ecdm_info.html
http://www.mcc.com/projects/env/resources.html
http://ecologia.nier.org/
http://ceq.eh.doe.gov/reports/1993/chap9.htm
http://www.time0.com/Platform/Utility_services/index.html

15

Life Cycle Assessment
Discussion and Industrial Applications

John L. Sullivan
Ford Motor Company, Dearborn, Michigan

1 INTRODUCTION

During the late 1980s and the decade of the 1990s, a heightened awareness and concern arose about the state of the world's environment. For instance, the media, environmental groups, and politicans both in United States and Europe frequently pointed to a growing solid waste problem. Often evening news programs would show clips illustrating waste disposal problems such as barges loaded with garbage with no place to dispose of their loads and beaches littered with wastes like hypodermic needles, serum bags, garbage, and paper. Also, plastic waste was more frequently being found in waterways, like the plastic rings used for beverage six-packs, some of it choking and snaring marine life. New York City was for a while dumping its waste in the Atlantic Ocean, but sometimes it shipped it to nearby states for burial. At the time, it seemed that we were running out of landfills. The good news is that out of the numerous solid waste concerns came a groundswell of sentiment to improve product and material recycling.

During this period, the notion of total product stewardship was advanced; it advocates that manufacturers should be responsible for their products over the entire product life cycle, that is, from cradle to grave. Also, at that time arose an important environmental improvement initiative for vehicles, called the partnership for a new generation of vehicles (PNGV) program. The objective of that program, a cooperative joint development agreement between the U.S. government, Ford Motor Company, General Motors, and Chrysler (now DaimlerChrysler) Corporation, was to develop vehicles with fuel efficiencies of 70 miles per gallon or more. The PNGV program arose from concerns about fossil fuel depletion and global warming due to carbon dioxide emission. Recently, global climate change concerns are becoming even more acute as more experts express their judgment that anthroprogenic carbon dioxide is affecting global climate. Indeed, international agreement has already radically limited the use of chlorofluoro carbons (CFCs) due to their role in stratospheric ozone depletion. Similar agreements have been proposed for carbon dioxide emission reductions in the Kyoto protocols.

Today, many manufacturers, including the electronics, automotive, and chemical industries, seek ways to improve the environmental performance of their products. They have looked for ways to extend product useful lifetime (a form of source reduction), recycle the materials of discarded products, refurbish and reuse old products, improve product efficiency (e.g., improved vehicle fuel efficiency), and substitute newer technology materials for traditional materials.

But are products that have been redesigned to improve their environmental performance really better? Sometimes product environmental improvement claims are misleading. The so-called ZEV, which is an electric vehicle, is a good example. Technically, the term stands for a zero emission vehicle, which to some means that it emits no (or little) CO, NO_x, SO_x, particulate, and hydrocarbons, whereas for others means it gives off nothing, including CO_2 or water. Clearly, it takes electricity to charge the batteries of these vehicles and that requires electric power plants, which generally produce electricity by the combustion of hydrocarbon fuels which in turn generate emissions. Perhaps it is better to refer to ZEVs as EEVs, that is, "emissions elsewhere vehicles" (Graedel 1995). This example is a case of an environmental improvement that moves emissions from the product use stage of the product life cycle to the fuel production part of the life cycle. Further, it illustrates how misperceptions can arise when a product is viewed as a "stand alone" entity rather than as a part of a product system. Hence, without taking a systemwide view of a product, an environmental improvement may not be as large as it appears. Indeed, it could even be negative.

There is a methodology that has been developed to provide a product systemwide perspective. It is called life cycle assessment (LCA) and is aptly suited to identifying burden shifting from one life cycle compartment to another. LCA is a cradle to grave environmental assessment methodology that considers all product environmental burdens (resource consumption and wastes/emissions generated) over the entire life cycle; from material production to part manufacture, product assembly, operation, servicing, maintenance, and end-of-life (EOL) disposition. LCA provides a needed big picture holistic perspective of a product system. Furthermore, it is the quantitative component of design for environment (DFE) and industrial ecology approaches to product and process environmental improvement.

In the sections to follow, LCA is described detailing many issues pertaining to its proper use. The development of the method and other approaches to it are also discussed. Also provided is a section on applications of LCA, including some for vehicle systems. Finally, a section is devoted to needed improvements of the method.

2 LIFE CYCLE ASSESSMENT

LCA is a methodology used to conduct environmental assessments of products over their entire life cycle, ranging from raw material extraction from earth to product manufacture, use, and recovery/disposal at product EOL. Since its inception (Fava 1991) the methodology has become increasing visible, particularly in the international environmental community. Some view LCA as a regulatory tool; others see it being used for ecolabeling purposes. In our view, LCA is best used for voluntary assessments conducted by industry on the overall environmental performance of their products. The strengths of LCA are to provide a macroview of the environmental performance of a product system and to monitor environmental burdens (consumed resources and generated wastes) of a product over its entire life cycle. The latter is particularly important, since sometimes an alleged environmental improvement has in reality shifted burdens from one part of the life cycle to another.

2.1 Background

The technical framework for LCA (Fava 1991) was first developed about a decade ago at a workshop held at Smugglers Notch, Vermont. The workshop was attended by 54 scientists and engineers, mostly technical experts on environmental issues from around the world representing government, public interest groups, the industrial sector, consultants, and academics. The roots of LCA go back to the early 1960s when there was an interest by some to elucidate the influence of population trends on raw materials and energy consumption. From those studies, some dire consequences were predicted, including global warming (due to waste heat—now known to be a regional problem) and ice cap melting. During this same general period, a number of fuel cycle studies were conducted to estimate the environmental and cost implications of alternative energy sources. Cradle to grave industrial energy analyses (Boustead 1979) were routinely conducted during the mid-1970s. In fact, resource and environmental profile analysis (Hunt 1974) was developed at that time and subsequently served as the basis for the development of LCA.

While the level of public sector concern about environmental issues tends to be cyclical, it became evident to the environmental science and engineeering community at that time that products can no longer be viewed solely in terms of the service they perform or their direct environmental burdens. Also needed for consideration are upstream and downstream environmental burdens. For example, in a vehicle LCA not only must vehicle fuel combustion burdens be considered but also those incurred in making the fuel. Hence, it became recognized that the principle advantages for conducting LCAs are as follows:

1. To provide a holistic view of the interaction between an industrial activity or product system and the environment;
2. To advance the understanding and general awareness of the interdependence of environmental consequences;
3. To provide decision makers with relevant information about potential environmental consequences of activities and to help identify environmental improvement opportunities for such activities.

2.2 Components of LCA

LCA, as per the Society of Environmental Toxicology and Chemistry (SETAC), is a four-component product environmental assessment methodology, including goal definition and scoping, inventory assessment, impact assessment, and improvement assessment phases. A depiction of the method is given in Figure 1. The foundation of the methodology is life cycle inventory (LCI) Analysis, which is an objective quantification of all energy and raw material requirements and generated emissions (air, water, and solid) incurred throughout the entire life cycle of a product, activity, or process.

Life cycle impact assessment (LCIA) is a process where the environmental loadings identified in the inventory are quantitatively or qualitatively attributed to environmental and human health effects. The final component of LCA is life cycle improvement analysis where various proposed product environmental improvements are assessed in terms of their inventory and

Figure 1   The SETAC depiction of the LCA.

impact performance and then compared to a base case. Improvement assessment is in reality an activity of DFE.

LCA describes a product system in terms of life cycle stages (Figure 2). All materials and energy inputs and outputs (flows) crossing the system boundary are quantified. Though not explicitly depicted in Figure 2, flows are also quantified for each unit process in the product system. In International Organization of Standards (ISO) terminology, the term elementary flow is often used. It refers to material and energy flows into a product system from the environment without previous human transformation or from the system to the environment without subsequent human transformation. Hence, while an intermediate product like plastic pellets moving from unit process A to process B is a flow, it is on the other hand not an elementary flow.

The ISO view of LCA is depicted in Figure 3. The ISO approach, also a four-component method, is seen as a more iterative process than the one shown in Figure 1. Further, instead of the improvement assessment phase of LCA, the ISO method has an interpretation component. In ISO, improvement assessment is considered just one of a number of reasons to conduct an LCA, and they would be outlined in the goal definition and scoping phase. Much of the methodological discussion to follow is based heavily on ISO 14040 series of standards. As seen in the sections to follow, an LCA begins with goal definition and scoping, ends with life cycle interpretation, and generates the information required to meet study objectives during the LCI and LCIA components.

Life Cycle Stages

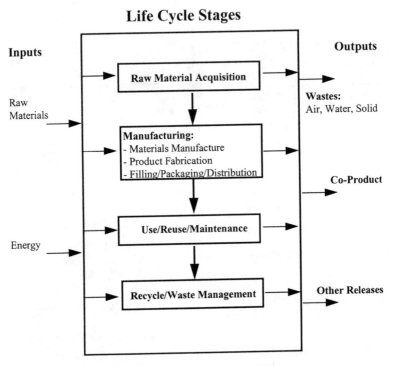

FIGURE 2   Life cycle stages of a product system.

2.3 Goal Definition and Scoping

The first component of LCA is goal definition and scoping. Though the original LCA framework (Fava 1991; Vigon 1992) recognized the need to conduct a goal definition and scoping exercise, it was not until later that it became a separate phase of LCA (Consoli 1993; Husseini 1994; Fava 1993; ISO 1997, 1998). At the beginning of an LCA, the goal of the study should be unambigiously stated along with the reason for conducting the study and the intended audience. For example, a manufacturer could elect to conduct an LCA on a product to determine its global warming potential, which is to subsequently serve as a benchmark for the environmental performance of some of the company's new products under design. The purpose should be stated in terms of the intended decision, data required, and level of detail. It is also important to distinguish between external or internal applications of study results. For example, an LCA study could be intended to influence public policy. There are many reasons to conduct an LCA, including pollu-

Life Cycle Assessment Framework

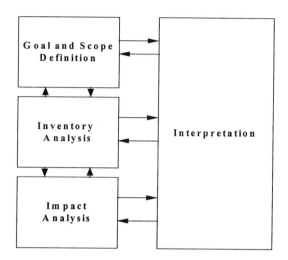

FIGURE 3   The ISO LCA framework.

tion prevention, comparisons of products and processes, site and technology selection, ecolabeling, strategic management, and product development and improvement.

An important part of the scoping process is to define the system and its boundaries, study assumptions and associated limitations, and data requirements. The scope of the study should be sufficiently broad and deep to support the study purpose. In defining the scope of a project, one also considers system function, the functional unit, allocation procedures, types of impact analysis to be used, data quality requirements, and a critical review, if any.

One of the most important definitions in an LCA is the functional unit (i.e., X gallons of soda to be packaged with Y pounds of 16 oz. beverage containers). Without definition of the functional unit, it is difficult to unambiguously attribute environmental consumption and loading to the product. In fact, the functional unit is not only important in LCA studies but also in cost analyses (activity-based costing, life cycle cost). The functional unit is a measure of the service performed. Following the definition of the functional unit, the system boundaries need to be defined. This is also an important step, for otherwise two studies on the same product could come out radically different. For example, if one wanted to compare two LCAs on automobiles and one study included the vehicle's infrastructure (roads, parking lots, garages, service stations, etc.) and the other did not, results from the two studies could not be directly compared without suitable adjustment.

During establishing the system boundaries, all germain life cycle stages, unit processes, and flows should be taken into consideration. This includes quantifying inputs and outputs of main production processes, distribution and transportation, production and consumption of fuels, and recycling. Figure 2 is a simplified diagram of a product system; typically, more detailed diagrams are required in practice. Developing process flow diagrams is indispensible as it highlights unit processes and their interrelationship. A general unit process is shown in Figure 4. Examples of unit processes include stamping steel, painting cars, injection molding plastics, cracking petroleum, and many others. The flow diagrams for each process with quantified flows leads ultimately to a more quantitative representation of the product system.

Next, the data categories need to be defined, again consistent with the goal of the study. An example of a list of data categories is given in Table 1, which are the categories used in the USAMP General Vehicle Life Cycle Inventory Study (Sullivan 1998). Generally speaking, data categories are chosen for both their environmental relevance and significance to the product system under discussion. For example, carbon dioxide and other combustion emissions are not a likely concern for a product that requires little energy to produce, such as a wooden stepladder, but they could be for one made of aluminum.

Not every input or output to a product system needs to be traced back to earth. In some cases, the omission of upstream (back toward earth) life cycle burdens of some aspect or component of the product system has little impact on system life cycle performance. For example, the sheet metal screws that hold the shrouding to a space heater are not likely to have an appreciable effect on the product's life cycle. The criteria usually applied to

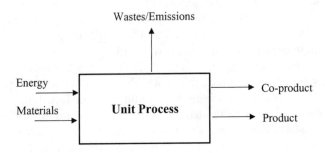

FIGURE 4   The general unit process.

TABLE 1 Data Categories for the USAMP/LCI Generic Vehicle LCI Study

1. Energy	4. Water emissions
Fossil	Dissolved solids
Nonfossil	Suspended solids
Process (electrical and nonelectrical)	Heavy metals
Transportation	Oils and greases
Feedstock	Other organics
Total	Phosphates and ammonia
2. Water consumption	5. Solid wastes
Ground	Total solid waste
Surface	Sanitary/municipal
3. Air emissions	6. Raw material consumed
Dust and particulates	All significant inputs as
Carbon dioxide	defined by the decision
Carbon monoxide	rules
Sulfur oxides	
Nitrogen oxides	
NMHC (includes halogenate	
hydrocarbons)	
Methane	
Acid gases: HCl and HF	
Pb	

the question of materials or product subsystem inclusion are based on mass, energy, or environmental relevance (ISO 1998).

Data quality critically influences the reliability and utility of LCA results. Unfortunately, LCI data tend to be quite variable. This is due to the use of data from different periods, from different geographic locations (was the aluminum made here or in Europe?), from a mix of technologies (e.g., was old or new manufacturing equipment used?) to make the same product system, and finally from facilities with different operating efficiencies. The indicators recommended by ISO 14041 to characterize the quality of LCI data, called data quality indicators (DQI), are precision, completeness, consistency, representativeness, and reproducibility. Precision is a measure of the variability of the data; completeness is the percentage of locations reporting data from the total number of possible reporting locations, consistency is an assessment of how uniformly the methodology has been applied to all components of the study; representativeness assesses the degree to which the study represents the true population; and finally reproducibility is an appraisal of the extent to which information on the methodology and the data permits an independent practitioner to reproduce study results. The first two indicators are quantitative and the remaining ones are qualitative. A more complete discussion of data quality issues can be found elsewhere (Fava 1992; Barnthouse 1998; Husseini 1994).

Because LCI data tends to be quite variable and generally not sampled very broadly, it is difficult to calculate means and standard deviations from such results. While qualitative DQIs are useful, precision is an important statistical quantity that provides decision makers with a measure of confidence in LCI results. Hence, some way to estimate means and standard deviations would be very useful. Fortunately, under development is a stochastic approach, which generates means and standard deviations from qualitative DQIs determined by experts (Kusko 1997; Kennedy 1996). In the approach, DQIs are developed for all the data elements of an LCI model to arrive at a mean and standard deviation for the product system as a whole. Qualitative DQIs alone are generally considered by most LCA practicaners as inadequate.

The final step in goal definition and scoping is to consider whether a critical review for the study is required. Reviews can be either internally or externally conducted. An external review is optional for internal studies but is considered mandatory in the ISO Standard (ISO 1998) for publicly disclosed studies making product comparative assertions. The purpose of the review is to certify that the study has been conducted in accordance with the ISO standards, to enhance its scientific and technical quality, and to help focus study goals, data collections, and critically screen study conclusions. In short, the objective is to enhance the understanding and credibility of LCAs, especially for interested parties and stakeholders. Due to the complexity of most LCAs, the numerous available methodological options, and the many data quality and availability issues, it is recommended that some kind of a review be conducted for all studies.

While goal definition and scoping is done early in a study, it is expected that the entire LCA process is quite iterative, as indicated in Figure 3. For example, it might be discovered that important data needed for one of the study's LCIA objectives is unavailable and cannot be inventoried. Hence, project objectives would need to be revised.

2.4 Life Cycle Inventory

As prescribed by ISO 14041, LCI is comprised of the following steps: gathering data, verifying to its quality and validity, properly allocating data to the product/process system in question, running the model to generate the inventory output, and interpreting and assessing the quality of results.

While data collection sounds simple, it does require preparation and is often difficult to collect. Especially for suppliers, the difficulty arises from concerns over revealing proprietary and/or confidential information that can being used against them by government, competitors, or the requesting entity. Also, there is difficulty in separating data for unit processes for multiproduct operations. However, as more industries adopt sustainable practices and want to be identified as doing so, the reluctance to provide LCI data should decrease. The data collection process is greatly facilitated by the flow charts and unit processes defined in the previous stage. Sometimes data has to be acquired from the literature. This is called secondary data, and when it is used it must be cited as such. For other details pertaining to data collection, see ISO 14041 (1998).

After the data have been collected, calculation procedures are applied to generate the results of the inventory. This is done for all defined unit processes, which are subsequently aggregated first by life cycle stage and finally summed to represent the functional unit of the product system. The environmental burdens list for the product is written by life cycle stage as follows:

$$\{\mathbf{B}\}_{tot} = \{\mathbf{B}\}_{mp} + \{\mathbf{B}\}_{pr} + \{\mathbf{B}\}_{op} + \{\mathbf{B}\}_{mn} + \{\mathbf{B}\}_{eol} \tag{1}$$

where $\{\mathbf{B}\}$ is the vector of emissions generated and resources consumed, and the subscripts tot, mp, pr, op, mn, and eol denote total, materials production, product manufacture, operation, maintenance, and end of life, respectively. Note that material production includes raw material acquisition, and all stages include their own transportation contributions. For example, the operational energy (E_{op}) of a wash machine includes the transportation of the coal to the power plant to produce the electricity to run the machine.

However, before calculating the final results, often times adjustments and validity checks need to be applied to the data. This is particularly true for fuels and electricity consumption where it is important to account for fuel cycle elementary flows. For example, the combustion energy (high heat value [HHV]) of gasoline is 47.2 MJ/kg of fuel burned, but fuel needed to produce that fuel is 8.9 MJ, bringing the life cycle energy (LCE) for the consumption of gasoline to 56.1 MJ/kg. LCEs for making a number of materials are given in Table 2. Some of the energy values have been reported before (Sullivan 1995).

Table 2 is a semicurrent listing of some LCE and total wastes estimates associated with making various materials. For some of them, depending on the age of the data, the variation in LCE might be as large as ±40% (Sullivan 1995); airborne waste values probably have similar variations due to the dominance of CO_2 emissions from energy consumption via combustion. Variations arise from two sources: (1) different technologies, old vs new or efficient versus inefficient, used in making a given material and (2)

TABLE 2 Estimates of Cradle to Gate Energy and Wastes for the Production of 1000 Pounds of Various Primary Materials

Material	Energy Consumption (Thousand BTU)	Solid Waste (lb)	Waterborne Waste (lb)	Airborne Waste (lb)
Aluminum sheet	86.6	2200	52	12,100
Brass	18.2	2700	10	2650
Cast iron	15.5	5100	8.5	2520
Copper wire	25.4	20,900	14.5	3800
Polyurethane foam	48.1	290	26.6	4700
Glass	7.0	98	2.5	1000
HDPE	36.2	83	6.6	1960
LDPE	39.7	140	7.7	2500
Lead	21.8	360	5.7	3050
Natural rubber	46.6	170	8.3	4200
PET	34.9	145	15.4	2800
Phenolic	38.6	73	13	2200
Polycarbonate	67.9	420	21	6270
Polypropylene	36.9	79	6.9	2040
PVC	31.7	270	7.9	3080
Synthetic rubber	64.6	138	10	4300
Steel	21.4	4600	17	3550
Zinc	24.4	451	7.5	3400

inconsistent application of LCI, particularly in setting system boundaries. Eventually, these variations will diminish at least in part due to the emergence of standardized life cycle practices.

An important data issue is to verify that the results from all processes have been appropriately allocated to products and coproducts. For example, in the production of caustic soda, chlorine gas is a coproduct that is sold to make other products. However, the allocation of chlorine's share of caustic soda life cycle burdens should include only common processes required to make both chemicals. Allocations to caustic soda should not occur for processes unique to chlorine and vice versa. Allocation can be a contentious issue. The ISO 14041 recommends avoiding allocation if possible or if necessary use mass, energy, or some other reasonable basis.

Examples of product input and output inventory data are given in Tables 3 and 4. These results are from the USCAR Generic Vehicle Life Cycle Study (Sullivan 1998); the functional unit is the five-passenger generic D-class family sedan with a lifetime drive distance of 120,000 miles.

An LCI ends with an assessment of the limitations of the analysis. Included in the analysis are a data quality assessment and a sensitivity

TABLE 3 LCI of the Generic Vehicle (Raw Materials Use)

Environmental flow: inflow	Units	Generic vehicle	Mat'ls production	Mfg. assembly	Fuel use	Maintenance & repair	EOL
(r) Bauxite (Al_2O_3, in ground)	kg	233	233	0.003		0.28	
(r) Chromium (Cr, in ground)	kg	0.94	0.94				
(r) Coal (in ground)	kg	2554	1034	635	776	99	11
(r) Copper (Cu, in ground)	kg	24	24	1.0 E-04			
(r) Ilmenite ($FeO.TiO_2$, in ground)	kg	0.98	0.33			9.9 E-05	0.045
(r) Iron (Fe, in ground)	kg	1486	1483	0.65		3.0	
(r) Lead (Pb, in ground)	kg	33	13	0.38		20	
(r) Limestone ($CaCO_3$, in ground)	kg	454	187	0.27	147	20	2.0
(r) Manganese (Mn, in ground)	kg	24	23	98		0.74	
(r) Natural gas (in ground)	kg	1827	470	219	1065	71	2.2
(r) Oil (in ground)	kg	16,842	419	86	16,133	169	35
(r) Olivine (in ground)	kg	8.6	8.5	1.0 E-05		0.0032	
(r) Perlite (SiO_2, in ground)	kg	2.5	2.4	0.057			
(r) Platinum (Pt, in ground)	kg	0.0015	0.0015				
(r) Pyrite (FeS_2, in ground)	kg	13	13			4.3 E-05	
(r) Rhodium (Rh, in ground)	kg	3.0 E-04	3.0 E-04				
(r) Sulfur (S)	kg	0.10	0.082	0.022		4.0 E-05	
(r) Tin (Sn, in ground)	kg	0.49	0.069	0.42			
(r) Tungsten (W, in ground)	kg	0.012	0.012			6.8 E-04	
(r) Uranium (U, in ground)	kg	0.044	0.014	0.009	0.019	0.0019	2.5 E-04
(r) Zinc (Zn, in ground)*	kg	18	18	0.051		4.3 E-04	
Iron scrap	kg	249	206			43	
Natural rubber	kg	25	9.1	9.4		16	
Raw materials (iron casting alloys)	kg	12	12				
Raw materials (unspecified)	kg	16	6.4			0.32	
Steel scrap	kg	492	444			49	
Water used (total)	L	80,413	62,941	9936	2080	5452	4.0

a Uranium use is for electricity production.

TABLE 4 LCI of the Generic Vehicle (Outflows and Energy Use)

Environmental flow	Unit	Generic vehicle	Mat'ls production	Mfg. assembly	Fuel use	Maintenance & repair	EOL
Outflow							
(a) Dust and particulates	g	55,301	27,414	8449	17,142	613,532	247
(a) Carbon dioxide (CO$_2$, fossil)	g	61,326,600	4,608,670	2,614,900	53,346,300	39,083	143,273
(a) Carbon monoxide (CO)	g	1,937,590	58,778	5938	1,833,109	4415	683
(a) Sulfur oxides (So$_x$ as SO$_2$)	g	137,380	31,163	15,253	86,234	2739	315
(a) Nitrogen oxides (NO$_x$ as NO$_2$)	g	255,703	12,986	8429	230,744	2148	806
(a) Nonmethane hydrocarbons	g	258,712	6930	14,142	235,321	3828	171
(a) Methane (CH$_4$)	g	69,578	13,810	5662	46,134	29	144
(a) Hydrogen chloride (HCl)	g	749	286	11	417	2.1	5.7
(a) Hydrogen fluoride (HF)	g	117	61	1.2	52	63	0.71
(a) Lead (Pb)	g	117	51	1.2	1.1	1053	
(w) Dissolved solids	g	7996	4771	1137	1018	512	
(w) Suspended solids	g	76,117	1993	2516	71,038	5.1	
(w) Heavy metals (total)	g	52	38	9.0	2.5 E-08	56	
(w) Oils and greases	g	7899	141	522	7172	7.2	
(w) Other organics	g	512	122	67	316	0.42	
(w) Phoslphates (as P)	g	16	7.4	7.8		12	
(w) Ammonia (as N)	g	2435	115	17	2829	277	
Waste (total)	kg	4376	2554	408	812	41	
Waste (municipal and industrial)	kg	430	23	70	8.3 E-05		
		17	58	0.0013	7.4	0.28	0.015
Energy reminder							
E (HHV) total energy	MJ	995,089	85,509	39,894	851,078	16,445	
E (HHV) fossil energy	MJ	988,881	81,677	39,088	849,894	16,075	
E (HHV) nonfossil energy	MJ	6212	3833	806	1184	373	
E (HHV) process energy	MJ	957,858	67,447	37,407	843,904	8355	
E (HHV) feedstock energy	MJ	25,964	16,667	961	319	8016	
E (HHV) transportation energy	MJ	11,270	1397	1527	6854	75	

analysis of all significant inputs and outputs and of choice of methodology. The question being addressed is whether the LCI results (inputs and outputs) support achieving the goal and scope of the study.

2.5 Life Cycle Impact Assessment

LCIA, the third phase of an LCA, is the process that imparts environmental significance to the output of an LCI. As seen in Tables 3 and 4, there is considerable output information derived from an LCI. While sometimes LCIs are used as a basis to make product decisions, this is generally difficult to do unless the data categories can be significantly reduced. LCIA is a way to do that. Afterall, the output from an LCI represents input and output data and not environmental impacts. In fact, ISO 14041 prescribes that an LCI study alone should not serve as a basis for product or process comparisons.

To demonstrate environmental significance, LCI data should be applied to an LCIA scheme. LCIA is a three-step process; as shown in Figure 5 LCI data are first binned in environmentally relevent categories in a process called classification. Next is a process termed characterization, where numerous LCI entries in each of the classification categories are further reduced to a single metric per category. However, this can only be done if scientifically supportable. Finally, in a process called valuation, all individual category values are reduced to a single environment metric for the entire product system. The valuation process is controversial and has been designated an optional element of the ISO LCIA Protocol (Barnthouse 1998). In the next few paragraphs, the LCIA steps are described in some detail.

LCIA has advanced considerably (ISO 2000; Udo del Haes 1999) in the last 5 years. There appears to be a general consensus on impact categories, which are listed in Table 5, along with the areas of protection that they represent. As seen in Table 5, all impact categories can be grouped into one or more areas of protection: human health, natural resources, ecological health, and artificial environment (Udo de Haes 1999). In each of the four areas of protection, there are specific impacts, sometimes called end points, which are desirable to avoid. For example, the loss or impairment of human life, loss of biodiversity, fishkill, and others are all end points. In general, each classification impact category can contribute to one or more end points.

Before continuing, some LCIA terminology should first be defined and clarified. The term stressor, often used in LCIA discussions, denotes conditions that may cause impacts, that is, harm to ecological and human health or depletion of resources. In short, stressors are either system inputs or

Classification   Characterization   Valuation

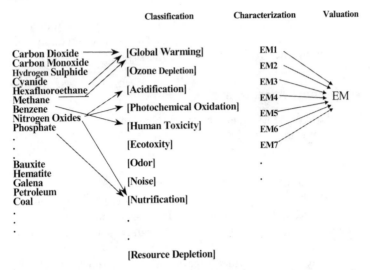

FIGURE 5   Depiction of LCIA classification, characterization, and valuation.

outputs; some examples are carbon dioxide, sulphur oxide emissions, toxic emissions, or an amount of consumed bauxite. There is typically a series of steps between a stressor and a category end point. For example, carbon dioxide emissions could lead to a global temperature rise, which could cause a rise in sea level, in turn causing coastal flooding resulting in loss of crops and habitat. These losses are end points. Resource depletion is also an impact category. There are three basic types of resources: reserve, stock, and flow. A reserve is a standing amount of a renewable or nonrenewable resource, a stock is a standing amount of a nonrenewable reserve, and flows are replenishable.

Also, associated with impact categories are the terms "category indicator" and "indicator result." A category indicator is a metric that represents the environmental mechanism by which the category is affecting the environment. For climate change, the indicator would be radiative forcing. On the other hand, the indicator result (or characterization) is an equivalent amount of a reference emission required to yield the same category indicator amount (e.g., radiative forcing) for all emissions of that category acting together. Since the science connecting end points to impact category results is generally not mature, common LCA practice is to assume an environmental relevence to the category and to use some intermediate indicator, called a midpoint, to represent the category. An example of a midpoint is the equivalent amount of carbon dioxide representing the net effect of all the warming gases in the climate change category for a particular product.

TABLE 5 Life Cycle Potential Impact Categories with Scales and Areas of Protection

Classification category	Spatial scale	Temporal scale	Human health	Ecological health	Resource depletion	Artificial environment
Climate change CI: radiative forcing IR: eq. CO_2	Global	C/D	X	X		X
Ozone depletion CI: ozone deplet. pot. IR: eq. CFC-11	Global	C/D	X	X	X	X
Human toxicity CI: YOLL, NOEC IR:			X			
Ecotoxicity CI: PAF, PNEC IR:	Reg/Lc	hr-yr		X		X
Photo-oxidation CI: PCOP IR:	Reg/Lc	hr/day	X	X	X	X
Acidification CI: Release of H^+ IR:	Cn/Reg	yr	X	X	X	X
Nutrification CI: IR:	Reg/Lc	yr		X	X	X
Abiotic resc. use					X	
Biotic resc. use					X	

CI, category indicator; IR, indicator result; Cn, continental; C, centuries; D, decades; PAF, potentially affected fraction; PCOP, photochemical oxidation potential; PNEC, potential no effect concentration; YOLL, years of lost life; NOEC, no observable effect concentration.

Generally, there are a number of gases in the global climate change category, such as carbon dioxide, methane, dinitrogen oxide, and numerous CFCs. Due to the physics of atmospheric radiation forcing (IR absorption and warming), each of these gases can be compared to an equivalent amount of carbon dioxide that absorbs the same amount of infrared radiation. Because each gas has a carbon dioxide equivalency factor, a single indicator result in terms of carbon dioxide can be computed to represent all the gases in the warming gas category for a product system. Though not fully developed for all impact categories, tables of characterization factors for various emissions in the impact categories have been developed (Heijungs 1992a,b). In the cases of stratospheric ozone depletion and global warming, the science is well established.

The final stage of LCIA is valuation, which in my view should be avoided. It is a subjective assessment where it is argued that X units of one impact is worth Y units of another. An approach often used to arrive at such relationships or weighting is willingness to pay; this is where respondants are asked what they are willing to pay to avoid certain consequences. In my opinion, reducing LCA results down to a single metric is ill conceived; instead developing a few metrics representing relevent categories is the more objective approach to LCIA. In the interpretation stage, decision makers can weigh the trade-offs.

LCA is a relative approach to environmental assessment and not an absolute one as in the case of ecological and environmental impact assessment (Barnthouse 1998). Ecological and environment risk assessment seek to measure or predict actual impact, including spatial and temporal considerations. LCIA, on the other hand, aims to make functional unit distinctions between products for the product systems as a whole. Generally, LCI data lack temporal and spatial information required to conduct environmental impact assessments (Barnthouse 1998). This has particular significance in considering regional, local, or global impacts.

The execution of an LCA involves only gathering as much data as required to meet the project objectives. ISO 14042 prescribes that data categories be selected for a study during goal definition and scoping. This process assures that the data gathering effort is consistent with study goals, including product/process environment evaluations. In short, LCIA drives LCI data collections and not vice versa.

LCIA is still under development, which is likely to continue into the forseeable future. A number of approaches to LCIA have been explored in the last decade, and a review of some them has been published elsewhere (French 1995).

2.6 Life Cycle Improvement/Interpretation

The last stage of the SETAC LCA framework is life cycle improvement assessment. It is the least developed, though some effort has been devoted to that end (Husseini 1994). The primary objective of the phase is to improve the environmental performance of the project or process in question. As mentioned above, this is in reality a DFE or industrial ecology activity, and in the most current LCAs, the improvement objective, if present, would be stated in the goal definition and scoping phase of LCA. Incidently, it is important to recognize that in typical product design assessment, engineering performance and cost are also considered and tend to be dominant. A matrix approach considering legal, cultural, cost, performance, and environmental factors has been advanced (Keoleian 1994) and applied to vehicle intake manifolds (Kar 1996) and instrument panels (Keoleian 1997).

The ISO standards (14040, 14041, and 14042) make no reference to improvement assessment but instead emphasize an interpretation stage. Since any environmental improvement motives are assumed to be a part of goal definition and scoping phase, this last phase of LCA focuses on identification of significant issues; evaluation of the study by completeness, sensitivity, consistency, and other checks; and conclusions, recommendations, and reporting. Identifying significant issues addresses the adequacy of study assumptions and methods, including allocation rules, category indicators, cutoff criteria, and selected impact categories. Generally, this means that the LCA results need to be presented in a form that shows where major contributions occur by life cycle stage (material product, end of life, etc.). This is particularly useful when product environmental improvement opportunities are being sought.

In the evaluation element of the life cycle interpretation stage, the practitioner completes a series of completeness, sensitivity, and consistency checks. In completeness checking, an attempt is made to ensure that all the required information has been gathered and appropriately used. In sensitivity checking, one evaluates the influence on the results by variation of assumptions, of methods, and in data. Finally, consistency checking determines if data gathering, models, assumptions, and methods have been consistently applied all along the life cycle or between LCAs of different product.

3 OTHER APPROACHES TO LCA

Above, the discussion was focused exclusively on LCA as advocated and developed by the SETAC and ISO efforts. In this section, we look at approaches that are variants of them. More complete discussions of these approaches can be found in the references cited.

3.1 EPS Method

The first alternative is the environmental priorites strategy (EPS) (Steen 1995, 1999a,b). EPS is a cradle to grave product environmental assessment method that methodologically is very similiar to the SETAC/ISO approach. In fact, the developer of the method has recently restructured it to be consistent with the ISO standards. There is a goal definition and scoping phase followed by an inventory data gathering process, which supports the impact categories included in the method.

In the EPS method, "impact categories" are identified in five "safeguard subjects": human health, ecosystem production capacity, abiotic stock resources, biodiversity, and cultural and recreational value. Some examples of impact categories used by EPS are life expectancy, severe morbidity, nuisance, wood production, crop production, fish and meat production and others. Consistent with the discussion above, these impact categories are actually end points. To each of these end points is attached a metric, like years of lost life, severe morbidity, and so on.

The emissions quantified in the inventory are assigned to the end points where pathways have been postulated to connect them. For example, EPS attributes effects in the all safeguard areas to carbon dioxide emissions. In just the human health category alone, EPS identifies four pathways that carbon dioxide presumably contributes to years of lost life (due to reduced life expectancy), namely through heat stress, starvation, flooding, and malaria, which are subsequently summed to a total. Since the EPS method is focusing on end points, a more balanced approach might also consider potential benefits of global warming.

EPS characterization factors express quantitatively the impacts of elementary flows on end points. Generally, each factor is comprised of five factors, i.e.,

$$f = F_1 \times F_2 \times F_3 \times F_4 \times F_5 \tag{2}$$

where F_1 is the valuation factor in monetary units of Euros (European units) per indicator value; F_2 is the extent of the flow interaction, like the number of extra persons expected to get chronic malaria; F_3 is the frequency factors usually set to 1; F_4 is the duration of the effect; and F_5 is a normalization factor, such as total global carbon dioxide emission. So EPS relates environmental impacts to all elementary flows, which are multiplied by their respective characterization factors, and finally summed to a single score.

$$\text{Score} = i_1 c_1 v_1 + i_2 c_2 v_2 + \cdots \tag{3}$$

where i is the elementary flow 1, c_1 is its characterization factor, and $v1$ is the valuation factor. In terms of Eq. (2),

$$c_i = F_{2i} \times F_{3i} \times F_{4i} \times F_{5i} \quad \text{and} \quad v_i = F_{1i} \tag{4}$$

It is F_1 that permits the scores to be summed, because F_1 relates end-point effects to a population expressed (via surveys) willingness to pay to avoid the effect. Hence, the final score is an economic unit.

EPS is an end-point focus environmental assessment approach. However, the connection made between the potential impact categories and EPS end points is conjecture and has not been demonstrated scientifically. On the other hand, LCA is a midpoint focused approach with impact categories as defined in Table 5 and category indicator results like equivalents of carbon dioxide and sulfur dioxide. Each of the LCA impact categories is believed to influence at least one of the human and ecological health end points and resource depletion. However, because the science connecting impact categories and end points is generally not well developed and certainly not quantitative, the usual practice of LCA, unlike EPS, makes no attempt to quantify end-point effects.

3.2 Eco-Indicator 95 Method

This method (Goedkoop 1995), advanced by M. Goedkoop of Pre'Consultants, is a "distance to target" method, where the measure of the seriousness of an effect is the difference between actual and target values for an emission. The method is also a single-indicator end-point approach. The basic Eco-Indicator 95 formula is

$$I = \sum_j w_j D_k \left(\sum_i \frac{E_i}{T_i} \right)_j \tag{5}$$

where w_j is a weighting factor for the seriousness of the three types of damage considered ($j = 1, 2, 3$), D_k is the critical damage level at the target level T_i, E_i is the contribution of the product life cycle to effect i, and T_i is the target value for effect I. The three types of damage considered are one extra death per million inhabitants per year, health complaints as a result of smog, and 5% ecosystem impairment (in the long term). After equating these three damages, the developers then set the weighting factors w_j to unity.

Rewriting Eq. (5) slightly differently, we have

$$I = D_k \sum \frac{E_i}{N_i} \frac{N_i}{T_i} \tag{6}$$

where N_i is the current extent of an effect. In fact, it is a normalization factor for effect "i," for example, total equivalent amount of CO_2 for the area of interest (country, world, etc.). The left-most quotient in Eq. (6) is the normalized effect score and the right-most quotient is the target reduction factor.

To compute an Eco-Indicator value for a product system, inventory emissions are classified into the effect categories given in Table 6 and applied to Eq. (6). National target levels are often not defined, but target reduction weighting factors N_i/T_i were estimated by the method developers for all of Europe.

The method has been recently updated to Eco-Indicator 99, which is more complex than its predecessor. The new method abandons the distance to target approach and focuses on a "damage function" approach.

3.3 Economic Input–Output

For application to LCA, economic input–output (EIO) analysis has been extensively developed by Lave and coworkers (Lave 1995; Cobas Flores 1996). They argue that in order to examine economy-wide environmental implications of a new product or a product design change, it is necessary to take into account the interdependencies of all affected sectors of the economy. Use of the EIO tables, including 519 sectors for the U.S. economy, was developed for LCA application. By coupling the toxic release inventory (a list of toxic chemical releases that companies are required to report) to the EIO tables, they generate an economy-wide listing of toxic environmental burden for products.

For the change in demand of a certain product, the EIO methods generate a list of direct and indirect outputs associated with the change. For example, to buy 1000 more television sets, direct effects including toxic emission would be increased by purchases from glass, electronics, plastic, and metal component suppliers and indirect effects would include the effects of the economic activity of the primary suppliers with their suppliers and so on.

Despite its economy-wide scope, EIO provides only sector information. For detailed environmental performance distinctions between products, a conventional LCA is best, especially if the objective is to initiate product or process improvements. Further, to run an EIO/LCA and distinguish between the environmental performances of PET and PP, two different polymers are not feasible, because all plastics appear undifferentiated in the plastic sector. Nevertheless, EIO/LCA is useful in comparing distinctly different products and is better equipped than LCA to characterize the environmental performance over the entire economy upon product change. However, what EIO/LCA captures that LCA does not is probably quite diffuse and therefore likely not very useful in product environmental improvement initiatives.

TABLE 6 Eco-Indicator 95 Classification Categories and Weighting Factors

Effect	Reduction factor	Criterion
Greenhouse effect	2.5	0.1°C per decade, 5% ecosystem impairment.
Ozone layer depletion	100	Probability of 1 death per year per million people.
Acidification	10	5% ecosystem impairment.
Eutrophication	5	Rivers and lakes; impairment of an unknown number of aquatic ecosystem; 5% ecosystem impairment.
Summer smog	2.5	Occurrence of smog periods, health complaints, particularly among asthma patients and old people. Also, occurrence of agricultural damage.
Winter smog	5	Occurrence of smog periods, health complaints, particularly among asthma patients and old people.
Pesticides	25	5% ecosystem impairment.
Heavy metals in air	5	Lead levels in children's blood, limited life expectancy and learning ability in an unknown number of people.
Heavy metals in H_2O	5	Cadmium content in rivers, ultimately also an effect on people.
Carcinogenic substances	10	Probability of 1 death per year per million people.

3.4   Threshold Inventory Interpretation Method (TIIM)

This method, developed by Franklin Associates (Hogan 1996), is a unique approach of conventional LCA in that it includes the spatial dependency of certain releases. This is typically not done in LCA. One of its underlying premises is that an emitted pollutant may or may not cause either a reduction of ecosystem or human health and hence only those that cause deterioration should be included. The method considers only atmospheric emissions, including those that have some combination of regional, global, and local effects. Representing local effects are the criteria pollutants, that is, lead, carbon monoxide, ozone, sulphur oxides, nitrogen oxides, and particulates. For global effects, carbon dioxide, carbon monoxide, and methane are included, including global warming and ozone depletion effects. The only regional effect considered is acid rain, which includes the pollutants of sulphur oxides, nitrogen oxides, particulates, ammonia, and hydrogen chloride.

The authors of TIIM advocate a three-step LCA process:

1.   Conduct a inventory of the product systems being compared;
2.   Perform a partial inventory interpretation/impact assessment of the LCIs of the product systems;
3.   If necessary, apply a TIIM analysis.

In short, for product environmental performance comparisons, an LCI should first be conducted, followed by an inventory interpretation stage. If results are sufficiently inconclusive, then TIIM should be applied. Otherwise, the environmental performance distinction between two products would be conspicious. For all practical purposes, the TIIM approach represents a standard LCA except that it focuses only on pollutants that have been scientifically proven to affect human and ecological health at the emission levels found. Of course, many pollutants are emitted that could be hazardous, but TIIM assumes that they are adequately managed by the regulatory system. A list of TIIM potential impact categories, subcategories, and emission categories are shown in Table 7. (Fluorochlorocarbons are usually associated with significant stratospheric ozone depletion, whereas methane and carbon monoxide normally are not. Since the fluorochlorocarbons have been mostly phased out, perhaps this potential impact category is no longer significant.)

Characterization models can be used in the global and regional emission categories, but the method does not recommend the use of such models, unless necessary. Also developed in the method are threshold emission factors (TEF), which are applied to all emissions in each impact category, i.e.,

$$\text{Emission } A_{\text{adj}} = \text{emission } A_{\text{phys}} \times \text{TEF} \tag{7}$$

TABLE 7 TIIM Potential Impact Categories, Subcategories, and Emission

Potential impact category	Potential impact subcategory	Emission categories	Effect
Ecosystem health	Greenhouse gases/ global warming	Carbon dioxide, methane	Global
	Ozone depletion	Carbon monoxide, methane	Global
	Acid rain	Sulphur oxides, nitrogen oxides, particulates, ammonia, hydrogen chloride	Regional
	Photochemical smog	Hydrocarbons, nitrogen oxides	Local
Human health	Irritant (eye, skin, GI tract)	Sulphur oxides, hydrocarbons, nitrogen oxides	Local
	Respiratory system effects	Carbon monoxide, sulphur oxides, nitrogen oxides, particulates, hydrocarbons	Local
	Cardiovascular system effects	Lead, carbon monoxide	Local
	Reproductive system effects	Lead	Local
	Behavioral effects	Lead	Local

For local effects, TEFs for criteria pollutants in local nonattainment areas (e.g., ozone concentrations is a particular areas exceed EPA standards) are assigned a value of 100% and in attainment areas 0%. Since the sum total of a particular emission in an LCI generally comes from numerous locations, then based on site-specific production capacities, a weighted average of the TEF for that particular emission is developed. The TEFs of all global emissions are assigned 100% as it does not matter where these emission are emitted. For the local effect of acid rain, a TEF of 100% is assigned to area considered at risk of acid rain effects and 0% otherwise. Incidentally, any particular pollutant may have more than one TEF. For example, sulphur oxides would have one for acid rain, another for an irritant, and finally one for respiratory effects.

The TIIM approach makes no attempt to determine the effects of an emission on end points; a simple loading model approach is taken to discriminate between product options, that is, less is better. As a matter of practicality and due to the generally wide variation of LCI results, during product comparisons only emissons results differing by 25% or more are deemed significant; otherwise, they are considered inconclusive. In a comparison of product options A and B, this difference for each emission category is computed from

$$\% \text{ Difference} = \frac{A - B}{(A + B)/2} 100 \tag{8}$$

The method works as follows: After the LCI is conducted, percent difference values are calculated for the emissions categories in a wide range of potential impact categories. In fact, in a conventional LCA conducted by Franklin, there are 24 potential impact categories, 14 for ecosystem health and 10 for human health, which is a more extensive list than for TIIM (Table 7). Collectively, these categories have 69 entries. Because generally each potential impact subcategory contains more than one emission, they developed a 25% significance criterion to declare an outcome to each subcategory. For example, the outcome of a subcategory is deemed "significantly less" if the system had no emission 25% higher than the other system, while at least one emission for the other system is 25% higher. If neither system has any emission 25% higher than the other or if each system has at least on emission 25% higher than the other, the results are deemed inconclusive. For a stage 2 analysis, if the overall results are inconclusive, a TIIM analysis is conducted. Here the purpose is to focus only on scientifically proven categories having ecosystem and human health impact. The number of potential impact subcategories is reduced from 24 to the 9 found in Table 7. The same rules of significantly less impact apply here also; the use of characterization factors for emissions in some impact subcategories is admitted.

4 APPLICATIONS OF LCA

The preceeding sections have presented an extensive discussion of the LCA methodology and a few of its variants. Despite its apparent conceptual simplicity, it is seen that in practice LCA is a somewhat complicated process to implement, though it becomes easier to use with experience. In this regard, a very useful reference for tips on how to conduct effective LCAs is the EPA LCI Guidelines and Principles document (Vigon 1992), which has a section entitled "Index of Guiding Statements and Principles." In the sections to follow, some applications of LCA are highlighted.

4.1 Aluminum Versus Thixomolded Magnesium Powertrain Castings

The first example application of LCA to be discussed herein is a study of die cast aluminum versus a hypothetical thixomolded magnesium powertrain structural component (Shen 1999). The goal of the study was to compare the life cycle environmental performance of these two parts, and the study scope included raw material production and product fabrication, use, and EOL disposition (recycling). The transportation of parts and components over the life cycle was also included. All consumed resources and generated emissions were collected that were consistent with the data categories of energy consumption, water use, air emissions, global warming potential, and acidification potential. The intended uses of the study were to provide auto manufacturers with a first look at an environmental performance evaluation of these two component systems and to provide an environmental burden listing for both of them, life cycle stage by stage.

The authors explored two approaches to making the magnesium from ore: ferrosilicon and the electrolytic processes. An issue of particular interest for the study was to explore the impact of an SF_6 free magnesium part production process, though it is still used during material production. Because SF_6 is a potent greenhouse gas 24,000 times greater than carbon dioxide, it is therefore desirable to avoid processes that use it.

Some key results of the study are shown in Table 8. The results are mixed. While the life cycle energy is lower for the magnesium parts, the aluminum part has the lowest GWP rating. This is due to the use of SF_6 in magnesium production. On the other hand, the magnesium parts have lower acidification and water use than the aluminum component.

These results are representative of more or less current magnesium and aluminum production practices. If, on the other hand, secondary aluminum is used and SF_6 use eliminated, the aluminum part LCE becomes 440 MJ and aluminum and magnesium part GWPs all decrease to 29, 31, and 36 kg of $[CO_2]_{eq}$, respectively. These results illustrate an outcome common to many LCAs—that is, LCAs often reveal both good news and bad news for the system under study. Stated otherwise, results are often mixed.

4.2 TIIM Applied to a Case Study of Three Fruit Juice Containers

Our second example LCA case study is an application of TIIM to a comparison of glass and plastic fruit juice containers. Detailed results for the study can be found elsewhere (Hogan 1996). The goal of the study was to determine the most environmentally friendly bottle option to deliver fruit juice from three possible bottle systems: one glass and two different plastic

TABLE 8   Selected Results for an Aluminum Versus Thixomolded
Magnesium Component : Per Part

LCA stage	Life cycle energy (MJ/part)	GWP[a] (kg of $[CO_2]_{eq}$)	Water (l/part)	Acidification (kg $[SO_2]_{eq}$)
Raw matrl acquis.				
Al	358	18.3	24	0.13
Mg-1	156	22.7	2	0.02
Mg-2	246	28.3	1.5	0.05
Prod. fabr.				
Al	45	1.3	9.8	0.01
Mg-1	78	3.8	2.8	0.04
Mg-2	78	3.8	2.8	0.04
Use				
Al	335	22	1.0	0.07
Mg-1	224	14.5	0.5	0.04
Mg-2	224	14.5	0.5	0.04
Total				
Al	744	41.6	34.5	0.2
Mg-1	487	42.2	6.0	0.11
Mg-2	571	47.9	5.5	0.14

[a]Global warming potential.

ones, denoted plastic A and plastic B. The scope of the study was cradle to grave for all primary flows into the system; omittted from the study were ancillary materials, such as lids, labels, and secondary packaging, and any associated burdens pertaining to their production or disposal. Details of the bottles are given in Table 9.

Even though all containers hold 64 ounces of fluid, for convenience the functional unit was defined as containers needed to supply 1000 liters of juice. For their stage 2 analysis (see Section 3.4), their list of potential impact categories included 14 ecosystem and 10 human health potential impact categories, which contains both waterborne and airborne wastes. A sampling from their stage 2 analysis is given in Table 10, which is a summary of five emissions in two potential impact categories.

If one can imagine this kind of listing for another 22 potential impact categories with a total of 69 emissions, it becomes evident that the LCI information can be overwhelming. Without effective reduction, this kind of analysis would be unworkable, particularly as it usually contains trade-offs. For example, it is clear from Table 9 that plastic A is better than glass

TABLE 9 Fruit Juice Container Details

Containers	Weight of container to deliver 1000 liters	Assumed recycling rate
64-oz glass bottle	372	19.4%
64-oz plastic A bottle	31	0%
64-oz plastic B Bottle	45	9.0%

in all global warming and ozone depletion emission categories, but the same is not true for plastic B versus glass.

To simpify the analysis, Franklin Associates summarized their stage 2 results even further, using their % difference and 25% significance criteria. They found that relative to glass, plastic A has "significantly less" impact in 21 impact categories and in two categories the results were inconclusive. However, for plastic B, it was better in only 5 categories, worse in 7, and inconclusive in 12 categories. Hence, they concluded that plastic A performed better environmentally than glass, but the results for the glass versus plastic B comparison were found to be inconclusive. Therefore, a TIIM analysis was applied to the plastic B versus glass bottle comparison.

After mapping the LCI data onto the TIIM potential impact subcategories and conducting an emission category loading analysis, it was found the plastic B relative to glass had significantly less impact in four subcategories, significantly more in one subcategory, and inconclusive in four. After applying a characterization model to the global warming emissions, the revised performance was less in five, more in one, and inconclusive in three. Though it is still not perfectly clear which is environmentally better, glass versus plastic B, a somewhat less complex comparison has resulted from the TIIM analysis.

The primary advantages of TIIM analysis is that it eliminates those emissions that have no scientifically proven effects on the environment or human health and reduces the number of emissions and potential impact subcategories that must be evaluated.

4.3 Life Cycle Inventory of the Generic Family Sedan

The life cycle inventory of the generic family sedan was a peer-reviewed study, conducted by the U.S. Automotive Materials Partnership (USAMP) in collaboration with the American Iron and Steel Institute, The Aluminum Association, and the American Plastics Council. The purpose of the study was to identify a suitable set of metrics to benchmark the environmental performance (not cost) of the "generic family sedan," in this

TABLE 10 Excerpts from Fruit Juice Inventory Interpretation Matrix

Category	Glass (kg/1000 L)	Plastic A (kg/1000 L)	Difference A—glass	Glass (kg/1000 L)	Plastic B (kg/1000 L)	Difference B—glass
Global warming						
Carbon dioxide	292	71.1	122%	292	186	44%
Methane	0.0038	9.9E-4	117%	0.0038	0.0084	-77%
Carbon monoxide	0.48	0.09	137%	0.48	0.64	-28%
Ozone depletion						
Methane	0.0038	9.9E-4	117%	0.0038	0.0084	-77%
Carbon monoxide	0.48	0.09	137%	0.48	0.64	-28%

case the 1995 Taurus/Lumina/Intrepid vehicle lines. The intended uses of the study were both internal and external. For the former, they are as follows:

1. To establish a database describing the environmental consumptions and releases of materials and processes used by USCAR members;
2. To improve understanding of the environmental implications of automotive manufacture, use, and disposition;
3. To facilitate assessment of the life cycle environmental inventory of alternative vehicle design options, to compare corresponding data sets, and to guide the evaluation of modifications for improvement;
4. To provide environmental input for use in strategic planning.

The external uses were to publish mutually agreeable interim and final reports (Sullivan 1998).

The scope of the study is a cradle to grave analysis of the generic family sedan; a flow diagram is given in Figure 6. The vehicle is a four-door six-passenger sedan with a 3.0-liter engine, a lifetime drive distance of 120,000 miles, a weight of 3380 lbs, a 10.7-second 0–60 mph time, and a metrohighway fuel efficiency of 23 mpg. For the study, the vehicle was divided into 7 systems and 17 subsystems. Others details of the study can be found elsewhere (Sullivan 1998). One of the omissions from the study is the vehicle infrastructure (i.e., service stations, roads, driveways, traffic signs, etc.). Vehicle storage and dismantling burdens and capital assests associated with aluminum, steel, and plastics production were also excluded.

Overall results of the study are found in Tables 3 and 4. Some conclusions of the study are as follows. The generic vehicle consumes 965 gigajoules of LCE, which corresponds to the consumption of 45,000 pounds of hydrocarbons (coal, oil, and natural gas) and the generation of 131,000 pounds of carbon dioxide. Further, the operational phase (fuel use) is dominant with respect to the other life cycle phases, consuming 85% of the LCE and generating 87% of the life cycle carbon dioxide, 94% of carbon

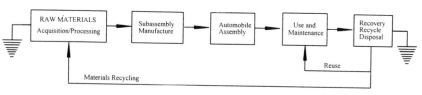

FIGURE 6   Major life cycle stages for the generic automobile.

monoxide, 90% of nitrogen oxides, 62% of sulfur oxides, and 91% of the nonmethane hydrocarbons. For water emissions, the operational stage generates 12% of dissolved matter, 93% of suspended matter, 90% of oils, and 94% of ammonia.

However, the material production and manufacturing stages also contribute significant burdens: 14% of consumed energy and for emissions 65% of particulates, 34% of sulfur oxides, 40% of hydrogen chloride, and 29% of methane. For water emissions, these stages are responsible for 90% of metals and 95% of phosphates but only 9% of oils. Regarding total solid waste, the material production and manufacturing stage contributes 68%, whereas the fuel use stage contributes only 18%. Solid waste is the only burden category where the EOL phase shows a significant contribution to the life cycle amounts, at 8% of the total waste production. Raw materials consumption and water consumption have also been tracked in this inventory.

A few other observations deserve comment. As expected, a large fraction of the life cycle air emissions are concentrated in the vehicle operation phase. Lead and particulate emissions are the exception. Lead emissions are related to battery production. Particulate emissions are associated with a wide range of material production and manufacturing processes. Solid waste is generated primarily in the materials production stage of the vehicle life cycle. EOL waste primarily from automotive shredder residue has a relatively small contribution to the total: as mentioned above about 8% of the total life cycle solid waste. Solid waste associated with gasoline production is more significant, accounting for 18%.

4.4 Influence of Vehicle Powertrain and Weight Changes on Life Cycle Performance

Vehicle weight reduction is thought to be an important way to improve the environmental performance of vehicle, through improved fuel economy. To explore the validity of this, LCA can be used to assess whether there are any important trade-offs associated with vehicle weight reduction via material substitution. For this, we focus on two published studies (Sullivan 1995, 1998). For those studies, the functional unit was defined as a vehicle system to provide transportation for a number of occupants over a lifetime distance of 120,000 miles, where the number of occupants is dependent on the vehicle system. All vehicles compared were functional equivalents. The studies were cradle to grave LCIs; not considered were vehicle infrastructure, some processing steps encountered in material transformation, and the ancillary materials associated with automotive production.

Almost always there is a benefit to weight reduction, though it is generally not appreciated that this benefit diminishes as the efficiency of the powertrain increases. In one study (Sullivan 1995) it was reported that LCE decreases with decreasing vehicle weight for both spark-ignited engine vehicles (SIV) and electric vehicles (EV), though in the case of the latter the benefit was much reduced. In fact, the authors found that the lightening coefficient for the SIV to be about 77,500 and 124,000 BTU per lb of weight saved, respectively, for virgin and 50/50 recycled/virgin aluminum substituting for steel, whereas for the EV the values are ≈ 0 and 51,900 BTU/lb per pound saved for the same two materials. In short, for the EV the rate of LCE reduction per pound of vehicle weight saved using the cited materials substitution is considerably less than that for the SIV.

The reason for this effect is clear. There is a trade-off between the operational energy saved in reducing vehicle weight and the production energy change when substituting lightweight materials for steel. Further, since the operational energy of a vehicle decreases with increasing powertrain efficiency, weight reduction induced energy savings decrease. This trade-off between production and operational energy changes upon weight reduction and/or change in powertrain is seen in Eq. (9):

$$\Delta E_{\text{tot}} \cong \left(\frac{\dfrac{E'_{\text{st}}}{C_{\text{st}}} - f \dfrac{E'_{\text{Al}}}{C_{\text{Al}}}}{1 - f} + \frac{dE_{\text{op}}}{dp_{\text{T}}} \right) \Delta p_{\text{T}} \tag{9}$$

where E' denotes the material production energies for steel (st) and aluminum (Al), p_{T} is vehicle weight, E_{op} is vehicle operational energy, f is the substitution factor (0.55 for aluminum replacing steel), and C_{ij} is the material production efficiency, typically around 0.95 for most material production systems. Equation (9) is the difference equation form of Eq. (1) for weight reduction via material substitution. This expression has been presented before (Sullivan 1995). Generally speaking, the first term in the brackets is negative, often sufficient in magnitude when lightweight materials replace steel to reduce the overall magnitude of the bracketed term. Note, the second term in the brackets is sensitive to powertrain efficiency and is around 110,000 BTU/lb and 60,000 BTU/lb for SIV and high efficiency powertrains, respectively.

Another example of this effect has been discussed for the new generation diesel engines, namely the compression ignited direct injection (CIDI) engines (Sullivan 1998). In that study, a D-class vehicle equipped with a CIDI engine realized a 40% reduction in LCE and a subsequent 900-pound weight save added another 7% reduction in LCE. On the other hand, a 900-pound weight save applied to the SIV D-class vehicle resulted in a 13% LCE reduction. All estimates are based on corporate vehicle simulation program

TABLE 11   Select Life Cycle Metrics for the SIV and CIV Comparison

Vehicle	Weight (lbs)	LCE-MBTU	GWP[a] (lbs)	LC-HC (lbs)	SO$_2$ (lbs)
SIV-1	3077	758	114,450	41,400	128
CIV-1	2938	456	63,150	26,600	95
CIV-5[b]	2112	404	56,130	24,100	97

[a]Global warming potential.
[b]50/50 virgin/recycled aluminum for steel.

(CVSP) runs and Ford's materials LCA database. The CVSP runs assumed that the weight reduction was done at constant performance (constant 0–60 mph time, gradeability, etc.) achieved by powertrain rematching (change bore and/or stroke).

An expression identical in form to Eq. (9) can be generated from Eq. (1) for LC-CO$_2$ and LC-HC. Some results are shown in Table 11. These results show that the benefits of the transition to high efficiency engines are considerably greater than those offered by weight reduction alone.

4.5   Production of Aluminum

The intentions of this study were to provide participating companies detailed inventories of various unit processes associated with aluminum production and to provide the USAMP Generic Vehicle Life Cycle Inventory Study (Sullivan 1998) comprehensive up to date LCI data on primary and secondary aluminum products in North America. A detailed report of the study is available from the Aluminum Association (1998). The uses of that study are similar to those listed above in Section 4.3 for the USAMP study. The goal of the study was to quantify, evaluate, and then identify opportunities to reduce the overall environmental consequences of the aluminum production system. Data was gathered in accordance with the ISO 14040 series of standards on LCA. The data categories are identical to those given in Table 1. The functional unit for the study was the system burdens required to provide 1000 kg of primary and secondary aluminum products, cast, rolled, and extruded.

The system boundaries of the study were cradle to gate, which includes extracting raw materials from earth (bauxite mining); material processes like anode production, alumina refining, primary ingot casting, and aluminum smelting; the manufacturing of extruded, cast, and hot and cold rolled aluminum; and finally recovery/recycling. This last stage includes consumer,

TABLE 12 Summary of Select Results from Aluminum Association's LCI Study For Aluminum Products

Category	Primary ingot	Primary cast	Secondary ingot	Secondary cast
Life cycle energy, MJ	186,300	208,900	11,690	37,511
Bauxite, kg	5090	5090	0	0
Treatment gases, kg	1	1	3	14
Total solid waste, kg	4590	5114	388	1190
SO_x, kg	84.5	93.2	4.5	1.45
CO_2, kg	11,600	12,700	616	1903
Dissolved solids, kg	18.1	21.2	1.42 E-3	3.12 E-3

Numbers cited are for 1000 kg of product.

manufactured, and shredded aluminum transport and shredding, decoating, and secondary ingot casting.

In my opinion, this study is one of the most comprehensive and complete LCIs conducted on a material that is available today in the public domain. Data were collected from 213 reporting locations representing 15 different unit processes; means and standard deviations have been reported for most of the unit processes covered in the study. For a more complete discussion of the methodology, including allocation procedures, deliberate omissions, and calculation procedure, the reader is referred to original report (Aluminum Association 1998).

A summary of select study results is given in Table 12. Casting refers to one of the type of aluminum products covered in the study; it follows ingot casting. It is clear from Table 12 that for shape cast aluminum production, the use of secondary aluminum consumes 82% less energy and generates 77% less solid waste, 85% less CO_2, and 98% less SOx than is the case for primary aluminum.

5 NEEDED IMPROVEMENTS TO LCA

The biggest challenge to routine successful application of LCA to product system environmental assessments is the availability of data. Indeed, one often hears that LCAs are too costly and time consuming to conduct. Up to the present time, LCAs have been conducted on a wide range of product systems, often being conducted under proprietary conditions. Very little has been done to gather these data in one location for public availability, as this type of LCI information has been generally considered too sensitive to be released by most manufacturers, due either to fear of the information being used against them by government and/or their customers or revealing too much about their processes and hence compromising their competitive edge. Nevertheless, some manufacturers have been rather forthcoming with LCI data pertaining to their operations. This seems to be truer of European manufacturers than North American ones. For instance, the European Plastics group (APME) has not only made available LCI data on a number of commodity plastics but also the processes used to make products from them. As mentioned above, the North American Aluminum Producers have also been forthcoming in releasing LCI data.

The development of a National or International LCI database would do much to improve access to LCI information and to reduce the time required to conduct them. However, probably a more systematic approach to developing this information is required. For example, instead of asking an automaker to reveal LCI information on the entirety of its assembly operation, which contains many processes, it might be more useful to identify the typical processes and materials required for automotive manufacture, acquire information on each of them in a gate-to-gate or cradle-to-gate fashion from tier 1 suppliers and auto manufacturers, and commit that information to a database. This should not be a difficult task. Afterall, there are not that many materials and manufacturing processes being used today to make automotive products. With such information from automotive and other producers, a manufacturer or supplier could piece together an LCI for any assembly or manufacturing operation, and the results can serve as a basis of comparison for their own product system or possible future products.

Supplier participation is crucial for successful application of LCA. However, suppliers often are neither equipped nor have the time to conduct such studies. If LCI information is going to be needed from suppliers, then they may often require help and guidance.

For purposes of reliability, LCI information submitted to a National database should be subjected to a peer review process. With readily available high quality LCI information, LCAs can be done quickly, reliably, and inexpensively. A set of trial LCAs sponsored by some authoritative organization could also be done to build confidence in LCA results.

Finally, further development of data quality indicators is necessary. For example, is there really a difference in the life cycle energy between two product systems where one has an LCE of 25 MJ and the other 28 MJ? Without appropriate statistical methods, it is difficult to tell. It appears that further development of the stochastic approach cited in section 2.3 could be very useful in this regard.

6 CONCLUSION

In the foregoing paragraphs, it has been shown that LCA is a comprehensive assessment process to evaluate the environmental performance of a product systems and not just the products over their entire lifetime. In fact, it has been seen that LCA encourages systems thinking. Certainly this kind of thinking is indispensible in industrial ecology initiatives. Since its inception, LCA has undergone continuous change, culminating with the recently completed ISO 14040 documents outlining a standardized approach. The major strengths of LCA are to identify environmental burden shifting from environmental compartment to another and to provide a holistic systemwide view of product systems. The scope of an LCA is generally cradle to grave, including raw material acquisition and production, part and product production and assembly, followed by product use, maintenance, repair, and EOL disposition. It is fundamentally a four-staged process: a careful definition of project goals and scope, inventorying resources consumed and wastes generated over the entire product system, assessing the environmental impacts of the burdens identified burdens, and interpreting the results in the context of study goals and objectives.

In the future, LCA will be more frequently used, particularly for assessing the merit of sustainability claims made about various product systems. Indeed, the holistic perspective of LCA makes it aptly suited for use in developing metrics to characterize sustainable development.

REFERENCES

Aluminum Association. 1998. Life Cycle Inventory Report for the North American Aluminum Industry. A report prepared by Roy F. Weston Inc. for the Aluminum Association, Publication AT2, November 1998.

Barnthouse L, et al. 1998. Life Cycle Impact Assessment: State-of-the-Art. The Society of Environmental Toxicology and Chemistry (SETAC).

Boustead I, Hancock GF. 1979. Handbook of Industrial Energy Analysis. John Wiley and Sons, New York.

Cobas-Flores E, Hendrickson CT, Lave LB, McMichael FC. 1996. Life Cycle Analysis of Batteries Using Economic Input-Output Analysis. Proceeding of the IEEE, 130–134.

Consoli F, et al. 1993. Guidelines for Life-Cycle Assessment: A Code of Practice, The Society of Toxicology and Chemistry (SETAC).

Fava J, et al., eds. 1991. A Technical Framework for Life-Cycle Assessment. The Society of Environmental Toxicology and Chemistry.

Fava J, et al., eds. 1992. Life Cycle Assessment Data Quality: A Conceptual Framework. The Society of Environmental Toxicology and Chemistry.

Fava J, et al., eds. 1993. A Conceptual Framework for Life-Cycle Impact Assessment. The Society of Environmental Toxicology and Chemistry.

French C. (Project Officer). 1995. Life Cycle Impact Assessment, A Conceptual Framework, Key Issues, and Summary of Existing Methods. EPA-452/R-95-002.

Goedkoop M. 1995. Eco-Indicator 95—Final Report, Weighting Method for Environmental Effects that Damage Ecosystems or Human Health on a European Scale. NOH (The National Reuse of Waste Research Programme), Report Number 9523.

Graedel TE, Allenby BR. 1995. Industrial Ecology. Prentice Hall, New Jersey.

Heijungs R, ed. 1992. Environmental Life Cycle Assessment of Products, Guide. NOH (The National Reuse of Waste Research), Report Number 9266.

Heijungs R, ed. 1992. Environmental Life Cycle Assessment of Products, Backgrounds. Leiden, NOH (The National Reuse of Waste Research), Report Number 9267.

Hogan LM, Beal R, Hunt RG. 1996. Threshold Inventory Interpretation Methodology: A Case Study of Three Juice Container Systems. Int J LCA 1:159–167.

Hunt R, Franklin W. 1974. Resource and Environmental Profile Analysis of Nine Beverage Container Alternatives. EPA Report 530/SW-91c, NTIS #PB 253486/5wp.

Hunt R, Welch R. 1974. Resource and Environmental Profile Analysis of Plastics and Non-Plastic Containers. The Society of Plastics Engineers, New York.

Husseini A, Kelly B. 1994. Life Cycle Assessment—Environmental Technology. Report Z760-94, The Canadian Standards Association.

ISO International Standard, ISO/FDIS 14040. 1997. Environmental Management—Life Cycle Assessment. Principles and Framework.

ISO International Standard, ISO 14041. 1998. Environmental Management—Life Cycle Assessment. Goal and Scope Definition and Inventory Analysis.

ISO International Standard, ISO 14042. 2000. Environmental Management—Life Cycle Assessment. Life Cycle Impact Assessment.

Kar K, Keoleian GA. 1996. Application of Life Cycle Design of Aluminum Intake Manifolds. SAE # 960410, Warrendale, PA.

Kennedy DJ, Montgomery DC, Quay BH. 1996. Stochastic Environmental Life Cycle Assessment Modeling. Int J LCA 1:199–207.

Keoleian G, Menerey D. 1994. Sustainable Development by Design: Review of Life Cycle Design and Related Approaches. Air & Waste 44:645.

Keoleian GA, McDaniel JS. 1997. Life Cycle Assessment and Design of Instument Panels: A Common Sense Approach. SAE # 970695, Warrendale, PA.

Kusko BH, Hunt RG. 1997. Managing Uncertainty in Life Cycle Inventories. SAE # 970693, Warrendale, PA.

Lave LB, Cobas-Flores E, Hendrickson CT, McMichael FC. 1995. Using Economic Input-Output Analysis to Estimate Economy Wide Discharges. Environ Sci Technol 29:420–426.

Shen D, Phipps A, Keoleian GA, Messick R. 1999. Life Cycle Assessment of a Powertrain Component: Diecast Aluminum vs. Hypothetical Thixomolded Magnesium. SAE #1999-01-0016, Warrendale, PA.

Steen B. 1995. Valuation of Environmental Impacts from Depletion of Metal and Fossil Mineral Reserves and from Emissions of CO_2. AFR Report # 70.

Steen B. 1999a. A systematic approach to environmental priority strategies in product development (EPS). Version 2000—General System Characteristics, Chalmers University of Technology, CPM Report 1999:4.

Steen B. 1999b. A systematic approach to environmental priority strategies in product development (EPS). Version 2000—Models and Data of the Default Method. Chalmers University of Technology, CPM Report 1999:5.

Sullivan JL, Hu J. 1995. Life Cycle Energy Analysis for Automobiles. SAE Paper # 951829, Warrendale, PA.

Sullivan JL, Costic MM, Han W. 1998. Automotive Life Cycle Assessment: Overview, Metrics, and Examples. SAE Paper # 980467, Warrendale, PA.

Sullivan JL, Williams RL, Yester S, Cobas-Flores E, Chubbs ST, Hentges SG, Pomper SD. 1998. Life Cycle Inventory of a Generic U. S. Family Sedan: Overview of Results USCAR AMP Propect. SAE Paper # 982160, Warrendale, PA.

Udo de Haes HA, Jolliet O, Finnveden G, Hauschild M, Krewitt W, Muller-Wenk R. 1999. Best Available Practice Regarding Impact Categories and Category Indicators in Life Cycle Impact Assessment. Int J LCA 4:1–15.

Vignon BW, Tolle DA, Cornaby BW, Latham HC, Harrison CL, Bogoski TL, Hunt RG, Sellers JD. 1993. Life Cycle Assessment: Inventory Guidelines and Principles. EPA/600/R-92/245, Washington, DC.

16

Life Cycle Engineering
A Tool for Optimizing Technologies, Parts, and Systems

Johannes Gediga, Harald Florin, and Peter Eyerer
University of Stuttgart, Stuttgart, Germany

1 INTRODUCTION

Products and services cause different environmental problems during the different stages of their life cycle. The life cycle engineering (LCE) tool aims to identify possibilities to improve the environmental behavior of technologies, parts, and systems under consideration. Therefore, it is necessary to systematically collect and interpret material and energy flows for all relevant processes. The whole life cycle of a system has to be considered to prevent the neglect or shift of possibly important environmental and economic aspects.

To support designers, engineers, and decision makers in making better more informed decisions, it is necessary to perform LCE studies, including economic assessments at a very early stage in product design. It is well known that the cost and environmental performance responsibility for a product is mainly in the hand of the designers.

To arrive at better informed decisions, tools are necessary that combine technical, economic, and environmental information. The decisions to be supported in different companies are mainly comparisons of technology and material options.

2 STATE OF WORK IN LCE

The starting point for the development of the LCE approach was the discussion about ecobalances for different packaging systems. Starting from this point, our department tried to work out a methodology from the view of an engineer and to adapt it to the decision process in the product development phase. For over 10 years the Institute for Polymer Testing and Polymer Science (IKP), University of Stuttgart has been investigating materials during their life cycle in the context of cooperation with the European automobile producers and their suppliers. Therefore, energy, raw material, emissions waste, and wastewater analyses have been made. In the foreground especially was the material selection, the usage phase, and the recycling of selected parts. From this point the methodology has been expanded from single parts to complex systems.

For the state of the work in the field of life cycle assessment, the following could be mentioned:

- After different studies at the IKP, a broad knowledge exists about environmental loads for the production of materials.
- Environmental knowledge about manufacturing processes is available.
- A mellow software tool, GaBi 3v2, is available and enables a fast modeling of life cycle assessments.
- Today, IKP has the chance to evaluate complex systems from the environmental and economic point of view.

3 LCE AS A DEVELOPMENT AND OPTIMIZATION TOOL

3.1 Task

A main goal for the development of the LCE approach is the integration into the development process of products and systems. In the following section it is demonstrated which method and evaluation techniques are used within the LCE approach.

Material and energy consumption and emissions and waste caused during production, usage, and recycling resp. landfilling of a product or system are raising costs and environmental loads. The amount of ecological and economic loads are already determined in the planning phase of a product. Revisions in the product design are feasible only with a high expenditure the later they are realized (Figure 1).

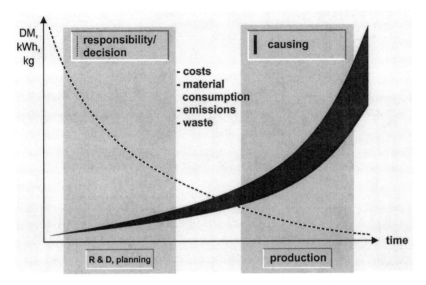

FIGURE 1   Economical and environmental responsibilities within the product development.

The experience within the LCE of the last 10 years certifies that the earlier the LCE is integrated into the development process, the higher the benefit with respect to the minimization of loads (cost and environment). Parallel to that, the following points have to be mentioned:

- Uncertainty in the result is rising. While this is accepted in a cost calculation, it is a main criticism point of all opponents of this approach in the environmental survey.
- Model calculations, based on the knowledge and experience from former studies with comparable issues on replacing specific considerations (i.e., modeling of the electricity consumption for the injection molding process as a function of mass of a part, the material itself, and the geometrical form of the part).
- The presentation of the results have to be in a condensed way to emphasize the important parameter for the decision process.

3.2 Methods

The planning process is divided into different phases. Figure 2 shows the integration of the LCA tool into the development process of the serial production. Depending on the phase of the state of the development process, there are different techniques (methods), assumption and system

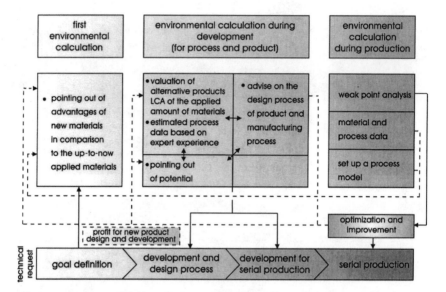

FIGURE 2   Integration of LCA in the development and design process.

boundaries (modeling, data origin), and evaluations (interpretation, presentation of the results) that can be applied to the LCE approach. In the following this issue is explained from the environmental aspect.

LCE for an already existing product will result in a list of environmental effects, which are caused during the production, use, and recycling of the considered product. A specific modeling in the software tool GaBi 3v2 and the detailed collection of material and energy flows over the product life cycle allow a weak point analysis and the preparation of optimization possibilities. This is used within the LCE approach if the environmental calculation takes place during the serial production of a part (process). There are three options to use the results:

1. After the error detection in the weak point analysis, the realization of improvements in the production follows. Basic changes and modification related to the product design or the production process itself are only feasible with extravagant expenses in this case.
2. The evaluation of the weak points regarding the optimization of products in the future. This specific analysis is carried out in the first and second phase of the development process.
3. The evaluation of the environmental calculation beside the serial production for the generation of a database for a nonspecific analysis is carried out in the first and second phase as well.

A further main issue of the LCE is the comparison of already produced parts or manufacturing processes.

With a specific LCE methodology, the evaluation methods in the above list are usable. Additionally is the comparison of the alternatives. The result of the comparison gives an answer to strategic questions regarding the product alternatives.

The LCE carried out in the development phase of products and manufacturing processes gives an estimation about the environmental loads connected to the planned application of products and processes:

1. A scenario analysis estimates potentials and risks for the application in the future.
2. Within a comparison, different alternatives can be evaluated.

In practice, the above-mentioned options are often used in a combined way. A comparison within an LCE is carried out for alternatives in serial production and other alternatives which are in the development or conception. The database for such studies is in some way specific because the already existing information of similar technologies is used for the environmental calculation of the considered product which is in the R&D phase. The following section shows the analysis of the results.

3.3 Combination of Technical, Economic, and Environmental Sectors

The starting point of LCE is the definition of the goal and scope of the intended study. It will be determined which product, process, or system with respect to the fixed system boundaries will be analyzed. In the second step, often very time intensive, the gathering of the necessary technical, economic, and environmental data is carried out. Afterward, a characterization of the data takes place. This means single materials (resources, materials, and emissions) will be classified due to the potential of the specific impact. Thereby a reduction of the extensive data amount from the inventory is achieved. The result is less amount of data in the form of impact potentials (i.e., acidification). A following interpretation step is much easier to perform. Figure 3 shows the LCE approach in form of a flow diagram.

In the evaluation step, the results of the three sectors (technical, economic, and environmental) are weighted and opposed. The goal is to determine the "best" alternative regarding all sectors. One method therefore is the life cycle efficiency analysis. The result of this approach is visualized with a 3D portfolio method (Figure 3).

On the basis of the modular multidimensional data collection, the LCE approach offers the opportunity to analyze weak points and detect

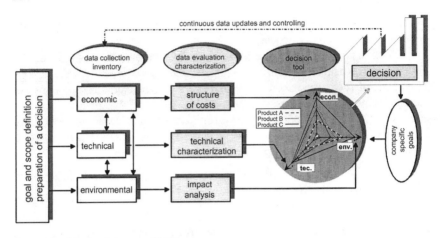

FIGURE 3   Methodology of LCE.

optimization potentials and thus a continuous optimization of the production process.

3.3.1   Economics Within the LCE Approach

For life cycle inventories (LCIs), usually only environmental data and parameters are of interest. Figure 4 illustrates what kind of data is needed for an LCI. These data not only include information about mass and energy flows for a special product or process, but also include economic information, because energy demand, for example, can be counted in [kWh] and in [DM]. This is also valid for material consumption, the products, possible byproducts, and wastes. Nearby, all flows are connected to economic numbers too. Regarding external costs such as CO_2 taxes, even emissions can be counted in economic units in future.

From this point of view, adding the economic parameters to the environmental once the chance for a more holistic decision support is easily possible. This approach is included in the methodology of LCE. Figure 5 shows what kind of additional economic data have to be collected for calculating the manufacture costs. First, information about working time must be collected. This includes the production time as well. With this information, a direct allocation of the direct labor costs on the product is possible.

To complete the manufacturing costs, information about machines must be available. Therefore, machine running times are necessary (machine [amount]) and the price component of the machines, including all costs resulting from the use of the machine. Figure 6 explains what kind of

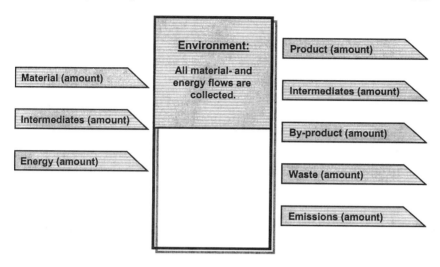

FIGURE 4   Environmental parameters within the LCE approach.

costs are included in the so-called machine set, the price component of the machines.

The "replacement value" in general is the starting point in the cost accounting. Additionally, the "estimated useful time" is needed. With this information the calculatory depreciation can be calculated. This parameter describes the wear of the machine resulting from the use. The "calculatory

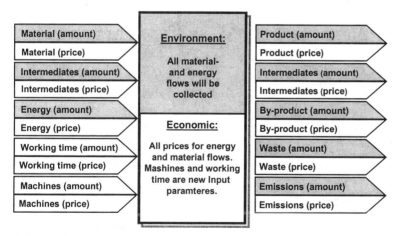

FIGURE 5   Environmental and economic parameters within the LCE approach.

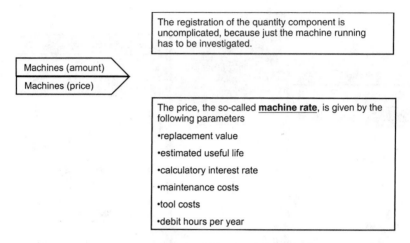

The registration of the quantity component is uncomplicated, because just the machine running has to be investigated.

Machines (amount)
Machines (price)

The price, the so-called **machine rate**, is given by the following parameters

•replacement value

•estimated useful life

•calculatory interest rate

•maintenance costs

•tool costs

•debit hours per year

FIGURE 6   Definition of the parameters included in the machine set.

interest" rate has to be calculated for the total operational capital (internal and loan capital). The "maintenance costs" include all costs that result from repair or service for the machines. The cost of wear of drills, cutters, and so forth is included in the so-called tool costs. Additional information is the estimated "debit hours per year."

In conclusion, with this data collection LCE enlarges the possibilities for analyzing and optimizing the life cycle or just single lifetime phases of a product enormously in comparison with a normal LCI. The environmental results can be used in product marketing, public policy, and product comparisons. Additionally, the economic investigation finds weak points and optimization potentials within the production chain, because this activity-based data collection support many of the so-called screws to regulate the manufacture.

LCE is a really important tool for the optimization of process chains and an effective support for decision makers, especially in an industry dominated by lean management and outsourcing. In the future evaluation methods for calculating external costs will be included, resulting, for example, out of taxes for emissions (i.e., CO_2 tax) or expenditures that are afforded by the national economy for the elimination of already caused damage of the environment.

4   LCE CASE STUDIES

This section considers various case studies. First, some environmental results of selected materials (metals and plastics) are shown in Section 4.1.

Section 4.2 shows a process comparison. Section 4.3 shows the difference between the production of the same product at different locations. Section 4.4 demonstrates a life cycle assessment (LCA) with the steps of goal and scope definition, inventory analysis, and impact assessment.

4.1 Material Comparison

Often, the designer of a new automobile uses traditional materials in their constructions, not knowing that there are other materials that can provide the same technical requirements by having reduced weight and/or having a more environmentally friendly production profile. Therefore, Figures 7–10 will give some environmental information about selected metals and Figures 11–14 about plastics that can be used for automotive applications.

4.2 Process Comparison

In the chemical industry, different production routes often exist for the manufacturing of plastics. One reason is that existing plants have to be used, although the process is not state of the art because it will cost too much to renew the plant. Also, new ways to produce the material are developed but the patent is the knowledge of just one single company. In this example, different possibilities for the production of propylene oxide (PO) are described (Figure 15) and results of the LCE study are presented.

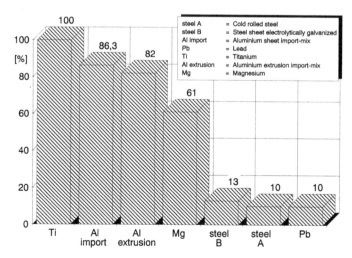

steel A	= Cold rolled steel
steel B	= Steel sheet electrolytically galvanized
Al import	= Aluminium sheet import-mix
Pb	= Lead
Ti	= Titanium
Al extrusion	= Aluminium extrusion import-mix
Mg	= Magnesium

FIGURE 7 Consumption of primary energy for the production of different metals, in relation to titanium.

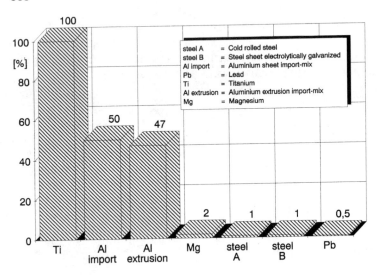

FIGURE 8 Appearing nitric oxides NO_x during the production of different metals in reference to titanium.

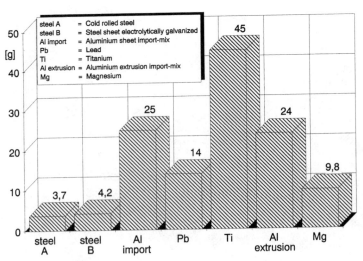

FIGURE 9 Sulfur dioxide SO_2 emissions into air during the production of different metals.

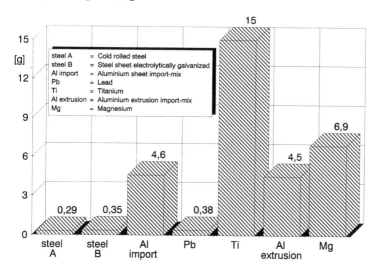

FIGURE 10   NMVOC emissions into air resulting from the production of different metals.

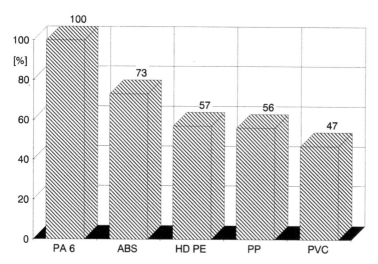

FIGURE 11   Consumption of primary energy for the production of different plastics, in reference to nylon 6 (PA 6).

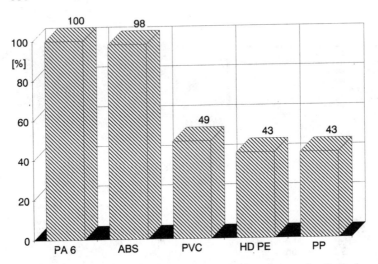

FIGURE 12 Appearing nitric oxides NO_x during the production of various plastics, in reference to nylon 6 (PA 6), German system boundaries 1993.

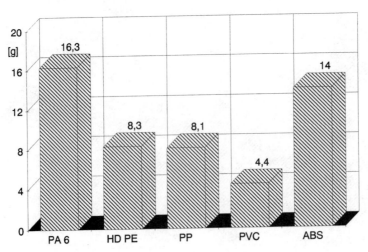

FIGURE 13 Sulfur dioxide SO_2 emissions into air during the production of different plastics.

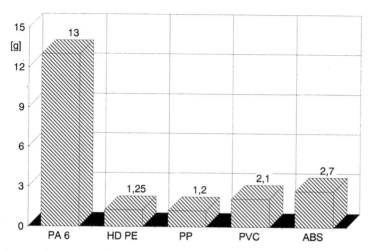

FIGURE 14   NMVOC emissions into air resulting from the production of different plastics.

PO is an important starting product for the manufacture of polyether polyol and propylene glycol, which is used in polyurethane and unsaturated polyester resins. Actually in Europe there is a production capacity of 1.4 million tons a year. In a project paid by the Enquête commission "Protection of humans and the environment," various process techniques for the synthesis of PO were analyzed, focusing on the protection of the environment.

Production of PO by using the conventional chlorine hydrogen process (CaO, NaOH) is as follows. PO cannot be produced by a direct oxidation like ethylene dioxide. Rather, the potential for chemical reactions of chlorine to be taken up by oxygen to the propylene molecule is used. Chlorine will be neutralized with hydrate of soda and is given to the wastewater as a strong thin aqueous solution as a kind of sodium chloride. The major byproducts are dichloro-propane dichlorodiisopropylether, which will be used for a energy utilization recycling or will be used as products by distillation. Additionally, neutralization with hydrate of soda calcium oxide is used to bring the chlorine out of the process (CaCl).

Production of PO by using the oxirane process with tertiary butyl alcohol as byproduct is as follows. In contrast, chlorhydrin (not chlorine) is used as reactive substance. With the formation of a hydrogen peroxide, an intermediate product, the oxygen is connected to the propylene molecule. As a byproduct, large quantities of tertiary butyl alcohol are produced. The wastewater is not contaminated by any considerable salt quantities.

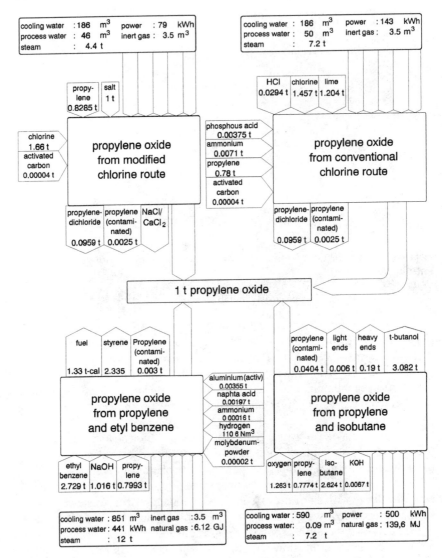

FIGURE 15 Alternative processes of PO production.

Production of PO by using the oxirane process with styrene as byproduct is as follows. Because of the use of ethyl benzene, the production of hydrogen peroxide intermediate is possible, which is needed for the oxidation of propylene. In this process there are bigger quantities of byproducts (such as styrene) produced too.

A comparison of the different production routes is possible only if the whole production chains are analyzed and if credits are given for the produced by products. In Figure 16 some results of the comparison of the chlorine hydrine route and the oxirane process are illustrated. Also shown is the difference in primary energy consumption, about 10%. For the CO_2 emissions, the results show a difference of nearly 50%. The reason is that the byproducts in the oxirane process are also responsible for some CO_2 emissions. Therefore, through an allocation procedure, the carbon dioxide emissions are divided into two streams: one for the PO and one for the byproducts.

Another example for the influences of different process technologies is given by the comparison of two production routes for the primary aluminum sheets. The first one describes the average production of primary aluminum which is representative for Germany, called "aluminum import-mix." The second production route describes the best available production process for primary aluminum which represents the best available technology, called "aluminum BAT." For both production routes the complete chain starts with the exploitation of the bauxite. Next are the steps of production of aluminum oxide (AL_2O_3), the electrolysis and the rolling plant, and finally the aluminum sheet is produced. All transports and all necessary products are considered. The production of primary aluminum

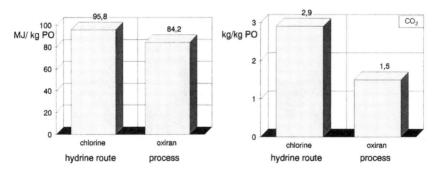

FIGURE 16 Comparison of the primary energy consumption and the CO_2 emissions of different processes for the production of PO.

does not take place at one location; therefore, country-specific power mixes are used for the energy supply.

Figure 17 shows the results of the use of primary energy, and it can be realized that the aluminum import mix scenario needs almost twice as much energy for the production of 1 kg aluminum sheet than the aluminum BAT scenario. The reason for this is the high amount of water power instead of power produced by burning fossil resources for the production steps electrolysis and rolling plant. Corresponding results are shown for the carbon dioxide and the other emissions into air.

4.3 Location Comparison

In this example the influence of the country-specific energy supply is demonstrated by the production of an air intake manifold (Figure 18). For this analysis exactly the same production route and the same material supplier are used. The only difference is that one production is located in Germany and the other one in Great Britain.

Foundations of this study were data from the industry, so it is shown that there are large differences in the production of the same part in different countries. LCE is not just a tool for analyzing parts; it also can be used for comparing various production locations.

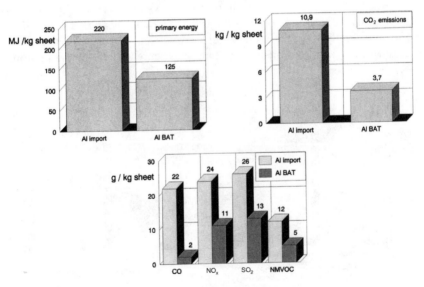

FIGURE 17 Results from the comparison of different primary aluminum production routes, like primary energy consumption and emissions into air.

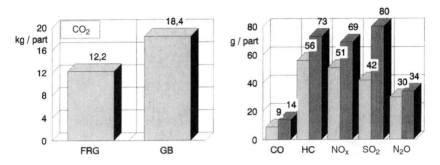

FIGURE 18 Atmospheric emissions caused by the production of similar air intake manifolds by different producers and at different locations.

4.4 LCA Case Study for Different Zinc Production Routes

This section shows the LCA for the comparison of two different production processes of zinc. The procedure is referred to the ISO 14040 series. The last step regarding the ISO standard, the interpretation step, is missing.

4.4.1 Goal and Scope

This LCA of zinc production with electrolysis and Imperial smelter was done to show the distinction in environmental profiles of the two different technologies. For both production routes the mining, benefication, roasting respectively sintering, and refining processes are considered. The main difference between the two zinc production routes is within the ore composition for the Imperial smelting furnace respectively the electrolysis. In the following paragraph, the system boundaries are described because these are the basic conditions for a LCA.

4.4.2 System Boundaries

The input material for the electrolysis process is a zinc–copper ore and for the Imperial smelting route a zinc–lead ore. The mining and benefication process is done for both ores in North America. That means country-specific electricity supply and energy carrier, needed for the processes, were used. Further, the transport from North America to Germany by bulk carrier is taken into consideration. Figure 19 shows the two production routes for the electrolysis and for Imperial smelting zinc.

The electrolysis and Imperial smelting plants are located in Germany. For this reason German electricity supply and energy carrier are used for the processes. A further assumption to compare the final product out of the two

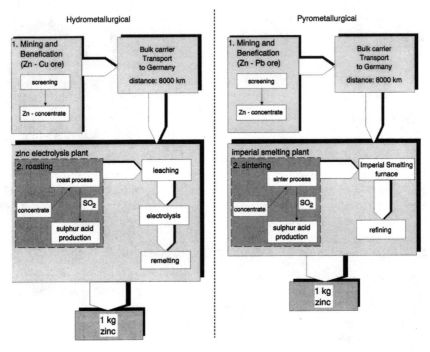

FIGURE 19   Production route of 1 kg zinc (electrolysis and Imperial smelter).

different processes (electrolysis and Imperial smelter) is that the zinc metal has the same material quality.

4.4.3   Data Quality

Most of the environmental process data used in this LCA is from industry. Other data used are from patent literature, literature, or calculations. The data given from industry are average values of measuring, done over a longer period of time. The data of the modules, like power generation, thermal energy, transports, and so on, are partly from industry, partly calculated, and partly from patent literature. Modules, like coke production, and other additional materials are mostly European average values from industry.

4.4.4   Inventory

In the following only a excerpt of the whole inventory is demonstrated. The nonrenewable resources and the primary energy demand, further the emissions, contributing to the global warming potential (GWP) and the acidification potential (AP) and selected heavy metals into air are represented.

Figure 20 shows the demand of primary energy for two different production routes of 1 kg zinc. The columns are divided in two parts. Every part demonstrates a country where the production takes place. One part represents the mining and benefication that takes place in North America, and the second part represents the metallurgical process step (sintering, roasting, electrolysis, leaching, smelting) that takes place in Germany. The high primary energy demand for the electrolysis (upper part of the left column) results from the high power consumption that is necessary for this step in the process. The lower primary energy demand of the Imperial smelting process in comparison to the electrolysis results out of low power consumption during the metallurgical production step for 1 kg zinc. The crude oil demand within this process is higher than in the electrolysis process. These depend on the combustion of light fuel oil during the sintering and Imperial smelting process. The high demand on electricity in the electrolysis process is the reason for the amount of brown coal, because up to 20% of the power generation in Germany is from brown coal. Hard coal in the Imperial smelting process is used as a reduction agent. For this reason the amount of hard coal is a little bit higher than for the electrolysis process, where hard coal mainly is used for power generation.

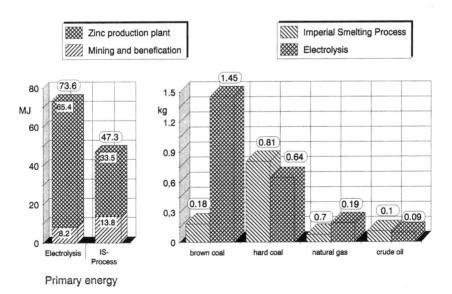

FIGURE 20 Comparison of the primary energy demand and nonrenewable resources for the two different production routes for 1 kg zinc.

In the following, selected emissions are shown, contributed to the different impact categories. The description of the formula for the determination of the impact potentials is demonstrated in Section 4.4.5.

Global Warming Potential. In Figure 21 on the left side, the GWP of the two production technologies is shown. On the right side, the chosen single emissions into air are demonstrated which are contributing to the GWP. In the electrolysis process route 67% of the CO_2 emissions are out of the power generation for the electrolysis itself; 31% of CO_2 is released during the leaching. In the Imperial smelting process, 34% of CO_2 emissions are released during the coke production and 28% released during the power generation. The remaining CO_2 are process specific emissions (e.g., combustion).

Acidification Potential. Emissions are demonstrated which are contributing to the acidification, a regional environmental impact problem. On the right side of Figure 22, the emissions are shown contributing to the acidification; 33% of the released SO_2 emissions are from the coke production. Another 25% of SO_2 emissions are emitted from the Imperial smelting furnace. The remaining SO_2 are mainly out of the sinter process and sulfur acid plant. The high HF and NH_3 emissions are to 80% released during the power generation. The higher amount of H_2S emissions into air are to 99% released during the coke production. Coke is used as reduction agent for the Imperial smelting process.

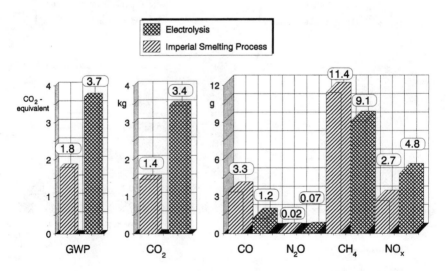

FIGURE 21 Comparison of the emissions into air which are responsible for an increasing GWP.

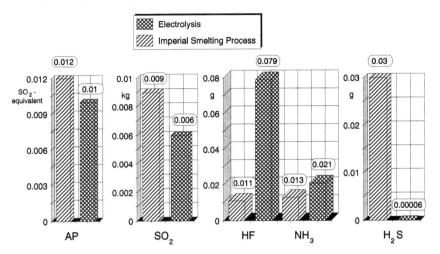

FIGURE 22   Comparison of the emissions into air which are responsible for an increasing AP.

On the left side of Figure 23, the whole amount of emitted heavy metals into air is demonstrated. Nearly 99% of the whole amount of emitted heavy metals within the Imperial smelting process route are directly from the smelting furnace. As an example, the process-specific zinc, lead, and cadmium emissions are shown on the right side of Figure 23. The reason of the higher amount of copper during the electrolysis production route is the zinc–copper ore, which is used in this process route.

4.4.5   Life Cycle Impact Assessment (LCIA)

The LCIA is the intermediate step between inventory and interpretation. The impact assessment shows the damage potentials of the different emissions. To achieve such an impact assessment, there are three steps necessary:

I. Definition of impact categories.   The quantification of environmental impacts of material and energy use is still not completely possible today. In accordance with the international discussion, four different classes of impact categories are defined:

    A. Global aspects
        1. Sustainability
            a. Renewable resources, nonrenewable resources
            b. Renewable energy carrier, nonrenewable energy carrier
            c. Water, ground
        2. Greenhouse warming
        3. Ozone depletion

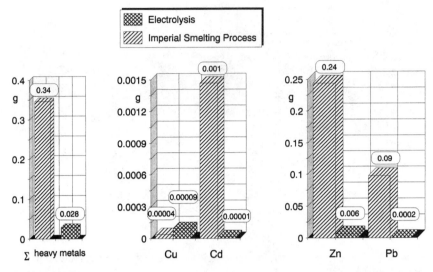

Figure 23 Comparison of selected heavy metals emitted during the production of 1 kg zinc.

    B. Regional aspects
        1. Acidification
        2. Landfill demand
    C. Local aspects
        1. Release of toxic emissions (human and ecotoxicity)
        2. Nutrification
        3. Photochemical ozone creation
    D. Not directly quantifiable categories
        1. Noise
        2. Smell
        3. Others

The above-mentioned list is the basis for this discussion. Research work is needed for the quantification of the different impacts. According to future research results, new categories can be added.

II. Classification. This process should identify emissions which are contributing to different impact categories. But it is necessary to pay attention in the case that a single emission can contribute to different categories.

III. Characterization. In the third step of the impact assessment, a statement over the considered environmental impacts will be achieved. Therefore, the contribution of the single emission to a special category

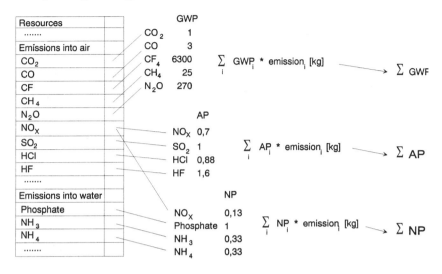

FIGURE 24   Calculation of the potential of impact categories.

has to be quantified. All the potentials of the single emissions have to be add up to the complete category potential. This step is very important for an LCA, because up to 200 and more outputs (emissions and wastes) are very difficult to handle. In Figure 24 the step summarizing the single emissions to one potential is shown.

Table 1 summarizes the potentials of two global, one regional, and two local aspects. Table 1 also demonstrates that zinc from the Imperial smelting process has higher impact on the environment than the zinc produced

TABLE 1   Examples for Different Impact Categories for Both Technologies

	Impact categories	Unit	Electrolysis production route	Imperial smelter production route
Global	Use of resources	1/kg a	0.00023	0.00023
	GWP	CO_2 equivalent	3.74	1.79
Regional	AP	SO_2 equivalent	0.010	0.012
Local	Human Toxicity (air)	HCA equivalent	0.019	0.034
	NP	Phosphate equivalent	0.00037	0.00072

with the electrolysis. The zinc produced with the electrolysis has only a higher GWP because of the high CO_2 emissions out of the power generation for the electrolysis.

As an example for the local impact, human toxicity into air and nutrification potential (NP) were chosen. A short explanation follows from which kind of emission the higher human toxicity of the Imperial smelting results. To the final value of human toxicity SO_2 emissions contributing with 34% and lead emissions into air with 46%. The heavy metal emissions in the Imperial smelting production route resulting up to 90% directly out of the Imperial smelting furnace.

4.4.6 Conclusion

For the final conclusion you have to consider different aspects. It is important to recognize that LCA should be only an additional decision tool beside the technical and economic aspects for decision making. In this case the environmental advantages, disadvantages, and environmental improvement options are shown for both technologies.

For the electrolysis production route, the advantage is the lower impact on the environment in comparison to the Imperial smelting process route in the shown categories except GWP. The disadvantages are that the high power demand during electrolysis is the reason for a high GWP and higher primary energy demand in comparison to the Imperial smelting process route and that jarosite has to be dumped in a separate place (hazardous waste).

For the Imperial smelting production route, the advantage is that cycle material can include impurities and the disadvantage is that environmental impact is higher than during the electrolysis process route, except GWP.

For environmental improvement options:

- The GWP could be decreased if the country specific power mix for the electrolysis contains more water power (more than 30%).
- The use of natural gas instead of light fuel oil during the leaching process would reduce the CO_2 emissions, which contributes to the GWP.
- The use of natural gas would also reduce the SO2 emissions, which are mainly responsible for the acidification, but would increase the NO_x emissions contributing to AP and NP.
- The substitution of light fuel oil with natural gas requires a closer inspection from the environmental point of view because of the contribution of NO_x to the acidification.

5 CONCLUSION

In Europe, North America, and Asia, LCA is used. They are interested in the environmental profiles of their products. And therefore LCI of ferrous metals, nonferrous metals, and plastics is the basis for LCA for products and systems.

Different conclusions can be summarized from different finished studies:

- The use phase can be considered as decisive with respect to energy and combustion emissions such as CO_2.
- For other emissions and environmental problems (e.g., heavy metal emissions, hydrocarbons) the material production or the processing are relevant or even dominating.
- Recycling partially has great significance.
- From the environmental and economic perspective there are no good or bad materials. The related design and production has to be the subject of the considerations.
- Adequate material designs and processes are extremely necessary to arrive at the best results. Lack of knowledge and missing experience are often the basis for inadequate designs or process solutions which lead to wrong decisions.

The optimization of products and systems makes a holistic consideration necessary. Overall improvements can also be achieved by single investments, economically and environmentally.

REFERENCES

Eyerer P. Ganzheitliche Bilanzierung—Werkzeug zum Planen und Wirtschaften in Kreisläufen. Berlin, Springer-Verlag, October 1996.
Society of Environmental Toxicology and Chemistry. A Technical Framework for Life Cycle Assessment. SETAC, Washington, DC, January 1991.
Society of Environmental Toxicology and Chemistry. Guidelines for Life Cycle Assessment: A Code of Practice. SETAC Europe, Brussels, 1993.
IPCC (Intergovernmental Panel on Climate Change). Climate Change 1994. Cambridge, University Press, 1995.
Heijungs R, et al. Environmental Life Cycle Assessment of Products. Leiden, October 1992.

17

Demanufacturing System Simulation and Modeling

**Reggie J. Caudill, MengChu Zhou,
Jingjing Hu, Ying Tang, and
Ketan Limaye**
New Jersey Institute of Technology, Newark, New Jersey

I INTRODUCTION

Over the last decade, significant research has been focused on demanufacturing to incorporate end-of-life considerations into the early product design stages and to develop optimal disassembly process planning algorithms (Johnson and Wang 1995; Penev and de Ron 1996; Tang et al. 2000). Case studies for recycling of computer monitors and television sets were presented (Boks et al. 1996; McGlothlin and Kroll 1995). The actions being taken by the electronics industry to automate and use robots for disassembly processes were also presented (Aqua and Dillon 1996). Group technology is used to classify products for cellular disassembly according to product use characteristics (Hentschel et al. 1995). A disassembly tool that supports environmentally conscious product design is reported (Srinivasan et al. 1997). The tool offers recommendations to product design based on four steps: product analysis, disassemblability analysis, computation of disassembly sequence and directions, and design rating. Demanufacturing complexity metrics and a design chart are developed to enhance the recyclability at the early design stage of a product (Lee and Ishii 1997). Disassembly sequence planning has also been addressed through time and cost indices (Li et al. 1995; Navin-Chandra 1993; Subramani and Dewhurst 1991; Suzuki et al. 1993; Tang et al. 2000). Relational structures of parts within the assembly, time, material, and regulatory database are combined with geometric CAD information for prescribed disassembly sequences (Miyamoto et al. 1996) and three-dimensional disassembly motion planning for extraction of selected components reversibly (Woo 1987; Zussman et al. 1993) and irreversibly (Lee and Gadh 1996). Gaucheron et al. (1998) proposed the global and local planning concept. Disassembly planning at the global level involves the planning of resource, reversible and irreversible methods, workflow, and tooling. The baseline estimate of dismantling time, energy, and waste flows is obtained and passed to the local planning level where the mating relationships between parts, precedence of part removal, and so on can be accounted for in terms of variation of disassembly task sequence, time, and energy. Heuristic approaches are developed and illustrated through car disassembly. Fastener planning in relation to disassembly has become a critical aspect of design for recycling and was addressed previously (VerGow and Bras 1994; Shu and Flowers 1995).

To identify weakness in the design and compare alternative designs, different evaluation methods are developed (Bras and Emblemsvag 1995; Kroll et al. 1996; Navin-Chandra 1993; Zhou et al. 1998). The rating scheme (Kroll et al. 1996) is based on the difficulty scores of each task. Activity-based-costing is proposed and used to estimate the cost incurred by different designs in the context of design for retirement (Bras and Emblemsvag 1995). Zhou et al. (1998) developed a method and the related algorithm to generate disassembly sequences and evaluate large products. First, a direction matrix is built from the product specification. Then all feasible sequences are generated. These sequences are then analyzed based on the cost information associated with the disassembly operations. The algorithms for whole and selective disassembly are developed and their complexity is analyzed. They are implemented in Java and C languages and tested for the product examples with a varying number of components from 10 to 100. These research results are very important in promoting and practicing design for environment, design for disassembly, and design for recyclability for new products. Equally important research is on the dismantling and demanufacturing of the persistently coming obsolete, worn-out, and malfunctioning products. Successful applications of the latter are most limited to few industries, for example, automotive, aircraft, computer monitors, and home appliances (Boks 1996; Guide 1996; McGlothlin and Kroll 1995; Gaucheron et al. 1998).

Currently, the basic assumption of the disassembly planners who treat the disassembly as an inverse problem of assembly is that the product is known a priori with certainty (Arai and Iwata 1992; Laperriere and ElMaraghy 1992). Navin-Chandra and Bansal (1994) formulated the disassembly and recovery process as a prize collecting traveling salesman problem. This formulation assumes a priori knowledge of the cost of decomposing the parts from a product and the fixed number of "cities" to be visited. When a specific part needs to be extracted from a worn-out product, a predictive disassembly plan problem can be formulated as a dynamic programming problem (Penev and de Ron 1996; Zülch et al. 1997). In their case the termination goal was fixed. It should be noted that solution methodologies available for a traveling salesman problem and a dynamic programming problem are all computationally intractable for complex systems. Modeling is one of the keys to study both assembly and disassembly processes. And/or graphs formulated by Homen de Mello and Sanderson (1996) are very useful to describe a product's topology and precedence relations for both assembly and disassembly planning. Heuristic search algorithms can be implemented to locate an optimal assembly task plan. Disassembly Petri nets are used to formulate a linear programming method to find optimal (dis)assembly plans (Suzuki et al. 1993). The method was, however, formulated under several assumptions. Two important ones are the terminal goal (i.e., fixed initial and final states) and no end-of-life values associated with each place (representing a component or subassembly). In dealing with worn-out products, the aim is to maximize recycled resources and minimize possible damage by the remainder that is landfilled, while economic factors are considered. Zussman, Zhou, and Caudill formally define a disassembly Petri net in terms of both its structure and functions associated with places and transitions (Zussman and Zhou 1999, 2000) by incorporating the end-of-life values, disassembly costs, and disassembly operation reliability. Thus the optimal process planning and execution can be accomplished. The results are demonstrated through an AT&T telephone and Motorola's car radio. Tang et al. (2000) recently made an in-depth survey of the present modeling and planning methodologies.

The above brief overview illustrates both the intense academic interest in modeling disassembly/demanufacturing operations and the significant strides made thus far in developing case studies for dismantling products. However, little research has been reported to help understand the system-level issues facing demanufacturers (Limaye and Caudill 1999). While system simulation has been especially important to evaluating production facilities (Carrie 1988; Hurrion 1986), these tools and techniques have not been applied widely to demanufacturing systems. The objective of this research is to develop and validate a computer-based tool to model, simulate, and evaluate the operation of electronic demanufacturing facilities. By taking a system-level perspective, the study provides a foundation for a comprehensive understanding of the interactions between facility management decisions, equipment choices, and alternative configuration layouts. The developed templates and simulation tool are very useful for demanufacturers to identify system bottleneck and improve the resource utilization and productivity. The simulation software serves as decision aid in determining some key operational parameters such as the number of workers and their responsibilities:

- The number of workstation and sort bins;
- Transportation requirements;
- Floor space allocation for storage, warehousing, and demanufacturing.

The outcomes of simulation include

- System throughput;
- Resource/worker utilization;
- Bottleneck identification;
- Hazardous-material/recyclable/solid-waste output characteristics;
- Cost and profit assessment.

To understand a demanufacturing system, we first discuss its activities and outcomes. As shown in Figure 1, a typical demanufacturing system takes discarded products collected from households or businesses and performs the following functions:

- Inspection of collected products;
- Staging of the workflow;
- Disassembly of products;
- Shredding of products or components;
- Separation into bins;
- Shipment of the recovered materials and components for further processing or use.

The system strives to maximize value recovery through reuse, remanufacture, and reengineering options while reducing residual disposal. The activities in demanufacturing are stochastic in nature, due to random variations in in-coming product streams, nondeterministic product conditions, and variable disassembly times.

System simulation techniques have contributed toward improving operational efficiencies of manufacturing facilities by assessing process bottlenecks, evaluating machine/worker utilization rates, and supporting decisions regarding efficiency improvements, changes in physical layout

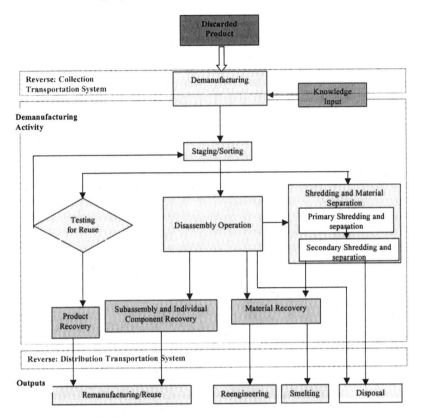

FIGURE 1   Model of a demanufacturing system.

and economics driving the business facilities (Carrie 1988; Hurrion 1986). However, very little research is focused on implementing similar techniques in the demanufacturing area. A major difference being that demanufacturing exhibits a one-to-many process-flow relationship, whereas manufacturing is a many-to-one.

Several system simulation software packages exist; however, these require users to learn the intimate details and nomenclature of a programmer and analyst. The objective of this research is to develop interfaces and specific logic modules related directly with demanufacturing activities without demanding that users learn the details of discrete-event modeling and simulation or the nuances of a programming language. The commercially available discrete-event simulation software Arena® (Arena 1998) has been chosen as the basis for this research because of its object-oriented approach, user-defined customized templates, and integration with Microsoft Visual

Basic® for formatted data input, customization of performance reports, and integration of an activity-based cost model. The next three sections discuss development of simulation modules, template designs, and activity-based cost modules.

2 MODULE DEVELOPMENT

The demanufacturing simulation tool is developed using modules, templates, and objects. A module is a single construct that may be selected from a template panel and placed in a model or can be built from the combination of existing modules in other template panels (Arena 1998). A template is a panel consisting of a group of objects called modules. The panel designed for a demanufacturer contains the objects necessary for facility layout, systematic workflow, and simulation of the facility.

Each object refers to a specific demanufacturing activity and uses detailed simulation logic behind its design to perform that activity. A user selects and locates these objects to layout the facility for a graphical representation of the demanufacturing operation. The construction of this interface provides fewer interactions with technical terms of the Arena software and concentrates more on demanufacturing terminologies for ease of operation.

As previously stated, demanufacturing is identified as a set of activities. The template panel groups these activities with a module representing each one of them as well some module doing the definition. The following fourteen modules have been developed:

- Simulation
- Incoming storage
- Inspection and testing
- Depart
- Storage and staging
- Signal control
- Disassembly workstation
- Disassembled part
- Subassembled part
- Shredder and separator
- Shipping and staging
- Transfer device
- Distance definition
- Bin

The primary modules are listed and discussed as follows.

2.1 Incoming Storage

The incoming storage module models the unloading dock for the whole facility to represent the arrival of product batches for demanufacturing. After arriving, the batches need to be unloaded and transferred into the facility. The user provides the following information: station name, the mean quantity of products, interval time of batches, and percentage of different products in the batch. In addition since the products need to be unloaded, there is a temporary storage area in this module. Thus the capacity of the storage area is also needed.

2.2 Inspection and Testing

The inspection and testing module represents the inspection stage in a demanufacturing system. The user provides inputs on unload time for the batch of products, name of inspection, maximum number of batches that can be handled at a time, number of operators, and the inspection time. Based on the condition of products, they can be transferred to three different destinations: remanufacturing area, resale area, and disassembly area. The percentage of the products being transferred to each destination also needs to be supplied.

2.3 Storage and Staging

The storage and staging module is designed for modeling the storage area of a facility, which requires name of storage, its capacity, and the percentage of products that go to the disassembly workstation as input. In both parallel and sequential operations, the products are organized and released according to preselected ranking rules. The module also gives users the ability to choose the transfer device, such as a transporter or conveyor.

2.4 Disassembly Workstation

The disassembly workstation is a primary production resource for demanufacturing systems. When a batch arrives at the workstation, it undergoes a time delay for tool setup at the workstation. After the setup operation, the batch undergoes another time delay simulating the actual disassembly process. The workstation changes its states as the underlying activity changes. The workstation has one of the following four states:

- *Processing:* This state indicates that actual disassembly operations are taking place.
- *Setup:* This is the proportion of time consumed in setting up tooling and other equipment and peripherals for the workstation.

- *Wait:* When an operator is not available at the workstation for disassembly but batches are there for processing, this state is assigned to measure the unproductive time of the workstation.
- *Idle:* When the operator and the workstation are available but no jobs are there to be run, then the workstation is idle.

All the states are predefined into the module definition, including their animation icons so that they can be used elsewhere in the model for generating percentage data on times devoted to each state.

User input is required for information on machine name, unload time for the batch of products, setup, and disassembly times as shown in Figure 2. The panel icon and operand definitions provide representation of the module on the template and reference of the user input into the logic, respectively.

2.5   Shredder/Separator

The shredder/separator operates in a semiautomatic sequential mode where entire batches of products and components are loaded into the hopper and, after multiple grinding stages and separation steps, basic "pure" materials are output into collection bins. This module requires user input for loading and processing time, shredder name, and number of bins for separated materials.

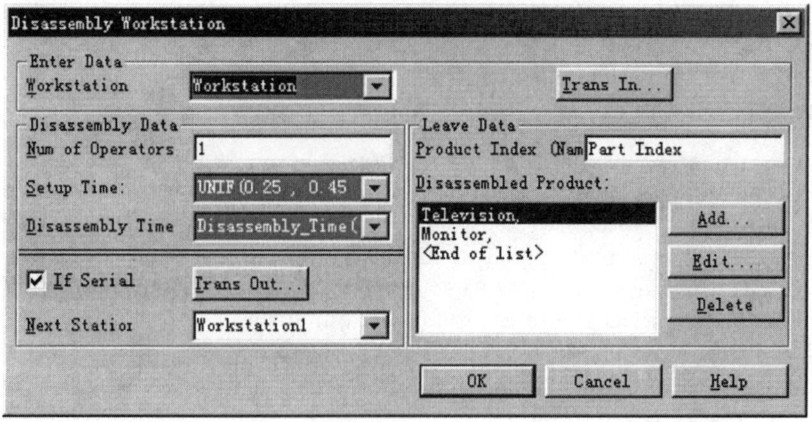

FIGURE 2   Dialogue window of a disassembly workstation module.

2.6 Bins

The disassembly plan determines the number of bins required for various recovered components or materials after disassembling or shredding a product. A bin is a container used for collecting the disassembled parts and components. The purpose of sort bins is to provide bulk handling (conveyer or forklift truck) of the disassembled parts to reduce the transfer times. The user screen gives access to specify the number of branches using the condition defined for that branch. For example, if a branch is modeled to operate only for a specific product (e.g., a monitor), then for other products that branch controls flow so as not to send materials to this designated bin. The logic provides a delay for loading the bin into a truck or conveyer. The user inputs are loading time, bin name, and capacity. The capacity of the bin affects the utilization of the disassembly workstation or operator because larger bins affect disassembly and transfer times. All the user inputs are referenced into the logic via operand definition window. The user view provides the animation icon for bins, which shows these bins being loaded with the materials, components, and assemblies from products.

3 TEMPLATE DESIGN

This section presents template design for demanufacturing system simulation using Arena functionality. Arena, as a commercially available discrete-event simulation software system, consists of the methods and applications to mimic the behavior of real systems. It is a convenient tool for bringing dynamic simulation and animation to business decision making and combines the ease of use found in high-level simulations with the flexibility of simulation languages and general-purpose procedural languages like the Microsoft Visual Basic programming system. As shown in Figure 3, standard Arena templates include Arena template, Siman template, and user-developed code in such languages as Visual Basic, C/C + +, and Fortran. Using them, designers can further develop commonly used constructs, company-specific processes, company-specific templates, and so on. They are further used for users to create templates at the highest level to achieve modeling and simulation goals. In template design, we need to consider at least four windows: logic, operand, panel icon, and user view.

3.1 Model Logic and Operand

The primary function of a module is to provide the actual logic to be performed during a simulation run. When users build a module in their own template, they define this logic just as they build a module, by placing

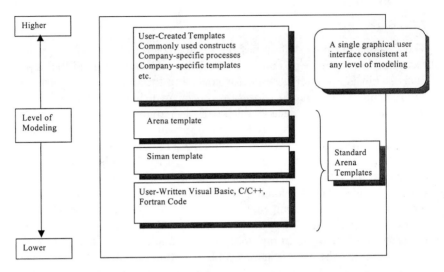

FIGURE 3 Abstraction levels of Arena.

and connecting modules from other templates. Figure 4 shows an example of the logic underlying a disassembly workstation module.

Each time this workstation module is placed in a model, this underlying logic is included. Thus, the entities that enter this module follow the underlying logic to do the operation based on the information that users input (like operation time, capacity of the workstation, how many operators at this workstation). Operands provide a user interface for complete description of the activity and its operational characteristics. An example is shown in Figure 5 that allows a user to input workstation name, number of operators, set up time, disassembly time, disassembled products name, and so on.

Double clicking the icon pops up a dialogue window that is defined by the operand window. Adding, deleting and editing functions are also included in this dialogue window as shown in Figure 2. When designing a module, we need to decide which parameters the user is permitted to change (e.g., the workstation name in our example) and which should be protected from any modification (e.g., the attribute name that is embedded in the logic window).

3.2 Panel Icon and User View

To utilize the demanufacturing module in an Arena model, the template panel has been attached to the template toolbar. Arena displays an icon drawn by the template's creator for each model, referred to as the module's panel icon. Figure 6 presents the panel icon of the developed template.

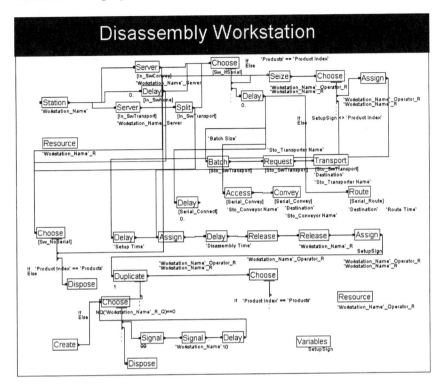

FIGURE 4   Module logic window.

When a modeler places an instance of a module in a model window, the graphic objects that are added to the window are referred to as the module's user view. It merely mimics a real workstation using an icon as shown in Figure 7. After setting up the above four windows, a simulation can be run animatedly. Furthermore, statistical analysis is integrated into all models, which provides comprehensive information to improve the system efficiency.

4   ACTIVITY-BASED COST MODEL

Demanufacturing costs depend on facility, operator, management strategies, and product characteristics. Variation of run time properties affects the operational cost, while variation of physical properties affects the fixed and variable costs. An activity-based cost model has been developed by identifying fixed and variable costs associated with each demanufacturing

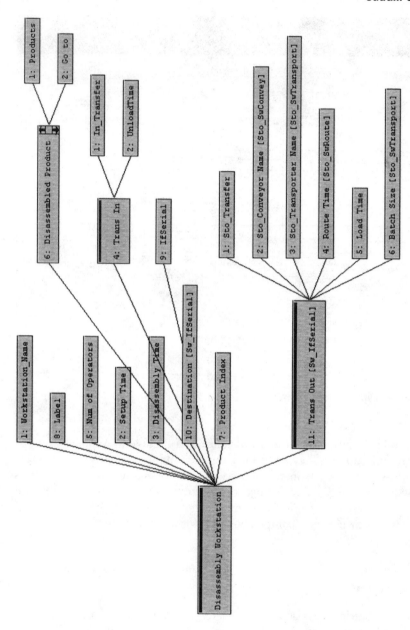

FIGURE 5 Operand window of the module.

Transfer	Support	Common	Utlarena	Demfg	Attach...

Incoming Do | Disassembl | Subassembl | Disassembly | Shredder&St | Bin | Inspection Ar | Shipping&St | Depart | Distance Def | Signal Contr | Simulate | Transfer Dev | Storage and

FIGURE 6　Panel icon of the template.

activity. The cost per product is the summation of variable costs per product representing the activities and fixed costs per product representing the investment cost for initiation of activities. Therefore,

$$\text{Total cost} = \sum_{i=1}^{m} \sum_{j=1}^{n} t_i a_{ij} c_{ij}$$

where m is the number of different types of products, n is the number of activities associated with product type i, t_i is the number of type i product, a_{ij} is the number of activities j of type i product, and c_{ij} is the cost of activity j of type i product. A summary of activities and cost drivers, including parameters required to calculate costs associated with each activity, is summarized in Table 1.

The following are major revenue categories for a demanufacturer:

• Resale of products (reuse);
• Sale of subassemblies and components (remanufacturing or reuse);
• Sale of recovered basic materials, such as metals, glass, and plastics (reengineering or smelting);
• Value of services rendered to customers to process/discard products in environmentally responsible manner and to recover proprietary or hazardous components from products, for example, batteries, mercury switches, or proprietary chips.

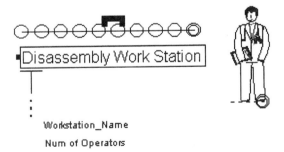

Workstation_Name

Num of Operators

FIGURE 7　User view of the module.

TABLE 1   Activities and Cost-Generating Parameters

Activity conducted	Cost-generating parameters
Disassembly of products	Disassembly times
Tool setup for disassembly	Disassembly times
Loading of the material or products Unloading of the material or products	Material handling times, type of material handling systems, times for labor used
Movement from loading dock to warehouse Movement from warehouse to inspection/staging area Movement from inspection/staging area to workstation Movement in between workstations Movement from workstation to collection/shipping area	Material handling times, type of material handling system, times for labor used
Disposal of fluff bins	Material handling times, types of material handling system, times for labor used, weight of disposed item, type of disposed item
Repair and maintenance of disassembly workstations	Down times

The marketplace drives the value of these materials and components; consequently, their values fluctuate depending on supply and demand. The total cost of operations and associated revenue reflect the bottomline business concerns of demanufacturers and can be used to compare the economic viability of alternative improvement options.

5   VALIDATION OF CUSTOMIZED MODULES

The objective of template development is to build an interface between users and a software engine that models the facility, performs the simulation, collects detailed operational data, and displays results. To verify the accuracy and validity of the simulation logic running behind the modules, model validation is performed using data collected from a typical small electronics demanufacturing facility.

5.1 Case Study Definition

The facility under study is a primary disassembler of electronics equipment with intent to ship the disassembled parts to outside vendors for further processing or resale. Electronic products coming into the facility range from televisions, computers, and monitors to microwaves and vacuum cleaners and even large machines like medical equipment, photocopiers, and mainframes. The following describes the facility and its operation:

- 10,000 sq. feet floor area out of which approximately 500 sq. feet area is actually used for disassembly (see Figure 8 for a layout).

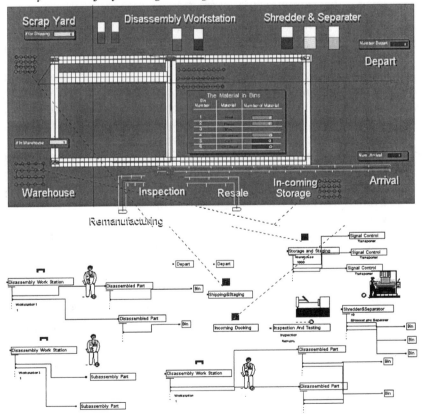

FIGURE 8   Animation snapshot showing facility layout.

- Staff includes a manager and two workers; each worker fetches batches from storage, disassembles products, and moves filled bins to the collection area for shipment. In addition, workers unload in-coming batches to the storage and staging area.
- Material handling equipment includes a single fork truck and manual movements.
- Disassembly workstations are arranged for single-worker parallel operation.
- Sort bins include containers for small motors, dirty steel, dirty aluminum, mixed copper, commingled plastics, wires/cables, CRTs, and circuit boards.
- Out of total products, 27% are televisions, 30% monitors, and 15% vacuum cleaners.

5.2 Data Collection and Validation

Data are collected on overall arrival of discarded products into the facility and throughput of recovered materials shipped as well as the flow of work through the facility in terms of job arrival/departure for each activity. The disassembly times recorded include the time required to transfer the disassembled part to the appropriate bin beside the workstation table. With these raw data, activities were stochastically characterized in terms of random variables with distributions fitted to the observed data.

Using these data, the system simulation tool with developed modules and templates was used to model and simulate current operations. Fifteen replications of this baseline were run to validate the modules and modeling structure. As indicated by the results summarized in Table 2, the simulation closely replicates the observed facility throughput.

Figure 9 shows the operator time allocation for existing operations. As seen, workers are involved in actual disassembly approximately 20% of their

TABLE 2 Summary of Observed and Simulated Data

Pallets coming into facility per month		Pallets processed per day	
Observed	Simulated	Observed	Simulated
129	127	3.8	4.3
226	249	4.3	4.8
113	114	3.5	3.0
Avg = 156	Avg = 163	Avg = 3.7	Avg = 3.9
	Error = 4.5%		Error = 5.4%

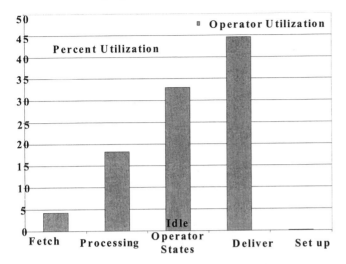

FIGURE 9 Operator time allocation for current operation.

time, while they are delivering the filled bins to the collection/shipping area approximately 45%. It can be observed that material handling—a non-value-added activity—consumes over 50% of the operator's time.

6 ANALYSIS AND RESULTS

After examining the simulation results of the current operation, some bottle-necks to the operation were observed immediately:

- Operator idle time is greater than the processing time. The potential cause may be unavailability of a transporter for the operator because of dispatching and unloading of the products at the storage area.
- Material handling time is greater than actual processing time sug-gesting to reevaluate the material handling activity.

The following suggestions were identified to remove these bottlenecks and improve flow:

- Revising the layout of the facility to reduce time for moving filled bins to the collection/shipping area.
- Increase size of the sort bins to hold more material leading to fewer trips by workers.

- Assign a separate worker to move bins while others continue to disassemble products.
- Addition of material handling equipment to reduce waiting for availability of the transporter.

6.1 Scenario Description

To illustrate the usefulness of the demanufacturing system simulation tool, the following two improvement scenarios have been selected to evaluate throughput, worker and workstation time allocation, and incremental cost/benefit trade-offs:

1. An additional operator is incorporated to receive incoming batches, stage work, and inspect products. This operator is not used for delivering the filled bins because the operator at disassembly workstation has to wait until the bins are relocated. If the additional operator is used to deliver the bins and disassembly operator is sent to get new bins, the time can be reduced but not significantly.
2. An additional operator and an additional forklift truck are incorporated for more efficient movement of materials and less delay for operators waiting for a transporter to be available.

6.2 Analysis and Comparison

These two scenarios were modeled and simulated with 15 replications to assess the effectiveness of the improvement options. Figures 10 and 11 represent the time utilization of the operators and disassembly workstations, respectively, for the baseline case and two improvement scenarios. Significant increase in time allocated to the actual disassembly process is achieved with reduction in idle time for both improvement options, with additional reduction in fetch time for scenario 2. With this better allocation of time, workers are more fully engaged in disassembly, as indicated by the improvements in workstation utilization, rising from 27% for the original baseline operation to 40% for the second scenario. These improvements result in overall increases in facility throughput. For a typical day, the existing operation processes 4.3 batches of electronic products—batch size of approximately 20 units. With the improved utilization of the disassembly workstation, scenario 2 processes almost twice as much, 7.8 batches per day, while the throughput of scenario 1 is 5.8. The difference in throughput for these two scenarios is due to the addition of a forklift truck so that conflicts between receiving incoming work and servicing workstations are avoided.

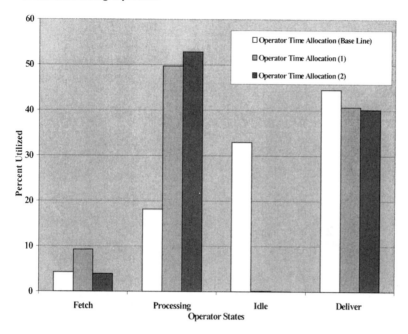

FIGURE 10 Operator time allocation for all three scenarios.

Table 3 summarizes the economic analysis for these scenarios. The following assumptions have been made:

- *Costs*: $8/hour labor rate, $875 annual capital expenditure for forklift truck.
- *Revenues*: Only income from recovered materials based on market values (Limaye 1999).

The study shows that the bottomline economics of the facility are improved as well. As observed for scenario 1, while the throughput is increased, the additional costs associated with the added worker exceeds the additional revenue generated. Whereas for scenario 2, the gain in revenue through increased throughput offsets the costs of hiring an additional worker and investing in additional equipment.

7 CONCLUSIONS

The results of this research demonstrate the usefulness of a demanufacturing system simulation tool to evaluate overall improvements in the operations of electronics demanufacturing facilities. Commercially available simulation

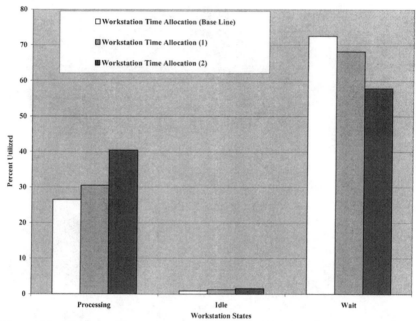

Disassembly workstation time allocation for all three scenarios.

software is fairly complex and requires extensive training; however, the customized computer-based simulation tool developed in this work provides an interface between the rigorous simulation nomenclature, structure and programming language, and the application domain of the demanufacturer. Future work will be desired in incorporating into the tool sophisticated scheduling algorithms and evaluation of end-of-life values based on the

TABLE 3   Relative Profit/Loss Summary for Scenarios

Scenario	Incremental daily labor and transportation cost ($)	Revenue ($)	Daily profit/ (loss) ($)	Daily output (batches)
Base line scenario	—	—	—	4.3
First scenario	64	34	(30)	5.8
Second scenario	65	89	24	7.8

real-time execution results (Tang et al. 2000). Applications of the tool to both education and industrial-scale demanufacturing systems will be pursued.

ACKNOWLEDGMENT

Supported by the Multi-lifecycle Engineering Research Center at NJIT through the R&D Excellence Program of the New Jersey Commission on Science and Technology and the Center's 45 industry members. In particular, we acknowledge the guidance and contributions of the Center's Industrial Technical Advisory Committee on Demanufacturing and Design-For-Environment.

REFERENCES

Aqua, E. N. and P. S. Dillon. 1996. Robotics and automation for the disassembly and recycling of electronic products. In Int. Symp. on Robotics & Manufacturing, pp. 257–262.

Arai, E. and K. Iwata. 1992. CAD System with Product Assembly/Disassembly Planning. Comput. Integr. Manufact. 12:41–48.

Arena Professional Edition Reference Guide. Systems Modeling Corp, Sewickley, PA, 1998.

Boks, C., W. Brouwers, E. Kroll, A. Stevels. 1996. Disassembly Modeling: Two Applications to a Philips 21″ Television Set. Proc. IEEE Intl. Symp. on Electronics and the Environment (Dallas, TX), pp. 224–230.

Carrie, A. 1988. Simulations of Manufacturing Systems. John Wiley & Sons, 1st ed, New York.

Caudill, R., et al. Multi-Lifecycle Engineering and Demanufacturing of Electronic Products. Symposium on Demanufacturing of Electronic Products, Deerfield Beach, FL, Oct. 1999.

Gaucheron, T., P. Sheng and E. Zussman. 1998. Hierarchical Disassembly Planning for Complex Systems. In Proc. 1998 Japan-USA Symposium on Flexible Automation, Ohtsu, Japan, July.

Gildea, L. F., M. D. Adicks and Battella. 1995. Front-End Shop Floor Control in a Remanufacturing Area. Proc. IEEE Intl. Symp. on Electronics and the Environment, Orlando, FL.

Guide, V. D. R., Jr. 1996. Scheduling using drum-buffer-rope in remanufacturing environment. Int. J Product. Res. 34:1081–1091.

Hentschel, C., G. Seliger, E. Zussman. 1995. Grouping of Used Products for Cellular Recycling Systems. Ann. CIRP 44/1:11–14.

Hurrion, R. D. (Ed.). 1986. Simulation Applications in Manufacturing. International Trends in Manufacturing Technology, IFS Ltd, UK.

Johnson, M. R. and M. H. Wang. 1995. Planning Product Disassembly for Material Recovery Opportunities. Int J Product Res 33:3119–3142.

Lambert, J. D. 1997. Optimal disassembly of Complex products. Int J Product Res 35:2509–2523.

Laperriere, L. and H. A. ElMaraghy. 1992. Planning of Products Assembly/ Disassembly. Ann CIRP 41/1:5–9.

Lee, B. H. and K. Ishii. 1997. Demanufacturing complexity metrics in design for recyclability. Proc. of IEEE International Symposium on Electronics and the Environment, San Francisco, CA, pp. 19–24.

Li, W., C. Zhang, H. Wang, and A. Awoniyi. 1995. Design for Disassembly Analysis for Environmentally Conscious Design and Manufacturing. Proc. ASME WAM, MED-2-2:969–975.

Limaye, K. 1999. System Simulation and Modeling of Electronics Demanufacturing Facilities. Masters Thesis, Dept of Mechanical Engineering, NJIT.

Kroll, E., et al. 1996. A Methodology to Evaluate Ease of Disassembly for Product Recycling. IIE Trans 28:837–845.

McGlothlin, S. and E. Kroll. 1995. Systematic Estimation of Disassembly Difficulties: Application to Computer Monitors. Proc. IEEE Intl. Symp. on Electronics and the Environment, Orlando, FL, pp. 83–88.

Miyamoto, S., T. Tamura and J. Fujimoto. 1996. ECO-Fusion, integrated Software for Environmentally-Conscious Production. Proc. IEEE Intl. Symp. on Electronics and the Environment, Dallas, TX, pp. 179–188.

Navin-Chandra, D. 1993. ReStar: A Design Tool for Environmental Recovery Analysis. Proc. 9th Intl. Conf. on Engineering Design, The Hague, Netherlands, pp. 780–787.

Navin-Chandra, D. and V. Vansal. 1994. The Recovery Problem. Int J Environ Conscious Design Manufact 3:65–71.

Penev, K. D. and A. J. de Ron. 1996. Determination of a Disassembly Strategy. Int J Product Res 34:495–506.

Shu, L. and W. Flowers. 1995. Considering remanufacturing and other end-of-life options in selection of fastensing and joining methods. Proc. of IEEE Intl. Symp. on Electronics and the Environment, Orlando, FL, pp. 75–80.

Srinivasan, H., N. Shyamsundar, and R. Gadh. 1997. A virtual disassembly tool to support environmentally conscious product design. Proc. of IEEE Intl. Symp. on Electronics and the Environment, San Francisco, CA, pp. 7–12.

Subramani, A. and P. Dewhurst. 1991. Automatic Generation of Product Disassembly Sequences. Ann CIRP 40/1:115–118.

Suzuki, T., T Kanehara, A. Inaba, and S. Okuma. 1993. On algebraic and graph structural properties of assembly Petri net. Proc. of IEEE Int. Conf. on Robotics and Automation, Atlanta, GA, May 2–6, pp. 507–514.

Tang, Y., M. C. Zhou, and R. J. Caudill. 2000. An Integrated Approach to Disassembly Planning and Demanufacturing Operation. In Proceedings of 2000 International Symposium on Electronics and the Environment, San Francisco, CA, pp. 354–359.

Tang, Y., M. C. Zhou, E. Zussman, and R. J. Caudill. 2000. Disassembly Modeling, Planning, and Application: A Review. In Proceedings of 2000 IEEE

International Conference on Robotics and Automation, San Francisco, CA, pp. 395–400.

VerGow, Z. and B. Bras. 1994. Recycling Oriented Fasteners—A Critical Evaluation of VDI 2243's Selection Table. Proc. 1994 ASME Advances in Design Automation Conf., DE-69-2:341–350.

Woo, T. 1987. Automatic Disassembly and Total Ordering in Three Dimensions. Proc. ASME WAM, PED-25:291–303.

Zhou, M. C., R. J. Caudill, and X. He. 1998. Evaluation of Environmentally Conscious Product Designs. In Proc. of 1998 IEEE Int. Conf. on Systems, Man, and Cybernetics, San Diego, CA, pp. 4057–4062.

Zülch, G., E. F. Schiller, and M. Schneck. 1997. Adaptive Dynamic Process Plans—a Basic for a Disassembly Information System. Proc. of the CIRP 4th Int. Seminar on Life Cycle Engineering, Berlin, Germany, pp. 400–412.

Zussman, E., A Kriwet, and G. Seliger. 1993. Disassembly-Oriented Assessment Methodology to Support Design for Recycling. Ann CIRP 43/1:9–14.

Zussman, E. and M. C. Zhou. 1999. A Methodology for Modeling and Adaptive Planning of Disassembly Processes. IEEE Trans Robot Automat 15:190–194.

Zussman, E. and M. C. Zhou. 2000. Design and Implementation of an Adaptive Planner for Disassembly Processes. IEEE Trans Robot Automat 16:171–179.

Zussman, E., M. C. Zhou, and R. J. Caudill. 1998. Disassembly Petri Net Approach to Modeling and Planning Disassembly Processes of Electronic Products. In Proc. of 1998 IEEE International Symposium on Electronics and the Environment, Oak Brook, IL, pp. 331–336, May 4–6.

18

DFE Materials and Processes

**Eero Vaajoensuu, Taina Dammert,
Markku Kuuva, and Mauri Airila**
Helsinki University of Technology, Helsinki, Finland

1 INTRODUCTION

In many industrial branches, addressing environmental aspects in product development and engineering design is still a relatively new issue. Even though interest in such an approach grew considerably during the 1990s, the possibilities to practically monitor, control, and affect the wide spectrum of environmentally significant product properties are not easily identifiable for designers. The primary focus is often on obvious issues (e.g., energy consumption or recyclability, highly important themes as such). However, in many cases a wider and more comprehensive total life cycle approach would often be called for. When introducing environmentally conscious engineering design approaches, it has thus been seen as useful to provide designers with a relatively general and comprehensive presentation of the basics of design for the environment (DFE). This includes goal definition, principles, definitions of terminology, rules and regulations, description of design tools and methods, services, information, and expert contacts available associated with DFE. When aiming for practical results and continuous improvement, it is also important to include DFE into the company environmental policy and to establish a DFE process integrated with the company's standard product development process, management system, and quality system.

Collection, edition, update, and distribution of this kind of basic information and subsequently providing the product developer with assistance in finding more detailed in-depth information on respective topics are important. These are topics of study and research at the Helsinki University of Technology (HUT) Laboratory of Machine Design's DFEE (Design for Environment in Electronics) research project (1998–2001). Use of the Internet as a medium in this context is of key interest and is being tested in the project (please visit http://www.machina.hut.fi/elvi/ for more information). This article presents essentials of DFE, driving forces, and some closely associated topics as identified in the DFEE project mentioned.

2 DFE PROCESS

DFE is by nature a general heading for several types of efforts aimed at reducing the environmental load of a product over its total life cycle (Figure 1). As such it includes many approaches complementing each other. A common denominator, however, is the general DFE process, which can roughly be divided into four stages:

- Identification of impulses for environmentally oriented actions;
- Analysis of environmental properties of own products and operations;
- Goal definition and action plan for improving environmental properties;
- Practical implementation of DFE.

In addition essential elements for the process are continuous feedback and integration of the approach into the standard operating practice of the company (Figure 2). The aim should be continuous improvement and adaptation of best available technology.

Reducing clearly identifiable environmental loads, singly, is the traditional DFE approach. An obvious example is efforts to reduce fuel consumption and emissions of combustion engines. The total life cycle approach, however, emphasizes the minimization of environmental loads over a product's total life cycle, including production of raw materials and energy, production of components, actual manufacture of product, distribution, usage phase of product, recycling, landfilling, incineration, and transportation. This approach may function as a tool to bring up important but perhaps not so apparent environmental issues and show connections between seemingly independent factors. Using the total life cycle approach principle does not necessarily require the use of numerical

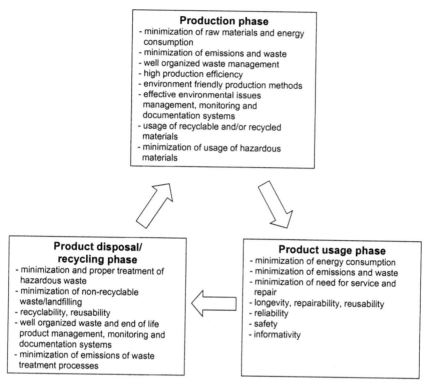

FIGURE 1   A simplified product life cycle model and some typical associated DFE goals for each stage.

life cycle analysis (LCA) tools; however, their application in the analysis stage is getting more popular. Total life cycle approach includes insight of the following:

- Products do not cause only immediate directly visible environmental loads—significant load types are usually included in the production stage, which may often be invisible to the consumer.
- All components and products have a past and a future. Proper handling of scrapped/discarded products is essential. Products do not cease to exist when they are thrown away.
- All different types of environmental loads and risks should be taken into account.
- Raising a product's overall ecoefficiency (=minimizing the environmental loads) is a central goal.

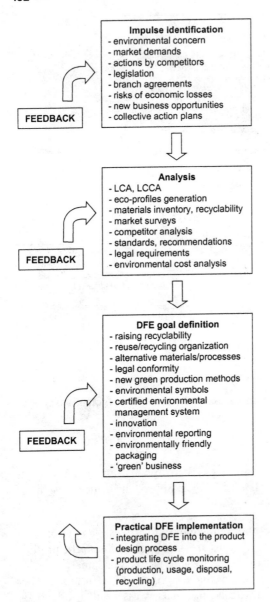

FIGURE 2   The general DFE process and some examples of issues associated with each stage.

- On a longer time scale, saving nonrenewable raw materials and energy resources is especially important.
- In many cases, materials recycling supports sustainable development.
- Product life cycles should be monitored and analyzed. Essential information on materials, production processes, and environmental product properties should be collected, updated, and kept available.

2.1 Impulses

General environmental concern has grown as consumption rates and associated cumulative environmental problems have become more apparent. In developed countries, consumer awareness has become more targeted on environmental issues. In some cases environmental properties of products have evolved into features used in competition between manufacturers. From a company's point of view, impulses for introduction of DFE thus include the following:

- Customer/market demands
- Initiatives and actions taken by competitors
- Legislation based requirements for the product and the production process
- Binding contracts made by branch organizations (or other similar local or national arrangements, e.g., to organize waste handling or recycling)
- Risk of economic losses (e.g., through taxes on use of certain materials, energy consumption, emissions/pollution, or other)
- New business opportunities identified for environmentally friendly products
- National and/or international action plans to reduce environmental load
- True individual environmental concern

Many of these impulses act simultaneously and may even be difficult to distinguish from each other. It is useful, however, before going in for more detailed analysis and development work to analyze these impulses and identify the primary ones in each case. Just building up a positive environmental company image using environmental reports, marketing methods, and so on, may serve its cause in some cases. But it may fail totally if the actual impulse requiring attention and action is upcoming legislation banning, for instance, some traditional production methods used. An essential part of functioning DFE is the ability to foresee upcoming changes, new requirements, and new technological breakthroughs and to be a forerunner when introducing new more environmentally friendly products and production methods. Environmental technology and legislation monitoring is a minimum requirement for each business today.

2.2 Analysis

After analyzing the action impulses, an evaluation of the current situation—environmental benchmarking—should be made. Essential methods that can be used in analyzing environment and DFE-related issues of a given product include the following:

- LCA and life cycle cost analysis methods, typically software supported
- Other (computerized) environmental properties modeling, inventory, and analysis methods—"eco-profile" generation
- Basic materials inventory
- Product recyclability determination
- Market surveys
- Competitor analysis
- Product analysis using standards and/or available requirement and recommendation lists as reference
- Evaluation of product conformity with legal requirements
- Company environmental cost reporting

LCA methods typically include four stages (a more detailed description is to be found in the ISO 14040 standard):

- *Goal definition and scoping.* In practice this usually includes system boundary definition, qualitative modeling of the product life cycle, where each individual stage is identified, named/numbered, and included in the total model.
- *Inventory analysis.* Quantitative input-output inventory of each stage, where all inputs and outputs (emissions) within the scope of the analysis are recorded for each stage is made.
- *Impact assessment.* Ecoprofile generation, where all inputs and outputs of each individual life cycle stage are combined to give a total picture of the overall environmental load.
- *Interpretation of results.* Analysis of the ecoprofile, evaluation of environmental effects caused by the loads.

LCA can be carried out manually but is by nature very work intensive and time consuming in most cases. Thus, several LCA software packages have been developed to make the application of the method easier and more efficient. In all cases it is important to understand that LCA results do not automatically give any technical answers to how the product should eventually be modified. Basically they just show the distribution of the environmental effects of each analyzed solution. Technical innovation is still the engineer's domain. It should also be noted that the question of how different types of environmental loads should be weighed against each other has not been conclusively answered, approaches and opinions may still differ from each other significantly. Here obviously more in-depth scientific research is called for. Some well-known LCA software packages, many of which include supporting databases, include (trademarks) the following:

- SimaPro (Pré Consultants, The Netherlands)
- LCA Inventory Tool (Chalmers Industriteknik, Ekologik, Sweden)
- PEMS, Pira Environmental Management System (PIRA International)
- The Boustead model (Boustead Consulting Ltd, United Kingdom);
- TEAMTM (The Ecobilan group, France)
- KCL-ECO (the Finnish Pulp and Paper Research Institute)
- Eco-Indicator 95 (Pré Consultants, The Netherlands)
- Eco-it 1.0 (Pré Consultants, The Netherlands)

Modified types of LCA-related software tools for environmental analysis with more product or property specific approaches have also been developed. In some cases these can be just relatively simply structured but extensive databases for materials inventories. In other cases the emphasis has been on easy operation, and the need for own materials inventories may be eliminated by relying on integrated databases containing information on essential and typical components, and processes. Analysis tools for specific issues (e.g., product structure and disassemblability analysis) or recycling profitability analysis, are also available. Some examples of software are (trademarks) as follows:

- EIME^{TM} (The Ecobilan group, France)
- EcoPurchaser^{TM} (the Swedish Institute of Production and Engineering Reseach, Sweden)
- DELTA environmental database 1.0 (Danish Electronics, Light & Acoustics, Denmark)
- DFE ver 1.1 (Boothroyd Dewhurst Inc., Wakefield, RI, USA)
- EcoScan® 2.0 (Turtle Bay, The Netherlands)

All computerized environmental property analysis methods require some training and expertise in general environmental issues before they can be effectively used. Not limiting the analysis just to separate individual cases is also suggested. When choosing an analysis tool, the company should look for one which is well suited for continued use in product properties monitoring and the use of which can be adequately supported by the company's other computerized systems, data structures, and so forth.

2.3 Goal Definition and Action Planning

Impulse and environmental property analysis of own products and production activities should next lead to a more defined DFE goal definition, some examples of which are (in no specific order) as follows:

- Increasing product recyclability
- Organizing product reuse and/or recycling
- Research for new alternative materials
- Elimination of hazardous materials
- Ensuring conformity with legislation
- Introduction of new environmentally friendly production methods
- Conformity with environmental symbol requirements
- Introduction of a certified environmental management system
- R&D—innovation—new environmentally friendly products
- Establishment and development of environmental reporting
- Environmentally friendly packaging
- Environmentally oriented business and consultation

Affecting the action planning, in the next section properties of materials and components with focus on electronics, electromechanical, and machine product applications are presented in more detail. Legislation and market-driven requirements with focus on European conditions are presented in Chapter 6. Recycling, reuse, and mechanical design are addressed in Section 4.

3 MATERIALS AND COMPONENTS SELECTION

In this section essential and typical environmental properties of some common materials and components are presented. The aim of the section is to show the reader typical environmental issues of these materials and products for a general understanding of what should be kept in mind when using them and setting targets for DFE. The presentation is not intended to be complete. In most cases obvious environmentally more friendly alternatives are difficult to find or may not yet exist. Thus, the presentation also shows many challenges for future research and development work.

3.1 Metals

Most common metals do not usually have markedly harmful direct environmental effects during their use. This can sometimes lead us to believe that their harmful environmental effects are negligible. It is easily forgotten that metal manufacturing can cause significant local environmental effects and consumes large amounts of energy. Some metals, however, are toxic for humans and the environment as such. Examples of these are chromium, nickel, and tin (see Table 2).

3.1.1 Energy

Manufacturing of metals consumes large amounts of energy (Table 1). Actual energy consumption depends on the particular plant and alloy produced. In copper production it is estimated that the energy consumption distribution for the different production stages is as follows: mining 20%, milling 40%, smelting 15%, and converting (and refining) 25% (1). Processing metal from scrap uses only about 4-40% of the energy needed to produce primary metal. This naturally depends on the purity of the scrap.

3.1.2 Coating Processes

Most coating processes for metals use toxic and hazardous materials. In addition, coating processes use significant amounts of energy. Sulfuric, hydrochloric, nitric, and phosphoric acids are commonly used for pickling. All acids are corrosive and harmful for the environment. Chromium, nickel, zinc, copper, and cadmium are used for plating. In the plating process, other chemicals are also used, including various cyanides. Some notable process-related details are as follows:

- Chromium baths used in the yellow chromium coating process contain chromium (VI) ions (Table 2).
- Cadmium is a probable human carcinogen (2).
- Cyanide is extremely toxic to humans, and inhalation exposure can be rapidly lethal. Long-term inhalation exposure may cause central nervous system effects (2).
- Nickel carbonyl may be used in refining processes (Table 2).

TABLE 1 Energy Consumption in Production of Common Metals

Metal	Energy required in primary manufacturing	Energy required in secondary manufacturing
Aluminum	222 GJ/ton	11 GJ/ton
Copper	112 GJ/ton	5-45 GJ/ton
Steel	20 GJ/ton	2 GJ/ton

Source: Ref. 1.

TABLE 2 Some Materials and their Environmental and Health Effects

Material	Environmental and health effects
Antimony and compounds	Respiratory effects, such as inflammation of the lungs, chronic bronchitis, and chronic emphysema, are the primary effects noted from chronic (long-term) exposure to antimony in humans via inhalation. Other effects in humans include cardiovascular effects (increased blood pressure, altered ECG readings, and heart muscle damage) and gastrointestinal disorders.
Arsenic and compounds (GaAs, GaP)	Inorganic arsenic compounds are toxic and carcinogenic to humans.
Cadmium and compounds	Cadmium is probable human carcinogen. It is also a cumulative toxicant in some organs such as the kidney.
Chromium	Chromium (VI) ions are toxic human carcinogens. Chromium (III) ions are much less toxic than chromium (IV) ions.
Cobalt and compounds	Acute (short-term) exposure to high levels of cobalt by inhalation in humans and animals results in respiratory effects, such as a significant decrease in ventilatory function, congestion, edema, and hemorrhage of the lung. Respiratory effects are also the major effects noted from chronic (long-term) exposure to cobalt by inhalation, with respiratory irritation, wheezing, asthma, pneumonia, and fibrosis noted. Cardiac effects, congestion of the liver, kidneys, and conjunctiva and immunological effects have also been noted in humans. Chronic exposure to high levels of cobalt via ingestion has resulted in cardiovascular effects in humans, with effects including cardiogenic shock, sinus tachycardia, left ventricular failure, and an enlarged heart. The State of California has determined under Proposition 65 that cobalt metal powder and cobalt (II) oxide are carcinogens (CCR 1996).
Lead and compounds (PbTe, PbS, PbSe)	Lead is a very toxic element, causing a variety of effects at low dose levels. Effects include brain damage, reduced growth of children, and kidney damage. Lead is classified as probable human carcinogen.

TABLE 2   (contd.)

Material	Environmental and health effects
Nickel and compounds	Respiration of nickel compounds increases the risk of respiratory organ cancer. Nickel carbonyl is the most acutely toxic form of nickel in humans. It is also a probable human carcinogen. Nickel refinery dusts and nickel subsulfide are human carcinogens. Nickel can cause allergic reactions in some people.
Tin and compounds	Some types of the organic tin compounds seem to weaken rats' immunity. Rats and mice exposed to organic tin compounds also had problems with reproduction and with the development of normal offspring.

Source: Ref. 2.

3.1.3   Emissions to Environment

Besides typical emissions from energy production (CO_2, NO_2, and SO_2), metal manufacturing also generates other emissions. These include heavy metals (lead, chromium, arsenic, and cadmium), metals (iron, copper, and nickel), oil, and dust. Emissions heavily depend on the metal manufactured and particular plant conditions.

Environmental laws and directives have forced the metal industry to start using new manufacturing technologies, which has led to decreasing emissions. Some emissions may even have decreased 90% from the 1960s to the 1990s (3).

3.2   Plastic Additives

Most polymers are not usable in technical applications without significant use of different types of additives. Depending on the polymer and its application, the additive concentration can be 1–80% (1). Thus, evaluation of environmental properties of plastics cannot be made without also addressing the additive properties.

Some additives are harmful to humans and the environment. Exposure to additives can occur through emissions from production and transportation of the additive; production, recycling, and incineration processes of the plastics; and by leaching from plastics during their use and in landfills.

Additives also usually make identification, classification, and recycling of plastic more difficult.

3.2.1 Stabilizers

A stabilizer increases a polymer's resistance to physical and chemical impacts and in that way increases the product's usability. Stabilizer concentration in thermoplastics is usually 0.1–5% (4). They are seldom used in thermosets.

Lead and organic tin compounds are used as heat stabilizers. Lead and its compounds are well-known environmental toxicants. Organic tin compounds may have effects on reproduction (Table 2). Also, cadmium compounds can be used as heat stabilizers. Their use is forbidden at least in Sweden because of cadmium's environmental effects. Heat stabilizers are used only in polyvinyl chloride (PVC) (4).

Some commonly used antioxidants are known or suspected to have effects on humans and the environment. Tetrakis and octadecyl are suspected to be accumulative. TNPP can contain free nonylphenol, which is harmful to human health and the environment. It is also suspected to be a hormone disrupter together with bisphenol A. Some tests have shown that bisphenol A also causes leukemia in tested animals. BHT is suspected to be slowly degradable, bioaccumulative, and toxic (4).

3.2.2 Plasticizers

Plasticizers make fragile and glasslike polymers tough and flexible. There can be tens of percents of plasticizers included in a plastic product. Typical plasticizers are phthalates, tricresyl phosphate, and chlorinated paraffin. Eighty to 85% of all manufactured plasticizers are used in PVC (4).

There are many different phthalates and their environmental effects are suspected to be different. In animal tests some of them are shown to cause cancer in relatively high concentrations. Some of them are also suspected to be hormone disrupters. Currently, phthalates are included in a European Community (EC) risk evaluation project that evaluates six phthalates used in PVC. The EC has temporarily banned the use of certain phthalates in products for children in 1999. Based on tests on animals it is presumable that some chlorinated paraffins can cause cancer in humans. They can also form toxic dioxin, furan, and PCB substances when they burn (4).

3.2.3 Flame Retardants

Substances used as flame retardants have a wide variety of chemical structures, chemical and physical properties, health and environmental properties, and modes of action (5).

Flame retardants have been under considerable debate in the 1990s, mainly because halogenated flame retardants can generate toxic dioxin and furan compounds when they are heated or incinerated. Also, a Swedish study (6) has linked PBDE compounds used as flame retardants to growing concentrations of tetra- and penta-BDEs in mother's milk. Growth has been exponential since 1972, with concentrations doubling every 5 years.

Besides the above it has been stated in the same study that some PBDE compounds cause liver cancer and abnormal offspring in test animals. It is also stated that PBDE compounds, along with PBB compounds act similarly to PCB in the environment. PBB compounds are fat soluble and may accumulate in food chains. They have been found in mammals, fish, and birds. PBB compounds also cause liver cancer and harm immunity. It is also stated in the same study that use of PBDE and PBB compounds should not be continued. The EC is likely to ban the use of some halogenated flame retardants in the future.

Antimony trioxide is used for increasing the fire-retarding effect of halogenated flame retardants. It is possibly carcinogenic to humans (7). Also, chlorinated paraffins can be used as flame retardants, some details about their environmental behavior are presented in Section 3.2.2.

3.2.4 Colorants

Traditionally, colorants are divided into three types: inorganic pigments, organic pigments, and dyes. Dyes in general make very good transparent colors that are very bright. Commonly used dyes include atzo dyes (yellow, orange, and red) and xanthene dyes (green, red, and orange) (1).

Pigments usually make nontransparent plastics. Some inorganic pigments contain heavy metals. Thus, production, use, and disposal are becoming more and more regulated. Commonly used pigments are, for example, titanium dioxide (white), carbon black (black), chromium oxide (green), and iron oxide (black, yellow, brown, and red) (1).

Traditionally, one of the biggest problems with colorants is their dustiness. Besides the contamination problem, inhalation can be a nuisance or hazardous, depending on the colorant involved. Several producers offer low dusting products or encapsulated products to improve this situation (1).

Metabolic breakdown of benzidine dyes releases a very toxic human carcinogen (2), benzidine. These dyes are no longer used in the United States and Europe. They are, however, still used in some other countries and can be imported from there (8).

Diarylide pigments break down to the probable human carcinogen (2) 3,3-dichlorobenzidine when heated above 200°C. Sometimes aliphatic amines are mixed to diarylide pigment. Aliphatic amines are irritating to the skin and eyes (8).

Cadmium is used in red colors; 80% of all cadmium-based pigments are used in plastics. Historically, cadmium pigments have been very important. This importance, however, has been decreasing continually because of the environmental issues associated with cadmium (Table 2) (1).

Lead is used in yellow and orange colors. Use of lead containing pigments has declined because of regulations restricting their production and use (Table 2) (1).

Carbon black has been defined as carcinogenic in some sources (9). Over 90% of this pigment is used in car tires where it is used as reinforcement at the same time.

Due to their environmental properties pigments containing mercury are no longer produced in the United States. However, they may still be used in some countries.

3.2.5 Blowing and Foaming Agents

Before CFC prohibition in the Montreal agreement, foamed plastics were made using CFC compounds. Now other compounds like pentane compounds and hydrocarbons are being used.

3.2.6 Reinforcements and Fillers

Reinforcements are used to increase the mechanical properties of plastic. Desired effects are increase of highest operating temperature, increase in durability, and tensile strength. Some commonly used reinforcements are glass fiber, boric fiber, and asbestos. Fillers (e.g., wood powder and cotton) are added to decrease the cost of plastic. Plastics with reinforcements and/or fillers are called composites. Composites are usually difficult to recycle.

3.2.7 Fungicides and Biocides

Most pure polymers can resist bacteria and mold. However, microbes can affect the additives in plastic. To prevent this, fungicides and biocides such as inorganic arsenic, bromine, and organic tin compounds are blended into plastics (Table 2).

3.3 Thermoplastics

3.3.1 Acrylonitrile Butadiene Styrene (ABS)

ABS is the most widely used engineering plastic. Typical flame-retardant ABS plastic (engineering plastic) contains 78.82% of ABS, 3.1% of antimony trioxide, and 18.8% of brominated flame retardant. Also, antioxidants and UV stabilizers are widely used in ABS plastics.

ABS can be manufactured from acrylonitrile, styrene, and polybutadiene or styrene acrylonitrile (SAN) and some other compounds. In all cases acrylonitrile and styrene are used. Acrylonitrile is a highly toxic probable human carcinogen, and styrene is a toxic possible human carcinogen (2). There may be residuals of these or other polymerization components such as emulsifiers, stabilizers, or solvents in ABS plastic. These compounds can leach from ABS plastic during its use. Recycling of ABS is difficult due to its many varied compositions.

3.3.2 Polycarbonate (PC)

PC is used for products like instrument panels, electrical connectors, telephone network devices, and outlet boxes. PC has an excellent stability, and only small amounts of additives such as stabilizers and flame retardants are used in some applications. Copolymer PC-ABC is widely used in car parts and in electronics industry.

PC is manufactured from the highly toxic substance (2) phosgene, which is derived from highly toxic (2) chlorine gas. The possible human carcinogen (2) methylene chloride and the probable human carcinogen (2) chloroform solvents are used in the production process. Also, a suspected hormone disrupter bisphenol A is used. Bisphenol A releases from polycarbonate when heated. Recently, a new chlorine, phosgene, and bisphenol A free production method has been developed (10).

3.3.3 Polyethylene PE)

PE is manufactured from ethylene, which is highly flammable but relatively harmless for the environment. It can be made either hard (high-density PE) or flexible (low density PE) without the use of plasticizers by different processing techniques. Cracking and incinerating PE may generate carcinogenic PAH substances. PE can be recycled or incinerated for energy recovery.

3.3.4 Polyethylene Terephthalate (PET)

PET is mainly used in packaging (transparent bottles, etc.). The main environmental concern about PET is additives such as UV stabilizers, flame retardants, and pigments. Also, heavy metal catalysts that are used in PET production may have effects on the environment. PET's recycling rates are relatively high compared to other plastics.

3.3.5 Polypropylene (PP)

PP is manufactured from propylene that is generated as a byproduct of petroleum processing plants. In production a chlorine intermediate process is usually used, though a viable nonchlorine process exists (10). Despite its widespread use, chlorine is highly toxic (2). Cracking and incineration of PP may generate carcinogenic PAH substances. PP can, however, be recycled or incinerated for energy recovery.

3.3.6 Polystyrene (PS)

PS is manufactured from styrene, which is toxic and a possible human carcinogen (2). Also, the toxic human carcinogen (2) benzene and toxic probable human carcinogen (2) 1,3-butadiene are used in the process. Probable human carcinogen (2) styrene oxide is released during processing of both styrene and PS. When PS burns, styrenes and PAHs may form. PS can be recycled, but its practical recycling rate is low.

3.3.7 Polytetrafluoroethylene (PTFE), Teflon

The main environmental issue regarding PTFE is hydrogen fluoride (HF), which is a toxic, corrosive, and ozone-destroying substance. HF can release in production of PTFE. When PTFE is heated above 400°C toxic perfluoroisobutyle can release. Above 690°C, PTFE burns but does not support combustion if the heat is removed. One of the combustion products is HF. PTFE cannot be incinerated for energy recovery, because HF will corrode incineration devices. PTFE is more expensive than most other thermoplastics.

3.3.8 PVC

Due to its low price and unique properties (easy to modify with additives), PVC is a widely used material in many applications. Its environmental impacts, however, have been under discussion since the 1970's. Environmental organizations have demanded that its use should not be continued. PVC defenders/producers claim that if its use stops, even worse environmental problems will appear.

PVC is manufactured from vinyl chloride that is highly toxic and a human carcinogen (2). Dioxins are released in the production of both vinyl chloride and PVC. In addition, less energy, raw materials, and fossil fuels are used to make PVC than other plastics. Also less CO_2 is released.

PVC always contains additives. Many of these additives are harmful to the environment. For example, phthalates used in children's toys have been recently under debate.

When PVC burns, the toxic gases dioxins and furans may form. Recycling of PVC is uneconomical due to its low cost and the fact that it cannot be mixed with other plastics. PVC can be incinerated for energy recovery only in special plants.

3.3.9 SAN

SAN is manufactured from the highly toxic probable human carcinogen (2) acrylonitrile and the toxic possible human carcinogen (2) styrene. SAN resins themselves appear to pose only few health problems. The main concern is that of toxic residuals (e.g., acrylonitrile, styrene, or other polymerization components such as emulsifiers, stabilizers, or solvents). Each component must be evaluated individually for toxic effects and safe exposure levels. These substances can leach from the plastics during their use.

3.4 Thermosets

3.4.1 Amino Plastics

Most amino plastics are manufactured from highly toxic probable human carcinogen (2) formaldehyde and either urea or melamine. In the uncured state the amino resin contains some free formaldehyde. Some formaldehyde may be released during cure and in some cases also after cure (e.g., from foamed insulations). Combustion or thermal decomposition of cured resins can produce hazardous substances, such as formaldehyde and extremely toxic (2) hydrogen cyanide.

3.4.2 Epoxy Plastics

The most widely used epoxy resins are diglycidyl ethers of bisphenol A (DGEBPA). Another resin, which is widely used in electronics, is the cresol-novolak epoxy. To facilitate processing and modify cured resin properties, other constituents may be included in the compositions: fillers, solvents, diluents, plasticizers, accelerators, and reinforcements. The main use of epoxy molding compounds in electronics is for encapsulation of solid-state devices such as diodes, transistors, and integrated circuits.

DGEBPA resins are derived from suspected hormone disrupter bisphenol A and epichlorohydrin. Epichlorohydrin is classified as probable human carcinogen and has been demonstrated to cause infertility (2). Bisphenol A is made of toxic phenol.

3.4.3 Phenolic Plastics

Reaction of phenol or mixture of phenols with formaldehyde forms the phenol-formaldehyde polymers (general phenolic thermoset). Phenol monomer is highly toxic, and absorption by the skin can cause severe blisters (2). Phenol is produced from toxic and known human carcinogen (2) benzene. Formaldehyde is highly toxic probable human carcinogen (2). As with all thermosets, phenolic thermosets are difficult to recycle.

3.4.4 Polyurethane

Polyurethane is manufactured from isocyanates and compounds containing a hydroxyl group. The most often used isocyanate is the extremely toxic possible carcinogen (2) toluene di-isocyanate. Tertiary amines used as catalysts are an extreme irritant (2). Thermal decomposition products of polyurethane includes extremely toxic (2) hydrogen cyanide.

3.5 Electronic Components

In this section environmental aspects of common electronics components are presented. This presentation is not intended to be comprehensive; thus, many chemicals and materials used in the components and their manufacturing processes may not be presented here.

The most typical hazardous materials used are italicized in this section, and their environmental impacts are presented in more detail in Table 2. Environmental issues of materials that are used only in one specific component type are presented in that context.

3.5.1 Active Components (Diodes, Transistors, and Rectifiers)

Active components are usually made of silicon (Si), *gallium arsenide* (GaAS) or germanium (Ge). In addition, very low temperature diodes are often made of *lead compounds* (PbTe, PbS, and PbSe) and very high temperature transistors are often made of indium phosphor or aluminum *antimony*. Schottky-diode barriers in active components can contain *nickel* (Ni) or *platinum arsenide* (GaP). In manufacturing processes any of the following hazardous substances can occur: *lead, mercury, chromium, cadmium*, cyanide, benzene, and toluene. Plastic packages are generally epoxy based and usually contain halogenated flame retardants and antimony trioxide. See sections 3.4.2 and 3.2.3 for details about their environmental issues.

3.5.2 Passive Components

Fixed wirewound resistors, wirewound potentiometers, and rheostats are made of nichrome (contains *nickel* and *chromium*), manganin (contains copper, manganese, and *nickel*), or constantan (contains copper and *nickel*). Fixed metal film resistors are made of nichrome, *tin-oxide*, or tantalum nitride. In addition, all of these components can contain *lead*.

NTC (negative temperature coefficient) thermistors can contain *cobalt* and/or *nickel*. PTC (positive temperature coefficient) thermistors can contain barium, strontium, and/or *antimony*. Inductors can contain *nickel*, *cobalt*, and/or *cadmium*.

Electrolytic capacitors usually contain manganese and can contain tributylamine, boric acid, and phosphoric acid. Tantalum capacitors can contain manganese oxide. Their crystals can contain *lead oxide* and packages can contain *cobalt*. Ceramic capacitors can contain *lead oxide* in termination glass and electrodes. Ceramic dielectric material can contain *cobalt*. Ceramic capacitors may also contain barium and/or *cadmium*.

3.5.3 Cables and Conductors

Most common cable and conductor materials are copper and aluminum; in addition, copper-plated aluminum and *tin-* or silver-plated copper is used. Most commonly used sheathing material in general applications is aluminum. *Lead-antimony* is used in high voltage power cables.

Typical insulator materials include polyethene, polypropene, PVC (most common), fluoric plastics (e.g., PTFE), and polyamide. Insulation materials are usually flame retarded so they can contain halogenated flame retardants and antimony trioxide. See section 3.2.3 for details about their environmental effects.

Cable manufacturing produces waste from extruding machine plastic runoffs, bottoms of stripwafers, and wire-end waste. In addition, small amounts of gaseous decomposition products are formed, mainly hydrocarbons and chlorine compounds (from PVC).

3.5.4 Batteries and Accumulators

In lead acid accumulators, the basic material of the electrodes is lead alloyed with other metals like calcium, *antimony*, and *arsenic*. Separators, the insulating sheets between the electrodes, are usually made of polypropylene. The electrolyte is a mixture of water and sulphuric acid or a gel compound.

The active electrode materials of NiCd accumulators are *nickel* hydroxide (Ni(OH)2) and cadmium hydroxide (Cd(OH)2). During charge nickel oxyhydroxide (NiOOH) and *cadmium* (Cd) will be formed. The electrolyte is concentrated aqueous solution of potassium hydroxide (KOH).

The exact raw materials of nickel metal hydride cells (NiMH) have not been announced by the producers. The positive electrode is usually *nickel* oxyhydroxide (NiOOH), negative electrode a metal alloy, and the electrolyte concentrated potassium hydroxide (KOH). NiMH accumulators of electric vehicles contain titanium, vanadium, and nickel alloys.

The cathode electrodes of lithium ion rechargeable cells are oxides of cobalt, manganese, or *nickel* ($LiCoO_2$, $LiMnO_2$, $LiNiO_2$). The anode is made of graphite. The electrolytes contain lithium salts in organic solvents like ethylene and propylene carbonate.

The electrodes of alkaline batteries are made of manganese oxide (MnO_2) and zinc (Zn), and the electrolyte is potassium hydroxide (KOH).

The mercury content of modern alkali-manganese batteries is less than 0.025 weight %.

The electrode materials of lithium batteries may be liquid or solid. Solid lithium iodine (Li-I_2) batteries have been in use to power pacemakers. The liquid electrodes are with the electrolyte of lithium based salts in organic solvents. The cathode may be a solid, dissolving, or liquid component. The solid cathodes are metal oxides or sulphides. A dissolving cathode may be gaseous sulphur dioxide (SO_2). A liquid cathode may be thionyl chloride ($SOCl_2$) and sulphonyl chloride (SO_2Cl_2).

The standard cell in Leclance batteries contains natural manganese dioxide (MnO_2) as the cathode, zinc as the anode, and ammonium or zinc chloride (NH_4Cl, $ZnCl_2$) electrolyte. Benefits of Leclance batteries are low price and wide range of battery types.

The materials of button cells are today zinc-silver oxide (Zn-Ag_2O) and lithium instead of mercury oxide. The electrolyte is potassium hydroxide (KOH).

In zinc-air batteries the active materials are zinc and air. The electrolyte is an alkaline aqueous solution of sodium hydroxide or ammonium chloride (NaOH, NH_4Cl).

3.5.5 Electromechanical Components

Electromechanical components like connectors, contactors, overload protectors, and relays are usually made of copper, tin, and brass (copper, zinc, lead). Relays can contain small amounts of *mercury*. Commonly used insulation materials are polyester, polycarbonate, phenolic plastics, and polyamide. Insulation materials can contain halogenated flame retardants and antimony trioxide. Coatings are often made of gold, silver, or palladium.

3.5.6 Displays

Usually, the environmental issues of flat panel display (plasma, electroluminescent, and liquid crystal displays) manufacturing are concerned with the process chemicals throughout photolithography steps, dopant, and process cases for deposition, wet and dry etchants, cleaning substances/techniques, and metallization processes. The proper waste disposal, emissions to air and water, and energy and water consumption may also cause environmental loads.

Typical materials and substances used in manufacturing processes of flat panel displays are as follows:

- Glass (e.g., soda lime)
- Color filters
- Developers

- Thinners
- Acetone
- Strippers
- Acids (e.g., HCl, HNO_3, H_2SO_4, HF)
- Ammonia
- Indium tin oxide
- Etchants (e.g., acids and gases such as SF_6, CF_4, Cl_2, BCl_3)
- Si_3N_4, PH_3, NH_3
- Metals (e.g., Al, Mo, Ta, Ti, Co)
- Trichloroethylene
- Deionized water
- Process cooling water

Emissions originating in the manufacturing process are as follows:

- Associated with energy consumption (e.g., clean room machinery, waste disposal equipment)
- Associated with air quality (volatile organic compounds and acid fume emissions from etching processes)
- Associated with global warming and ozone depletion (releases, e.g., from etching and cleaning processes)
- Solid wastes

In addition, these processes consume large quantities of water (e.g., process cooling water, deionized water production). There are no established recovery and recycling practices for end-of-life flat panel displays.

In cathode ray tubes (CRT) the main raw materials are glass sand (SiO_2), soda ash (Na_2CO_3), limestone ($CaCO_2$), feldspar, and other fining, coloring and oxidizing agents (e.g., K_2O, MgO, ZnO, BaO, PbO). The screen is coated with a luminescent material (phosphor), typically zinc sulfide. The neck and funnel consist of leaded glass containing mainly silicon oxide and *lead oxide*. Other environmentally harmful materials include barium, phosphor, strontium, beryllium, and cesium. A typical 28-inch cathode ray tube display contains about 1 kg of *lead*. The percentage of lead in glass parts of a typical CRT is as follows: panel 0–3%, funnel 24%, neck 30%, and frit 70%.

Information on CRT manufacturing processes can be found from net site of the EPA DFE computer display project: http://www.mcc.com/projects/env/projects.html

CRTs are usually difficult to recycle. Before recycling the glass parts of a CRT must be separated and the coatings removed. The current recycling processes are expensive and must be carried out in special plants.

3.5.7   Printed Circuit Boards

Various chemicals used in the manufacturing processes of printed circuit boards include sulfuric acid, hydrofluoric acid, formaldehyde, phenols, toluene, benzene, and cyanide. Many of these are hazardous for humans and the environment. In addition, manufacturing consumes large amounts of water.

Laminate materials can contain cyano-acrylates or phenol-formaldehyde and finishes can contain *lead* or *nickel*. Printed circuit boards are usually flame retarded with mixture of brominated flame retardant and antimony trioxide (see also Section 3.2.3)

3.5.8   Solders and Adhesives

Typical tin-lead solders contain ca. 60% *tin* and ca. 40% *lead* by weight. Surfaces are cleaned with soldering flux, which contains colophony resin and isopropanol. Use of lead-containing solders could be banned in the EC in the future.

Lead-free solders are usually *tin*-based. For example, Sn/Ag3.8/Cu0.7 (tin/silver/copper), Sn/Cu/*Sb*/Ag (tin/copper/antimony/silver), Sn/Bi/Ag/Cu (tin/bismuth/silver/copper), Sn97/Cu3 (tin/copper), 96Sn/4Ag (tin/silver), and 95Sn/5*Sb* (tin/antimony) formulas are used.

Many materials can be used to make glues. Most common glues used in electronic industry are epoxy based (see also Section 3.4.2).

4   RECYCLING, REUSE, AND MECHANICAL DESIGN

As described earlier, DFE may, and in its practical forms usually does, include several approaches complementing each other. As an example of an individual, more specified, and relatively clearly defined target setting product recyclability, reusability, and design for recycling (DFR) is addressed next. As with DFE, DFR also should be included in the general product development process and not addressed as a separate issue. Practical feedback from actual reuse and recycling processes is of key importance. It should be noted that "recycling" can—depending on context— refer to both materials recycling and product/component reuse, even if the primary interpretation should be materials recycling.

4.1   Choosing the DFR Approach

When going in for DFR it is important that a principal goal definition is made and the efforts are given sufficient organizational support. Real results are very difficult to obtain if the company does not officially recognize the importance of DFR alongside with other strategic product properties and features. Two basic levels of approach can be identified:

- Product recycling is made easier using technical solutions anticipating the most likely recycling processes to be used for the product.
- In addition, the company actively participates in building up, running, and financing recycling practices and operations.

The latter, of course, requires essentially higher commitment and investments. Efforts in this direction are usually driven by legislation or market requirements. However, examples of recycling operations strictly based on business principles also exist (e.g., metals recycling). Actually running recycling operations, however, is not a prerequisite for DFR. Well-coordinated DFR can substantially help manufacturing companies to cope with the more strict recyclability and recycling requirements of the future. It is also good to remember that recycling as such should perhaps not be the target; its meaning lies in that it should essentially lessen the product's environmental load on a total life cycle basis, which is not automatically the case. Successful elimination of the actual need to recycle some difficult materials by eliminating them from the structure in the planning stage is in a way ideal DFR.

4.2   Recycling Practices and Technologies Used

To formulate more detailed DFR requirements, information about the recycling practices and technologies existing or to be used is essential. To optimize product properties for a certain recycling treatment, it is important to know

- Who carries out the disassembly and the recycling?
- What parts/materials are reused/recycled?
- Where and in what kind of conditions does recycling take place?
- What methods/technologies are used?
- What tools, identification methods, etc. are available?
- Are there some economic restraints which must be taken into account?
- What are the cooperation parties (collection, transport, etc.)?
- What are the identified recycling/recyclability requirements (targets) now and what will they be in the future
- How is recycling monitored, measured, and on what basis are statistics made?

All this information should be used to formulate a realistic recycling strategy for the product in question to function as reference in DFR goal definition. For instance, manual disassembly, which is often given high priority in many DFR theories, may actually be of small importance if the recycling practice and subsequent strategy does not include this phase. This is actually a typical case for products treated by mechanical shredding and materials separation. The idea of DFR is to raise the efficiency of the recycling process so that a higher percentage of all materials in the product can be successfully recycled.

4.3 Detailed DFR Requirements

When defining detailed DFR requirements it is important to understand how differences in the recycling strategy affect the requirement profile. As stated earlier it is especially important to understand the difference between recycling and reuse. Recycling in this context means processing and utilization of materials available from products to be scrapped. Reuse means using products or components as such, perhaps slightly modified, and repaired but essentially in their original form, primarily for their original purpose. Typical basic recycling strategies include the following:

- Product/component reuse
- Product/component reuse—modified, repaired, and/or conditioned
- Product/component reuse—for alternative/secondary purpose
- Systematically organized component reuse as spare parts or as such in production
- Materials recycling without any prior separation processes
- Materials recycling after manual disassembly
- Materials recycling after mechanical shredding and separation
- Materials recycling after manual disassembly, mechanical shredding, and separation
- Incineration—energy recovery
- Chemical recycling—special processes
- Elimination of need to recycle—biodegradability—minimization of landfilling

It is easy to see that the terminology as such can be problematic and thus it is always important to clearly define what is actually meant by the term "recycling." In everyday discussions, the risk of confusion is particularly apparent.

Refined methods (often computer supported) to analyze product disassemblability have been developed. These may be well suited for fine tuning the recyclability properties but also simpler recommendable basic DFR indicators are

- Recyclability percentage—%-ratio of recyclable materials in the product. Note the importance to clearly define what is regarded as recyclable and the difference between theory and actual praxis.
- Disassembly time—the time required to manually disassemble the product according to specifications set in the recycling strategy. Most reliably obtainable in practical tests.
- Simple yes/no type indicators—e.g., includes hazardous waste X, requires special treatment, material symbols used, etc.

Existing infrastructure also plays a major role in product recyclability evaluation. Theoretically, fully recyclable products may be nonrecyclable in markets where collection and recycling infrastructure and systems do not exist.

4.4 Practical Mechanical Design Guidelines

General design methodologies, which automatically produce ideally recyclable and/or reusable products including all essential structural details regardless of market conditions, existing infrastructure, and so on, do not exist. Key properties defining recyclable and reusable products are given in Figure 3. These properties naturally can be achieved by a large number of alternative technical solutions. Subsequently, the success of DFR should be evaluated based on how the product structure supports recycling and reuse in practice in different conditions. Separate individual theories are often optimized for special conditions and certain strategies only. Lists of good general DFR principles and DFR checklists have been developed, however. These design principles supporting DFR typically include the following:

- Standardization
- Modularization
- Structural simplification
- Minimization of number of parts
- Timeless design
- High quality
- Informativity
- Serviceability
- Availability of spare parts

If the chosen recycling strategy includes product and component reuse, properties like serviceability and disassemblability should be prioritized. Here it is important that parts or structures to be reconditioned and reused do not get damaged when disassembled and handled. Supporting design rules include the following:

	Materials and components identification made easy by using markings and/or other coding systems	
Recyclable materials and materials combinations optimized for recycling processes and/or easy separation		Included information to ensure proper and safe product handling and treatment at all life cycle stages

Materials separation friendly product structure for easy separation of recyclable, hazardous, and waste fractions	Disassembly friendly product structure for easy disassembly of reusable components, service, and repair

FIGURE 3   Key properties of recyclable and/or reusable products.

- Use self-explanatory product structures e.g., structures for which the disassembly (as well as assembly) order, tools needed, joint types etc. are obvious and/or clearly identifiable/visible.
- Favor joint types which can be fastened/opened with typical standard tools.
- When possible, favor simple product structures without overlapping part arrangement. Parts to be reused or serviced should ideally be disassembled without having to disassemble other parts prior to this.
- Avoid structures which have to be adjusted/measured/set each time they are disassembled/reassembled.
- Avoid structures which require measurements with external equipment in assembly/reassembly.
- Group components logically according to disassembly order, common recycling method, other treatment required, or e.g., component value.
- Ensure accessibility and visibility.
- Ensure sufficient space for tools used in service and disassembly.
- Unify assembly/disassembly directions.
- Use joining methods easy to disassemble, especially avoid methods which include elements which are broken when disassembled,
- Minimize number of joining elements,
- Standardize types and sizes of joining elements (e.g., screws).
- Use structural symmetry or integrated mechanical guides (forms) to ensure correct reassembly.

- Ensure changeability/serviceability especially of components subject to mechanical wear (bearings etc.).
- Include suitable integrated forms for special disassembly tools when use of these is unavoidable.
- Avoid use of free rotating counter nuts which may be difficult to handle.
- Avoid shrink fittings in structures often to be serviced/disassembled
- Avoid gluing in structures to be disassembled.
- When possible use interlocking parts, which eliminate the need for separate joining elements.
- Reserve sufficient material thickness in areas subject to mechanical wear.
- Make evaluation of component condition easy (include wear markers etc.).
- Avoid complex structures subject to dirt and corrosion.
- Practice corrosion engineering. Use functional coatings when needed but take into account their possibly negative effect on material recyclability.
- Especially ensure corrosion protection and functionality of joints intended to be opened during service.
- Ensure sufficient lubrication.
- Make filling and draining of liquids easy.
- Do not integrate too large parts to avoid the need to replace large structures because of minor damages only.
- Support product longevity by keeping spares available and reasonably priced.

If the chosen recycling strategy does not include product and component reuse, but primarily materials recycling, properties such as use of recyclable materials, including material identification possibilities and use of easily separable structures are to be prioritized. Here it is not so important that parts or structures remain intact, so more rough disassembly methods can be and are used. A product can also be optimized for mechanical shredding and separation processes. Supporting design rules include the following:

- Unify materials used in the product (e.g., avoid metal parts in a plastic-dominated structure and vice versa).
- When possible unify the material of joining elements and other small components with the dominating material type.
- Use materials for which practical and economic recycling practices exist. Most materials are theoretically recyclable (e.g., many plastic types) but recycling services do not exist (e.g., because of poor economic feasibility, technical difficulties, or logistics problems).

- Avoid material combinations that can cause problems in the recycling process.
- Avoid composites, laminates, and integral structures in which materials are combined in a difficult to separate way.
- When possible use materials in pure form—avoid unnecessary coatings etc.
- Check effects of eventual coating materials on the basic material recycling process.
- When possible use single material integral structures.
- Ensure easy materials separability.
- Ensure materials identification possibilities—use relevant material symbols.
- Ensure easy possibilities to (manually) disassemble relative large components manufactured from materials that differ from the rest of the structure before the mechanical shredding and separation process.
- Ensure identification and correct treatment of hazardous materials possibly included.

As it is easy to see some of the design rules presented may contradict the rules in the reuse-oriented DFR rule list presented earlier. This makes it even more important to ensure that the recycling strategy has been planned well enough before going in for detailed design.

In both reuse- and recycling-oriented DFR, ensuring sufficient and relevant environmental information for all associated parties is highly important. Information distribution methods may vary, including printed material, information attached directly to the product, Internet services, and phone services.

- The consumer needs user's manuals, service manuals, warnings, prohibitions (to avoid danger), instructions how to deal with the product when it is to be scrapped, material symbols when some disassembly is intended to be carried out by the consumer, and service information.
- The waste/scrap handling company needs information on product material content, possibly hazardous materials included, correct handling storage and transportation methods, recycling processes available, and special arrangements.
- The service/repair company needs service manuals, spare parts lists and so on.
- The recycling company needs information on product material content, possibly hazardous materials included, correct handling, storage, and transportation methods.

5 CONCLUSION

The guiding principle of DFE in all product groups is to state environmental properties as essential requirements equal with traditional and obvious technical requirements such as strength, functionality, efficiency, safety, and ergonomics starting at the very beginning of the product planning process. Evaluations and choices on this basis are then to be made at every decision-making stage. DFE should be seen as a complementing, not as an alternative or rival, approach to existing product planning methodologies or practices. The general quality, usability, and safety of the product should not be allowed to be affected in a negative way.

DFE is most likely to succeed when it is carried out as an integral part within the normal product design process supported by long-term commitment to research and innovation. Adding favorable environmental properties afterward to an otherwise complete design is difficult or even impossible. Crucial basic decisions where DFE principles can be applied are typically made at the beginning of the engineering design process. This also applies to DFR as an essential element of DFE (11).

6 ABBREVIATIONS

The following list includes explanations for essential chemical substance abbreviations used in the text (list not exhaustive).

BDE	Bromodiphenyl ether
BHT	Butylated hydroxytoluene (cas: 128-37-0)
Bisphenol A	4,4′-Isopropylidenediphenol (cas: 80-05-7)
CFC	Chlorofluorocarbon
Octadecyl	3,5-Bis(1,1-dimethylethyl)-4-hydroxybenzenepropanoic acid octadecyl ester (cas: 2082-79-3)
PAH	Polyaromatic hydrocarbons
PBB	Polybrominated biphenyl
PBDE	Polybrominated diphenyl ether
PCB	Polychlorinated biphenyl
TBBA	Tetrabromobisphenol A
Tetrakis	3,3-bis(1,1-dimethylethyl)-4-hydroxy-, 2-(3(3,5-bis(1,1-dimetylethyl)-4-hydroxy-phenyl)-1-oxopropoxy)-methyl)-1,3-propandiol benzenepropanoat (cas: 6683-19-8)
TNPP	Nonylphenol phosphite (cas: 26523-78-4)

7 TERMINOLOGY

The following list includes explanations for some essential DFE-related terms applicable in this article (list not exhaustive). Some caution is advisable since international terminology interpretation and definition may slightly vary, depending on context.

- *Best available technology (BAT)*. The best-known, existing technical solution/alternative by applied criteria.
- *Design for environment (DFE)*. General heading for efforts aimed at designing and producing environmentally more benign products.
- *Design for recycling (DFR)*. General heading for efforts aimed at designing and producing optimally recyclable/reusable products according to chosen strategy.
- *Design for service (DFS)*. General heading for efforts aimed at designing and producing easily serviceable and repairable products.
- *Life cycle analysis (or assessment) (LCA)*. Environmental impact analysis for a given product's life cycle including qualitative life cycle (environmental) model generation, quantitative input output inventory, ecoprofile generation for the total life cycle model (combined presentation, all stages included), ecoprofile analysis, and optionally also combination of different types of loads using weighing factors to produce more easily comparable indicators. LCA is typically carried out using software and data base support.
- *Life cycle cost analysis (LCCA)*. Cost analysis of the total life cycle of a given product. Typically used to analyse usage phase associated costs (e.g., energy, service, and repair) for alternative solutions. If environmental loads generate actual costs (e.g., fees, taxes) they are also included in LCCA.
- *Materials recycling*. Recovery, separation, and processing of material fractions from scrapped products into materials reusable in industrial production. Type and quality of recycled material may vary depending on basic material properties, fraction purity, process available, and so on. Some fractions can also be used to form mixed recyclates.
- *Product and/or component reuse*. Functional reuse of products or components principally in their original form. Repair, minor modifications, and secondary applications may be associated with reuse.
- *Product life cycle (commercial, marketing)*. The time cycle including product development, market introduction, production/marketing and eventual production discontinuation of a product/model. Characterized by investments made into and profits gained from a given product.

- *Product life cycle (environmental).* The total physical cycle completed by an individual product during its existence including associated materials and energy production, actual production phases for product and its components, the usage phase, and the disposal/recycling phase. Characterized by physical inputs and outputs for single product or given quantity.
- *Product recyclability.* Percentage ratio of recyclable materials (materials recycling) included in product. Exact definition of 'recyclable' varies, especially theoretical and practical (actual market area and infrastructure specific) interpretations may differ markedly, case by case caution is advised.
- *Recyclate.* Recycled material
- *Recycling/recovery rate.* Percentage ratio of end of life products being recycled/recovered. Case by case caution is advised since monitoring methods, associated definitions, and other condition specific properties may vary.
- *Total life cycle approach.* Emphasizing the total (combined, cumulative) load of a product life cycle in environmental effect analysis instead of focusing on single loads/load types only.

REFERENCES

1. Kirk-Othmer Encyclopedia of Chemical Technology; 4th edition. New York: Wiley-Interscience Publications, 1991–1998. Vol. 1–22.
2. U.S. Environmental Protection Agency. Health Effects Notebook for Hazardous Air Pollutants. http://www.epa.gov/ttn/uatw/hapindex.html
3. O. Kolehmainen, ed. Suomen ympäristön tila. Sisäasiainministeriö, Helsinki, Finland, 1982, 328 p.
4. Tillsatser i plast—Slutrapport från plastadditivprojektet, The Swedish National Chemicals Inspectorate. Stockholm, Sweden, 1995. 80 p. KEMI rapport 15/95.
5. The flame retardants project—Final report. The Swedish National Chemicals Inspectorate. Stockholm, Sweden, 1996. 89 p. KEMI rapport 5/96.
6. Phase-out of PBDEs and PBBs. Report on a Governmental Commission. The Swedish National Chemical Inspectorate. 15 March 1999 http://www.kemi.se/aktuellt/Pressmedd/1999/flam_e.pdf
7. National Institute for Occupational Safety and Health. International Chemical Safety Cards. http://www.cdc.gov/niosh/ipcsneng/neng0012.html
8. HM Smith. Safety, health and environmental regulatory affairs for colorants used in the plastic industry. Coloring Technology for Plastics, no. 52 1999, pp. 87–98.
9. CalEPA Air Resources Board Toxic Air Contaminant Summary. http://www.scorecard.org/chemical-profiles/html/carbonblack.html

10. PVC Plastic: A Looming Waste Crisis. Greenpeace, April 1998. http://
     www.greenpeace.org/~comms/pvctoys/reports.html
11. M Kuuva, M Airila. Conceptual Approach on Design for Practical Product
     Recycling. Design for Manufacturability—An Environment for Improving
     Design and Designing to Improve our Environment. ISBN 0-7918-1269-3, p.
     115–123, New York: ASME, 1994.

19

Application of Ecodesign in the Electronics Industry

Ab Stevels

Philips Consumer Electronics, Eindhoven, The Netherlands

1 INTRODUCTION

Environmental care in industry has already existed for many decades. In the early 1960s, the detrimental effect of emissions to air, water, and soil was recognized as a global scale, and since that time legislation, regulation, and voluntary programs have been started to abate pollution. For more than 20 years main focus has been on production processes and hence on industry sectors involved in basic production (chemicals, materials like steel, paper, etc). Environment was seen as part of industrial engineering; solutions to environmental problems were sought in "end of pipe" cleaning through investment in installations.

In 1987 the Brundtland report (1) for the first time called attention to the fact that products (the result of production processes) can also cause substantial environmental loads. Product embodiments use sometimes scarce resources and can contain environmentally relevant substances as well. Packaging, packaging waste, and transport to the users can contribute considerably to the overall life cycle burden of products. For products using consumables like water, gas, and electricity, this holds in an even more outspoken way for the so called user phase. Finally, the end-of-life phase is relevant as well (recycling of discarded products, adverse environmental effects of landfill, and incineration).

Due to the very nature of its products, environmental issues in the electronics industry started to get more attention in the early 1990s. Improvement programs focused (and still do) on prevention, that is, reducing environmental effects up front by appropriate product specification and design. In this way product management and development groups got involved next to the production departments. Because products, once produced, move or potentially can move all over the world, environmental product issues have a global character in contrast to production/manufacturing issues, which are primarily local/national.

Authorities and consumer groups were the first to move after the awareness phase. In various countries around the world the first draft legislation on electronics products started to appear in the early 1990s. Test magazines started to include environmental paragraphs in their tests reports. Reaction of the industry was primarily fairly cautious; compliance with legislation/regulation and preventing bad test scores ranked high on the agenda. Basically this a *defensive* attitude; in this stage (1992–1996) environment was therefore primarily seen as a cost rather than as an opportunity to enhance business. Around 1995, the electronics industry started to realize that environmental and economic interests run to a large extent parallel:

- Resource reduction (energy use, materials, packaging) means also cost reduction.
- Reduction of disassembly times means mostly also reduction of assembly times.
- Reuse of subassemblies, components, and materials is cheaper than buying new ones.

This gave an enormous impetus to *cost-oriented* environmental programs. A new type of program, *customer-oriented* or *proactive*, was started by several companies as of 1997–1998. The basic idea was to increase market share through offering environmental benefits (which are communicated in terms of financial, immaterial, and "emotional" benefits as well) to the customer.

In Section 2 the general characteristics of the defensive, cost-oriented, and proactive approaches are discussed and elaborated on. In Section 3 examples are given of a typical defensive activity: setting up a basic environmental organization, mandatory rules, and establishing chemical content of electronic products. In Section 4 examples of cost oriented activities are addressed: environmental management systems (ISO14001), energy reduction, and packaging reduction. In Section 5 examples of proactive activities are presented: an Eco Vision Program, environmental benchmarking, and a strategy for environmental communication.

The examples given in Sections 3, 4, and 5 are from the author's practice at Royal Philips Electronics, Product Division, Consumer Electronics. Activities there have developed starting from a defensive approach in 1992 to which in later stages cost-oriented programs (from 1995 and onward) and the current proactive approach have been added.

In section 6 the "cultural" effects of the introduction of these programs are discussed, both in terms of successes and of items further to be improved. On basis of this experience a general model for integration of ecodesign into business has been developed. This model has turned out to be widely applicable outside the electronic industry. Details about this management model are given in Chapter 24 of this book.

2 GENERAL CHARACTERISTICS OF ENVIRONMENTAL APPROACHES IN THE ELECTRONICS INDUSTRY

The general picture of the environmental approach in the electronics industry is summarized in Table 1. In practice, individual companies in the electronics industry operate environmental affairs in a way which is a mix of the approaches shown in Table 1. The exact structure of the mix both depends on external and internal factors. External factors include geography (regions, countries of the world where business is done); product characteristics (environmental potential); customer awareness (both private and professional customers); position in the market (competition); and position in the supply chain (power, leverage). Internal factors include business focus (i.e., ambition and ethics); management style; and availability of skills.

Because integration of environment in the (electronics) industry is also a cultural process, the three approaches have also a sequence in time. From this perspective the defensive approach is to be seen as a minimum approach to start with and to be done by all companies. Based on the experiences built up in this phase, further steps can be taken to introduce the cost related green programs. For instance, a proactive approach can be developed. Practice has shown that jumping directly to the proactive mode of operation fails in the market. When the defensive items are not appropriately addressed, such programs are very vulnerable.

The drivers (item 1 in Table 1) are strongly geography dependent; generally speaking, legislation and customer awareness are best developed in Europe, liability and cost reduction are most important in the United States, and resource reduction is highest on the agenda in Japan. Management style strongly influences items 2–4 of Table 1 (management and organizations): Centralized organizations operating top-down can

TABLE 1 General Characteristics of Environmental Approaches in the Electronics Industry

Item	Defensive approach	Cost oriented approach	Proactive approach
Driver	Legislation/regulation	Money/cost	Market/customer
Management	Environmental declaration	Policy	Vision
	Command and control	Projects	Integrated in the business
Main objectives	Comply	Improve with respect to previous generation	Be better than competition
Organization	Formal structure	Delegated responsibility	Management of processes
Core processes	Manufacturing	Product creation process	Chain management
	Suppliers (purchasing)		
Control	Afterward	Built-in	Upfront
Activities	Substances reduction	Material reduction	Designs with lower cost for user
	Standby energy reduction	Energy reduction	Green designs which are easier to operate or fun
	Take-back of discarded products	Reduction of (dis)assembly time	Durable products
	ISO 14001 (partly)	ISO 14001 (partly)	Products with emotional benefits (green image)
Supporting tools	Checklists	Manual	Greening your business handbook
	Chemical content tool	Packaging reduction tool	Ecoindicator software
	Environmental weight calculation	Energy reduction tool	Benchmark tool
		End-of-life cost analysis tool	STRETCH creativity tool

Training	How to comply	How to reduce	How to integrate with business
Communication to the outside world	Compliance beyond minimum	Environmentally friendly but not more expensive	Green and other benefits combined
Language of communication	Environmental ("scientific green")	Reduction of resources	Perceived green
Main benefits delivered	Green and societal benefit	Green and company benefit	Green and customer benefit

move swiftly in the defensive approach, whereas decentralized ones with a bottom-up culture do well in proactive approaches. This is also because for such an approach, tailormade solutions depending on product characteristics have to be developed.

Items 5 and 6 (processes and control) depend externally strongly on the position of the company in the supply chain and internally on the business focus. The activities (item 7) to be done have a strong relation with products characteristics and with the customers. Products of complex nature, with substantial volume, weight and energy consumption, mostly have the highest potential for resource and cost reduction. Especially in professional markets such an activity will be highly rated. Environmental tools for cost related activities (item 8) are the ones which are most easy to develop and to operate. The some holds for training (item 9).

In the field of environmental communication (items 10–12) there are clear distinctions. The electronics industry is perceived as high-tech and professional and is therefore well positioned to perform good in compliance and in cost reduction. Especially in societies with a high income per capita, (brand) image plays a tremendous discriminating role in the markets. Being seen as a "caring" company (through a proactive approach) is of primary importance in this field.

3 EXAMPLES OF A DEFENSIVE APPROACH

3.1 Organization of Environmental Responsibility in a Global Electronic Company

To make corporate environmental goals visible and deployable, one of the members of the Group Management Committee, preferably the President and CEO, should be responsible for environmental affairs. By nominating a green standard bearer it is clear that the company takes green issues very seriously and wants to integrate them in all operations. At the corporate level support to the chief environmental officer should be given by a Corporate Environmental Office/(CEO).

An appropriate headcount in the electronics industry is approximately one person per US$ 5 billion of revenue. The task of this CEO is

- To develop the corporate policies, strategies, and programs;
- To handle external affairs (legislation, communication);
- To monitor progress of company programs.

A replica of the corporate structure should be made at the divisional and business group level. At the divisional level a member of the Senior Management Team should be responsible for environment. Support at divisional level is to be given by an Environmental Competence Center (ECC); the headcount of such an ECC should be on the order of 1 person per 2 billion US$ of revenue. The tasks of such an ECC are as follows:

- Support of the divisional environmental steering team
- Making of divisional programs, roadmap
- Support of implementation at business groups
- Ensuring availability of know-how and supporting tools
- Training and audit

At the business group level environmental matters should be handled by a member of the Management Team. Support is to be given by a Divisional Environmental Manager (1 person per billion US$) of revenue and line of business/plant Environmental Managers. Most of the persons in last category will be part-timers, located in the quality or health and safety departments. The main tasks of the divisional environmental managers are supporting implementation and reporting progress.

The structure sketched above shows that in the electronics industry, environment is seen as a line responsibility. Integration of environmental issues in the normal operations (also if a company still has a defensive approach) should have a high priority. In this respect, environment will follow developments as have happened with quality issues. Started as something separate to be addressed by specialists, it has now become fully integrated in the tasks of all employees.

3.2 Mandatory Rules

To ensure a minimum of environmental care in all operations, companies should have minimum mandatory rules. Application of these rules should be checked on product release and/or in manufacturing operations reviews. For the electronics industry these mandatory rules include

- *Banned substances.* Brominated flame retardants of certain types, heavy metals (Cd, Hg, etc.), ozone depleting chemicals, organic solvents and liquids (PCB, PCT).
- *Availability of environmental information.* Energy consumption, environmentally relevant substances (see also Section 3.3), and recyclability.
- *Packaging.* Material application, printing inks.
- *Marketing and labeling of products and/or product parts.*
- *Customer information.* For environmental optimal operation and disposal of discarded packaging and products.
- *Batteries.* Marking, how to handle.

The precise formulation of the mandatory rules varies from company to company. Some of them stick strictly to fulfilling legal requirements and have regional policies if requirements differ. Others go beyond the minimum and mostly have global mandatory environmental rules.

3.3 Chemical Content of Electronic Products

Knowing the chemical content of electronic products is important to fulfill actual legal requirements. It will also be helpful in anticipating future developments. It is crucial to start elimination well in advance of passing of laws, because finding alternatives will involve a lot of work. Some substances will not be legislated in the future, for instance, because a scientific basis for forbidding them is not available. However, using such substances ('the suspects') could do harm to the brand image of the company.

The vehicle used by Philips Consumer Electronics (PCE) to find out about chemical content is the so-called chemical content questionnaire (see Appendix 1). This questionnaire has been sent to all components and materials suppliers. This action included many hundreds of suppliers all over the globe and some 20,000 code numbers. Apart from the list, the supplier gets an accompanying letter explaining the procedure. It is essential to make clear that if in any category the supplied items exceed the threshold limits in the list, this does mean that PCE wants to starts improvement actions with the supplier and does not want to terminate relationships. On the contrary, it is stressed that we *want to know* the chemical content of our products and we *want to improve* our products in close cooperation with the supplier.

The answers given by the suppliers are processed by specialists of the ECC. When information has been considered to be complete, the component/material concerned is given a so-called environmental indicator (E.I.).

E.I. = 9   Component/material contains no environmentally relevant substances. *Fully released.*

E.I. = 6   Component/material contains environmentally relevant substances but no Philips banned substances. There are no good alternatives. *Temporarily released.*

E.I. = R   Component/material contains environmentally relevant substances (there are good alternatives) or component/ material contains Philips-banned substances. In both cases: *Rejected.*

The results of this environmental classification are communicated to the organization through updates of the Environmental Design Manual and a

TABLE 2   Chemical Content of Printed Wiring Board GFL-V2

	Number (%)	On weight basis (%)	Target % (number)	Target % (weight basis)
Total number of components, 3637				
Chemical composition known	95	98	99	99.5
of which fully released	64	87	80	92
Temporarily released	36	13	20	8
Released rejected		0.2	0.0	0.0
Total	**100%**	**100%**	**100%**	**100%**

computer database to which all S&V/CE development groups are connected.

In the product creation process the environmental performance is
checked at milestones. In so-called product cross-sections the chemical of
the product is described in terms of fully released, temporarily released, and
rejected components/materials. When rejects are still present, the milestone
cannot be passed.

A physical example of a chemical content project has been the work on
the composition of printed wiring board in GFL-V2 (in 1997). This board
has been used for several years in the mid range of televisions (21 and 25
inch). See Table 2 for environmental indicators.

After the determination of the E.I. it was concluded that the design of
GFL-V2 still contained some rejected components. Moreover the percentage of temporarily released components and materials is still pretty high
(36% and 13%, respectively). On basis of this information, the decision was
taken to reduce the number of rejected code numbers to zero and to reduce
the temporarily released ones to 20% (number wise) and 8% (on weight
basis). This project was successfully executed before release in the beginning
of 1998.

4   EXAMPLES OF A COST-ORIENTED APPROACH

4.1   The Environmental Opportunity Program of Royal Philips Electronics

This program was introduced in 1996 as a follow-up of a period in which
defensive attitudes were dominant. The main items are given in Table 3. In
the program a clear distinction is made between the mandatory Corporate

TABLE 3   The Environmental Opportunity Program of Royal Philips
Electronics (1996–1998)

Corporate part
  All factories EMS certified (ISO 14001 or EMAS)
  25% energy reduction in all operations
  15% packaging reduction in all operations

Product division part
  Ecodesign according to business needs
  Supplier requirements
  Creation of internal, external network
  Active participation in legislation, regulation discussion

part and the part at the discretion of the product divisions. In practice the
corporate part was the dominant one, with energy saving and packaging
reduction as the carriers for the ISO 14001 program. As is explained in
Section 4.2, starting with the cost savings side of ISO 14001 rather than
with the more formal part offered many advantages in practice. In this way
environmental management systems become a logical result from integrated
practice oriented activities instead of a set of upfront stand alone items.

4.2   Energy Saving in Manufacturing Operations

Energy saving in manufacturing operations has been treated as the core
platform on which the ISO 14001 certification was to be obtained. This
means that these projects have been organized in such a way that they fit
in both "upstream" and "downstream" ISO 14001 activities. Table 4 shows
that on basis of practical experiences in the factories, the ISO elements, as
far as not yet present, are to be organized or built as structures. Experiences
has shown that this "carrier" approach is very effective indeed.

   To create the platform for energy saving actions, a so-called energy
potential scan (EPS) has been carried out in many consumer electronics
factories. This EPS is in fact making a detailed and systematic inventory
of all energy flows in the production system. Data collection sheet formats
were organized in such a way that these could used both for upstream and
downstream activities. A general observation has been that the very fact that
comprehensive data are brought together in one concentrated form makes
awareness, creativity, and effectiveness in saving energy enormously stimu-
lated.

   The results of such an EPS is a list of prioritized options to save
energy, both in terms of its environmental effect and in terms of pay back
time. The items to be prioritized strongly depend on the location (need for

TABLE 4   Energy Saving as Core Activity in ISO 14001

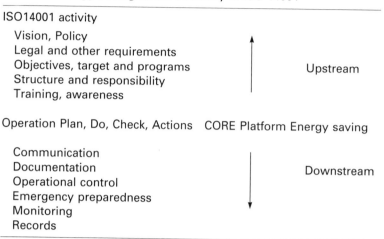

ISO14001 activity

Vision, Policy
Legal and other requirements
Objectives, target and programs   Upstream
Structure and responsibility
Training, awareness

Operation Plan, Do, Check, Actions   CORE Platform Energy saving

Communication
Documentation   Downstream
Operational control
Emergency preparedness
Monitoring
Records

air conditioning/heating in winter), the type of products manufactured (assembly, processing), degree of automation, and so on. So execution of locally tailormade action plans is necessary. On average for Royal Philips Electronics energy reduction programs have brought savings of US$ 40 M/ year with an average pay back time of investments in 2 years.

4.3   Packaging Reduction

The packaging of products has a multitude of functions. Apart from its protection function, it can also play a role in handling, communication of messages to the customer, and creating brand image. These items should be mapped out in detail before starting reduction actions. This should prevent that such "add-on functions" of packaging disappear in the process.

A first step in packaging reduction is getting facts. In the PCE case these include the following basics:

- Integral environmental load and cost of packaging and transport
- Ratios (see also Section 6.2): packaging weight/product weight and packaging volume/product volume.
- Environmental weight ratio. This is a number taking into account material application, substances in packaging, and recyclability.

These members (and subsequent simulations) are used to establish the main strategies for packaging reduction. These include

TABLE 5 Integral Environmental Load and Costs of Transportation

	Integral environmental load (%)	Integral cost (%)
Packaging material	48	42
Packaging operation	<1	3
Transport	45	43
Storage	1	12
End of Life	6	<1

- Material reduction (works out on integral load and weight ratio);
- Increase of amounts of recycled materials (affects environmental weight);
- Volume reduction (works out on integral load and volume ratio);
- Material replacement (affects environmental weight);
- Improving fragility (shock resistance) of the product or matching fragility better with the packaging concept (works out on all categories).

For audio products, integral environmental load and costs have been established for products manufactured in Asia and sold in Europe (Table 5). Table 5 shows that for both environmental load and costs the potential is approximately equal for material reduction and for volume reduction.)Also the data in Table 8 point in the same direction.)

Fragility measurements showed that in fact the packaging was over-dimensioned, especially in the respect of the EPS buffers. In the execution the volume reduction strategy was to one to be preferred. Design avenues for material reduction were derived from the benchmark (see Section 6.2). The total effort yielded a reduction of environmental load and integral costs of 8%, of which 6% is to be attributed to volume reduction and 2% to weight reduction.

5 PROACTIVE APPROACH

5.1 Philips Eco-Vision Program (1998–2002)

The formulation of the Philips Eco-Vision program as a proactive approach to environmental issues was a result of several paradigm shifts:

- Environment is business rather than a technicality.
- Environmental benefits as perceived by other stakeholders are key rather than scientific calculations of environmental gains.
- Best environmental care means when compared with competition.

• Understandable communication of environmental results is just important as achieving the results themselves.

The current Eco-Vision program is presented in Table 6. The cornerstone of the program is communication of top achievement in green as embodied in green flagship products to customers and other stakeholders. These achievements are to be realized through management of the cross-functional processes around creation, production, and marketing/sales of these products (see also Chapter 24 of this book).

In the creativity phase, environmental benchmarking (where do you stand with respect to the competition) is a key element, this is described further in Section 5.2. In Section 5.3 an example is given of communication about green flagships.

5.2. Environmental Benchmarking

The relation of environmental benchmarking with ecodesign is shown in Figure 1. To do a proper benchmarking, the system boundaries should be well defined and the functionality of the products to be compared should be as identical as possible (2). Also a list of items to be benchmarked should be available; this list contains the items which will be used later

TABLE 6 The Eco Vision Program of Royal Philips Electronics

Products (per line of business)
 Green focal areas in product communication
 Green flagships in 1998
 X% of products fully ecodesigned in 1999
 Y% of products fully ecodesigned in 2001
 15% packaging reduction in 2000 (ref. 1994)
 $Y > X$ to be determined by each division

Manufacturing (reference 1994)
 35% waste reduction in 2002
 25% water reduction in 2002
 Hazardous substances reduction in 2002
  Category I, 98%
  Category II, 50%
  Category III, 20%
 25% energy efficiency in 2000 (stretch to 35% in 2002 to be decided upon)

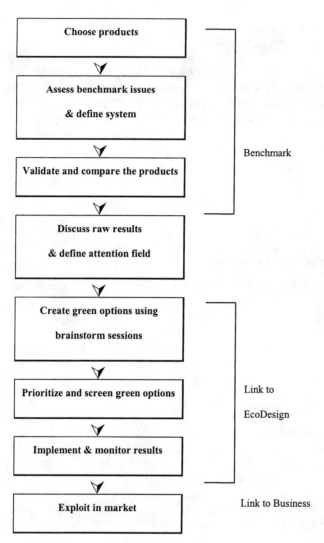

FIGURE 1   Relation of environmental benchmarking and ecodesign.

in communication with stakeholders. For this purpose, the Eco-Vision program has defined five focal areas:

- Energy consumption;
- Weight/material application;
- Packaging and transport;

- Substances in particular hazardous substances;
- Recyclability.

For electronic products, benchmarking items have been divided in five groups as well (Table 7). The example given in Table 8 on products on the market in 1997 demonstrates that environmental benchmarking can be very powerful, both in terms of generating data and ideas for further improvements but also for product positioning in the five focal areas. Table 8 compares the main properties of traditional audio systems . The four products selected consist of a tuner/amplifier, a double tape deck, a compact disk changer, and two loud speakers and have approximately the same functionality and features.

The results of Table 8 show that in spite of the fact products of this type are already on the market for more than 10 years, there are substantial differences in almost all focal areas and categories. Apparently the companies active in this field have completely different design strategies and using them to environmental objectives is relatively new.

For PCE (Audio Group) the above-mentioned benchmark results showed clearly the strategy how to develop green flagship products, that is, bringing products to the market which have a superior environmental performance with respect to the competition. This strategy included the following:

- Keep the lead in standby energy consumption.
- Increase the small lead in energy consumption in the operational mode.
- Stay among the best in weight issues.
- Reduce packaging weight and volume so that it becomes at least on par with competition.
- Drastically improve disassembly times.
- Since energy consumption is a major contributor to the life cycle impact score, the lead in the score will be automatically kept.

5.3 Environmental Communication

5.3.1 Green Communication at the Company Level

Green communication at the company level should particularly contribute to enhancing the brand image. Putting more green "into the brand" can be realized by

TABLE 7 Benchmark Items

Energy

Energy consumption off mode
Standby mode
Operational mode
Energy consumption of subassemblies
Energy consumption for user scenario's
Battery life and costs

Materials
Weights of plastic applications

Weights of metal applications
Weights of subassemblies, speakers
Weights and surface area of printed wiring boards
Weights of cables and wiring

Packaging
Packaging material weights
Packaging volume
Packaging weight/product weight
Packaging volume/product volume
End-of-life costs of packaging

Substances
Number of weight of suspect
Components, subassemblies
Recylability
(Calculated) disassembly times
Estimated material recycling efficiencies
Estimation of end-of-life cost

Life cycle calculations
Environmental impact (eco-indicator) of the various life cycle phases
Environmental impact (eco-indicator) of total life cycle for various user scenarios

TABLE 8 Example of Benchmarking Results for Audio Systems

Benchmarking items	Product of		
	Competitor 1	Competitor 2	Competitor 3
Energy (W)			
Standby, 2	11	8	12
Operation, 21	22	30	23
Tuner, 20/25	31/28	18/50	23/34
CD, 25/27	25/28	31/60	26/34
Tape decks, 23/24	22/27	31/43	25/34
Weight			
Parts total, 4300	4100	4600	6200
Transformer, 1800	1800	2100	2800
Sound system, 6887	9988	5612	9453
Packaging			
No. of boxes, 1	2	2	1
Weight (g) total, 2895	2607	1804	3401
Packaging weight/ product weight, 0.17	0.14	0.12	0.15
Volume ratio box/ product volume, 2.06	1.89	2.02	2.56
Environmental weight ratio; 0.91	0.95	0.98	1.11
Disassembly time (sec)			
Total, 160	90	100	150
Of which due to screws, 90	40	50	90
Life cycle score (mPT); 600	1200	1600	1300

- *Leadership.* Top management shows visible involvement in green; communication of environmental vision; and visible proactive attitude in trade associations.
- *Programs.* Communication of corporate environmental programs (see Sections 4.1 and 5.1), communication of awards, ISO 14001 certificates obtained, having a supplier requirement program; and communication of successes obtained in the programs.
- *Making documentation available.* Examples in the Philips Electronics case are green product brochure, (public) ecodesign guideline book *From Necessity to Opportunity*, the environmental annual report, and Internet green homepage.
- *Sponsorship.* Examples in the Philips Electronics case are sponsoring of Chair in Ecodesign (Design for Environment) at Delft

University of Technology; sponsoring of environmental conferences; and sponsoring of cleaning up the Antarctic.
- *Hardware.*   Green flagship products (see Section 6.3.2).

5.3.2   Green Communication at the Product Division Level

Green communication at the product division level should be directed
toward the methods and tools that are applied to ensure that the "green"
products brought to the market are really outstanding with respect to competition (or to conventional products). In the case of PCE particular attention is paid to

- Communication of the benchmarking method used (see also Section
  6.7).
- Explaining what the life cycle principle means in terms of combining the five focal areas: energy, materials, packaging and transport,
  substances, and recyclability.
- Explaining the ecodesign procedure followed (see Chapter 24 of this
  book).
- Communicating what a green flagship means and presenting these
  products.

5.3.3   Green Communication a Specific Product

This type of communication refers to scores in the focal areas. Table 9
shows an example for PCE data for the Audio System type no. FW870C,
a product which was launched in the market in September 1999. This product
duct was developed on the basis of the benchmarking results presented in
Section 6.2 and the design strategy resulting thereof.

6   EFFECT OF ENVIRONMENTAL APPROACHES ON THE ORGANIZATION

6.1   Defensive Approach

PCE started with a basically defensive approach of environmental issues in
1992–1993. Basic elements of the environmental program were at that time
formulation and deployment of an environmental declaration; setting up of
an environmental organization (see Section 3.1); formulation and monitoring of mandatory environmental (design) rules (see Section 3.2); and start of
an environment on banned and environmentally relevant substances (see
Section 3.3).

The effect of organizing this defensive program in the organization is
summarized in Table 10. An immediate result of implementation of the
program was that a strong environmental awareness was created. In spite

TABLE 9 Green Communication Focal Areas for Audio System FW870C

Focal area	Unit of	Message
Energy	kWh/$	Over the life cycle of the product, energy cost of the product is $35 lower than for the best competitor
Weight	kg	The weight of the product is 15% lower than of the best competitor (this saves resources)
Packaging	kg	The packaging weight is now 5% lower than of the best competitors (this saves resources)
Substances	Concentrations	N.A.
Recyclability	%	Now better than the competition
Life cycle performance	Ecopoints %	This product has a life cycle impact score which is 35% lower than the average of competitors

of urgent cost cutting and restructuring efforts taking place at the time of introduction, "green" got a solid place on the business agenda.

CE was from the very beginning perceived as a caring company, seen as one of the first companies in the electronics industry to take real action. In particular, this was achieved by sending letters to all suppliers about the chemical content program and by communicating this to the outside world, in particular to authorities in various countries.

A further advantage was in the systematic approach; through the presence of a network of full-time (in only few cases) and part-time (in most cases) environmental managers throughout the organization, all kinds of environmental information were gathered and improvement actions

TABLE 10 Effects of Defensive Approach

Good	To be improved
• Awareness created	• Perception as threat by organization
• Action taken, first mover	• Benefits for company doubted
• Environmental managers in place	• "This is technical"
• Collection for information	• Philips, what is in it for me?
• Program further developed	

going beyond the mandatory program we started. In such a way, the basis for further development of green activities was created.

All these positive effects could not prevent that pending legislation was seen as a threat; benefits for the company were doubted. The introduction of the program coincided with a major restructuring and this proved to be a serious handicap. Because the content of the program had chemical content as the core, it was seen as technical and highly specialized in nature. This mindset had to be turned into a perception of "green" as a business item in later stages.

6.2   The Cost-Oriented Approach

The environmental opportunity program (1996) (see Section 4.1) widened the scope of the environmental opportunities substantially. Effects on the organization are in Table 11.

The obligation to put an operational Environmental Management System in place according to the internal ISO 14001 standard confronted the business groups for the first time in a systematic way to deal with the environment. Of particular significance was that an ISO standard is a global one and as such is much more appealing to a global business as consumer electronics than national or regional (draft) legislation.

By linking energy savings activities in manufacturing directly to ISO 14001 through introduction of appropriate organizational structures and

TABLE 11   Effects on Organization of Environmental Opportunity Program

Good	To be improved
• Business groups systematically confronted with environmental concerns through EMS	• Sometimes bad experience with ISO 9001, why should ISO 14001 be better?
• Clear cost saving through saving energy	• "It works in the factories, why not for products"?
• Clear cost saving through reducing packaging	• Ecodesign manual too static
• Ecodesign taking off, manual in place	• LCA turns out to be difficult, need for more practical approach
• Supplier requirements	• Business rationale, resistance from purchasing
• Internal and external network built	

reporting formats, the cost-saving potential of ISO 14001 could be made very visible.

A similar effect was reached through the packaging reduction programs. These got an environmental flavor by increasing the amount of recycled materials used and by eliminating to a large extent expanded polystyrene (which is perceived as an environmentally unfriendly material).

With the momentum created in this way, ecodesign in general got more attention. Also the presence of appropriate metrics and supporting tools contributed towards this effect.

The good environmental image of Philips also had a positive effect on supplier relations. Requirements were accepted as an opportunity to learn and to improve rather as a threat and source of cost increases. The internal and external networks were further strengthened by the environmental opportunity program. Through its performance, authorities took proposals and initiatives seriously, although this did not always work out in final regulation.

In spite of all successes there were still items to improve. Implementation problems (in spite of the practical approach through savings side rather than addressing the more formal part first) with ISO 14001 occurred in the situation where business units or factory locations had mixed negative experiences with the ISO 9001 quality programs. Also, savings in factories (utilities) turned out to be easier to achieve than in product design itself.

This was in large extent due to the fact that the ecodesign (= design for environment) tools and manuals were formulated in environmental rather than in business language so that it was sometimes difficult to get the message across and to boost creativity. This was enhanced by the fact that in spite of all savings potential realized, a clear strategy to exploit environment also upstream (suppliers) and downstream (in the market) was not yet in place.

Environment stayed in the technical domain, sometimes even resisted by other internal stakeholders. This was enhanced by external events; authorities took in this period still a formal attitude toward the electronics industry, that is thinking in principles rather than solutions. Particularly in Europe this was believed as not justified and as unfair by the industry.

6.3 The Proactive Approach

The Eco-Vision program (1998; see Section 5.1) created a tremendous shift in mindset in the organization. The fact that the President and CEO introduced Eco-Vision personally contributed substantially to its success. Soon it turned out that an introduction of the first green flagship products led to

TABLE 12   Effects of the Proactive Approach

Good	To be improved
• Vision, strategy, roadmaps in place	• Deployment to improve
• Environment integrated into business	• Need to keep it separate in the beginning of process
• Broad-based actions—fantastic results	• A lot of strain on the support organization
• Ecodesign works well	• Consolidation into concepts, feasibility remain hard to fight
• Green marketing put into practice	• Special strategy needed to circumvent prejudice

increases in market share ($+2\%$), price premiums (an average $+3\%$), and a lower bill of materials (approx. -5%). Therefore, the outcome—also in the cultural sense—has been very positive as set forth in Table 12.

On a strategy level, environment was integrated into the business, in vision, strategy, and roadmaps. This got further support by results obtained in practice (see above). In fact Eco-Vision's success strained the support organization; not all initiatives, requests, and questions could be adequately handled in the beginning because of the volume of all the work involved. It also turned out that in spite of all integration efforts in later stages, environment should be kept separate in the very beginning of the design process. In this stage green creativity is basic issue. Putting in too much day to day business issues here turned out to be very disadvantageous for "thinking out of the box," and specifically for environmental thinking.

Special attention needs to be paid to deployment as well. The basic mindset of many employees in the beginning of the program was that environment is a threat rather than an opportunity for the organization. Only the communication strategy of joint environmental and other benefit can overcome this.

APPENDIX 1

Philips Consumer Electronics List of Environmentally Relevant Substances

Component (family):   Supplier:

Supplier type number: Date:

Component weight (excl. packaging):

Compound	Threshold conc., ppm (mg/kg)	Tick off if actual conc. > threshold	Actual conc., ppm (mg/kg)
Antimony and—compounds	10	☐	
Arsenic and—compounds	5	☐	
Berylium and—compounds	10	☐	
Cadmium and—compounds	5	☐	
Chromium (hexavalent)	10	☐	
Compounds		☐	
Cobalt and—compounds	25	☐	
Lead and—compounds	100	☐	
Mercury and—compounds	2	☐	
Metal carbonyls	10	☐	
Organic tin compounds	10	☐	
Selenium and—compounds	10	☐	
Tellurium and—compounds	10	☐	
Thallium and—compounds	10	☐	
Asbestos (all types)	10	☐	
Cyanides	10	☐	
Benzene	1	☐	
Phenol (monomer)	10	☐	
Toluene	3	☐	
Xylenes	5	☐	
Polycyclic aromatic hydrocarbons	5	☐	
CFCs and halones	0	☐	
Acrylonitrile (monomer)	25	☐	
DMA, (N,N)-dimethylacetamide	10	☐	
NMA, (N)-methylacetamide	10	☐	
DMF, (N,N)-dimethylformamide	10	☐	
NMF, (N)-methylformamide	10	☐	
Diethylamine	10	☐	
Dimethylamine	10	☐	

Compound	Threshold conc., ppm (mg/kg)	Tick off if actual conc. > threshold	Actual conc., ppm (mg/kg)
Nitrosamide	10	☐	
Nitrosamine	10	☐	
Ethylene glycolethers and—acetates	10	☐	
Phthalates (all)	25	☐	
Formaldehyde (monomer)	40	☐	
Hydrazine	10	☐	
Picric acid	10	☐	
PBBE, polybrominated biphenyl ethers	10	☐	
PBB, polybrominated biphenyls	10	☐	
PCB, polychlorinated biphenyls	1	☐	
PCT, polychlorinated triphenyls	10	☐	
Pentachlorophenol	10	☐	
Dioxines	0	☐	
Dibenzofurances	0	☐	
Other halogenated aromatic Hydrocarbons	20	☐	
Epichlorohydrine (monomer)	10	☐	
Vinylchloride (monomer)	1	☐	
PVC and PVC blends	1000	☐	
Other halogenated aliphatic hydrocarbons	10	☐	

REFERENCES

1. The Brundtland Commission. Our common future. Report of the World Commission on Environment and Development. Oxford University Press, Oxford, UK, 1987.
2. AJ Jansen, ALN Stevels. The EPass Method: A Systematic Approach for Environmental Product Assessment. Proc. CARE Environment '98. Vienna, November 1998, 313–320.

20

The Economics of Disassembly for Material Recovery Opportunities

Michael R. Johnson and Michael H. Wang
University of Windsor, Windsor, Canada

1 INTRODUCTION

Two groups today are increasingly interested in understanding and improving the economics of product disassembly to facilitate product stewardship and reverse logistics strategies. The first group, recyclers, are presented with the problem of dismantling products which were designed as long as 15 years ago, without thought to their ultimate disposal. Many of these designs make reclamation and disassembly of products a costly and unprofitable process. The second group is a growing number of manufacturers and designers who are interested in the implementation of product stewardship programs and the ultimate costs associated with material recovery opportunities (MROs). An MRO is defined as an opportunity to reclaim post-consumer products for recycling, remanufacturing, and reuse.

The life cycle of a product can be defined as stages of possible environmental detriment which may occur at each of the following stages: raw material extraction (i.e., birth of material), processing, transportation, manufacturing, use, and final disposition (Figure 1). Product stewardship is becoming a popular notion that the producer of the product or service is

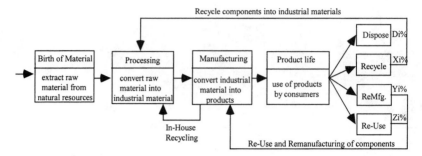

FIGURE 1   Material/product life cycle.

made responsible for taking care of its environmental impacts throughout the product's entire life cycle. MRO not only provide the means to divert materials from disposal but also to reduce the virgin material requirement, thereby diminishing potential upstream impacts of material extraction and processing. Now that it has been briefly demonstrated that MRO provide clear advantages to the environment, the next question is what methods in engineering, design, and manufacturing can help to evaluate the economics associated with product stewardship and the disassembly of products for recycling, remanufacturing, and reuse ?

Only recently have manufacturers and recyclers questioned the profitability of recovering durable products. For the most part, however, these groups have resorted to using heuristics for analyzing the breakdown of products and the associated costs. As identified at a local disassembly facility in Canada, the process of disassembly is haphazard at best. This often means that once high-value components are identified from within products, product disassembly continues (regardless of its profitability) until such materials are recovered. As the product dismantling process continues, other economic questions surface:

- How may the recovery process itself generate the highest possible return of investment?
- Is there a particular disassembly sequence that will maximize the return?
- Is it better to recover only specific components rather than all components?
- What design characteristics facilitate ease of disassembly and how are they to be employed?

All the above questions may be answered by using a thorough analysis of the disassemblability of a product.

2   ECONOMIC CONSIDERATIONS

Today, the actual dollar value derived from disassembly and recycling seems to be rather precarious in an economy subject to low virgin material costs and a changing market demand for materials. However, times are changing, and worldwide trends such as the current product stewardship legislation in Germany seem to indicate the need for an emphasis on material recovery rather than continued disposal. The idea of "cradle to reincarnation" design is continually being advanced in both research and industrial organizations. With this in mind, we propose that an economic assessment of disassembly is required for designers and decision makers to diagnose the feasibility of MRO.

Background information on the economics of dismantling products is discussed in the next section. Issues such as the value of postconsumer products, disassembly costs, and disposal costs are known to fluctuate considerably depending on the product's materials and disassembly process and inevitably impact on the profitability of demanufacturing.

2.1   Material Value

As discussed by Simon et al. (1992), all product components, parts, and subassemblies exhibit some form of material value at the end of the product's useful life. They define material value as the amount a customer would pay for the disposed item. In other words, material value may be defined as the actual gain, or dollar value, from recovered materials at the end of the product's life.

This definition could also be viewed from the manufacturer's point of view to consider how much would the manufacturer pay for the item at the end of the product's life. The advantage of this perspective is that a quantitative assessment may be used to compare the manufacturer's virgin material cost with the cost to upgrade (or reprocess) the reclaimed material to a quality level of the virgin material. In all situations, the material value will vary depending on the MRO used, the state of the product on disposal, the cost of disposal, and other external costs associated with reprocessing materials.

For some materials, such as precious metals, material values are more easily defined. For example, five German companies have joined forces to create a system for recycling postlife catalytic converters. Their goal is to recover the highest proportion of the precious metals (high platinum and rhodium content) from the converters and return the stainless steel casings to the material cycle as a secondary raw material. As early as 1987, BMW recognized the potential for reclaiming valuable materials from used catalytic converters. However, for other materials, material value is not as easily defined and may become rather ambiguous. To help designers assess such values the following list of cost factors should be considered:

1. Total reprocessing/upgrade costs including transportation, sorting, cleaning, disassembly, and reprocessing. Often the design of the product may render a component or part nonrecoverable. For example, in situations where parts are painted, it becomes very difficult to justify recovery because of high reprocessing costs. Thus, design has an important part in making material values recoverable after consumer use.

2. Product/component(s) disposal state: The shape of the product on disposal must be assessed. Consideration of the product's use by consumers must be assessed. For example, some materials are exposed to certain environmental conditions (ultraviolet or corrosive exposure, etc.) during its use that may substantially deteriorate its material value.

3. Availability of recycling and transportation infrastructures.

4. Disposal fees.

All these factors may be used to assess the value of postconsumer products by comparing the value of the used components with respect to virgin materials costs. For example, if one was considering recycling, the reprocessing cost is used to assess material value by comparing the cost to improve the quality (i.e., the cost of upgrade) of materials with respect to the price of virgin materials. The material value should be a reflection of economic benefit gained from recovering postconsumer products.

2.2 Economic Analysis of the Disassembly Sequence

In this section, the economics of disassembly with respect to the sequence of removing parts are addressed. At one extreme, as the product is disassembled piece by piece, no material value is gained (i.e., the material reclaimed has no value). At the other extreme, each component removed has a positive material value and the total valued reclaimed increases higher and higher. However, in actuality, a product will consist of a variety of material values ranging from positive to negative values. A negative material value may arise in a number of circumstances; materials or parts of the product are toxic and require special handling disposal methods; one or more components have a low material value and existing landfill costs are high; or it may occur in instances where legislation has required recycling and large costs have been incurred to comply with such regulations.

Figure 2 represents the economic importance of the disassembly sequence. The product shown in this graph consists of a range of positive

DISASSEMBLY COST VS. DISASSEMBLY SEQUENCE

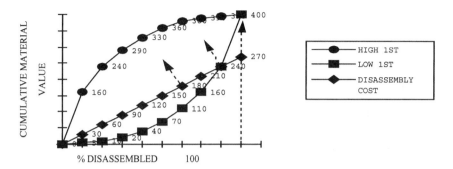

FIGURE 2   Importance of the disassembly sequence.

material values. The two curves represent that disassembly is constrained by the arrangement of components and can reach its goal of complete product disassembly through a wide range of disassembly sequences. In fact, the range between the two curves represents the range of feasible disassembly sequences. The upper curve will result if materials are removed through successive steps from the highest to the lowest material values. Likewise, the lower curve results if materials are removed through successive steps from the lowest to the highest material values. The middle line represents the linear relationship of the constant labor rate ($/hr) and time. If materials are to be removed following the lower curve, the total material value recovered in the disassembly process may be rendered uneconomical especially in a situation of high labor rates. Therefore, to remain economically viable, it is important to remove high value items first and then remove material with diminishing values.

In fact, if we represent material value as recovered at the instance when the component is released from the product, then the material value curve increases in steps, as shown in Figure 3. We define the point "X" in Figure 3 as the target value where economic benefit is maximized. Under the assumption that the disassembly sequence continues in a fashion of "high value materials removed first," this point may also be represented in the disassembly process as the disassembly step or instance which results with the onset of a negative or zero benefit.

In this fashion, disassembly rules can be generated to assess the point at which the disassembly process should stop, to minimize costs and maximize returns of reclaimed materials:

LABOR STEP SEQUENCES VS. DISASSEMBLY COST

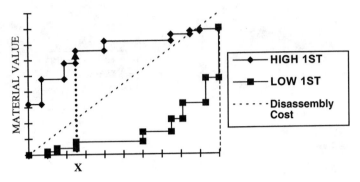

FIGURE 3   Gradient value of material recovery.

Rule 1: If total profit loss margin (PLM_T) > 0, continue disassembly.
Rule 2: If total profit loss margin (PLM_T) ≤ 0, stop and consider disposal

The above two rules are used in the situation where the disassembly sequence progresses in a perfect order of reclaiming high material values to low material values. The disassembly process halts when a negative or zero PLM value is encountered. However, in reality, the order of disassembly sequence is constrained by the arrangement of components and the choice to remove parts in such a perfect order is highly unlikely. In this case, rule 2 is oblivious to the fact that higher value material may exist further down the disassembly hierarchy. This presents the idea that all disassembly sequences must be generated and assessed in order to optimize the disassembly sequence.

The next section discusses some of the different approaches developed in research to evaluate and optimize the economics of product disassembly for MRO.

3   RESEARCH CONTRIBUTIONS

Research over the past 5 years has flourished on evaluating the disassembly of end-of-life products for demanufacturing. For all the research completed in this area, researchers have tended to consider various factors including or excluding various disassembly characteristics. Table 1, by no means exhaustive, provides an overview of the researchers and their assumptions made for evaluating disassembly. The following is a brief description of the various

TABLE 1 Disassembly Analysis: An Overview of the Various Authors and Their Assumptions

Authors	[1] G	[1] NG	[2] E	[2] S	[3] Y	[3] N	[4] Y	[4] N	[5] Y	[5] N	[6] EOL	[6] S
Subramani and Dewhurst (1991)												
Laperrierre and ElMaraghy (1992)												
Navin Chandra (1995)			?	?			?	?	?	?		
Zussman et al. (1994)			?	?			?	?	?	?		
Johnson and Wang (1998)					?	?						
Spicer and Wang (1995)			?	?			?	?	?	?		
Huang and Wang (1995)			?	?			?	?	?	?		
Penev and de Ron (1996)			?	?			?	?	?	?	?	?
Woo and Dutta (1991)			?	?			?	?	?	?	?	?
Chen et al. (1994)			?	?			?	?	?	?	?	?
Vujosevic et al. (1995)			?	?			?	?	?	?	?	?

See text for explanation.

assumptions that correspond to the numbered column headings found in Table 1:

[1] The product's three-dimensional structure: Some researchers have avoided this due to the added complexity. Thus in Table 1, "G" stands "yes," the authors have considered the geometric interfaces between parts; and "NG" stands for "no," the authors have not considered it.

[2] The search space: Some researchers have simplified by the problem by not considering the entire search space. The trade-off of not considering all possible disassembly paths is an obvious reduction in complexity and search efficiency. "E" stands for entire, and "S" stands for simplification.

[3] Precedence constraints: Some parts must be removed before others. "Y" stands for "yes," it has been considered; and "N" stands for "no," it has not been considered.

[4] Successive disassembly operations may have distinct benefits due to changeover characteristics (i.e., similar setups, tooling, etc. between operations). "Y" stands for "yes," it has been considered; and "N" stands for "no," it has not been considered.

[5] Yield: Some disassembly operations lend themselves to a greater release of materials than other operations. "Y" stands for "yes" and "N" stands for "no," it has not been considered.

[6] The purpose of the model itself: Many researchers have not considered its use (or have not discussed it). Two common categories were found: (1) "S" stands for serviceability and (2) "EOL" stands for end-of-life considerations.

And finally, a question mark denotes the fact that it was not obvious what assumptions were made.

Subramani and Dewhurst (1991) developed a method for the generation of the disassembly sequence from a serviceability point of view. They use a relational model to represent the geometric relationships between parts. This model consists of four entities: parts, contacts, attachments, and relations. They proposed the use of a disassembly diagram that holds information in terms of the disassembly direction of a part with respect to their contacts or obstructions.

Laperriere and ElMaraghy (1992) developed a method for integrating assembly sequence generation with various aspects of disassembly. They used an artificial intelligence search algorithm to expand the directed search graph. A number of criteria are used to optimize the resulting assembly plans: stability, number of reorientations, concurrency of operations (which they defined as parallelism), and grouping of similar operations.

The last criterion is particular interesting. It is also defined as clustering. The clustering of similar tasks into successive operations in the assembly sequence may significantly reduce the number of setups and tool changes. For example, consecutive pressing of different parts on a base may sometimes be performed simultaneously in a single operation (Laperriere and ElMaraghy 1992).

Zussman et al. (1994) used principles from engineering design combined with a graphical technique to evaluate all possible disassembly paths. They applied utility theory to model the objectives of a company regarding end-of-life considerations and to take into account the uncertainties about future development of economic and technical conditions. Essentially, an optimal disassembly path is selected as the path that demonstrates the highest utility (i.e., to the designer or company using this technique).

Johnson and Wang (1998) developed a method for the optimization of disassembly sequences using a two-commodity network formulation. These researchers used a number of criteria to reduce the search space prior to sequence generation: (1) material compatibility for end-of-life considerations, (2) material designed for disposal, and (3) identification of "clustered" tasks (i.e., successive disassembly operations which reduce setups and tool changes). This approach is explained further in a later section of this chapter.

Navin Chandra (1995) developed an algorithm to generate disassembly sequences for material recovery. This research is based on the Pareto-Optimal A* methodology applied to what is called the Opportunistic Traveling Salesman Problem.

Spicer and Wang (1995) developed a method for generating optimal disassembly plans using genetic algorithms. This method can handle large problems but has been based on a simplification of the problem. Each part was considered to be a single entity. There was no concept of parts coming together to form subassemblies. In other words, the product was modeled as if it were a collection of n parts, each indivisible.

Vujosevic et al. (1995) developed an algorithm for the identification of a disassembly sequence for the maintainability of mechanical parts based on tool selection, time and cost analysis, and human factors. Alternate algorithmic approaches have also been carried out by Woo and Dutta (1991) and Chen et al. (1994).

Penev and Ron (1996) designed a "disassembly strategy" for disassembly sequence generation utilizing the method of dynamic programming. They combined the theory of graphs (using and/or graphs) and dynamic programming to derive an optimal disassembly sequence based on the economics of material recovery.

4 ECONOMIC MODELS FOR DISASSEMBLY

One of two actions must be taken within the disassembly process: reclaim the component (i.e., for a specific MRO) or dispose of it. At the forefront of this decision, three decision elements should be considered:

1. The option of recovery, its costs and benefits;
2. The present disposal cost of the component;
3. The possibility that disposal of the component is the best alternative but disassembly is still required because of the need to recover a valuable part which is attached to it; the cost of disassembly and disposal.

This section is devoted to developing economic indices to assess the possible trade-off considerations between the above three elements.

Consider the following terms and their corresponding values:

Cd_k	Disassembly cost for the kth component (\$)
Cp_k	Disposal cost for the kth component (\$)
$PLM_{RECOVERY}$	A decision index for recovery: profit/loss margin of reclaiming component k (\$)
$PLM_{DISPOSAL}$	A decision index for disposal: profit/loss margin of disposing component k (\$)
PLM_{FINAL}	The resultant PLM generated in the cost analysis (\$). Used as input in the disassembly sequence generation
n	Total number of components reclaimed
m	Total number of components within product
$d1$	Total number of components disassembled but disposed of
$d2$	Total number of components disposed of (without disassembly)

Upon disassembly, a specific component k is defined as having an individual reclamation value (Rv_k) of

$$Rv_k = mv_k \times \text{weight}(\text{wt}_k) \times df \tag{1}$$

where mv_k represents the material value (\$/unit weight) of the kth component and df is the depreciation factor specified by the user which falls between 0 and 1. The depreciation factor accounts for the condition of the component upon disposal. For example, some materials are exposed to certain environmental conditions (ultraviolet or corrosive exposure, etc.) during its use which may substantially deteriorate the material value.

The disassembly cost associated with removal of the kth component can be represented as

$$Cd_k = t_k \times C_L \tag{2}$$

where t_k is disassembly time for the kth component and C_L is assumed to be the labor rate (in \$/unit time).

The disposal cost associated with landfilling the kth component can be represented as

$$Cp_k = Cp_R \times wt_k \tag{3}$$

where Cp_R is the current disposal rate (\$/unit weight) and wt_k is the weight of the kth component.

Within the previously defined parameters, the profit/loss margin (PLM) of recovering the kth component for material recovery can then be expressed as

$$\text{PLM}_{\text{RECOVERY},k} = Rv_k - Cd_k + Cp_k \quad (k = 1, 2, \ldots, n) \tag{4}$$

where Cp_k is the savings from not paying the disposal cost of the kth component. The disposal cost (i.e., Cp_k) may be quantified as a tangible benefit to the manufacturer when recovery is implemented (assuming that the manufacturer has taken responsibility for product disposal). The values of Eq. (4), Rv_k, Cd_k and Cp_k, are positive for all m components. From a recycler's perspective, the disposal cost (i.e., "Cp_k") associated with Eq. (4) is actually the revenue associated with the initial act of recovery. This value is also known as the tipping fee that is paid to the recyclers from the liable party (i.e., manufacturer). In both cases, the revenue of recovery is based on the total reclamation value and the disposal cost.

The only other alternative is disposal. The profit/loss margin of disposing the kth component can then be expressed in one of the following two forms:

$$\text{PLM}_{\text{DISPOSAL},k} = \begin{cases} -Cp_k & \text{Disposal cost only} \\ & (k = 1, 2, \ldots, d1) \quad (5) \\ -Cp_k - Cd_k & \text{Disposal plus disassembly cost} \\ & (k = 1, 2, \ldots, d2) \quad (6) \end{cases}$$

Equation (5) represents the disposal cost of the kth component. Equation (6) represents the disposal cost plus the disassembly cost of the kth component. This may occur when disassembly is still required because of the need to recover a valued part that is attached to it.

Equations (4), (5), and (6) are quantitative indices of the three decision elements (which were listed at the beginning of this section) to consider in product disassembly analysis. The disassembly process may be optimized using these decision elements to assess the trade-off between recovery and disposal of individual components.

5   A FRAMEWORK FOR ECONOMIC PRODUCT STEWARDSHIP

This section provides a brief overview of the research carried out by the present authors to establish a methodology for evaluating the disassembl-ability of consumer product for MRO. Here we define product "disassembl-ability" as the ability to optimize the design and disassembly process for removal of specific parts or materials in a manner which will simultaneously minimize costs and maximize the material value to be reclaimed.

Economic indices, which are developed in the previous section, pro-vide an indicator of the profitability of material and component recovery compared to alternative options such as disposal or storage. In our research, this is identified as a level 1 disassembly analysis and the outcome of this analysis will include the following:

1.   Cost estimates are generated for recovery versus disposal of indi-vidual components.
2.   Unprofitable disassembly operations are abandoned from further analysis.
3.   Material reclamation values are maximized from the available material markets.
4.   An optimal economic index (called the PLM_{FINAL}) is affixed to each disassembly operation under consideration in the schedul-ing process. This index can be used as input into a level 2 analysis of product disassemblability (i.e., the optimal disassembly sequence generation level).

Figure 4 outlines two levels of this approach used to evaluate and generate the most economically viable disassembly sequence. Using the results of the first level, a profit matrix is formed at the start of level 2 to represent the profits between disassembly operations. The formation of the profit matrix is known as the PLM (or profit-loss margin) matrix formation. Next, criteria for reducing the profit matrix and hence the search space are applied. Three criteria are established to reduce the search space and facil-itate recovery opportunities: (1) material compatibility between adjacent parts and components, (2) grouping multiple adjacent parts together for disposal, and (3) disassembly efficiencies that may occur due to concurrent disassembly operations. Finally, the reduced matrix forms the objective

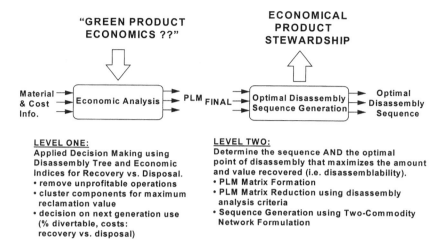

FIGURE 4   Two levels of disassembly analysis methodology.

function of a linear programming formulation that generates the optimal disassembly sequence.

6   CONCLUSIONS

This chapter presented an introduction into the economics associated with disassembly of products for material reclamation. The fundamental parameters of evaluating the economics of product disassembly were discussed, including material values, disassembly costs, and disposal costs. Also discussed was the relationship of these economic parameters with respect to the sequence of disassembly operations and the need to determine the point in the disassembly process that will maximize profitability. An overview of today's research contributions was discussed as well as the various assumptions considered by different researchers. Finally, the last two sections of this chapter developed economic models to demonstrate the relationship between the economic parameters of disassembly and a framework for economical product stewardship.

REFERENCES

Ashley S. Designing for the Environment. Mech Eng March, 1993, pp. 52–55.
Chen SF, Oliver JH. Parallel Disassembly by Onion-Peeling. ASME Conference: Advances in Design Automation, 1994, 69(2), pp. 107–115.
Huang T, Wang M. Optimal Disassembly Sequence Generation using Neural

Networks. Proceedings of the Fourth International Congress on Environmentally Conscious Design and Manufacturing, 1996, Cleveland, Ohio, pp. 231–238.

Johnson MR, Wang MH. Economical Evaluation of Disassembly Operations for Recycling, Remanufacturing, and Reuse. Int J Product Res 36:3227–3252, 1998.

Laperriere L, ElMaraghy HA. Planning of Products Assembly and Disassembly. Ann CIRP 41:5–9, 1992.

Marks MD, Eubanks CF, Ishii K. Life-cycle Clumping of Product Designs for Ownership and Retirement. ASME Design Theory and Methodology Conference, 1993.

Navinchandra D. Design for Environmentability. ASME Design Theory Method 37:119–125, 1991.

Penev KD, Ron AJ. Determination of a Disassembly Strategy. Int J Product Res 34:495–506, 1996.

Schmaus T, Kahmeyer M. Design for Disassembly—Challenge of the Future. Kahmeyer, Fraunhofer Institute, Germany.

Simon M. Design For Dismantling. Professional Eng November, pp. 20–22, 1991.

Simon M, Fogg B, Chambellant F. Design for Cost-Effective Disassembly. Chambellant, Manchester Polytechnic, UK, 1992.

Spicer A, Wang M. Optimal Disassembly Sequence Generation for Complex Products: An Artificial Intelligence Approach. 3rd Intl. Congress on Environmentally Conscious Design and Manufacturing, Las Cruces, NM, 1995.

Subramani AK, Dewhurst P. Automatic Generation of Disassembly Sequences. Ann CIRP 40:115–118, 1991.

U.S. Congress, Office of Technology Assessment. Green Products By Design: Choices for Cleaner Environment. OTA-E-541. Washington, DC: U.S. Government Printing Office, October 1992.

Vujosevic R, Raskar R, Yetukuri N. Simulation, Animation, and Analysis of Design Disassembly for Maintainability Analysis. Int J Product Res 33:2999–3022, 1995.

Zussman E, Kriwet A, Seliger G. Disassembly Oriented Assessment Methodology to Support Design for Recycling. Ann CIRP 43:9–14, 1994.

21

Product Disassembly and Recycling in the Automotive Industry

Sanchoy K. Das

New Jersey Institute of Technology, Newark, New Jersey

1 INTRODUCTION

In terms of both mass per unit and the total disposal volume, automobiles represent the single most important class of products that are disposed of by consumers. There are approximately 170 million automobiles currently on U.S. roads, and it is expected that a minimum of 8 million automobiles will be disposed of every year. It is estimated that currently about 4 million vehicles are processed each year by recyclers. The global upward trend in auto production volume has not yet shown any significant signs of leveling off, and as a result we can expect the auto disposal rate to continue to increase in the near future. As a consequence of these trends, the disassembly and recycling of automobiles is an area of significant interest in life cycle analysis. Fortunately, the disassembly and recycling of autos is relatively well evolved when compared to other product wastestreams. We find there are two primary reasons for this: there is a good market for reclaimed auto parts and components and over 60% by weight of an automobile is steel, a material that is highly amenable to recycling. As a testament to steel recycling, consider the fact that in 1998 the auto recycling industry generated enough recycled steel to manufacture 13 million new cars. Today typically, about 25% by weight of every new car is made of recycled steel. Even so, more than 24% of every disposed automobile does end up in a landfill site. Further, changes in material composition and stricter landfill regulations pose several new challenges to the auto disassembly and recycling industry.

The auto disassembly and recycling industry serves as a model for all other product industries. Our ability to economically build and evolve this industry is an indication that such life cycle processes can be established for other product groups. In the last few years significant transformation has occurred in automobile design, and this is evident in the material composition of new vehicles and the assembly structure of the vehicle. Given that the typical life expectancy of an automobile is between 10 and 15 years, we can therefore expect new recycling challenges within the next decade. In this chapter we first review the current industrial practice in automobile recycling. We then discuss key issues in the future development of automobile recycling. Our consensus is that we must inevitably develop and achieve the economic operation of high volume automobile disassembly lines. We discuss existing approaches that have been proposed for the construction of such facilities. Finally, we propose a prototype schematic for the design and analysis of high volume automobile disassembly lines.

2   CURRENT AUTO RECYCLING PRACTICE

Auto recycling is a relatively large industry; for instance, in 1997 the Association of Automobile Recyclers estimated the industry had total revenues in excess of about U.S. $8 billion. The industry tends to be highly localized, in that facilities operate on a regional basis. In the United States there is on average at least one auto recycling facility in every regional county. The primary reason for this is that transportation costs need to be minimized to maintain profitability in this relatively low margin industry. Figure 1 illustrates the current structure of the industry and the associated disassembly supply chain. The primary value adding components in this chain are the disassembly facilities. Both technologically and economically, disassembly provides the greatest opportunities in the future recylcability of automobiles.

The two main stages in the processing of the disposed vehicle are disassembly and shredding. As shown in Figure 1 the disassembly facilities can be divided into the following two distinct groups. Note that many facilities are a hybrid of the two groups. The first group is *part scavenging recyclers*, in which a disposed vehicle is scavenged on an as-needed basis. Here the disposed vehicle is left on a lot. When a customer requests a part

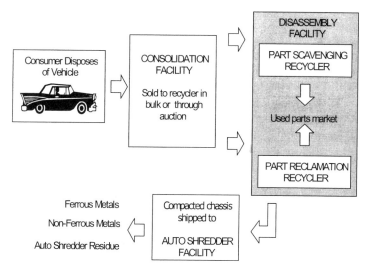

FIGURE 1   Structure of the automobile recycling industry.

from the vehicle, a disassembly crew will go to the vehicle and retrieve the specific part. Rarely is the vehicle in such cases brought into a disassembly bay. After a period of time when it is projected the vehicle has little additional value left, it is shipped to the shredding facility. Such facilities are commonly referred to as "junkyards" since they specialize in vehicles that are over 10 years old. The second group is *part reclamation recyclers*, in which a facility disassembles all the marketable parts from an entering vehicle. These parts are then inventoried and ready for off-the-shelf retail sale. The vehicle is typically brought into a disassembly bay, which is configured as a manual single station line. These facilities specialize in late-model vehicles, and consequently the value and overall market for the parts is much larger. The more advanced facilities have computer databases that forecast precisely which parts of the vehicle are in demand, and the disassembly operation will focus on these parts.

In both cases the entering vehicle will go through a pretreatment process, which is vital to proper recycling. This includes the drainage of operating fluids such as engine and gear oils, engine coolant, refrigerant, and gasoline and the removal of the gas tank, battery, and tires. It is critical that the draining and collection of fluids is performed correctly to prevent any groundwater contamination. Disassembly facilities are often cited by local environmental agencies for fluid seepage. The average volume of operating fluids in a car is approximately 19 liters and the detailed composition is shown in Table 1. In 1997 an estimated 11 million gallons of fluids were recovered from disposed vehicles.

In addition to the above fluids there are several components that are packed with grease (e.g., continuous velocity joints), and these too must be carefully handled to avoid environmental hazard. The common approach to fluid drainage is gravity, but often facilities will use vacuum pressure pump to accelerate the drainage. The evacuated fluids can be resold if sufficient volumes are collected. The majority of older vehicles contain ozone damaging refrigerants, and these require special handling during evacuation. Government regulations require a trained and licensed technician to evacuate the air-conditioning systems. The used refrigerant is of value and may be sold to a licensed buyer for reuse.

The recycling of rubber from used tires remains one of the key challenges in the auto recycling process. More than 275 million discarded tires are dumped annually into the environment in the United States alone and more than 5 million a day globally. So far there have been only a few commercial solutions to deal with rubber from used tires. A handful of companies are collecting used tires from the disassembly facilities and then recycling the scrap rubber into a wide variety of industrial and consumer products and materials. This process involves processing the scrap rubber into metal-free chips for civil engineering and alternative fuel applications and for use in crumb rubber production. Crumb rubber powders can be used in the manufacture of new tires on a limited basis and also to market asphalt rubber concrete (ARC) and asphalt rubber membrane (ARM). The concept of adding crumb rubber to the asphalt mix to improve the qualities of the paving material is now a proven and accepted technology. Some of the benefits attributed to ARC and ARM are significant road noise reduction and improved braking performance. Ground tires are often used as a limited fuel source in power plants. A disassembly facility will often attempt to sort out those tires with minimum wear. These relatively good quality

TABLE 1 Fluids in an Automobile

Fluid	Volume (L)
Engine oil	2.6
Transmission oil	1.3
Final drive oil	1.1
Steering gear oil	0.8
Engine coolant	2.8
Gasoline	10.4

tires are sold domestically or exported for reuse. Large quantities of old tires are stored in holding facilities with little projected utility.

Reclaimed parts represent the greatest value component in auto recycling. Depending on their condition and sales potential, parts and components are dismantled, reconditioned as required, and then sold to customers. These parts could include whole front and rear ends, body panels, engines, transmissions, alternators, and wheels. Today's auto recycler uses an elaborate electronic system to inventory, sell, and decide what to dismantle. On average, a used part is 50% cheaper than a new part of like kind and quality, so when parts are available and used, the cost of repairs and therefore the cost of insurance is kept to a minimum. In the United States, it is estimated that by utilizing used parts, the auto recycling industry saves an estimated 85 million barrels of oil that would otherwise be used in the manufacture of new replacement parts.

Table 2 lists the parts that are commonly recovered from disposed automobiles for reuse and/or refurbishment. Engines and transmissions represented the largest items by mass that are recovered, and these require special handling. For passenger cars the engine is typically dropped from the bottom and therefore requires a lift to raise the vehicle. Conversely, in trucks and sport utility vehicles the engine is raised from the top and there-

TABLE 2 Reusable Parts Commonly Recovered from Automobiles

Part name		Part name	
1.	Air-conditioning compressors	18.	ABS parts
2.	Alternators	19.	Blower motors
3.	Brake boosters	20.	Brake master cylinder
4.	Body panels	21.	Crossmember or K-frame
5.	CV joints	22.	Dash pad
6.	Doors and door panels	23.	Drive shafts and axles
7.	Fenders and bumper covers	24.	Electronic control modules
8.	Engines	25.	Fuel pumps
9.	Grills and metal trim	26.	Headlights and light clusters
10.	Ignition systems	27.	Mirrors
11.	Radiators	28.	Radiator/condensor fan motors
12.	Starters	29.	Steering column
13.	Steering gear/rack and pinion	30.	Strut
14.	Transmission/transaxle	31.	Water pump
15.	Wheels (steel and aluminium)	32.	Window crank handles
16.	Window motors	33.	Window regulators
17.	Windshields	34.	Wiper motors

fore requires a hoist. The disassembly facility will maintain a computerized database that identifies which parts are in demand in each vehicle model, and typically only these are retrieved. One drawback is that the disassembler is provided little to no information from the manufacturer on how to disassemble a part. Most facilities are therefore dependent on the experience of their workers.

At the end of the disassembly process, the remaining vehicle chassis is prepared for shredding by compacting. Shredders are larger facilities and can handle a considerable volume of vehicles. As a result typically there will be one shredder that is supplied by many disassembly facilities. The shredding process, which handles one car every 45 seconds, generates three output material streams: iron and steel, nonferrous metal, and fluff (fabric, rubber, glass, etc.). The iron and steel shred are magnetically separated from the other materials. The steel shred is then transshipped to a steel mill for recycling. The nonferrous metals (about 6% by weight) contain mostly aluminum, copper, and zinc and are forwarded to different recyclers. Auto shredders annually generate 3–5 million tons of fluff, commonly referred to as auto shredder residue (ASR).

The ASR material has little utility and is usually landfilled. ASR contains fibrous textiles, polyurethane foams, plastics, rubber, and a wide variety of light metal contents. This material may be saturated with oils and engine coolants and miscellaneous dirt and stone. ASR is a nonhomogenous material, and the composition changes from hour to hour. Research groups are investigating approaches for potential reuse of this material, with plastic reclamation offering the greatest opportunity. The recycling of automotive plastics, which constitute about 20–30% of ASR by weight, is not currently practiced on a commercial basis. The components are not recycled for several reasons, including a lack of a cost-effective technology for recovering plastics, potential contamination of the plastics with other materials from the ASR stream, and the absence of a strong market. The process is of particular importance in Europe, where the waste from end-of-life vehicles (ELVs) must be reduced by 40% by the year 2005 under European Union directives.

3 DISASSEMBLY ECONOMICS

The process of disassembly attempts to break-up a product into several pieces, with the expectation that the pieces together have a net value greater than the discarded product. But unlike an assembly process, the net value added percentage in disassembly is significantly less. This implies that for a disassembly activity to be profitable, the labor time, equipment needs, energy needs, skill needs, and space requirements must be relatively small.

The overall economics of the disassembly process is still not well understood. There is much reported research on the life cycle analysis (LCA) of products (Carnahan and Thurston 1998). These LCA models attempt to capture all costs associated with production, usage, and disposal of the product. To capture the associated economics, we can model disassembly as a multistep process defined by a disassembly plan. A disassembly plan is described by the sequence of processing steps, the part or parts worked on in each step, and the part portions, parts, and subassemblies remaining at the end. At the end of each step there are three possible outcomes: a reusable part is reclaimed, a part or subassembly is reclaimed and forwarded for material recycling, or the accessibility of other parts within the assembly is improved. During each step the disassembly worker performs one of two classes of activities: unfastening processes, where a fastening device is removed in an operation, which reverses the assembly fastening action, disassembly processes, which includes all other activities that facilitate the separation of the product into its parts.

The first step in the economic analysis is to project what the valuable outputs of the disassembly process for a given product will be. This is done by a disassembly planner who is familiar with the bill of materials of the product and the fastening and joining structure of the product. Let $i = 1$, ..., N be the commodity material output bins in the facility and $j = 1, \ldots,$ M the reusable parts reclaimed from the product. For automobiles this set of parts is represented by the list shown in Table 2. Then we introduce the following notation to represent the potential value streams from vehicle disassembly: W_i is the net generated weight of bin i output per unit product, V_i is the market value of bin i per unit weight, and R_j is the reuse value of reclaimed part j. Note that the part reuse value is net of any cleaning, refurbishment, and inspection costs. Estimation of these variables can be done from experience data or historical norms. For instance, in the case of personal computers, we can expect that 45% of the aluminum content will enter the aluminum high bin, 25% the aluminum mix bin, and 30% the general waste bin. More accurate numbers are derived from developing a detailed disassembly process plan for the product. Pnueli and Zussman (1997), Johnson and Wang (1995), Spicer et al. (1996), and Penev and Ron (1996) among others have suggested methods for deriving a product disassembly plan.

There are four primary cost elements involved in automobile disassembly. First, there is the direct labor time associated with the operation. There are several approaches by which this can be done. In an effort to standardize disassembly operation times, Dowie and Kelly (1994) conducted a series of disassembly experiments with simple operations. They recorded times ranging from 0.2 to 2.5 seconds for a wide variety of operations, including screw removal, cutting, and snap-fit release. These exclude any access issues or execution problems. They expected the real-life times to be 10 to 20 times more. Kroll (1996) developed a method for estimating the disassembly time using work measurement analysis. Vujosevic et al. (1995) also used a work measurement tool to estimate disassembly times in developing a simulator for maintainability analysis. Alternatively, a few products can be disassembled and a labor estimate derived using time and motion analysis.

The second cost element is the disassembly effort. This includes the associated tooling and fixturing needs, part accessibility, worker skill and instructions, process hazards, and force requirements. Das et al. (2000) proposed a multifactor model for estimating the disassembly effort index (DEI) for a product, and this is amenable to our economic analysis. They computed the DEI score as a function of seven factors: time, tools, fixture, access, instruct, hazard, and force requirements. The DEI score for each step is defined in the 0 to 100 range, with zero indicating no effort. This range is assigned on a weighted basis to each of the seven factors. This is a common practice in multiattribute utility theory and is often referred to as the amalgamated utility model. To assign the range, they surveyed and studied different disassembly facilities. The final allocation was as follows: time, 25%; tools, 10%; fixture, 15%; access, 15%; instruction, 10%; hazard, 5%; and force, 20%. For each factor, an independent utility was formulated, using the assigned range as anchors. Figure 2 illustrates a scoring card for implementing the proposed estimation scales. For each step, the user enters the appropriate score from the scale. The scores are added to give the DEI score for that step. Finally, the DEI score for all steps are summed to give the overall DEI score. Clearly, the greater the number of steps the higher the DEI score will be.

The third cost element is the vehicle cost plus the logistics cost. The logistics cost includes the cost to collect the vehicle from the point of disposal and process it through the consolidation facility. Typically, this cost is equivalent to the price a disassembler will pay for the disposed vehicle on receipt. The fourth and final cost is the inventory carrying cost associated with the disassembly activity. This includes the cost of the occupied space and the associated regulatory costs. Disassembly facilities located in high population density areas tend to have high carrying costs. Note that the disposed autos are not amenable to stacking and thus occupy their direct footprint in real estate. One reason why the inventory cost is important is that typically a facility will delay the disassembly of the vehicle until there is sufficient demand for its parts. We introduce the following notation to represent these costs.

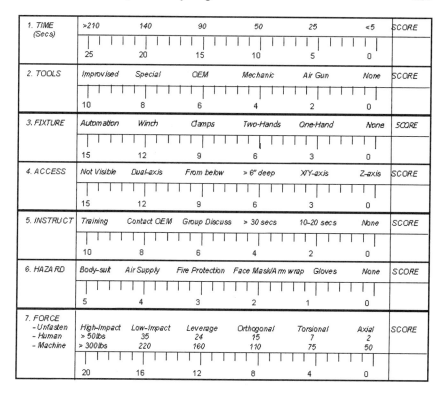

FIGURE 2   The DEI scoring card (Das et al. 2000).

α The labor cost rate per unit time including all direct time-related costs

The overhead or effort cost rate per unit index

μ The inventory carrying cost for the vehicle

C The purchase and logistics cost associated with the product

T the time to disassemble the vehicle

τ The total time the vehicle is at the facility (from receipt to transport to shredder)

The DEI score for the product

Ψ The cost to landfill the nonrecylable and nonreusable content

Observe that β is representative of the indirect and overhead cost and can be estimated by calibrating past cost performance of the facility with the

DEI scores. For instance, if the indirect and overhead cost for five previously processed products can be estimated, then we can compute the DEI score for these five and subsequently derive a trial β for each. The average of these would then provide a calibrating β. The profitability of the disassembly operation then is a function of the associated costs and values generated. We propose the following measure for the disassembly return on investment.

$$\text{Disassembly return on investment } = R_d = \frac{\sum_i W_i V_i + \sum_j R_j - \psi}{\alpha T + \beta \Delta + C + \mu \tau} - 1$$

For a given scenario one could determine a threshold level for R_d and that could be used to decide whether disassembly is an attractive alternative. Note that R_d represents the gross profitability of the disassembly activity. While there are no available surveys of auto disassembly facilities, our research indicates that most facilities will not accept a disposed vehicle if the expected $R_d < 0.7$. In such cases the vehicle will go directly to a shredder. Typically, a facility will maintain an expert database which enables it to reliably project the R_d value and hence make a decision to process the vehicle. There is a growing trend in the industry to develop facilities which focus primarily on high R_d value vehicles, that is $R_d > 1.2$. Such facilities do extensive levels of disassembly to retrieve a large number of reusable parts. Recently, the Ford Motor Company has entered into collaborations with these facilities so as to setup a supply chain for its automobile repair operations.

Observe from the above equation that R_d is very sensitive to τ when μ is high. Since the market for specific parts changes constantly, the disassembler must evaluate R_d in a dynamic decision making environment. Figure 3 illustrates this sensitivity for an example case. For the example $\Sigma_i W_i V_i = \$132$, $\Sigma_j R_j = \$870$, $\alpha = \$12/\text{hr}$, $T = 8\,\text{Hr}$, $\beta = 0.04$, $\Delta = 2132$, and $C = \$455$. For this example we derive R_d as a function of τ for the three cases of μ equal to $\$16$, $\$21$, and $\$26$ per month. We see from Figure 3 how the change in μ can significantly effect the disassembly decision. Consider the case when the threshold $R_d = 0.4$, then a facility with $\mu = 16$ can hold the vehicle in disassembly for as long 8 months and still meet its target. On the other hand, the facility with $\mu = 26$ only has a 5-month holding window. Using this concept a comprehensive model for deriving the purchase price C of a disposed vehicle as function of demand and disassembly value could be derived.

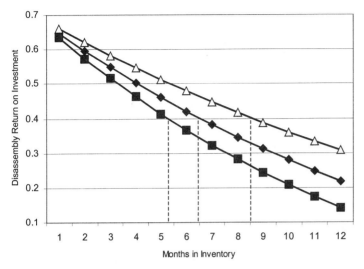

Months in Inventory

FIGURE 3   Effect of the holding cost on the disassembly return on investment.

4   KEY ISSUES FOR THE FUTURE

There are several issues that we believe will determine the future of the auto recycling process. Since automobiles represent a significant portion of the consumer disposal stream, it is imperative that we address their recyclability. Our research leads us to believe that at least four of these issues can be labeled as key and must be addressed to achieve a balance between the rate at which automobiles are being disposed and the rate at which we can recycle them. We introduce and discuss each issue in the following subsections.

4.1   Product Design for Disassembly

Efficient and cost effective disassembly is greatly dependent on the design of the product. The future state of auto recycling will therefore be determined by design related efforts we implement today. Fortunately, manufacturers are making significant attempts to incorporate design for disassembly in their development efforts. Multiple developmental efforts with this objective are already underway in all parts of the globe. In Japan, for instance, auto manufacturers are collaborating with relevant industry partners, including materials processors, parts suppliers, and disassemblers, to develop new technologies and practices that will make ELVs 95% recyclable by 2015 for the industry as a whole. The aims are to dramatically reduce the amount of plastics and glass that are disposed of in landfills and to generate energy from the incineration of plastic and rubber automotive parts.

As another example, BMW's Z1 Roadster, launched in 1991, has plastic side panels that come apart like the halves of a walnut shell. One of the lessons learned from that program is that glue or solder in bumpers should be replaced with fasteners so that the bumpers can come apart more easily and the materials more easily recycled. BMW is also changing instrument panels. In the past these were made of an assortment of synthetics glued together. Now BMW uses variations of polyurethane, foam, and rubber so the panel can be better recycled. In 1991 about 80% by weight of the Z1 was recyclable, and BMW is aiming for 95% by year 2005. A similar initiative is underway in the U.S. auto industry, which has targeted that by 2002 all new cars will have less than 10% material content that is not recyclable. This target includes both material recycling and the use of waste material as power generation fuel. As a note of caution, just because a material is recyclable does not mean it is easily recycled. Many materials can only be recycled if a sufficient mass of relatively pure content can be collected. In a complex assembly this implies that a considerable degree of disassembly must be achieved to cross the purity threshold.

Das and Matthew (1999) reported on the definition of material output bins for electronics demanufacturing. Here we expand their approach to develop a list of material output bins that should be the target output for an auto disassembly facility. Impurity levels are one of the most widespread roadblocks to more widespread recycling. The majority of materials recycling processes are limited in the amount of impurity they can accept. The level of impurity determines the quality of the recycled material and its market price. With present recycling technology a large part of the purification process must be accomplished by physical disassembly and sortation. The list was developed from a survey of existing disassembly facilities. A total of 21 bins are recommend, and these are listed in table 3.

4.2 Material Recovery and Recycling

A key issue for the future is to create efficient markets for the materials that can potentially be reclaimed from disposed automobiles. The design of automobiles has changed quite significantly in the last two decades. The primary material change has been in the increased use of plastics and in particular the use of thermoset polymers. The recycle value of these materials tends to be low. Further, the presence of these new materials tends to compromise the efficiency of the steel reclamation process. Currently, the separation and recycling costs often make the recyclate more expensive than virgin materials. To achieve their target, auto manufacturers must develop

TABLE 3 Material Output Bins for an Auto Disassembly Facility

Number	Output bin name (Max impurity)	Notes
1	Rubber (8%)	Tires and weather stripping. Ground tires can be made into tiles or carpet back.
2	High carbon steel (10%)	Wheel rims. Steel rims have higher carbon content than other steel parts.
3	Steel mix (20%)	Brake system and suspension components. This commodity has lower quality steel and is mixed with plastic and other materials.
4	Aluminium mix (14%)	Transmission and radiator. The aluminum in these parts is contaminated with other materials.
5	Aluminum high (2%)	Engine blocks and alloy wheels. High-quality aluminum has good reusability.
6	Copper mix (22%)	Wiper motor, alternator, starter, and wires. The impurity is usually steel and plastic.
7	Nylon (4%)	Carpeting can be recycled into plastic products.
8	Platinum (2%)	Primarily from the catalytic converter.
9	Battery	Battery refurbished and sold.
10	PVC plastic (6%)	Pillars, floormats, and door panels. Can be shred and remade into counter tops and tiles.
11	ABS plastic (5%)	Dashboard.
12	Plastic mix (6%)	Miscellaneous plastic components.
13	Textiles/fabric	Headliners and seat covers.
14	Glass	Glass from windows. Recycled into bottles.
15	Windshield	Can be recycled into PVB and plate glass.
16	Engine and transmission oil	Can be sold as a commodity.
17	Engine coolant	Engine coolant is recovered and sold as a commodity.
18	Gasoline	Gasoline is recovered and sold as a commodity.
19	Steering/gear oil	Can be sold as a commodity.
20	Freon	Freon is recovered and sold as a commodity.
21	High-density foam (10%)	Seat cushions. Reused for packaging.

new engineering materials that are easier to recycle. At the same time they must make it easier for the disassembler to identify and sort the different types of plastics. An important consideration given the fact that when mixed together, dissimilar plastics frequently produce an inferior recycled product. To this end, Nissan, Toyota, Honda, Mitsubishi, Mazda, Subaru, and other Japanese manufacturers have all adopted systems for marking automobile parts. Bumpers and other large plastic parts are stamped in multiple places, making it easier to identify the material even in the shredded state. In addition, manufacturers are beginning to consolidate the types of plastics they use to further improve the properties of recycled materials. The trend is to favor thermoplastics and liquid crystal polymers, which return to their original structure when heated. Similarly, plastic bumper material can be made into construction filler.

The Vehicle Recycling Development Center (VRDC) is a joint research project of Chrysler Corp., Ford Motor Co., and General Motors Corp. to develop automotive recycling technology. The VRDC is managed by the Vehicle Recycling Partnership, one of the research teams under the U.S. Council for Automotive Research. Other collaborators in the VRDC include the Automotive Recyclers Association, the American Plastics Council, and the Institute for Scrap Recycling Industries. Since becoming fully operational in January 1994, the center's efforts have focused initially on plastics and fluid recycling.

The removal of fluids quickly and cleanly is a key issue in automobile disassembly. THE VRDC has set a goal of 20 minutes for the draining of all fluids, while minimizing any risk to the environment. The VRDC plans to develop industry standard design guidelines for fluid caps so that the same fluid removal equipment can be used on all vehicles. Moreover, a drain plug might be added to automatic transmissions so that the transmission fluid could be drained rather than vacuumed. The VRDC is investigating a variety of methods for reprocessing the reclaimed fluids into their original form. Progress has been made, for example, in reclaiming antifreeze. The center currently is testing a recycling machine which analyzes the antifreeze for its basic components and regenerates it into a reusable condition. Likewise, the center has demonstrated how windshield washer fluid can easily be saved and reused in vehicles. Equipment to remove and refine brake fluid also is being tested.

Plastics present the most formidable recycling challenge, since each type of plastic must be identified and evaluated separately. The VRDC observes that there are as many as 15 different types of plastics in the instrument panel alone. In addition, a variety of bonding methods ranging from molding to gluing is used to assemble the instrument panels. This further complicates the sorting process. The VRDC is testing equipment that can help identify the various kinds of plastics. In addition, the center is cataloging all plastic components as each vehicle is dismantled. The information gained from this process is being used to develop a computerized database for the disassembly industry. In the future, a disassembler will be able to input the part taken off the vehicle and a computer-generated worksheet will identify the material that was used to make the part. It then can be assigned to the appropriate recycling bin. In addition to identifying materials, the VRDC is pursuing opportunities to expand the recycling infrastructure to more automotive materials.

Research and development funds also are being used to study the appropriate recycling of air bags and how to reduce the amount of lead in new car models. Even though manufacturers use recycled automotive plastics to build a variety of products, a certain amount still ends up in the landfill as shredder waste. Vehicle designers are taking recycling into account by including the increased use of one-piece molds and common materials in bumpers, seats, interior panels, and carpets.

4.3 Reduction and Disposal of ASR

The disposal or recycling of ASR is one of the most important issues in the automobile recycling industry. While the material complexity of ASR combined with its nonhomogenous structure make it difficult to do anything with it, a few research groups have made some progress in their quest to find efficient cost-effective methods to recycle the ASR or "fluff." Researchers at Argonne National Laboratory are working on finding new ways to recycle fluff (Jody et al. 1993). They are testing a technically promising and potentially economically viable process to recycle polymers and other constituents from ASR. The value of the recovered plastics could be very competitive in a variety of applications; one of the first might be polyurethane foam for carpet padding. The Argonne technique for recycling ASR begins with the mechanical/physical separation of the ASR into several fractions: polyurethane foam, which is separated and cleaned; iron-oxide-rich "fines," which may be used by the cement industry as a source of iron oxide; and a plastics-rich stream, from which Argonne dissolves and recovers heat-formed plastics (thermoplastics). The solvent used in the process is regenerated and continuously recycled. The process utilizes advanced froth flotation technologies.

Reduction of ASR requires that the source materials are removed during the disassembly process and subsequently feed to commercial recycling streams. If fines, foam, and plastics are recycled, as much as 75% of landfill waste could be eliminated. In Japan the auto manufacturers association and Nissan have teamed with Kobe Steel, Ltd., and Nakadaya Co.,

Ltd., a large shredder operator, to build a pilot incinerator to safely dispose of unusable waste and convert the energy into distilled gas for other uses. Argonne researchers have developed new methods for removing and cleaning seat foam, which currently is a costly difficult process. Results to date show the recycling methods are an economical way to produce quality materials, and the lab is currently building a prototype system to clean foam more efficiently and in a continuous manner. The polyurethane foam material in modern car seats can be recycled into residential carpet backing but not into new seat foam again. The VRP is evaluating the use of automotive seat foam in carpet padding and sound suppression padding in future production vehicles.

4.4 High Volume Auto Disassembly

The successful future of automobile disassembly lies in our ability to build facilities that are capable of processing several vehicles per day. The only way this can be realistically done is to design and build automobile disassembly lines. The ideal disassembly line will structurally replicate the activities of an auto assembly line. The key benefits of a high volume disassembly line are as follows:

- Achieve economies of scale;
- Division of labor, which leads to lower labor costs and better accessibility;
- Can use assembly technologies more readily;
- Analytical understanding of the entire disassembly operation;
- Greater degree of disassembly—which leads to a smaller carcass;
- Potential to reclaim a wider range of parts and materials.

While there are no such disassembly lines currently in operation, several approaches and associated equipment have been proposed. These are briefly reviewed here.

4.4.1 Car Recycling Systems

Car Recycling Systems (CRS) has developed a system for the large-scale disassembly of old cars. The system can process any model of car; further, it is possible to rapidly switch from one car model to another. The first CRS line went into operation in 1989, and today CRS has lines operating in The Netherlands, Belgium, Germany, and the United States. The lines permit high-yield low-waste production via a continuous transport system; step-by-step dismantling is carried out at six workstations promoting a fast work rate. Each workstation is equipped for the removal of specific materials. Stage by stage, the wreck is dismantled in the most efficient order, starting with the glass, followed by plastics and rubber, and finishing with the engine, axles, and wiring, as sources of aluminum, iron, and copper. All that is left behind is a bare body shell, which is removed as scrap. If the material thus obtained is supplied separately—type by type—and in a constant stream, they become an interesting source of raw materials for customers. Bumpers, for example, can be ground into granulate from which a new generation of plastic can be manufactured. The operations in the CRS line are divisible into the following five stages:

1. *Stage 1: Removal of contaminated and hazardous substances.* In a separate environmental cell, the battery, fuel, oil, coolant, windscreen washer fluid, airbag, and air conditioning are first removed from the wrecks. This process prevents hazardous situations or contamination arising during further processing. It also promotes a safe and pleasant working environment. Subsequently, the wrecks are transported to initial storage, necessary for the buildup of stocks required for serial dismantling.

2. *Stage 2: The car is hoisted onto the trolley.* The wrecks are placed on a special transport trolley and taken to the start of the CRS line, where a crane system with a special car grab places them on the rail transport wagon on the line. In stage 1, the windows, doors, bonnet and boot cover, window rubbers and bumpers, seats, dashboard and interior upholstery, front and rear lights, outer fittings, and any objects left behind on the car by its last owner are systematically removed from the wreck. This stage is completed in the first three workstations.

3. *Stage 3: Rotating the car 180 degrees.* The wreck is rotated through 180 degrees, using a secured tilting mechanism. Designed by CRS, the tilting mechanism is equipped with catch trays for any loose materials, which fall from the car during tilting. The operators work on movable work platforms, mounted on both sides of the car, at working height. In this tilted position, the mounting points for the engine, gearbox, and axles can be easily released, and the exhaust system removed. The wreck is then returned to its original position.

4. *Stage 4: Final workstations.* In this stage, the shock absorbers are unscrewed to separate the heavy components from the chassis. At the final workstation, all remaining parts are removed from the chassis, including the wiring, the heater unit, the radiator, and items such as the windscreen wiper fluid tank. A final check is also carried out to determine whether the dismantling process has been fully completed, leaving a bare and empty body shell.

5. *Stage 5: Disposal of materials.* All dismantled materials are collected in separate containers along the dismantling line. This results in material flows intended for transport to processors. One of these waste flows is scrap. The spare body shell in principle can be transported directly

to the steel industry for reuse, once it has been compressed into the correct format using a crusher, for example. Due to this working method, it is no longer necessary to use a shredder installation.

4.4.2 De Mosselaar System

A Netherlands-based company, De Mosselaar BV, has been developing a semiautomated dismantling center for auto disassembly. In their line the vehicle is progressively disassembled in six stages. The line is designed to provide operators with side access, bottom access, and top access to the vehicle. A major component of this line is a large hydraulic fixture that can rotate the vehicle 180 degrees along its primary axis. This makes the bottom of the vehicle accessible to an operator in the work-up or work-down position. There is reportedly only one De Mosselaar line installed in the United States at a Maryland facility. The design itself does not assume any particular disassembly sequence or process plan, and it is left to the disassembly facility to make these decisions. It is therefore not possible to estimate the humanpower required to operate the line or the level of disassembly attained. The company projects a fully operational line will process between 25 and 40 cars per day.

4.4.3 Autoline System

Another Netherlands-based company, Autoline Recycling Equipment BV, has developed at least two solutions. The first of these is a sequential dismantling line, which is built around a series of lifting units. The unique feature of this line is that the handling or lifting equipment is suspended from an overhead rail. Each lift units has its own hydraulic arms and is electrically driven. A user could design a line with as many stations as needed in a modular fashion. The system provides the following advantages:

- The floor space is kept clear of handling equipment.
- The design is flexible both in the of number of stations and the distance between stations.
- The vehicle is lifted easily without the use of a forklift truck.
- Operators can dynamically adjust the vehicle height to the desirable level.

The stability with which misshapen vehicles are lifted and moved, though, is not clear. A key drawback is the cost, since each unit is equipment by itself. Further, return of the units to the start of the line is another problem.

The second solution targets the single station automobile disassembly facility. The company has developed a rollover hoist that is similar in structure to a cylindrical axis robot. Using a remote control device the operator

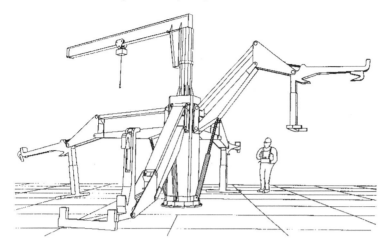

FIGURE 4 AutoLine rollover hoist for vehicle disassembly.

can use the hoist to lift the vehicle and then rotate it into any desired position. Lifting is done via hydraulic cantilever arms that grip the roof of the vehicle while at the same time supporting the bottom. The hoist also is modular in that as many as four lifting arms can be added. Figure 4 illustrates a four-arm hoist.

5 A PROTOTYPE DISASSEMBLY LINE

The future of successful automobile disassembly is greatly dependent on developing high-volume low-cost disassembly lines that effectively reclaim all reusable parts and separate all materials. We propose here the schematic design of such an automobile disassembly line. We believe that this schematic can be used by any disassembler to construct a line with the supporting product handling technology and material/part redistribution logistics. This design was developed in collaboration with a major consortium of automobile disassemblers based in New Jersey. To collect the design data, the research team conducted a direct observation of the disassembly process at several facilities. From this we constructed a list of disassembly activities and an inventory of the required tools. In conjunction we also identified and classified the retrievable assets and recyclable materials that are reclaimed. Part of these data was obtained by studying the history of part sales. Our discussions with industry professionals in combination with our direct observations led us to the following principles in designing the line:

- Most single-station manual operations are quite efficient, and so we must try to extrapolate that approach to a line.
- Lifting equipment complicates the line both in terms of handling and fixturing.
- Vehicle must be fixtured from the bottom and in the rectangle between the wheels.
- Fixturing footprint must be constant across models.
- Handling pallet must be easily returned to the start of the line.

The key design questions are in how many stations the vehicle should be disassembled and what is done at each station. Our approach was to use classical assembly line design theory to create a 5- to 15-stage disassembly line, including any divergence points and branches. Key features of the proposed design are as follows:

- The disassembly line consists of 12 workstations in total. The main line consists of 9 workstations and 3 branch workstations.
- The target for the end product is a steel-only chassis.
- Disassembly cycle time is 17 minutes, with several stations having a slack time of 2 to 3 minutes. This implies that 28 vehicles will be disassembled during an 8-hour shift.
- Proposed direct humanpower for the line is 16 workers.
- The direct disassembly labor time per vehicle is 4.5 hours. To benchmark this time consider that at existing facilities the partial disassembly typically requires between 5 to 8 hours of labor time.

The conveying mechanism is a towline conveyor that pulls the custom-designed pallets on a rail track.

Figure 5 summarizes the details of the proposed schematic design. Each station box in Figure 5 identifies the disassembly process activity, the output bins where the reclaimed materials are sent, and the operator accessibility required for the operations. The bin numbers refers to those listed in Table 3. The accessibility numbers are as follows: 1, back; 2, top; 3, front; 4., bottom; 5, left side; and 6, right side. Table 4 lists the parts that are removed at each station. An important feature of this schematic that only four stations (1, 5, 6, and 7) require access from the bottom. We expect station 1 activities will be done before loading the station onto a mechanized line. Hence, since the remaining three stations are in sequence, the vehicle has to be raised only once.

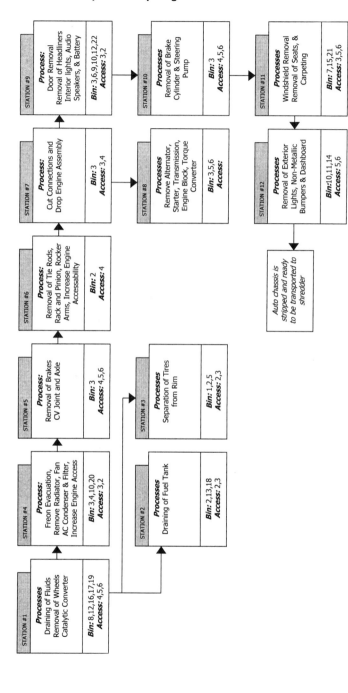

FIGURE 5 Automobile disassembly line station details.

TABLE 4   Parts Reclaimed in Each Station

Station No.	Parts
1	Catalytic converter
2	Fuel tank
3	Rims, tires
4	Radiator fan, air-conditioning condenser and compressor, air filter, freon
5	Calipers, brakes, CV joints, axles
6	Rack and pinion system, tie rods, rocker arms
7	None
8	Alternator, starter, transmission, engine block, torque converter
9	Doors, headliners, interior lights, audio system, battery
10	Brake cylinder, steering pump
11	Windshield, seats, carpets, computer
12	Exterior lights, nonmetallic bumpers, dashboard

6 SUMMARY

The automotive recycling industry plays a necessary and crucial role in the efficient ecological disposal of inoperable motor vehicles. In addition to conserving natural resources, automotive recycling plays an important role in reducing air and water pollution and solid waste generation. The effective and efficient disassembly of discarded automobile is critical to achieving a sustainable material reuse policy. Since automobiles represent such a large portion of the industrial production stream, any development in the recycling will have significant effect on global environmental policies. Our research shows that considerable progress has been made and continues to be made in this regard, and the auto disassembly industry, particularly in Europe and the United States, is preparing for the challenges of the future. The main themes for the success are design for disassembly, increased material reclamation and recycling of plastics and fluids, reduction of ASR, and the development of high volume disassembly facilities.

REFERENCES

Carnahan J, Thurston D. (1998) Trade-off modeling for product and manufacturing process design for the environment. J Indust Ecol, 2:79–92.

Das SK, Matthew S. (1999) Characterization of material outputs from an electronics demanufacturing facility. Procs IEEE International Symposium on Electronics and the Environment, Boston, pp. 75-81.

Das SK, Yedlarajiah P, Narendra P. (2000) An approach for estimating the end-of-life product disassembly effort and cost. Int J Product Res 38:657–673.

Dowie T, Kelly P. (1994) Estimation of disassembly times. Unpublished Technical Report, Manchester Metropolitan University, United Kingdom.

Gungor A, Gupta SM. (1997) An evaluation methodology for disassembly processes. Comput Indust Eng 33:329–332.

Jody BJ, Daniels EJ, Bonsignore PV, Brockmeier NF. (1993) Recovering recyclable material from shredder residue. J Metals 46: 40–43.

Johnson M, Wang M. (1995) Planning product disassembly for material recovery opportunities. Int J Product Res 33:3119–3142.

Kroll E. (1996) Application of work-measurement analysis to product disassembly for recycling. Concurrent Eng Res Appl 4:149–158.

Penev K, de Ron A. (1996) Determination of a disassembly strategy. Int J Product Res 34:495–506.

Pnueli Y, Zussman E. (1997) Evaluating end-of-life value of a product and improving it by redesign. Int J Product Res 35:921–942.

Spicer A, Zamudio-Ramirez P, Wang M. (1996) Automotive applications of disassembly modeling. The 4th International Congress on Environmentally Conscious Design and Manufacturing, July , Cleveland, OH.

Vujosevic R, Raskar R, Yetukuri N, Jyotishankar M, Juang S. (1995) Simulation, animation, and analysis of design disassembly for maintainability analysis. Int J Product Res 33:2999–3022.

22

Environmentally Friendly Manufacturing

Joseph Sarkis
Clark University, Worcester, Massachusetts

1 INTRODUCTION

Organizations are beginning to feel a myriad of pressures for managing their environmental sustainability. Reactive and remedial measures of the past may not be adequate to develop and maintain a competitive advantage, especially one that would incorporate the natural environment. As such, the breadth of what organizations can do with respect to the natural environment has increased greatly within the last decade. Research to support these efforts has also seen a great surge. The reason is very clear: The environment is important to people, organizations, and countries, and corporations should heed, and maybe even lead, efforts that focus on industry's relationship to the environment.

One place within the organization that can support these corporate environmental sustainability efforts is the manufacturing and operations function. The set of processes and supporting activities that are central to an organization's value chain are those that encompass the production and manufacturing function. It is the manufacturing function, from a systems perspective, that takes input from the environment, transforms this input, and generates the necessary outputs that will satisfy the customer. The possibilities for improved environmental performance occur at each of

523

these stages. The viewpoints of the roles of the organization and the manufacturing function can be developed from both an internal (micro) and external (macro) perspective. We shall show the various implications and developments from both perspectives. There are many linkages and necessary overlaps when viewing the organizational system from these general and detailed perspectives.

The chapter includes the following sections:

1. An industrial ecology perspective: The roles of various functions within and external to the organization are discussed. Various industrial ecology and ecosystem relationships are described. An overall viewpoint, focusing on the value chain perspective, provides the foundation for green supply chains. This section provides an external macro viewpoint of manufacturing's role.

2. Corporate strategy and the manufacturing function: How the natural environment is integrated within corporate strategy as well as its influence and relationships are presented. Various strategic models and literature and its impact on the manufacturing function are presented. For example, topics such as ecocentrism and sustainability are briefly discussed. These topics rely heavily on the management literature and management theory. This section alters the focus of the chapter to a top-down, but internal perspective of the manufacturing's relationship to the remainder of the organization.

3. Dimensions of manufacturing strategy and environmental friendliness: This section covers the major dimensions of manufacturing strategy based on product, process, and technological categories. A review is presented of literature in this area and how they view manufacturing strategy and its linkage to environmental friendliness. From a product perspective, we focus on the linkages between procurement, design, and marketing and how they influence the manufacturing product issues. A process perspective will look at issues related to process design, environmental management systems, closed loop manufacturing, and business process reengineering, to name a few. The technological perspective will link up implications of various technologies to manufacturing including information, manufacturing, and control technology. A summary subsection in this part of the chapter links up the various dimensions and how they may overlap with various manufacturing practices such as life cycle analysis (LCA), total quality environmental management, and green supply chains. This section focuses on some of the internal operational microaspects of environmentally friendly manufacturing.

4. Emerging and evolving issues: This final section summarizes of the latest evolving issues related to the virtual enterprise, information technol-

ogy, globalization, the strategic focus, and environmental performance measurement.

2 INDUSTRIAL ECOLOGY

Lowe (1993) defined industrial ecology as "a systematic organizing framework for the many facets of environmental management. It views the industrial world as a natural system—a part of the local ecosystems and the global biosphere. Industrial ecology offers a fundamental understanding of the value of modeling the industrial system on ecosystems to achieve sustainable environmental performance."

An industrial ecology (ecosystem) can be described by three levels of relationships (Jelinski et al. 1992). The first is a linear materials flow system, where the input is unlimited materials and resources, which are transformed by the ecosystem component with unlimited waste. This type of system clearly assumes unlimited resources and capacity for formation of unlimited waste; it is not a realistic system. A second level is what has been defined as a quasi-cyclic materials flow ecosystem or type II ecology. In this system energy and "limited" resources flow into a transformation cycle composed of a number of ecosystem components with limited waste flow occurring as an output. The final system has energy flowing in with all ecosystem components linked together and no waste outflows. The third level is a utopian and idealistic level that is completely sustainable as long as energy flows into the system.

A type II ecosystem model as an example of industrial ecology is shown in Figure 1 (Sarkis 1994). Within this system the elements include materials extractors and growers that supply the raw materials for the man-

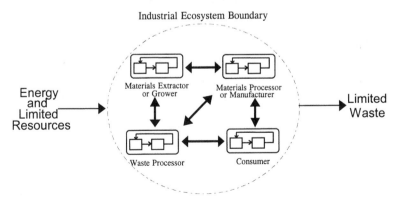

FIGURE 1 A generic industrial ecosystem and its components.

ufacturers and processors whose outputs are consumed by the consumers. All these elements produce outputs that are transformed by waste processors that may be used by the other ecosystem components.

A few communities and industrial parks are beginning to model this industrial ecology. In this situation, wastes of one manufacturing plant may be useful material for another manufacturing plant (Dwortzan 1998). A number of proposals for U.S. ecological industrial parks of networked manufacturing facilities are part of the President's Council for Sustainable Development Task Force on Eco-Industrial Parks. Lowe (1993) and Starik and Halme (1994) described an operational industrial ecosystem modeling healthy interaction between business and the environment at Kalundborg, Denmark. This industrial ecosystem consists of a web of multidimensional recycling among an electric power-generating plant, an oil refinery, a biotechnology production plant, a plasterboard factory, a sulfuric acid producer, cement producers, local agriculture and horticulture, and district heating utilities. Some U.S. state government economic planning agencies (for example, the Texas Natural Resource Conservation Commission and its Departments of Economic Development) when convincing organizations to locate and expand in their regions identify and build relationships with suppliers and customers in areas that will help them in their recycling and environmental efforts. Additionally, they provide documentation and a database of sources of various waste material that may be used as recycled material in manufacturing. This database has both regional and national sources. These databases once they are widely adapted and available will be integral to design of products.

2.1 Green Supply Chains

The organizational process manifestation of the industrial ecosystem is environmentally conscious supplier-customer relationships that form the green supply chain. One characterization of this type of supply chain is shown in Figure 2. Here we see the various elements of the green supply chain (Bloemhuf-Ruwaard et al. 1995; Lamming and Hampson 1996; Narasimhan and Carter 1998; Sarkis 1995b; Walton et al. 1998). The development of this green supply chain model is based on the value chain of an organization (Porter 1985). Hartman and Stafford (1998) also defined the green supply chain and its practices as "enviropreneurial" value chain strategies. The focus is to gain competitive advantage and improve financial performance from these strategies.

Figure 2 shows a number of environmental issues as well as operations within a typical supply chain. The primary focus in this figure is the management of products and materials that flow through the supply chain and

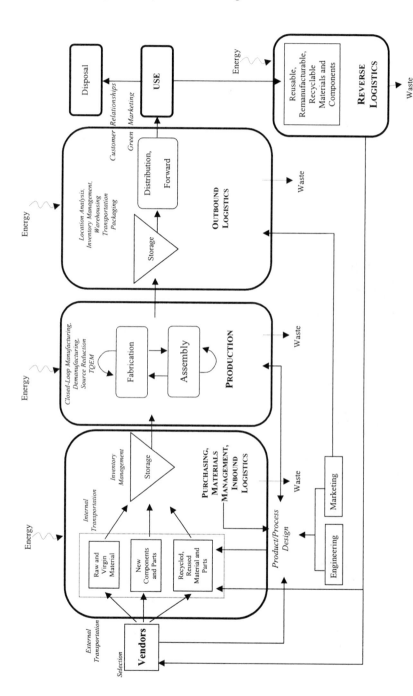

FIGURE 2 Operational functions and environmental practices within the supply green supply chain.

relationships among the various functions. Looking at this chain we see that vendors (who have their own internal and external supply chains) supply the necessary materials. The materials may include raw and virgin material; parts and components from original manufacture; and recycled, reusable, or remanufactured materials. There are environmental implications depending on the focus of acquisition activities at this stage. Vendor supplies have environmental implications. For example, a recycled material from a distant location may not be as environmentally sound as a virgin material from closer locations. The policies for selection of vendors are a central issue for purchasers (Noci 1997; Walton et al. 1998). These materials are then stored and may be managed under the auspices of the purchasing function. Each of the major functions will be profoundly impacted by the design of the product and the process. This impact needs to be managed by including the various functions and vendor(s) in the design of these processes and products. Included in design issues will be such topics as LCA and design for the environment concepts.

The production function is composed of assembly and fabrication. Within this area environmental issues such as closed-loop manufacturing, total quality environmental management (TQEM), demanufacturing, and source reduction all play some role. This element of the green supply chain is the focus of this chapter, and in a later section we explore this function in more detail with its linkages to organizational strategy made clear.

Outbound logistics (Wu and Dunn 1995) includes such activities and issues as transportation determination, packaging, location analysis, and warehousing, as well as inventory management (for finished and spare parts good items). Marketing's influence is clear here with customer relationships that need to be maintained, as well as living up to green marketing developments and presentations (Roberts and Bacon 1997). The "use" external activity is the actual consumption of the product, a situation where product stewardship plays a large role (Dutton 1998; Hart 1995). At this stage, field servicing may occur, but from an environmental perspective the product or materials may be disposed or return to the supply chain through the reverse logistics channel. Within this channel, the product can be deemed to be reusable, recyclable, or remanufacturable. The reverse logistics function (Carter and Ellram 1998; Fleischman et al. 1997; Pohlen and Farris 1992) may feed directly back to an organization's internal supply chain or to a vendor, starting the cycle again. Each major supply chain activity consumes energy and generates some level of waste. Reduction in energy usage and waste generation are issues that need to be addressed throughout the supply chain.

3 CORPORATE STRATEGY AND THE MANUFACTURING FUNCTION

Whereas the green supply chain and industrial ecology focus on the role of an organization and its functions along a horizontal and external process, the foci of the issues presented here are on a vertical/horizontal and internal relationship. The discussion in this portion is guided by the structure presented in Figure 3. In this figure we see the various external natural environmental demands and requirements, including industrial ecosystems, ecocentric theory and product stewardship, and green consumers. Internal to the organization are the strategic elements which include the overall organizational strategy and the manufacturing functional strategy (other functional strategies can be viewed as support elements). At the next level are the operational and tactical elements of the manufacturing interface. These elements are driven by product and process design and include a number of other manufacturing-oriented elements. These form the core of

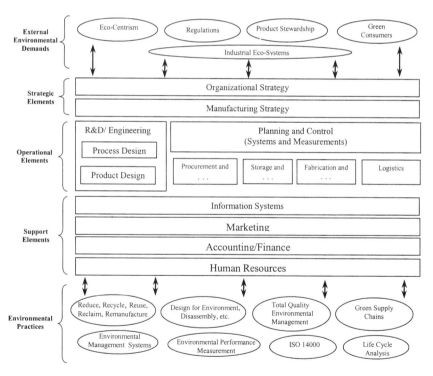

FIGURE 3 Manufacturing organization functions and environmental influences.

the next section and its discussion. In this section we focus on some strategic elements.

3.1 Evolving Natural Environment Management Theory and Corporate Strategy

Organizational theory and perspectives are driven by social norms and viewpoints. Much of the ethical debate and roles of organizations with respect to external entities and communities has evolved from corporate social responsibility work. To this end, much of strategic management theory has been driven by fundamental economic theory. This most traditional theory is espoused by economist Milton Friedman (Friedman 1970), who essentially stated that it is the ethical and social responsibility of firms to make profits for their shareholders. This viewpoint has come under attack by a growth in a viewpoint that firms should be focusing on the broader community of "stakeholders," which is espoused in stakeholder theory (Freeman 1984). The stakeholders in this situation would include anyone that is directly impacted by the operations of an organization, including shareholders, employees, customers, and communities. There have been extensions to stakeholder theory to incorporate the environment (Starik 1995). They argue that nonhuman nature should be included among the stakeholders considered by organizations. More recent effort in this area has included consideration of firm operations from the perspective of natural resources (Hart 1995) and ecocentric management (Shrivastava 1995a). Another level of analysis as presented by Gladwin et al. (1995) espouses the "sustainability" theory, which focuses on economic, environmental, and social measures for organizational responsibility. Some will argue that all these theories are anthropocentrically (human centered) and are not biocentrically (environmentally centered) theories (Bucholz 1998).

These theoretical developments are forces that have caused manufacturing industry to reevaluate how they develop their corporate and manufacturing strategy. In fact, a recent survey of 1000 U.S. manufacturers by *Industry Week* (Miller 1998) found that 90% have environmental strategies and 80% believe they have environment friendly operations mechanisms. To further explain these relationships, we now look at how these two are linked and at some guidelines, as set forth by the literature and practice.

3.2 Manufacturing Strategy and Strategy Development

Manufacturing strategy provides a vision for the manufacturing organization based on the business strategic plan. It consists of objectives, strategies, and programs, which help business gain or maintain a competitive advantage (Schroeder and Lahr 1990). Manufacturing strategy's role in manufacturing enterprises has become more visible in recent years. Initially, it was viewed as an operational consideration and planning was short term and internally focused within the manufacturing function. Skinner (1968) was an early proponent of integrating manufacturing strategy with the overall business strategy. Dimensions of manufacturing strategy include product, process, and technological factors and substrategies (Adler 1989; Hayes et al. 1988; Samson 1990). Each of these dimensions is also closely linked to various environmental issues facing manufacturing enterprises.

Environmentally friendly manufacturing incorporates the planning, development, and implementation of manufacturing processes and technology that minimize or eliminate hazardous waste, reduce scrap, are operationally safer, and can design a product that is recyclable, and can be remanufactured or reused (Weissman and Sekutowski 1991). To show how the manufacturing strategy can begin to incorporate these factors, a look at the development of standard corporate strategy is warranted.

The standard approach to development of a corporate strategy and the role it plays in a corporation as defined by Porter (1980) is to first determine a mission for the organization. The mission then drives objectives formation which gets interpreted into strategies policies for the overall organization. This is defined as the mission, objectives, strategy, and policy (MOSP) setting process. The driver for the MOSP process is the consideration of the external competitive environment of the organization. This is accomplished by focusing on competitors' strengths and weaknesses and from these, determining the firm's opportunities and threats (SWOT). To help in carrying out a SWOT-MOSP environmental strategy development process, a general open systems model to carry out the linkage of external and internal interactions for the planning and control of corporate environmental strategy has been developed by Klassen (1993). This approach provides insight into the characteristics of the internal and external processes that interact to form the strategies for an organization. Interacting with the manufacturing strategy are other functional strategies, which are also constrained by the overall business strategic plan. These strategies must filter down to the various functional objectives which will drive the functional strategies. A detailed outline of the functional strategies and their interfaces among each other as well as the linkages with the corporate policies and strategies is well established in the literature (Hax and Majluf 1991; Meredith 1987; Noori 1990; Stahl 1989).

These functional strategies (manufacturing, marketing, finance, etc.), must clearly fit within and not conflict with the overall business strategies. Thus, it is imperative that a corporate strategy must include as one of its dimensions environmental consciousness from all functions within the organization, not just manufacturing. Each of these functional areas will also have a profound impact on the development and implementation of a manufacturing strategy. As evidence of the importance of explicitly and proactively including and addressing environmental issues in corporate strategy are the external pressures from various sources. For example, marketing strategy may drive product development as defined by customers. The linkage is clear and critical.

3.2.1 Functional Environmental Strategies and Linkages to Manufacturing

The market focus and their primary external pressures have been defined as "green consumerism." Some of the key players and drivers of green consumerism include environmental groups, green voters, the business-to-business sector, retailers, employees, and shareholders (Ottman 1992). One of the controversies in green marketing and customer relationships is whether customer interest in environmentally sound products relates to actual purchase. Various studies have shown that interest is usually higher than actual purchase. This argument can be made for either individual consumers or corporate and industrial buyers. Even though this issue has been shown to be an individual consumer phenomena (Mandese 1991; Schlossberg 1990; Yam-tang and Chang 1998), the extension to corporate buyers needs a more complete evaluation. Earl and Clift (1999) argued that individual consumers and business-to-business consumers (e.g., purchasing agents) have similar attitudes, actual behavior, and understanding of purchasing green products. They argue against the theory supported by Karna and Heiskanen (1998) that business consumers are more aware of environmental and green product characteristics and issues than individual consumers. Whether or not these consumer groups have equal levels and understanding of environmental issues with regard to purchasing products will influence organizational marketing strategies and practices.

Financial strategy can be important in a number of ways ranging from evaluation and justification of various internal environmentally friendly programs to acquiring external capital investment because of environmental practices to improved environmental performance associated with these practices. Even though the literature is mixed on the issue of financial performance's relationship to environmental performance (e.g., Cordiero and Sarkis 1997; Hart and Ahuja 1996; Klassen and Whybark 1999; Russo and Fouts 1997), some literature (Klassen and McLaughlin 1996; Porter and van der Linde 1995; Schmidheiny 1992; Shrivastava 1995b) has proposed that when integrated strategically, environmental management enhances business performance, but others (Cordiero and Sarkis 1997; Walley and Whitehead 1994) have empirically and anecdotally indicated that environmental performance may be hindering organizational performance. The results have been mixed for a number of reasons, but the primary reasons can be attributed to the type of performance measures used and the length of time before performance is measured and whether the environmental programs had proactive or reactive characteristics.

4 DIMENSIONS OF MANUFACTURING STRATEGY AND ENVIRONMENTAL FRIENDLINESS

There are a number of ways to define manufacturing strategy and its dimensions. At this stage of the discussion the dimensions set the stage for operational and tactical issues. The use of the major areas of operational manufacturing functions will provide an idea of the type of work and issues involved at this level. The areas include procurement/acquisition, inventory management, fabrication and assembly, logistics, and planning and control. Reviews in the literature of environmentally friendly manufacturing at all levels of analysis (i.e., strategic and operational) have been increasing. Some of these include works by Darnall et al. (1994), Gungor and Gupta (1999), Gupta (1995), Hanna and Newman (1996), Narasimhan and Carter (1998), Sarkis (1995a, 2001), and Sarkis and Rasheed (1995). Various practices related to environmentally friendly manufacturing are observable in a wide variety of organizations and industries. A summary of exemplary practices is described in Table 1.

4.1 Procurement and Acquisition

We have discussed some of the issues related to material and product acquisition when within the green supply chains and green purchasing discussion. But there are a number of additional issues that can be addressed. Material, supplier selection, practices, and supporting mechanisms are among these issues.

One of the major difficulties in the procurement process is to manage the suppliers and materials sources needed for environmentally friendly manufacturing. There is a great amount of uncertainty associated with the supply availability. For example, in remanufacturing and demanufacturing environments, Guide and Srivasatava (1997) pointed to a number of issues that lead to uncertainty in the supply, as well as in the overall production environment (discussed in the upcoming sections):

- Probabilistic recovery rates of parts from the inducted cores;
- Unknown conditions of the recovered parts until inspected;
- The part matching problem components, along with common ones;
- The added complexity of a remanufacturing shop structure;

TABLE 1   Exemplary Industrial Practices for Environmentally Friendly
Manufacturing.

American Tool Corporation—Process and product substitutability by replacing the
deionization system with a reverse osmosis system at their DeWitt facility,
eliminated the need for all chemicals for regeneration of plating process waters.
This system enabled them to control costs, while achieving higher quality
production and reduce waste (Hearns 1999).

Champion International—invented the bleach filtrate recycle (BFR) process, a new
technology that can reduce discharge levels of dissolved organic materials and
advance the development of a closed-loop system that would eliminate all process
effluents from the pulp bleaching process (Hartman and Stafford 1998).

Compaq (formerly Digital Equipment Corporation)—Has a recovery facility for
electronics parts where purchasing agents also serve as marketers and salesmen.
They have developed relationships with a number of organizations that send their
electronics components and products there for demanufacturing and organizations
(such as Envirocycle Inc.) that use their outputs as inputs for their products
(Sarkis et al. 1998)

Crown Cork and Seal's—Minnesota plant involved production workers as the
centerpiece of its efforts to eliminate waste and reduce toxins. They formed teams
of workers to focus on environmental issues, including recycling teams and quality
improvement teams whose objectives focused on an environmental problem and
hazardous waste generation. From these efforts, the plant achieved reductions in
solvents, air emissions, and solid waste disposal (Florida 1996).

Disney—Built an on-site material recovery facility (MRF), which began handling
recyclables from the Walt Disney World Resort. The MRF handles more than 45
tons of paper, plastic, glass, steel, aluminum, and cardboard daily, representing an
average recycling rate of more than 30% of these materials. Other used equipment
and excess items are sold to Cast Members or auctioned to the public.

Kennedy Die Casting—In Worcester, Massachusetts has three separate internal
closed-loop waste water systems. Each is unique to its manufacturing centers,
depending on the level of contamination and the use of the water. One process
includes a simple heat capturing and cooling water closed-loop system. In this
system the water has no direct contact with the material or solvents. Another
closed-loop wastewater treatment process is more complex with the use of settling
tanks, skimmers, and centrifuges to help decontaminate the water for reuse (Sarkis
2000).

Novartis—Conducts benefit-risk analysis for its core manufacturing processes. The
results are broadly disseminated as publicly available information. Employees are
made responsible for contributing to better environmental performance. For
auditing purposes the company has developed an objective reporting system called
SEEP which allows the company to obtain real time accurate information on 90%
of emissions from the company worldwide (Narasimhan and Carter 1998).

TABLE 1   (Contd)

Uzin Georg Utz Gmbh and Co. (cement and tile manufacturer)—Focuses its effort on investment in advanced production technology. They view technology as a strategy to improve environmental effectiveness of its production function. Uzin's major processes are physical and mechanical in nature (not chemical); thus, the cleaning and maintenance of these products is of environmental concern to the organization (Narasimhan and Carter 1998).

- The problem of imperfect correlation between supply of cores and demand for remanufactured units;
- Uncertainties in the quantity and timing of returned products.

These uncertainties and complexities can be better managed through the procurement and acquisition function. Unfortunately, without a good supplier market and ample supplies that can be provided by these markets, volatility and uncertainty is almost guaranteed. That is why close management of the supply chain is critical.

The relationship and types of products and services that are offered need to take a different viewpoint. For example, DuPont has developed a partnership with Ford Motor in which DuPont's payments are based on the number of cars that are painted. This creates an incentive for the two companies to use paint as efficiently as possible (Denton 1998). Digital, Corning Asahi, and General Electric is another partnership to use recycled materials (Sarkis 1995b), where the partnership was formed to use material that had been returned and whose byproducts could be used for new products.

The relationship between marketing, design, engineering, and manufacturing needs to be managed very carefully. One of the important aspects of materials management is a thorough LCA of materials and products (Young and Vanderburg 1994). This will be necessary for design for the environment principles. The development of inventories and databases is critical to this endeavor. Unfortunately, for procurement specialists, there are few if any of these tools that are easily available for various industries. There have been some tools developed but at a relatively high level from a macro industry perspective (Horvath 1999) or from a specific company perspective (Nagel 1998). General tools available across industries that can be very specific to certain products are not easy to find. Horvath's work is web-based where organizations can determine life cycle impacts for a number of products. This type of easy availability requires flexibility in the system and thus has to necessarily be at a generic level of analysis.

In managing the purchasing relationship with suppliers, a number of additional issues have arisen with respect to supplier selection. As we have mentioned, third party certification is one issue related to this selection process. ISO 14000-like guidelines are one set of guidelines to aid in this area. Cascio et al. (1996), pointed out that to become ISO 14001 certified, it would not be necessary for an organization to

- Request information from the supplier or contractor about activities performed, chemicals used, waste generated, potential releases to the atmosphere, and other actual or potential environmental aspects and impacts experienced by them.
- Impose its own Environmental Management System on suppliers and contractors.
- Visit a supplier site to ensure that applicable legal requirements are being met.
- Demand that suppliers and contractors have a registered EMS.

ISO 14000 currently has some limitations in this area. Whereas ISO 9000 standards (for quality) mention external relationships with customers and suppliers, these relationship issues are not explicitly mentioned in the ISO 14001 standards.

The role of purchasing and procurement staff and employees for an organization that carries out disassembly or demanufacturing also becomes altered. For example, Sarkis (1998) pointed out that purchasing agents in this environment can serve as salespeople. The reason for this is that some products which are at the end of their life (e.g., computers) may require disposal, but some organizations are capable of taking these disposed materials and actually make them profitable by demanufacturing them into useful materials or remanufacturing them into new products. Of course, the role of procurement would be altered, because this may become a service that is offered to organizations, whereas instead of just waiting for materials to arrive, proactive corporations seek the materials and may even share in the profits with potential "customer-suppliers."

4.2 Storage and Inventory Management

Inventory management has a number of implications from an environmental perspective. Taking a narrow view of the internal just-in-time philosophy (little JIT) as coming under inventory management, where the goal is to minimize waste, it naturally works well when an organization is seeking to be environmentally friendly. The management of inventory, if done efficiently in a "lean" way, is one of the win-win opportunities available. Not only can there be an elimination of waste from obsolescence or spoilage, but the energy and waste costs of maintaining facilities to manage this inventory is lessened. Of course, a poorly managed inventory system can cause numerous problems, critical of which is customer satisfaction.

Inventory modeling of recycled parts and materials has been around for over two decades (Cohen et al. 1980). The inventory issues relative to raw materials, work-in-process, and finished goods are all impacted by environmentally focused source reduction, process management, and systems. As stated previously, the greater uncertainty associated with material coming into the system makes for higher variability and fluctuations, making inventory management in an environmentally friendly firm (if that firm completes remanufacturing and disassembly and relies on recycled and returned materials) a difficult proposition.

Gorgun and Gupta (1999) argued that additional concerns with the quality and reliability of the inventory from returned and recycled materials is another complexity added on to inventory management. Noting these limitations, the use of some traditional models, such as the economic order quantity approaches, can still be used in the evaluation of remanufacturable or repairable products. Some of the work in this area includes deterministic and stochastic modeling as well as continuous and periodic review models (see Inderfurth 1997; Korugan and Gupta 1998; van der Laan et al. 1996a,b, 1997). Development of decision support systems for disassembly inventory planning are also under continuous development (e.g., Rudi and Pyke 2000).

4.3 Fabrication and Assembly

Fabrication and assembly are essentially the manufacturing processes traditionally considered the shop floor of a manufacturing organization. The types of processes can be traditionally grouped into project, job-shop, batch, assembly line, or continuous flow systems (Hayes et al. 1988), each with its own strategy. Not much work has been completed to determine the influence that these type of systems will have on environmental issues and what potential environmental issues emerge. For example, cost reduction and efficiency are characteristics of assembly systems, which may yield relatively superior environmental performance, but system inventory and possibility for scrap may increase, thus possibly yielding inferior environmental performance.

Another issue, in the literature and practice, is that assembly has been addressed somewhat more than fabrication, especially when the disassembly literature is considered within the assembly scope. Defabrication has rarely, if at all, been addressed, since it is difficult to take an item and realter its physical or chemical form back into another form which can be reused later.

To evaluate some of the fabrication and assembly issues, we shall consider the implications of various aspects of major environmentally focused recovery practices and processes such as reduction, reuse, remanufacturing, and recycling.

Reduction will require integration of ideas for waste minimization. This practice is not a big gap for organizations that are seriously practicing quality initiatives. The total quality movement necessarily overlaps and encompasses many environmental initiatives of organizations (Isbell 1991; Mannion 1996). The ultimate goal of zero defects fits well with the concepts of zero emissions for organizations. Continuous improvement, a major tenet of TQM efforts, has also been integrated with environmentally conscious business practices. In fact, the new ISO 14000 standards have, as a foundation, the plan, do, check, act stages of Deming's continuous improvement cycle (Hemenway and Hale 1996). By reducing wastes, efficiencies tend to increase, and win-win situations are more prevalent. The difficulties arise when the benefits of efforts to reduce wastes are not outweighed by the costs of developing, planning and implementing these reduction efforts.

Recycling, remanufacturing, and reuse are all differing levels of the general term recycling. Recycling practices are not only issues that have to focus on internal process capabilities but external process capabilities. Developing and integrating internal recycling processes may require significant investment in technology. As a quick overview, Freeman (1992) and Kryger and Dyndgaard (2000) mention methods for on-site reuse and recycling for a few industries. Some examples include printing; use a vapor-recovery system to recover solvents; textiles, use ultrafiltration system to recover dye stuffs from waterwater; and tape measure, recover nickel-plating solution using an ion-exchange unit. In addition, some companies have flourished with the implementation of newer closed-loop type process technologies.

In closed-loop or "zero-pollution," manufacturing's goal is to reuse any wastes or byproducts within the manufacturing system, a microindustrial ecosystem. The success of a closed-loop manufacturing system requires both prevention (e.g., substitution) and reuse capabilities. This was evident at Hyde Manufacturing company (a small manufacturing enterprise located in Southbridge, MA) where substitutes were found for two major hazardous chemicals. A fluid management program that reuses coolants was an example of how they maintained a closed loop manufacturing system (Hasek 1997). Another example of closed-loop manufacturing is the Ciba-Giegy dyestuffs in Indonesia (Leake and Kainz 1994). The decision, in this example, was whether to build a dye treatment facility that costs more than the manufacturing plant or to reuse the rinse water. It was decided to reuse the rinse water. The amount of dye in the rinse water alone helped to pay off the cost of the environmental improvements within 2 years.

Wastewater closed-looped production processes are being increasingly adopted by industry, primarily due to the ease of incorporating such systems and their accrued benefits. One unique example involves a customer-supplier relationship. A partnership between Caterpillar and BetzDearborn (Anonymous 1999) helped develop a dewatering process that produced a solid product—a concentrated sludge. To minimize the plant's environmental impact still further, Caterpillar and BetzDearborn purified the treated chrome rinse water to produce demineralized water that could be returned to the plating line. As a result, up to 80% of all wastewater now is reused. Chrome tank solutions from the demineralized water can be recycled indefinitely. Economic savings were also substantial in this situation.

Remanufacturing and reuse materials recovery processes also require some refurbishing and disassembly process capabilities. Processing equipment that is capable of cleaning and maintaining products is one of the first requirements. Disassembling products are also necessary for recycling materials that arrive from the after-market of these products. Currently, most disassembling technology and capabilities focusing on the processing of materials and products, are relatively manual. The development of automated systems (which, in parallel with, design for disassembly practices) will become necessary as these markets and volumes increase. Currently, flexible machining systems are designed for new product flexibilities or "forward" manufacturing capabilities. These systems may eventually require development of not only assembling a variety of products but disassembling and reassembling these similar products. Flexibility definitions for processing equipment would need to include the additional dimension of disassembling capabilities. Much of the emphasis in this area has been on model development for disassembly planning and modeling or on design for disassembly approaches. Actual process technology development for disassembly is rarely mentioned in practitioner or research literature (Boks and Templeman 1998; Nagler 1999, Penev and DeRon 1996).

4.4 Logistics

The logistics portion of the manufacturing function is meant to focus on internal materials movement rather than external relationships that have been evident and discussed under green supply chain. The materials movement can also include packaging and its requirements. The discussion of internal materials movement and its impact on environmental practices is not well established. The use of JIT internally may impact how much inventory is held but also the amount of movement of materials through the system. Smaller batches and rapid replacement requires significant energy and movements. The trade-offs at this level must be evaluated. Planning and routing for disassembly systems can also be an issue for internal logistics. The packaging and materials may be relegated to the use of recycled containers, especially internally, rather than packaging that needs to be destroyed as it is sent from one department to another.

4.5 Planning and Control Models, Systems, and Measurement

Planning and control models for manufacturing have received a significant amount of attention. We have already discussed some inventory control and management type models, especially for disassembly purposes. A number of models for managing in this environment are also evident, including routing, scheduling, and sequencing issues (Gupta and Taleb 1994; Lambert 1997; Yokota and Brough 1992). These may be core aspects of control systems for production activity control, which are part of the major control systems in manufacturing.

The major control systems in manufacturing are material resources planning and enterprise resources planning (ERP) systems. These systems will require integration of a number of environmental characteristics, especially in a remanufacturing/disassembly environment. The research and practice in this area has been quite limited (Guide et al. 1996; Panniset 1988). One such issue is the integration of reverse bills of material that will aid in managing inventory disassembly of products (Krupp 1993). The planning and forecasting for material flows into a system will also be an issue (Guitini 1998; Krupp 1992). The uncertainty with the type of supply for replacement or recyclable parts is greater than that of virgin materials. The diversity of "suppliers" in this type of environment is greater since it is heavily dependent on the variety of customers. Another reason for the uncertainty is due to the immature reverse logistics channels in most manufacturing industries. Completing master scheduling plans for materials that organizations have little control over is also a concern (Guide et al. 1996).

Another area to manage manufacturing operations include benchmarking and performance measurement schemes. Environmental benchmarking of operations will be critical from a total environmental management approach. Some resources do exist; for example, there is an existence of a government-supported environmental management benchmarking database and study that help many organizations gather information on best environmental manufacturing practices (Department of the Navy 1997). As more proactive environmental programs develop, additional information and sharing will necessarily occur. Incorporating these data and these sources into organizational environmental management is an issue that needs to be addressed. The work on performance measurement for environmental purposes has been strategic but does include some operational environmental performance measurement development (Azzone and Tyteca, 1998). A number of these measures will be very important for informing stakeholders (customers, community, etc.) through environmental reports which will be used for external benchmarking purposes (CERES 1999).

Tracking systems for environmental wastes and toxicities of wastes from manufacturing systems would be necessary to provide a more accurate portrayal of environmental costs. Polariod's environmental accounting and reporting system is one such example (Nash and Nutt 1992; Stark 1993). In this system, supervisors and managers in different production plants keep records of the various levels of wastes or hazardous materials that come into and flow out of a system. The wastes are grouped among different levels and recorded. The use of this system helps keep track of the types of wastes, which in turn can provide data for estimation of potential liabilities, costs of disposal, and other life cycle costs.

Investment in technology and programs incorporates a number of measurement/control issues. For example there exists a need for further integration of environmental factors into decisions for project selection, equipment evaluation, and product development. Total cost analysis (White 1995) has been proposed as an approach to more accurately introduce environmental factors (many of which are very uncertain and intangible) into the capital budgeting decision process. White (1995) shows the uncertain impact of not considering, or the inability to locate and integrate, environmental costs into traditional financial analysis tools such as net present value and payback. The use of traditional evaluation techniques may not be enough. Environmental factors are not only intangible and uncertain but are necessarily strategic. These characteristics need to be brought into any decision making framework. The development of these tools (much less the actual application) has not been very pervasive. Presley and Sarkis (1994) have attempted to further the analysis by incorporating factors such as the above-mentioned costs for justification of environmentally conscious design and manufacturing projects. The integration of standard financial analysis tools with more advanced multiattribute decision tools is an approach that has been put forth by a number of researchers (see Sarkis 1998, 1999). The increasing complexity of these types of analyses still makes it difficult to actually apply in manufacturing organizations. Integrating environmental factors and the necessary organizational infrastructure to acquire this information is also in its infancy, especially in the design process.

4.5.1   Productivity Analysis

A tool that can prove useful in performance measurement of productive units is data envelopment analysis (DEA). This tool helps to determine the relative efficiency of units by utilizing a multifactor productivity measure based on a number of inputs and outputs. Even though the literature is relatively sparse on these issues, there are a number of papers that have focused on broader environmental issues. DEA-based modeling and research that have incorporated environmental factors do exist. Example macroeconomic DEA-based works that include some form of environmental factors evaluation include efficiencies of energy alternatives (Criswell and Thompson 1996), use for forest management (Kao et al. 1993), industrial regional development (Karkazis and Boffey 1997), country productivity evaluations (Lovell et al. 1995), site location (Thompson et al. 1986), industrial productivity and emissions (Yaisawarng and Klein 1994), and opportunities for environmental improvements in the agricultural industry (Ball et al. 1994; Poit-Lepetit et al. 1997). Some researchers have also applied DEA to evaluation of pollution efficiency for individual plants and/or firms using emission data (see Haynes et al. 1993, 1994; Sarkis and Cordeiro 1998; Tyteca 1996).

Most of the macroeconomic examples from the literature focus on the industry or national level of analysis, with a focus on a set of factors of which environmental issues were not necessarily central issues. The focus of much of this research was evaluating or ranking various organizations within industries and how shifts can be made to improve overall environmental performance. The focus on these papers was from a policy perspective rather than corporate managerial and competitive dimensions. The only work that explicity evaluates environmental performance using DEA from a competitive perspective was by Sarkis and Cordiero (1998). It seems that within this relatively comprehensive review, there is ample room for external organizational application of DEA on environmental issues from a competitive perspective. Internal use of DEA as a managerial tool or for research study from an environmental perspective is lacking.

Internal benchmarking using DEA can incorporate internal measures of performance such as scrap rate, process emissions, and other measurable items as outputs, with capital costs, number of employees, and environmental staff as possible inputs. The use of this technique for operational measurement and benchmarking has significant potential.

4.5.2 Process Change Management and Organizational Influences

Benchmarking and other approaches for control of processes use measurement to influence how an organization completes its changes in these processes. Some of these process changes incorporate the issues of outsourcing and reengineering. The relationships of these organizational change (restructuring) issues with environmentally friendly manufacturing have not been studied (Kasperson et al. 2000). A number of issues do arise in this situation, part of which include the downsizing (laying off management and staff) for competitive reasons. For example, the impact of newly reengineered processes and their environmental influences have not been studied or whether these new processes are meant to address environmental issues or the traditional issues of efficiency. The outsourcing issue is a strategic purchasing decision. Part of the issues that arise from this situation arise from whether a supplier has appropriate environmental systems in place at least as equal to those of the original manufacturer. Usually a smaller manufacturer is the supplier of outsourced manufacturing services, and they may not have the resources to monitor or operate in an environmentally friendly way. In addition, downsizing staff may require outsourcing of environmental auditors and staff. The lack of internal experience and expertise may hurt an organization's environmental performance, but external expertise may prove to be better. These issues need to be evaluated from an environmental perspective, whereas many times they are evaluated from a time, quality, or efficiency perspective.

5 EMERGING ISSUES IN ENVIRONMENTALLY FRIENDLY MANUFACTURING

The area of environmentally friendly manufacturing has gained substantial momentum in recent years. Even though there is a significant amount of practitioner and research work that still needs to be completed to understand and develop environmentally friendly manufacturing programs, a significant foundation has been built. There are a number of issues that are still arising and a number of issues that still need to be addressed. In this section we look at some of these issues. Many of them have broad implications to all organizational functions, not just manufacturing, and we shall make clear the manufacturing relationships. We only touch upon some of these emerging issues. Additional issues are always arising and engineers and management must be ready to anticipate them or respond quickly to them when they arise.

5.1   Virtual Enterprises

Virtual enterprises are new organizational forms that are emerging due to increased competition, decreasing product life cycles, and developments in new technologies. These new organizational forms are meant to rapidly form to take advantage of a given market opportunity (Goranson 1999). The virtual enterprise is formed from a set of separate organizations who integrate their core competencies to address this market opportunity.

A number of environmental issues are of concern here. Some have been mentioned earlier but have to be reconsidered within the context of the virtual enterprise. For example, the responsibility for managing the production or aspects of the production function may require selection criteria that includes environmental characteristics. The product that flows through this system will require an LCA over a number of companies. The responsibilities of this product and its materials before, during, and after its delivery will need to be made more explicit. Since the virtual enterprise is meant to diffuse after its formation, product responsibility and stewardship (not to mention issues that deal with environmental regulations and possible reactive regulatory issues) must be dealt with.

A set of standards that overlap organizational boundaries, since the processes are expected to cross and re-cross organizational boundaries in a weblike network relationship, need to be managed. The standards need to be part of a clearly defined environmental management system. How this environmental management system integrates and manages the production function across organizational boundaries is a serious issue that needs to be addressed. Even the relatively new practices associated with green supply chains may have to be altered to be most effective in this environment.

5.2   Information Technology

Related to the virtual enterprise is the increase integration and use of information technology within the manufacturing function. Interorganizational information systems (e.g., internet, electronic data interchange) and e-commerce issues related to the environment, especially manufacturing related information, have not been addressed. The replacement of movement of information, rather than materials and products, will have profound influence on environmentally friendly manufacturing. The role of information technology in this environment can also focus on internal management issues related to manufacturing ranging from databases on substitutable materials and processes that manufacturing engineers can have access to for incorporating the most environmentally sound technique or material.

We have also discussed the increasing role of ERP systems in organizations. This means that environmental data will be readily available throughout the organization. The environmental management system will also need to be supported by this information technology ranging from monitoring of processes to helping in decision making and policy setting by the organization.

5.3 Globalization

The globalization of competition and organizations has caused a situation where there are opportunities and problems with respect to environmentally friendly manufacturing. The globalization issue has allowed for organizations to find the locations that may not be as environmentally restrictive in terms of regulations. This may cause some organizations to be more lenient in their environmental goals. This characteristic is a traditional problem similar to finding the lowest labor costs and exploiting these differences for competitive advantage. Yet voluntary programs do provide possible opportunities for proactive organizations. For example, organizations that have ISO 14000 standards in place (in less restrictive locales) may be at an advantage for gaining relationships and business from proactive multi-national corporations seeking to incorporate green supply chain issues. Vendor selection in green supply chain issues may be more difficult from an international supplier perspective. For example, if one of the criteria is how well these suppliers perform on meeting legislative regulations, the results can prove deceptive. What is acceptable in one country may be penalized in other countries. Maintaining a true green supply chain is more difficult in this perspective.

On the other side of this coin is the difficulty, with globalization, in meeting the various regulations as set forth by different countries. The use of standards may be helpful from this respect, where an ISO 14001 registered organization could be assumed to meet local environmental regulations.

5.4 Strategic Focus

Manufacturing's role in organizational strategy has been well established. Environmental strategy is beginning to become more pervasive in organizations. Linking these two strategies can prove to be a potent competitive weapon for organizations. Yet the traditional mindset that manufacturing should focus on cost reduction and environment on responding (and fighting) government regulations have to be put aside. The proactive approaches which require manufacturing to be a full partner in corporate strategy and the natural environment as a "win-win" proposition are a philosophy that is pervasive in the organization. Elevating environmental issues (similar to quality issues) to strategic decision making and systemic levels is necessary.

To do this there needs to be a grasp and support of environmentally friendly manufacturing by all functions and employee levels.

Another important strategic focus of environmentally friendly manufacturing is the diffusion of environmental management systems throughout the organization. These systems are meant to address concerns that are both short term and long term, across functions and organizations. These systems need a strategic perspective and will support an organization's strategic focus.

5.5 Performance Measures

We discussed the issue of performance measures as part of the systems and controls of the organization. Yet this area is one of the critical areas that organizations and researchers have had a difficult time incorporating. Measurement is important for a number of reasons, including such areas as environmental reporting, TQEM, LCA, environmental management systems, and selection of green suppliers.

Characklis and Richards (1999) pointed to two major questions pertaining to environmental performance measures: how to improve existing ones and what new metrics are needed. Within the improvement question, they mention that there is a need for addressing comparability and increasing life cycle coverage. The shift to sustainability has required additional measures to be integrated with environmental measures and comparability among them has become more difficult. In addition, data such as that offered by the Toxic Releases Inventory database but relative comparability across industries may be difficult to address. Varying toxicity levels and media implications have caused any data to be difficult to rationalize as appropriate for measuring performance. The determination of the actual impact of these metrics to environmental performance is also very difficult to achieve. Better modeling approaches are required to provide better estimation. Another issue is the normalization of the performance metrics across various stakeholders. The extension of performance metrics across the supply chain and product life cycle has also been viewed as an important issue to address.

6 CONCLUSION AND SUMMARY

The issues in environmentally friendly manufacturing are as broad as those of manufacturing strategy due to the relative importance of this issue for the long-term survival and prosperity of the organization. The topics presented here have described the role of environmentally friendly manufacturing within the broader organization. The visionary corporate strategy integrates environmental and manufacturing issues and relies on these issues to develop competitive strategies. There are internal strategic and operational considerations as well as external relationship issues that management, staff, and employees need to be made aware of. Researchers and practitioners are all attempting to address the many concerns of environmentally friendly manufacturing. This field is in its relative infancy, and many ideas still need to be investigated and generated, and many lessons learned.

REFERENCES

Adler, P.S. (1989) Technology strategy: guide to the literatures. In: Richard S. Rosenbloom, ed. Research in Technological Innovation, Management and Policy, Vol. 4. JAI Press Inc., Greenwich, CT, 1989, pp. 25–151.

Anonymous (1999) Buyer/supplier teamwork helps minimize environmental impact. Purchasing 127 (8), pp. 111.

Azzone, G., and Manzini, R. (1994) Measuring strategic environmental performance. Business Strategy and the Environment, 3, pp. 1–14.

Ball, V.E., Lovell, C.A.K., Nehring, R.F., and Somwaru, A. (1994) Incorporating undesirable outputs into models of production: an application to U.S. agriculture. Cahiers d'Economique et Sociologie Rurales, 31, pp. 59–73.

Bloemhuf-Ruwaard, J.M., van Beck, P., Hordijk, L., and van Wassenhove, L.N. (1995) Interactions between operational and environmental management. European Journal of Operational Research 85 (2), pp. 229–243.

Boks, C., and Tempelman, E. (1998) Future disassembly and recycling technology: results of a Delphi study. Futures 30 (5), pp. 425–442.

Bucholz, R.A. (1998) Principles of Environmental Management, 2nd Edition. Prentice Hall, New Jersey.

Carter, C.R., and Ellram, L.M. (1998) Reverse logistics: a review of the literature and framework for future investigation. Journal of Business Logistics 19(1); pp. 85–102.

Cascio, J., Woodside, G., and Mitchell, P. (1996) ISO 14000 guide: the new international environmental management standards. McGraw-Hill, New York.

CERES (1999) Sustainability Reporting Guidelines: Exposure Draft for Public Comment and Pilot Testing. CERES-Global Reporting Initiative, Boston, MA.

Characklis, G.W., and Richards, D.J. (1999) The evolution of industrial environmental performance metrics: trends and challenges, Corporate Environmental Strategy 6 (4), pp. 387–394.

Cohen, M.A., Nahmias, S., and Pierskalla, W.P. (1980) A dynamic inventory system with recycling. Naval Research Logistics Quarterly, 27, pp. 289–296.

Cordeiro, J. and Sarkis, J. (1997) Environmental proactivism and firm performance: evidence from industry analyst forecasts. Business Strategy and the Environment, 6 (2), 104–114.

Criswell, D.R., and Thompson, R.G. (1996) Data envelopment analysis of space and terrestrially-based large-scale commercial power systems for earth. Solar Energy 56, pp. 119–131.

Darnall, N., Nehmann, G., Priest, J., and Sarkis, J. (1994) A review of environmentally conscious manufacturing theory and practices. The International Journal of Environmentally Conscious Design and Manufacturing 3, pp. 49–58.

Denton, T. (1998) Sustainable development at the next level. Chemical Market Reporter, 253, pp. 3–5.

Dutton, G. (1998) The green bottom line. Management Review 7 (9), pp. 59–63.

Dwortzan, M. (1998) The greening of industrial parks. MIT's Technology Review 100, pp. 18–19.

Earl, G., and Clift, R. (1999). Environmental performance: what is it worth?" In Greener Marketing: A Global Perspective on Greening Marketing Practice, edited by M. Charter and M.J. Polonsky. Greenleaf Publishing, Sheffield, England, pp. 255–274.

Fleischmann, M., Bloemhof-Ruwaard, J., Dekker, R., van der Laan, E., van Nunen, J.A., and Van Wassenhove, L.N. (1997) Quantitative models for reverse logistics: a review." European Journal of Operational Research 103 (1), pp. 1–17.

Florida, R. (1996) Lean and green: the move to environmentally conscious manufacturing. California Management Review 39 (1), pp. 80–105.

Freeman, H.M. (1990) Hazardous Waste Minimization. McGraw-Hill, New York.

Freeman, R.E. (1984) Strategic Management: A Stakeholder Approach. Pitman, Boston.

Friedman, M. (1970), The social responsibility of business is to increase its profits. New York Times Magazine, September 13, pp. 122–126.

Goranson, H.T. (1999) The Agile Virtual Enterprise: Cases, Metrics, Tools, Quorum Books. Westport, Connecticut.

Guide, V.D.R. Jr., and Srivastava, R. (1997) Repairable inventory theory: models and applications. European Journal of Operational Research 102, pp. 1–20.

Guide, V.D.R., Srivastava, R., and Spencer, M.S. (1996) Are production systems ready for the green revolution? Production & Inventory Management Journal 37, pp. 70–76.

Guitini, R. (1998) Forecasting in remanufacturing. Fifth International Conference on Environmentally Conscious Design and Manufacturing, Rochester, NY.

Gungor, A., and Gupta G. (1999) Issues in environmentally conscious manufacturing and product recovery: a survey. Computers & Industrial Engineering 36, pp. 811–853.

Gupta, M. (1995) Environmental management and its impact on the operations function. International Journal of Operations and Production Management, 15, pp. 34–51.

Gupta, S.M., and Taleb, K.N. (1994) Scheduling disassembly. International Journal of Production Research, 32 (8), pp. 1857–1866.

Hanna, M.D, and Newman, W.R. (1995) Operations and the environment: an expanded focus for TQM. International Journal of Quality and Reliability Management, 12, pp. 38–53.

Hart, S.L. (1995) A natural-resource-based view of the firm, Academy of Management Review 20, pp. 986–1014.

Hart, S.L., and Ahuja, G. (1996) Does it pay to be green: an empirical examination of the relationship between emission reduction and firm performance. Business Strategy and the Environment 5 (1), pp. 30–37.

Hartman, E.L., and Stafford, E.R. (1998) Crafting "enviropreneurial" value chain strategies through green alliances. Business Horizons 41 (2), pp. 62–72.

Hax, A.C., and Majluf, N.S. (1991) The Strategy Concept and Process: A Pragmatic Approach. Prentice Hall, Englewood Cliffs, NJ.

Hayes, R.H., Wheelwright, S.C., and Clark, K.B. (1988) Dynamic Manufacturing: Creating the Learning Organization. The Free Press, New York.

Haynes, K.E., Ratick, S., Bowen, W.M., and Cummings-Saxton, J. (1993) Environmental decision models: U.S. experience and a new approach to pollution management. Environment International, 19, pp. 261–75.

Haynes, K.E., Ratick, S., and Cummings-Saxton, J. (1994) Toward a pollution abatement monitoring policy: measurements, model mechanics, and data requirements. The Environmental Professional 16, pp. 292–303.

Hearn, T. (1999) A win-win proposition. Industrial Distribution, 88 (11), p. 110.

Hemenmay, C.G., and Hale, G.J. (1996) The TQEM-ISO14001 connection. Quality Progress 29 (6), pp. 29–32.

Horvath, A. (1999) Supply chain environmental assessment of the telecommunications sectors. Proceedings of the 1999 IEEE International Symposium on Electronics and the Environment, Danvers, MA, May 11–13.

Inderfurth, K. (1997) Simple optimal replenishment and disposal policies for a product recovery system with lead-times. OR Spektrum 19, pp. 111–122.

Isbell, T.S. (1991) The Backside of TQM: How Waste Indicates Haste. Industrial Engineering, pp. 42–45.

Karna, A., and Heiskanen, E. (1998) The challenge of "product chain" thinking for product development and design: the example of electrical and electronic products. ournal of Sustainable Product Design 4 (1), pp. 26–36.

Kao, C., Chang, P.L., Hwang, S.N. (1993) Data envelopment analysis in measuring the efficiency of forest management. Journal of Environmental Management 38, pp. 73–83.

Karkazis, J., and Boffey, B. (1997) Spatial organization of an industrial area: distribution of the environmental cost and equity policies. European Journal of Operational Research, 101 (3), pp. 430–441.

Kasperson, R., Kasperson, J.X., Gray, W.P., Jiusto, S., and Sarkis, J. (2000) Industrial restructuring and corporate risk management. Proceedings of the Annual Meeting of the Society for Risk Analysis / Europe, Edinburgh, Scotland, May 14-17.

Klassen, R.D. (1993) Integration of environmental issues into manufacturing. Production and Inventory Management Journal 34 (1), pp. 82–88.

Klassen, R. and McClaughlin C. (1996) The impact of environmental management on firm performance. Management Science 42 (8), pp. 1199–1214.

Klassen, R.D., and Whybark, D.C. (1999) The impact of environmental technologies on environmental performance. Academy of Management Journal, 42, p. 599.

Korugan, A., and Gupta, S.M. (1998) A multi-echelon inventory system with returns. Computers and Industrial Engineering 35 (1-2), pp. 145–148.

Krupp, J.A.G. (1992) Core obsolescence forecasting in remanufacturing. Production & Inventory Management Journal, 33, pp. 12–17.

Krupp, J.A.G. (1993) Structuring bills of material for automotive remanufacturing. Production & Inventory Management Journal 34, pp. 46–52.

Krut, R., and Munis, K. (1998) Sustainable industrial development benchmarking environmental policies and reports. Greener Management International, p. 87.

Kryger, J., and Dyndgaard, R. (2000) Introduction to cleaner production. In A Systems Approach to the Environmental Analysis of Pollution Minimization, edited by Jorgensen, S.E. pp. 87–100. Lewis Publishers, Boca Raton, FL.

Lambert, A.J.D. (1997) Optimal disassembly of complex products. International Journal of Production Research, 35 (9), pp. 2509–2523.

Lamming, R., and Hampson, J. (1996) The environment as a supply chain management issue. British Journal of Management 7, pp. 45–62.

Leake, M., and Kainz, R. (1994) Environmental enhancement: an agile manufacturing perspective. Agility Forum Paper AR94-03, Bethlehem, PA.

Lovell C.A.K., Pastor J.T., and Turner J.A. (1995) Measuring macroeconomic performance in the OECD—a comparison of European and non-European countries. European Journal of Operational Research 87, pp. 507–518.

Lowe, E. (1993) Industrial ecology—an organizing framework for environmental management. Total Quality Environmental Management, 3, pp. 73–85.

Mandese, J. (1991). New study finds green confusion. Advertising Age 62 (45), pp. 1–2.

Mannion, F. (1996) Enhancing corporate performance through quality-driven pollution prevention. National Productivity Review 16 (1), pp. 25–32.

Miller, W.H. (1998) Cracks in the green wall. Industry Week, 247, pp. 58–65.

Meredith, J.R. (1987) Implementing the automated factory. Journal of Manufacturing Systems 6, pp. 1–13.

Nagel, M.H. (1998) Environmental supply-line engineering: eco-supplier development coupled to eco-design—A new approach. Bell Labs Technical Journal 3, pp. 109–123.

Nagler, B. (1999) Reintroducing Remanufacturing. Machine Design 71 (22), p. 78.

Narasimhan, R., and Carter, J.R. (1998) Environmental Supply Chain Management. The Center for Advanced Purchasing Studies. Arizona State University, Tempe, AZ.

Noci, G. (1997) Designing "green" vendor rating systems for the assessment of a supplier's environmental performance. European Journal of Purchasing & Supply Management 3 (2), pp. 103–114.

Noori, H. (1990) Managing the Dynamics of New Technology. Prentice-Hall, Englewood Cliffs, NJ.

Ottman, J. A. (1992) Green consumerism: the trend is your friend. Directors & Boards 16, pp. 47–50.

Pannisett, B.D. (1988) MRP II for repair/refurbish industries. Production and Inventory Management Journal 29, pp. 12–22.

Penev, K. D., and de Ron, A. J. (1996) Determination of a disassembly strategy. International Journal of Production Research 34, pp. 495–506.

Poit-Lepetit, I., Vermersch D., and Weaver R.D. (1997) Agriculture's environmental externalities: DEA evidence for French agriculture. Applied Economics 29 (3), pp. 331–338.

Pohlen, T.L., and Farris, M.T. (1992) Reverse logistics in plastics recycling. International Journal of Physical Distribution & Logistics Management 22(7), pp. 35–47.

Porter, M.E. (1980) Competitive Strategy. Free Press, New York.

Porter, M.E. (1985) Competitive Advantage: Creating and Sustaining Superior Performance. The Free Press, New York.

Presley, A., and Sarkis, J. (1994) An activity based strategic justification methodology for ECM technology. The International Journal on Environmentally Conscious Design and Manufacturing, 3(1), pp. 5–17.

Roberts, J.A., and Bacon, D.R. (1997) Exploring the subtle relationships between environmental concern and ecologically conscious consumer behavior. Journal of Business Research 40 (1), pp. 79–89.

Rudi, N., and Pyke, D.F. (2000) Product Recovery at the Norwegian National Insurance. Interfaces 30 (3), pp. 166–179.

Russo, M.V., and Fouts, P.A. (1997) A resource-based perspective on corporate environmental performance and profitability. Academy of Management Journal 40, pp. 534–559.

Samson, D. (1991) Manufacturing and Operations Strategy. Prentice Hall, Brunswick, Australia.

Sarkis, J. (1995a) Manufacturing strategy and environmental consciousness. Technovation 15 (2), pp. 79–97.

Sarkis, J. (1995b) Supply chain management and environmentally conscious design and manufacturing. International Journal of Environmentally Conscious Design and Manufacturing 4 (2), pp. 43-52.

Sarkis, J. (1998) Evaluating environmentally conscious business practices. European Journal of Operational Research 107, pp. 159–174.

Sarkis, J. (1999) A methodological framework for evaluating environmentally conscious manufacturing programs. Computers and Industrial Engineering 36 (4), pp. 793–810.

Sarkis, J. (2001) Manufacturing's role in corporate environmental sustainability: concerns for the new millennium. International Journal of Operations and Production Management, forthcoming.

Sarkis, J., and Cordeiro, J. (1998) An investigation of the relationship between environmental and financial performance of organizations. Conference Proceedings of Performance Measurement—Theory and Practice, Edited by A.D. Neely and D.B. Waggoner. Cambridge, U.K., pp. 255–262.

Sarkis, J., Liffers, M., and Malette, S. (1998) Purchasing operations at Digital's computer asset recovery facility. In Greener Purchasing: Opportunities and

Innovations, edited by T. Russel. Greenleaf Publishing, Sheffield, England, pp. 270–281.

Sarkis, J., and Rasheed, A. (1995) Greening the manufacturing function. Business Horizons 38 (5), pp. 17–27.

Schroeder, R.G., and Lahr, T.N. (1990) Development of a manufacturing strategy: a proven process. In: J.E. Ettlie, M.C. Burstein, and A. Fiegenbaum eds. Manufacturing Strategy. Kluwer Academic Publishers, Boston, pp. 3–14.

Schlossberg, H. (1990) Canadians are serious about their environment and ours, too. Marketing News, March 19, p. 16.

Schmidheiny, S. (1992) Changing Course: A Global Business Perspective on Development and the Environment. MIT Press, Cambridge, MA.

Shrivastava, P., (1995a) Ecocentric management for a risk society. Academy of Management Review 20, pp. 118–137.

Shrivastava, P. (1995b) Environmental technologies and competitive advantage. Strategic Management Journal 16, pp. 183–200.

Skinner, W. (1969) Manufacturing—Missing Link in Corporate Strategy. Harvard Business Review 47, pp. 136–145.

Stahl, J.F. (1989) Manufacturing strategic planning: key to competitive strength. CIM Review 6 (1), pp. 14–20.

Starik, M. (1995), Should trees have managerial standing? Toward stakeholder status for non-human nature. Journal of Business Ethics 15, pp. 207–208.

Porter, M.E., and van der Linde, C. (1995) Green and competitive: ending the stalemate. Harvard Business Review 73, pp. 120–134.

Thompson, R.G., Langemeier, L.N., Lee, C.T., and Thrall, R.M. (1990) The role of multiplier bounds in efficiency analysis with application to kansas farming. Journal of Econometrics 46(1-2), pp. 93–108.

Tyteca, D. (1996) On the measurement of the environmental performance of firms— a literature-review and a productive efficiency perspective. Journal of Environmental Management 46, pp. 281–308.

Tyteca, D. (1998) Sustainability indicators at the firm level,. Journal of Industrial Ecology 2 (4), pp. 61–77.

van der Laan, E.A., Dekker, R., and Salomon, M., (1996) Product remanufacturing and disposal: a numerical comparison of alternative control strategies. International Journal of Production Economics 45(1-3), pp. 489–98.

van der Laan, E.A., Dekker, R., Salomon, M., and Ridder, A. (1996) An (s,Q) inventory model with remanufacturing and disposal. International Journal of Production Economics 46, pp. 339–350.

van der Laan, E.A., and Salomon, M. (1997) Production planning and inventory control with remanufacturing and disposal. European Journal of Operational Research 102, pp. 264–278.

Walley, N., and Whitehead, B. (1994) It's not easy being green. Harvard Business Review 72 (3), pp. 46–52.

Walton, S.V., Handfield, R.B., and Melnyk, S.A. (1998) The green supply chain: integrating suppliers into environmental management processes. International Journal of Purchasing and Materials Management 34 (2), pp. 2–11.

Weissman, S.H., and Sekutowski, J.C. (1991) Environmentally Conscious Manufacturing. AT&T Technical Journal 70, pp. 23–30.

Wu, H.J., and Dunn, S.C. (1995) Environmentally responsible logistics systems. International Journal of Physical Distribution & Logistics Management 25(2), pp. 20–38.

Yam-Tang, E.P.Y., and Chan, R.Y.K. (1998) Purchasing behaviours and perceptions of environmentally harmful products. Marketing Intelligence & Planning 16(6-7), pp. 356–362.

Yaisawarng, S. and Klein, J.D. (1994) The effects of sulfur-dioxide controls on productivity change in the united-states electric-power industry. Review of Economics and Statistics 76, pp. 447–460.

Young, S.B., and Vanderburg, W.H. (1994) Applying environmental life-cycle analysis to materials. JOM: The Journal of the Minerals, Metals & Materials Society 46(3), pp. 22–27.

Young, P., Byrne, G., and Cotterell, M. (1997) Manufacturing and the environment. International Journal of Advanced Manufacturing Technology 13, pp. 488–93.

Yokota, K., and Brough, D.R. (1992) Assembly/disassembly sequence planning. Assembly Automation 12 (3), pp. 31–38.

23

Design for End-of-Life Strategies and Their Implementation

Ab Stevels
Philips Consumer Electronics, Eindhoven, The Netherlands
Casper Boks
Delft University of Technology, Delft, The Netherlands

1 INTRODUCTION

When compared with the standard design process, one of the most salient new elements in ecodesign is that businesses now have to start thinking about what happens to the product after it leaves the factory. Additionally, businesses must develop a scenario for what happens to the product when the user no longer wishes to use it; in other words, what happens in the product's end-of-life phase? This chapter deals with the subject of optimization of end-of-life systems.

The main reasons why product developers must concern themselves with the end-of-life phase when developing a product are the following. First, there is an increase in both magnitude and diversity of waste problems in affluent societies. Society is also becoming aware that waste and recycling are an important concern and that therefore our natural resources, materials and energy, should be more efficiently and in a sustainable manner. This concern has lead several countries to legislation initiatives, particularly in Western European countries, both on a European Union level [1] and in individual member states [2–4], but also in Asia [5] and in some individual states of the United States, particularly Massachusetts, Minnesota, and California. These legislation initiatives focus in particular on producer responsibility and product take-back, controlled treatment of discarded products, and recycling targets. In response to that, the industrial world has expressed the wish to achieve environmental improvement without imposing an unnecessary financial burden on society. In recent literature [6, 7] frameworks for doing so have been discussed. And as the front runners that have proactively started to take end-of-life considerations into account have shown, responsible product stewardship will often yield business opportunities related to marketing, cheaper assembly, image profiling, and second-hand markets.

What is an end-of-life system? The end of life of a product refers to all that can happen to a product after the initial user has discarded it. Questions to be answered are as follows:

- Is the product taken back and reused?
- Are useful components removed from the product for reuse or are only the materials reused?
- Is the entire product incinerated or directly dumped on a rubbish tip?

Partly because of the developments in legislation on manufacturers' responsibilities and the increased level of environmental awareness, all entrepreneurs must be able to answer these questions. The view taken by management with regard to the end-of-life system of the product is a strategic consideration as well. It is dependent on consumer behavior, infrastructure, and local and international legislation. This subject is further elucidated in section 2. It is necessary to determine on a product level what end-of-life system is technically and economically the best. This is further explained in section 3. Subsequently, it is the designer's job to redesign the product in such fashion that it is optimally suited for the chosen end-of-life strategy. The relevant design strategies are discussed in section 4. This chapter is concluded by section 5, which focuses on how the optimization of the end-of-life system can be quantitatively underpinned.

2   THE DEVELOPMENT OF END-OF-LIFE SYSTEMS

2.1   Preferences Between Different End-of-Life Options.

Before the organization of an end-of-life system is discussed, it is shown from an environmental perspective what general preferences exist when it comes to the end-of-life processing of discarded goods. Ranked highest is product life-time extension (Figure 1). When it can be prevented that the user discards a product in the first place, for instance, by extending its technical or economical life, any initial environmental impact from collection or processing the product can be forestalled. Also, the "investment" in materials is better used.

Another preferred solution is to strive toward reuse of the product as a whole, either for the same or a new application. If this is not possible, the reuse of subassemblies and components through remanufacturing and refurbishing should be considered. Next in the order of preference is material recycling. Material recycling processes exist in many different shapes and forms (see section 3.4 for descriptions and references), but when it comes to determining the order of preference for end-of-life destinations, three applications can be distinguished:

- Recycling in the original application (primary recycling).
- Recycling in a lower-grade application (secondary recycling). In this case the quality of the recycled material does no longer meet the

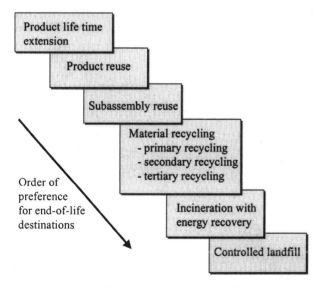

FIGURE 1   Order of preference for end-of-life destinations.

specifications of the original application, but secondary applications are still possible. A well-known example is the application of secondary plastics into park benches.

- Recycling of plastics into raw materials by means of thermal or chemical decomposition (tertiary recycling). This is the decomposition of long plastic molecules into elementary raw materials, which are subsequently reused in refineries or for the production of petrochemicals. Tertiary recycling is also referred to as feedstock recycling, by means of pyrolysis, hydrogeneration, gasification, or chemolysis.

If recycling is not possible, three more options remain for end-of-life treatment. These are, in order of environmental preference,

- Incineration with energy recovery (quaternary recycling, also called thermal recycling). This involves the incineration of nonreusable materials by using energy generation technology and good flue gas purification.
- Incineration of nonreusable materials without energy generation technology but with flue gas purification.
- Depositing residual material in a controlled fashion as solid waste at a landfill site.

In most countries, law prohibits incineration without flue gas purification and uncontrolled dumping.

In general, from an environmental perspective, the end-of-life strategies should be preferred in the sequence as outlined above. However, sometimes recycling or safe incineration with energy recovery is to be preferred to reuse or remanufacturing. Examples of such cases are as follows:

- The energy consumption of some electric appliances (e.g., television sets) has been reduced to such an extent over the past 10 years that it no longer makes sense to promote the reuse of these appliances older than 10 years.
- For materials containing toxic organic compounds, safe incineration (= destroying) is better than recycling (if at all possible).

2.1.1 Cascade Recycling

Which of the above environmental preferences apply to specific end-of-life systems is determined on the basis of the so-called cascade concept. This is the aim to keep a material in the most high-grade application possible for as long as possible. An example how the cascade concept applies in the chemical industry is the implication that the high-grade raw material, in this case natural gas, is not used to heat a physical space directly but is first used several times before being used for that purpose:

1. The natural gas is first used to produce plastics.
2. These plastics are then used in high-grade products.
3. After these products have been disposed of, the plastic is recycled and used in a low-grade product.
4. After this product has been discarded, the plastic is again used as packaging material.
5. This packaging material is then converted, e.g. by feedstock recycling, into low-grade fuel.
6. This fuel is finally used to generate energy.

2.2 Step-by-Step Plan for Developing an End-of-Life System

It is necessary to make a clear choice for the most suitable end-of-life strategy. The design tasks for optimizing life extension, product reuse, component reuse, and material reuse are not the same: They can sometimes even counteract one another. This is especially true when two other main options for ecodesign, the reduction of material and reduction of energy consumption, are simultaneously taken into consideration. Furthermore, a choice on strategic level has to be made, because the implementation of an end-of-life strategy could influence the entire organization.

When thinking about improving a product's end-of-life system, several different aspects need to be taken into account. These aspects are all contained in a so-called step-by-step plan for developing an end-of-life system:

- *Action 1.* Draw up a profile of the product's current end-of-life system (see section 2.3).
- *Action 2.* Analyze the main masons why the users dispose of the product (see section 2.4).

This step-by-step plan is further discussed in the subsequent paragraphs.

In the work of Rose et al. [8, 9] a compact method, the end-of-life design advisor, was developed. This approach uses statistical analysis on significant product characteristics such as wear-out life, design cycle, and number of materials to determine the preferred or current end-of-life strategy.

2.3 Action 1: Draw Up a Profile of the Product's Current End-of-Life System

The development of a product end-of-life strategy entails a fair deal of information collection. First, information should be gathered regarding the product in general.

1. *Who owns the product?* A product can be owned by individual persons or by organizations, offices, and institutions (the so-called professional customers). This question is important since professional customers often purchase products in bulk and dispose of them in bulk as well. They will usually have signed an agreement with the manufacturer, and agreements of this nature are today starting to contain an increasing number of stipulations as to what happens at the end of the product's life. Individual owners do not usually have such an agreement and they usually purchase only one or few products via the retail trade, not directly from the manufacturer.

2. *What kind of ownership applies?* A product can be owned, rented, or leased by the user. If the product is rented or leased, then the user has a different attitude toward aspects such as maintenance, return of the product (in actual fact this is disposal of the rented product), and so on.

3. *What is the price?* Generally speaking, the price of a product gives a good indication for its value on a second-hand market. For example, the second-hand market for cars is huge and varied while the market for color televisions is much smaller and more restricted. Also, when a product's price is lower, more often the price for the second-hand product will be determined on the basis of demand.

4. *How big is the product?* The size of the product has an enormous influence on the return logistics. Products that fit into a bin-liner are put out for refuse collection. It then costs a great deal of expense and effort to have them recovered. Medium-sized products that are too big for a bin-liner can be transported by the consumer. Apart from collection depots, these products often end up in attics or cellars. Large products that are difficult to transport are often given to the dealer (or his/her counterpart) upon the purchase of a new product.

5. *What is the average life of the product?* Products with a short lifetime for the first consumer (e.g., 2–5 years) are more likely to be eligible for high-grade reuse than products with a lifetime of 15 to 20 years. When these products are discarded they are technically obsolete, out of fashion, or use too much energy.

6. *What is the product's weight?* The heavier a product and the more material it contains, the more financially attractive material recycling becomes. This leads to one of the ecological design dilemmas: From the

perspective of material reduction and energy saving it is not preferable to design a product using a large amount of material. The prevention of waste and emissions should have preference over recycling, as recycling is really an end-of-pipe solution.

2.4 Action 2: Analyze the Main Reasons Why the Users Dispose of the Product

When designing an end-of-life system one must be aware of the reasons for the product's disposal. This information is necessary to establish the options for improving the product's performance and thus postpone disposal. The following questions are relevant here.

1. *Is the product disposed of because of specific technical defects?* The product may have a specific component or connection, which is difficult or very costly to repair in the event of a defect; this gives a direct design clue for life-time extension. It could also be a case of wear and tear (i.e., a gradual deterioration of performance) which is much more difficult to counteract.

2. *Is the product sensitive to trends?* Some products are very sensitive to changes in fashion. If a product goes out of fashion, it no longer gives the same satisfaction to the first user and may therefore be discarded. When resold at a lower price, it might be attractive to second users, however.

3. *Are there new products on the market which offer more features?* When products offering more features become available, the consumer soon values the "old" product to be no longer satisfactory.

2.5 Action 3: Determine Which Legislation and Regulations Affect the End-of-Life System

When drawing up end-of-life scenarios it is essential to know which legislation or regulations apply, or will apply, with regard to the product's reuse and disposal. The following information should be collected:

1. To what extent is the manufacturer responsible for the end-of-life phase?
2. Does a take-back obligation already exist for discarded products?
3. How must the costs of returning and processing the product be financed?
4. What rules and prices apply with regard to product reuse, material recycling, and the incineration and dumping of residual waste?

Legislation has been, or soon will be, enforced in several European countries (Denmark, Germany, the Netherlands, Austria, Sweden, and Switzerland). This will give substantial form to the so-called producer responsibility and will extend all the way to the organization of end-of-life systems in the future. The intention of the legislation is that manufacturers/ importers of products will systematically take products back and ensure that they are processed in an ecologically sound fashion. Take-back should preferably not incur any costs for the last user but can for instance be financed through a price increase of new products (the so-called internalizing principle) or adding a fixed recycling fee to the price.

In the countries mentioned above, negotiations are conducted between the government and industry as to how this will be materialized. Given the complexity of this subject matter, plus the financial deficit in most product chains, reaching consensus is difficult anywhere. The Netherlands was one of the first countries to adopt producer responsibility and take-back legislation as of January 1, 1999.*

2.6 Action 4: Contact the Suppliers

Discussion with the suppliers is the next essential step in drawing up end-of-life scenarios. Discussions with suppliers can improve dealing with end-of-life issues in two ways:

- Thanks to their expertise and industrial possibilities, suppliers can, like OEMs, usually achieve subassembly/component/material reuse considerably more efficiently and cheaper than the manufacturer of the product can.
- By reducing or eliminating substances, which are harmful to the environment, in the deliveries of suppliers, the environmental impact and the costs can be reduced in those cases where it is impossible to prevent the generation of waste. Similar aspects have been placed on the agendas of many companies when discussing orders.

---

* As a consequence of the new Dutch legislation on producer responsibility and product take-back, VLEHAN and FIAR (branch organizations for suppliers of white and brown goods) have initiated as of January 1, 1999 a collection system for discarded white and brown goods. The collection system is based on several waste collection streams, stemming from municipals and from shops ("old for new"). A cooperation of manufacturers and importers will take care of all collected appliances, which will be recycled to the greatest possible extent. The system will be financed by a fund under joint supervision of industry and the authorities. Its main financial input is contributions for removal, paid as a surcharge by customers when buying a new product.

2.7 Action 5: Establish How Discarded Products Can Be Collected

The return of discarded products can be organized in three different ways:

- Adopting a system whereby the consumer hands in the product to a recycling station or municipality/county-run center;
- Adopting a system whereby the municipality/county picks up the goods from the last user or by private refuse collection services;
- Adopting a return system via the retailers: the product is traded in for a new product, the consumer or retailer is given a "return premium," or a money deposit system is introduced.

More expensive products like cars are often "traded in" or used to bargain for a discount on the purchase of a new one. These traded in products can usually contribute to the realization of more environmentally friendly end-of-life strategies. Ultimately, all these products end up in the waste circuit where it was last used.

2.8 Action 6: Determine Who Is Going To Recycle or Process the Product

The processing of collected end-of-life products can be done either by the original manufacturing company or it can be contracted out to specialized recycling firms. The in-house processing of discarded products is often given preference. The reason is that the quality of the products, components, or materials can then be safeguarded. These are subsequently brought back into the industrial process after being upgraded. Nevertheless, in-house processing will generally be limited to processing waste and rejects from in-house production processes since virtually all businesses want to stick to their core activity. The choice for a processing firm to process end-of-life products depends on the end-of-life strategy that has been chosen:

- *Reuse*: In the event of product reuse it is important that the manufacturer knows how much expert knowledge is available in the (repairs) firm as the product will bear the manufacturer's name in its second life too, not the name of the firm that has repaired it.
- *Remanufacturing/refurbishing*: If it is a case of remanufacturing or refurbishing (repairing/upgrading a product to a new or almost new condition), then it is important that the processing firm guarantees quality and volume. Also access to secondary markets—often on other continents—plays a significant role.
- *Recycling*: Processing for material reuse sets entirely different requirements. The main concern is the requirements in connection

with available technology and sufficient size of scale. For instance, to process car wrecks effectively a shredder plant is needed with a capacity of 50,000–1,000,000 ton/yr. The high-end engineering plastics without flame retardants used for television cabinets can be recycled profitably on a line, which has a capacity of 2000 ton/yr (this corresponds with 300,000 TVs of the present generation). Ecologically sound and cost-neutral processing of printed circuit boards calls for a processing line with a capacity of 2500 kg/day.

- *Incineration*: If the products will end up in incineration plants, either in whole or in part, then it is important to know what the terms are for acceptance of the product. Inquiries should be made whether the product will be accepted for incineration as domestic refuse or if it will be classified under a special category of waste (involving a higher incineration charge). This is illustrated in Table 1, which shows a typical overview of additional charges to be paid for incineration of waste in a Dutch waste treatment plant. In general, it is advisable to ask various processing firms for quotes for end of-life costs for products that still have to be developed. An incoming quote often leads to a discussion as to how end-of-life costs can be reduced. In turn, this usually generates meaningful hints regarding product designs.

2.9 Action 7: Select the Most Efficient End-of-Life System

The answers to the foregoing questions concerning a specific product can now be used to formulate scenarios for the route the product will follow at the end of its life-time. We recommend establishing a main scenario and drawing up several alternatives with the uncertainty of future developments in mind. This helps determine which design actions are the most significant ones for the end-of-life system. The scenarios will usually not be complete or accurate because knowledge in the field of end-of-life systems is still being developed and not all the relevant knowledge and data are freely available. This does not need to be a drawback; the main thing is to establish the most suitable option.

Several future developments must be kept in mind when optimizing the end-of-life system. In Boks and Tempelman [10] several of the following developments were investigated:

- Users will think twice before they discard products, which means they will keep products as long as they perceive that the products have value.

TABLE 1 Additional Charges for the Composition of Waste in a Dutch Waste Treatment Plant

		Additional charge in connection with the composition of waste (as per 1.1.95, per ton, excl. VAT and eco-taxes)			
Component	Percentage	Additional charge	Component	Percentage	Additional charge
Fluor	0.1–0.5%	$ 53	Mercury	10–20 ppm	$ 53
	0.5–2.0%	$ 103		20–50 ppm	$ 103
	Every 2% extra	$ 103		Every 50 ppm	$ 103
Bromium	0.5–2.0%	$ 53	Chloride	1–2%	$ 24
Silicon	Every 2% extra	$ 103	Zinc	2–5%	$ 53
Sulphur			Lead	5–10%	$ 103
Phosphorus			Copper	Every 5% extra	$ 103
Arsenic			Nickel		
Antimonium			Cobalt		
Silver			Chromium		
Tin			Manganese		
Iodine	500–1000 ppm	$ 53	Vanadium		
Cadmium	Every 1000 ppm extra	$ 103	Tungsten		
			Molybdenum		
PCBs	10–50 ppm	$ 390	Sodium	1–2%	$ 33
	> 50 ppm	$ 1595	Potassium	Ever 1% extra	$ 53

ppm, parts per million.

- Legislative bodies will develop more legislation and regulations. Manufacturer responsibility will be imposed to a greater or lesser degree.
- A more effective processing industry will emerge.
- The technological options available will increase, especially in the field of mechanical processing of wastes.
- The market for recycled materials will improve.
- The incineration and especially dumping (i.e., landfill) of waste will be subjected to more regulations and become more expensive.
- Suppliers will become progressively more involved in integral chain management.
- Environmental impacts and the resulting costs will become easier to quantify.

Given the above developments it is recommended that more than one scenario is drawn up and that wide margins are allowed in corporate decision making. All companies should develop several scenarios and include "extremes." Future developments such as a drastic scaling down (e.g., all discarded products are exported) or the inverse, an increase in scale (a total stop on exports), should also be taken into account. This also holds for the consequences of a drastic changes in costs (incineration and dumping become considerably more expensive, energy becomes more expensive, a slump in the price of secondary materials due to the greater supply, etc.). Scenario analysis such as this facilitates understanding the extent to which the end-of-life design improvements under development will be future proof.

3 STEP-BY-STEP PLAN FOR AN END-OF-LIFE ANALYSIS ON PRODUCT LEVEL

After making a basic choice on a strategic level for a certain end-of-life system (i.e. deciding on one of the options from Figure 1), that system should be developed in greater detail. A step-by-step plan is set out to determine the end-of-life destination on product level.

In Figure 2 a chart is given that can be helpful in determining likely end-of-life destinations of products or product parts. For example, it becomes clear whether or not some parts can be recycled. On the basis of this assessment, the management policy, and the interactions between these two, design guidelines for redesign can be given. The various decision points in the chart will each be examined in the following sections.

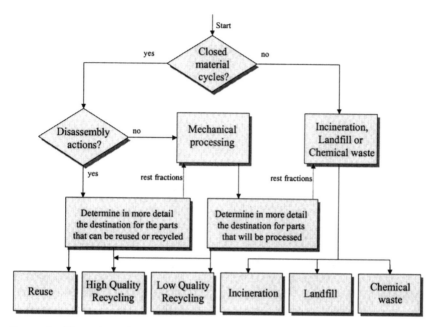

FIGURE 2 Flow chart for determining product end-of-life destinations.

3.1 Closed Material Cycles

The entry to the chart is the strategic choice for a certain end-of-life system as described in section 2. This choice determines whether or not the product will be kept in the loop and therefore roughly what the end-of-life system will be. After this the consequences for the reference product will be looked at.

3.2 Is Product Disassembly Applicable?

Disassembly is done for two reasons. First, it is important to obtain the purest form of secondary materials possible. Second, hazardous substances must be isolated so they do not contaminate the material fractions or exert too great an influence on the environmental results or on the financial return. Disassembly is done by hand and is therefore costly. In Western Europe the total industrial rate is approximately U.S. $0.5 per minute (including wages, overheads, housing). Such rates imply considerable restrictions on the amount of material that can be retained from disassembly. Table 2 shows the minimum amount of material that must be detached per minute on the basis of the proceeds from secondary material and recy-

TABLE 2   Grams of Material That Must Be Separated per Minute

Material	Amount of material in grams to be separated to achieve cost-neutral disassembly	Material	Amount of material in grams to be separated to achieve cost-neutral disassembly
Gold	0.055	PEE	250
Palladium	0.035	PC, PM	350
Silver	3.1	ABS	800
		PS	1000
Copper	300	PVC	4000
Aluminium	700		
Iron	50,000	Glass	6000

cling costs. These figures are based on the so-called full industrial rate. If the work is carried out within the framework of subsidized job-creation schemes or outside the developed world, then the limits are naturally based on a lower material weight. This is also the case if there is consensus in society that recycling must go much further than being based on principles of a market economy (mandatory recycling quota of product categories).

Given the possibility that disassembly in the future might become a partly automated process and that the options for processing elsewhere in the world will become more liberal, the "limits" indicated in Table 2 should not be interpreted strictly. When establishing the end-of-life strategy, wider margins will need to be taken into account than those shown in Table 2.

In terms of semiprecious metals (copper, aluminium) and plastics, only those components which contain 300 to 1000 g of the relevant material are eligible on the basis of these economic data. This is very important: Design attention will need to be given to components weighing between 300 and 1000 g if these products are to be constructed in such a way that they will be easy to disassemble.

To estimate the probable disassembly time of a product, it is advisable to establish so-called standard times by carrying out disassembly tests. These should relate not only to the dismantling activity itself but also to so-called secondary activities such as preparing the product for disassembly, gathering the tools needed for the job in hand, turning or moving the product, removing materials, and breaks. In Table 3 a number of standard disassembly times are given. For more information on disassembly modeling, see Boks et al. [11].

For a company it is economically attractive to disassemble products by hand when the disassembly costs of useful parts (for reuse or recycling) are smaller than the end-of-life costs. These end-of-life costs are determined by

TABLE 3 Examples of Average Standard Disassembly Times

Disassembly operation	Time required (s)
Screw (instantly accessible)	6.5
Screw (obstructed access)	10.5
Click (instantly accessible)	3.5
Click (obstructed access)	7.5
Glued joint	12.0
Nut/bolt combination	11.5
Welded joint (per point)	7.5
Welded joint (per surface)	18.5
Slide	3.0

the revenues of the recycling of materials or the reuse of parts minus the costs made for the removal of the remaining parts. The decision whether or not to disassemble will be taken on the basis of a rough estimation of the entire product. When the choice is made to disassemble the product, end-of-life destinations can be determined in more detail.

3.3 Determine in More Detail Which Parts Can Be Reused or Recycled

Parts can be reused when the technical life of the product is longer than the economical life. Furthermore, it is important that there is a market for these parts. On the one hand one has to look at the current market, whereas on the other hand it should be estimated if there are possibilities for future markets. In most cases a part cannot be reused when it fulfills an aesthetic function.

High quality material recycling can be achieved when parts are made of monomaterial and did not get a surface treatment. When parts are not made of monomaterial, 80% of all precious metals can be recycled with high quality and 75% of aluminium and ferrous can be recycled with low quality. Also, 80% of all thermoplastics (if they consist of no more than two types, are 90% of the fraction, and have no pollution) can be recycled with low quality. The leftover fraction and the complex parts that do not meet the requirements can be considered for mechanical processing.

3.4 Consider Whether the Product Is Fit for Mechanical Processing

Material not detached in the disassembly process usually ends up in a combined form on so-called mechanical processing lines. This processing consists in actual fact of two steps: shredding and separation. Shredding is essential to break down the different sorts of material and achieve a reduction in volume. Shredding occurs in hammer mills, cutting mills, or by means of cryogenic milling. Separation is achieved in several different separation stages. First, ferrous metals can be separated by magnetic separation. After magnetic separation, often an additional separation process is necessary because for shredded motors, coils, or trafos complete separation cannot be achieved using magnetic separation only. This is because copper parts remain attached to iron parts in the shredding process. Subsequently, aluminum is separated by means of eddy-current processes. Eddy current processes separate materials, which have different conductive properties using an alternating magnetic field.

A variety of other separation methods are then used to separate other fractions. These can be copper-content materials, mixed plastics, and precious metals. Dry mechanical separation techniques are wind sifting (used for separation of nonferrous metals and plastics), gas cyclone separation, vibration methods, fluid beds, and sieves (air-based or other). The advantage of these techniques, above wet mechanical separation techniques, like hydrocyclone separation, flotation, and sink/flow separation, is that there is then no problem of wastewater (although dust can also be problematic). For more information on separation techniques, see previously published references.

The separation abilities of these plants have been improved enormously over the past few years through the incorporation of return-flow circuits. Nevertheless, there is (still) no universal mechanical processing plant able to handle a wide range of products. Consequently, all plants are in one way or another specific to a certain product category. An example of a processing line is shown in Figure 3.

3.5 Determine in More Detail Which Parts Will Be Suitable for High and Low Quality Recycling

Of product parts containing ferrous materials, 95% of the magnetic material can be made suitable for high quality recycling by means of magnetic separation processes. High quality recycling of other plastic parts can be achieved when the material consists of more than 95% of thermoplastics of one kind. When this 95% consists of two materials, the part can be recycled with low quality. Parts which contain less than 95% of thermoplastics can be considered for other separation techniques. Of these parts 60% of the precious metals can be separated and recycled with high quality; 80% of the aluminium and 90% of the ferrous can be recycled with low quality. Other

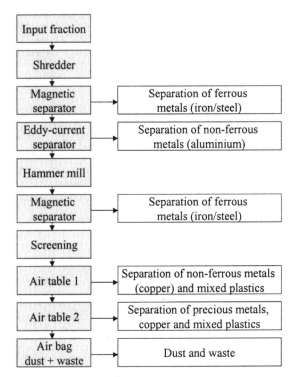

FIGURE 3   A processing line for miniaturization and separation of consumer electronics.

plastic parts and the rest fraction can be considered for incineration and landfill.

3.6   Will the Parts Be Incinerated, Dumped, or End Up As Chemical Waste?

When a choice between incineration and landfill has to be made, the caloric value is considered. High caloric value materials produce energy during incineration, and therefore this is preferable to landfill. Metals produce 0 MJ/kg while plastics produce 40 MJ/kg. Therefore, the best solution for plastic parts, if they cannot be recycled, is incineration. One should roughly calculate this value for the materials. When the caloric value is higher than approximately 8 MJ, incineration is preferred.

  If the product is not stripped of its toxic parts, the entire product has to be treated as chemical waste and therefore end up in higher waste incineration tariffs. Separation of parts will be attractive when the costs of dismantling and processing of toxic parts minus the economic value of the remaining nontoxic materials are lower than the costs of processing the entire product (see also Step 2: Motives for Disassembly).

3.7 Results of the Flow Chart

Once we have looked in detail at the consequences (on the level of product parts) of strategic decisions in the field of take-back, reuse, and recycling, feedback to this strategic level can be achieved. The results from the chart, in particular the feasibility of disassembly and mechanical processing, can be of influence on the decision whether or not to close material loops. When this feedback has taken place, priorities within the end-of-life system can be set. Subsequently, the product has to be optimized according to the chosen end-of-life system. Guidelines on how to do this are described in the next section.

4 OPTIMIZE THE PRODUCT ACCORDING TO THE END-OF-LIFE SYSTEM

In the following paragraphs special attention is given to the recycling of a number of specific materials, toxic materials, metals, plastics, glass, and ceramics.

4.1 Design Guidelines for Recycling Metals

Metal waste and high metal-content fractions can generally be recycled in metal forges. The presence of other metals is not so much an obstacle in the way of processing but can cause the fraction to be rejected, or at least it will influence the value of the fraction negatively. Table 4 shows the compatibility matrix for metals and for a number of substances what kind of pollution is a limit for metal refinery. Incompatible metals in scrap can—as long as they are adulterated with other more pure scrap—be included in the process. However, this will result in a substantially lower yield price for any batch treated that way. Copper fractions are often heavily polluted with mixed plastics. In itself this does not have to be troublesome since the plastics will burn during the melting process and even contribute toward the caloric value. However, it is essential to verify that the copper smelter in which the material will end up is fitted with a good flue gas purification unit to ensure that the flue gases released from burning the flame retardants in the plastics will not pose a problem.

TABLE 4   Compatibility of Metals

Target material	Penalty elements	Metal refinery limit
Copper	Mercury	50 ppm
	Bismuth	300 ppm
	Fluorine	150–300 ppm
	Cadmium	1000 ppm
	Chlorine	1000 ppm
	Antimony	2000 ppm
	Nickel	5000 ppm
	Nickel + cobalt	5000–1000 ppm
	Zinc	5%
	Aluminium	25%
Magnetic iron	Zinc	2000 ppm
	Copper	5000 ppm
	Tin	5000 ppm
Aluminium	Copper	5%

4.2 Design Guidelines for Recycling Plastics

As we can see from the compatibility matrix for thermoplastics in Table 5, very few plastics are compatible with each other. In practice we see that virtually only those plastics taken from the disassembly process are pure enough to be eligible for material recycling; even then very strict design conditions must be met. Additives such as fillers, fibreglass reinforcement, and flame retardants make plastics recycling virtually impossible.

Plastics resulting from mechanical processes are usually too mixed and contain too many pollutants (including flame retardants). This kind of fraction automatically ends up in incinerators. Recycling of so-called thermoset plastics can be problematic as well. Recycling by means of dissolving/leaching material would seem plausible but is as yet unavailable on an industrial scale. For the time being incineration is the only option. Over the past few years other processes have been under development to make better use of plastics waste than the recovery of energy alone by means of incineration:

- The use of plastic waste as a reducing agent in blast furnaces (substitute for coke).
- The previously referred to tertiary recycling (or feedstock recycling):
  - Pyrolysis: The decomposition of molecules by heating them in a vacuum whereby the energy required is derived from the plastic

TABLE 5 Compatibility Matrix for Plastics (From VDI)

	PS	SAN	ABS	PA	PC	PMMA	PVC	PP	PE, LD/HD	PET	Thermosets
PS	+	−	−	−	−	0	−	−	−	−	−
SAN	−	+	+	−	+	+	+	−	−	−	−
ABS	−	+	+	−	+	+	0	−	−	−	−
PA	−	−	−	+	−	−	−	−	−	+	−
PC	−	+	+	−	+	+	+	−	+	−	−
PMMA	0	+	+	−	+	+	+	−	−	+	−
PVC	−	+	0	−	+	+		+	−	−	−
PP	−	−	−	−	−	−	+		−	−	−
PE, LD/HD	−	−	−	−	+	−	−	−		+	−
PET	−	−	−	+	−	+	−	−	+		−
Thermosets	−	−	−	−	−	−	−	−	−	−	−/0/+

itself. This leads to gaseous or liquid hydrocarbons, which can then be processed further in refineries as raw material.

- Hydrogenation: By heating plastics in the presence of hydrogen the polymer chains are decomposed into oil which can also be processed further as raw material for use in the petrochemical industry.
- Gasification: Plastics are heated in the presence of air and oxygen. The synthesis gas thus formed, consisting of carbon monoxide and hydrogen, can then be used for the manufacture of methanol or ammonia. It can even be used as a reducing reagent for the manufacture of steel in blast furnaces.
- Chemolysis: By implementing solvolytic processes, such as hydrolysis, alcoholysis, or glucolysis, polyesters, polyurethanes, and polyamides can be decomposed to their monomers. The original plastics can then be remanufactured by means of polymerization.
- Partial incineration and partial conversion into organic compounds ("Schwelbrennverfahren" or the Siemens process).

Generally speaking, these processes are characterized by their large scale (needing a large amount of material) while it is still under discussion where these processes should be positioned really high on the list of environmental priorities. Nor have the ultimate costs been established yet.

4.3 Design Guidelines for Recycling Glass and Ceramics

There are also compatibility problems with glass (Table 6). Engineering glass from cathode ray tubes and LCD monitors cannot be recycled in combination with bottle glass or window glass and can consequently be recycled in the manufacturer's specific glass smelting furnaces only. Considering that engineering glass has to meet extremely stringent requirements, recycling is as yet problematic; however, it has started (i.e., Mirec, the Netherlands, and Envirocycle of Pennsylvania).

Glass and ceramics are mutually incompatible; most ceramics are also mutually incompatible. With a view to the production method (i.e., powder sintering), ceramics recycling is out of the question in practice.

Because of the problematic recycling options, other methods are being sought for the reuse of glass and ceramics. These are usually low-grade applications, which also are not without their environmental drawbacks, such as

- The use of glass and ceramics as an additive for building materials and for road construction (possibly leaching in the long run);

TABLE 6 Compatibility of Glass and Ceramics

	Bottle glass	Window glass	Drinking glass	TV (screen)	TV (cone)	TV (neck)	LCD (screen)	Ceramics
Bottle glass	+	–	–	–	–	–	–	–
Window glass	+	+	+	–	–	–	–	–
Drinking glass	+	0	+	–	–	–	–	–
TV (screen)	0	0	–	+	0	–	–	–
TV cone)	–	–	–	–	+	+	–	–
TV (neck)	–	–	–	–	+	+	+	–
LCD (screen)	0	0	–	0	–	–	–	–
Ceramics	–	–	–	–	–	–	–	–/0

- The use of lead-content glass as an additive in the lead ore smelting process (also produces slag material);
- Use as raw material for low-grade applications (e.g., foam glass/ glass wool).

The drawback here is that, as well as glass, oxides like lead oxide and barium oxide are introduced into the material flows or the physical environment. Here, the leaching properties are more detrimental than those of oxides in pure glass.

4.4 Recycling Toxic Materials

When dismantling products containing toxic materials, preference should always be given to the option of preventing waste and emissions. This means designing the product in a way that it contains such materials, components and subassemblies that make disassembly on grounds of "chemical" content not necessary anymore. If the prevention option is not feasible, then environmentally harmful substances, such as those used in components, must end up in a metal fraction which is processed by slag treatment and/or flue gas purification. This makes it possible to "control" the amount of toxicity. Although it can result in a higher cost for the fraction in question, it can still be lower than the cost involved in separate disassembly and further processing of the concentrated fraction.

4.5 Safe Incineration

The incineration of materials (plastics, wood, cardboard, paper) is generally lower on the list of environmental priorities than product, component, and material reuse. One major exception is the destruction of environmental pollutants during incineration (i.e., the destruction of flame retardants).

Incineration is usually not complete, and a residual product is left behind (slag). Metal can sometimes be recovered from this slag. How this residual slag is treated in an ecologically responsible manner is one of the most important cost factors for determining the incineration rate to be charged. This applies to an even greater extent for the purification of flue gases released during incineration, especially aggressive chloride and bromide-based gases. Combined with organic residues, these compounds are the basis for dioxin production. This, thanks to the technological improvements and more careful processing in modern incinerators, is now more nearly under control and flue gases today meet the extremely strict standards that have been set. Obviously, the highest rates are charged for those waste categories which demand the most intensive flue gas purification processes (Figure 1).

4.6 Design Guidelines for Disposal of Waste

Waste disposal by landfill in the United States and Western and Northern Europe is today subjected to strict control. Risks are being eliminated to quite a considerable extent through rigid acceptance policies. One important aspect is that in densely populated countries like Germany and the Netherlands, and in several parts of the United States as well, a shortage of landfill capacity is emerging. The costs, which are currently less than those for incineration, will increase considerably in the near future and are expected (for domestic refuse for instance) to overtake in the end those for incineration. It must be taken into account that the large-scale burial of waste will, in the next 5 to 10 years, no longer be possible in large parts of Europe and the United States. This is the basic reason why government policy focuses not only on reuse but particularly on reducing the amount of waste to be buried.

5 QUANTITATIVE UNDERPINNING TO OPTIMIZE THE END-OF LIFE SYSTEM

By following the priorities discussed in section 2 and by elaborating on them by implementing the steps in sections 3 and 4, in most cases a sound picture of the realistic end-of-life strategy shall be obtained. This can be underpinned in the quantitative sense by adopting different methods: a life cycle analysis, an analysis of the life cycle costs, and/or an analysis of the end-of-life costs.

5.1 Life Cycle Assessment

Although literature describing life cycle assessment (LCA) methods is abundant, unfortunately the LCA methodology has as yet been unable to generate complete environmental profiles for end-of-life problems. This is primarily due to a problem of a methodological nature. What are the system boundaries for recycling? Second, it is not yet clear how we should deal with embedded toxicity. How should toxicity, which is released into the environment, for example, through leaching from buried waste, be included and on which time scale (e.g., 1, 10, 100, 1000, 1 million years)? Third, it is not clear how "emissions" formed by the physical presence of a rubbish mountain (which does not fit in any landscape) should be included. A fourth problem is that reliable data on the environmental impact of collection systems, separation systems, and recycling processes are still inadequate.

In Table 7 very approximate figures are given for the reduced impact of recycled materials. The figures are based on ecoindicator values for the environmental impact of materials [16]. Transport energy for collection is

TABLE 7   Environmental Impact of Recycled Materials Relative to That of Primary Materials

Secondary materials	Environmental impact as fraction of impact of primary materials
100% recycled glass	0.8
100% recycled iron	0.4
100% recycled plastic/paper/cardboard	0.4
100% recycled copper	0.25
100% recycled aluminium aluminium	0.1

included. It is evident that the use of recycled material has a beneficial effect on the environment.

5.2   End-of-Life Costs

When making life cycle cost calculations it will appear that the end-of-life costs make up for only a small part (1–7%) of the total costs. The first suggestion of this, that the end-of-life costs are insignificant, is incorrect. This is because one must not only look at the absolute size of the amounts but particularly at the degree to which they can be influenced. There is a very big difference here in comparison with the other items of expenditure: Whereas ways to reduce the costs of production and transport, for example, have always been sought (and indeed many options to achieve this have already been found), end-of-life expenditure has only been focused on in recent years.

It is important to calculate the end-of-life costs on the basis of price quotations from service providers in the field of logistics and recycling/processing. The company itself must also build up an insight into the end-of-life costs. Experience has taught that this generates a great deal of understanding of the subject and stimulates environmental design improvements.

Data must be gathered on rates charged for waste and incineration while the value of the secondary material must also be known. For metals, scrap prices are related to the prices quoted on the London Metal Exchange. For plastics the value of high-grade secondary material is approximately 60–70% that of the new price. Furthermore, it is helpful to gain knowledge on recovery grades for applicable end-of-life processes. How data such as these can be used to understand in more detail the correlation between design and end-of-life costs is explained in for example Boks et al. [17] in which also the use of product end-of-life scenario analysis is explained.

REFERENCES

1. Second Draft: Proposal for a Directive on Waste from Electrical and Electronical Equipment. European Commission Directorate General XI: Brussels, 7 July 1998.

2. Draft proposal on take-back of white and brown goods (Ontwerpbesluit verwijdering wit- en bruingoed, concept). Staatscourant 5.8, 24.03.1998 [in Dutch].

3. Regeringens Proposition: Hantering av uttjänta varor i ett ekologisk hållbart samhälle—ett ansvar för alla. Proposition 1996/97:172 [in Swedish].

4. Ordinance on avoidance, reduction and processing of waste from used electrical and electronical equipment (Verordnung über die Vermeidung, Verringerung und Verwertung von Abfällen gebrauchter elektrischer und Elektronischer Geräte (Elektronik-Schrott-Verordnung—ESV)). Draft from 11.07.1991 [in German].

5. Nagel, C., Nilsson, J., and Boks, C. European End-of-Life Systems for Electrical and Electronics Equipment. Ecodesign'99, February 1999, Tokyo.

6. Klausner, M., and Grimm, W.M. (1999) Integrating product take-back and technical servies. In the Proceedings of the IEEE International Symposium on Electronics and the Environment, Danvers MA, May 11–13, 1999.

7. Stevels, A.L.N., Ram, A.A.P, and Deckers, E. Take-back of discarded consumer electronic products from the perspective of the producer—conditions for success. Journal of Cleaner Production 7:383–389, 1999.

8. Rose, C. M. and Ishii, K. Product End-of-Life Strategy Categorization Design Tool. Journal of Electronics Manufacturing [Special Issue on electronic product reuse, remanufacturing, disassembly and recycling strategies], 9:41–51, 1999.

9. Rose, C. M., Beiter, K. A., and Ishii, K. Determining of End-of-Life Strategies as a part of Product Definition. 1999 IEEE International Symposium for Electronics and the Environment Conference, Danvers, MA, May 1999, ISBN 0-7803-5495-8, pp. 219–224.

10. Boks, C.B. and Tempelman, E. Future Disassembly and Recycling Technology. Futures 30:425–442, 1998.

11. Boks, C.B., Kroll, E., Brouwers, W.C.J., and Stevels, A.L.N. Disassembly Modeling: Two Applications to a 21" Philips Television Set. In the Proceedings of the 1995 IEEE International Symposium on Electronics and the Environment, May 1996, Dallas, Texas.

12. Nijkerk, A.A. Handbook of Recycling Techniques. N0EY353293/0710, Dutch National Research Programme for Recycling of Waste Substances (NOH/NOVEM), Den Haag, NL, ISBN 90-9007664-6, 1995.

13. Arola, D.F., Allen, L.E. and Biddle, M.B. Evaluation of mechanical recycling options for electronic equipment. In the proceedings of the IEEE International Symposium on Electronics and the Environment, Danvers, May 11–13, 1999.

14. Zhang, S., and Forssberg, E. Metals recycling from electronic scrap by physical separation. Proceedings of Care Innovation'98, November 1998, Vienna, Austria.

15. Ram, A.A.P., Deckers, J.M.H., and Stevels, A.L.N. Recyclability of Consumer Electronics—Design for Non Disassembly. Proceedings of Care Innovation'98, November 1998, Vienna, Austria.
16. Goedkoop, M. De Eco-indicator 95, Randleiding voor ontwerpers (The Eco-indicator 95; manual for product developers), NOH report 95 10, Utrecht, NL, ISBN 90-72130-78-2 [also available in English], 1995.
17. Boks, C.B., Stevels, A.L.N. and Ram, A.A.P. Take-back and recycling of brown goods. Disassembly or shredding and separation? In the proceedings of the 6th International Seminar on Life Cycle Engineering, Kingston, Canada, June 1999.

24

Integration of Ecodesign into Business

Ab Stevels

Philips Consumer Electronics, Eindhoven, The Netherlands

1 INTRODUCTION

The need to achieve sustainable development poses an enormous challenge for society. This chapter highlights the way in which companies can contribute toward this goal. Basically, this is through environmentally friendly design (ecodesign, sometimes also called design for environment [DFE]). This seems to be a technical issue in the first place. In this chapter it is explained, however, that ecodesign needs to be integrated with all other business processes ("integration") to have real effect.

The Brundtland Commission report [1] brought the concept of sustainability to the attention of a wide audience. In subsequent years various actors became more and more involved (Figure 1). Although business had started to implement environmental care in production processes already several decades earlier, the attention to products and their ecodesign have evolved only recently (1992–1995).

First, activities were chiefly of defensive nature, that is, directed toward compliance with legislation and toward preventing bad image in the press. Emphasis was therefore on eliminating banned substances, cleaner production, recycling of packaging, and power management for standby

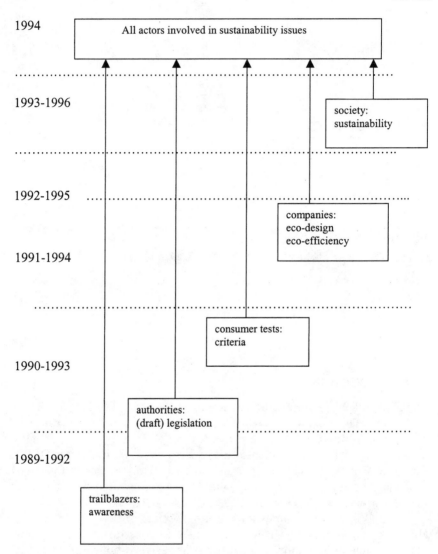

FIGURE 1 Involvement of actors in sustainability/ecodesign issues.

mode. Shortly after, it was discovered that many ecodesign activities can lead to direct cost reductions, for instance:

- Reduction of weight and volume of packaging (lower material and transport cost);
- Use of recycled material (lower material cost);
- Reduction of material use, electronic miniaturization (low material and component cost);
- Reduction of disassembly time (lower end-of-life cost, lower assembly cost).

Most recently, the proactive approach is gaining momentum. The idea behind this is to increase sales by bringing to the market products having both a good environmental performance ("societal benefit") and a good performance to the individual customer ("customer benefit"). This approach makes that these "green AND" products are not only attractive for buyers with environmental awareness but also for those having a more critical attitude toward green issues.

The processes underlying the integration of defensive, cost-oriented, and proactive ecodesign into business are identical. The idea generation for ecodesign ("green options") asks

- Where do we stand? Environmental product analysis and green supplier assessment;
- Where do we want to go? Brainstorms on future concepts.

The exploitation of ecodesign results asks

- What are we going to tell? Environmental communications;
- How are we going to sell? Environmental marketing and sales.

Before discussing these items in detail, first the strategic dimensions of integration of ecodesign (DFE) into business is discussed.

2 ENVIRONMENT AS PART OF VISION, POLICIES, AND STRATEGIC ANALYSIS

Enhancing a business by appropriately addressing environmental issues requires that these should be addressed at all levels of the operations. It is of paramount importance (see also Chapter 19) that environment is also addressed at the companies strategy level. Figure 2 gives a schematic overview of the processes involved and shows that the making of environmental strategies and roadmaps is rather adding environmental chapters to existing ones than something essentially new. Below, the key concepts of Figure 2 are explained in more detail.

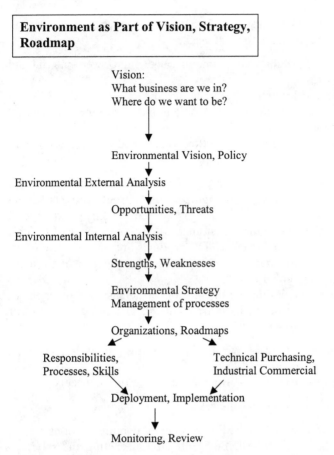

Environment as Part of Vision, Strategy, Roadmap

Vision:
What business are we in?
Where do we want to be?

Environmental Vision, Policy

Environmental External Analysis

Opportunities, Threats

Environmental Internal Analysis

Strengths, Weaknesses

Environmental Strategy
Management of processes

Organizations, Roadmaps

Responsibilities,   Technical Purchasing,
Processes, Skills   Industrial Commercial

Deployment, Implementation

Monitoring, Review

FIGURE 2 Environment as part of vision, strategy, and roadmap.

2.1 Vision

An environmental *vision* is to be formulated by the president and CEO of a company. It should define the scope of the business and overall environmental goal to be reached in the future. The vision should make clear to the employees and to external stakeholders where the company stands and where it wants to go in environmental terms. An example of an environmental vision is given in Table 1. Key words in this vision are as follows:

- *Leading.*   This introduces an accountability element that is evaluation of the company performance in comparison of that of competitors.

TABLE 1 Environmental Vision

Philips shall be the leading ecoefficient company in the lighting and
electronics industry.

Cor Boonstra, President and CEO, Royal Philips Electronics

- *Ecoefficient.* Containing ecological and economic improvement.

2.2 Policy

An environmental *policy* defines the principles based on which environmental programs and activities should be operated. An example of an environmental policy is given in Table 2. Key words in this policy are as follows:

- *Development.* All activities should make the company move into
  more sustainable direction.
- *Prevention.* Clear preference to address environmental issues up
  front (at the start of product/process development) rather than at
  "end of life" (remediation at end of life cycle/process).
- *Total effect.* This introduces the life cycle principle in all considerations. This means that environmental gains achieved for specific
  environmental categories (e.g., energy, material, recyclability, toxicity) should be checked for possible losses in other departments.
- *Open contact.* Companies have also a societal responsibility and
  should as such contribute with their expertise and their data to
  sustainability of the society as a whole.

2.3 Internal and External Analysis: Environmental Trends

The environmental external and internal analysis can be done in complete
similarity with SWOT analysis as usual in strategy making. The external

TABLE 2 Environmental Policy

- Sustainable development
- Prevention is better than cure
- The total effect on the environment counts
- Open contact with authorities

analysis has to be executed both with reference to the current situation and the future developments. The important items are as follows:

- Customers, market developments
- Legislation, regulation
- Competition
- Technological developments
- Availability of skills and techniques

Environmental trends with customers and in markets are summarized in Table 3. The main drivers behind these trends are that customers have developed from passive to active stakeholders:

- More awareness that environmental costs have to be directly and indirectly paid by users
- Still increasing emotions/political fallout about environmental disasters

Main developments in legislation/regulation are given in Table 4. An important issue to be analyzed in this category is differences in priorities in various parts of the globe. This is both relevant for style of operation (legislation vs. voluntary programs) and priority in items addressed (energy, chemicals/substances, take-back) and way of enforcement. This is particularly relevant because products of today become more and more "global" products. Fragmented legislation/regulation might seriously jeopardize developing optimal green products.

Competition items are given in Table 5. The ongoing trends in this table clearly show that in environmental issues the proactive approach is gaining momentum with respect to a compliance oriented ("defensive")

TABLE 3   Environmental Trends: Consumers

- Awareness consumer
- Increasing
- Diverging interest
- OEM requirements
- Public sector purchasing requirements
- Increasing green labeling
- Energy consumption
- Labeling schemes
- Blue Angel
- TCO '95
- Market is becoming the driver instead of regulation

TABLE 4 Environmental Trends: Legislation/Regulation

- Legal/regulation requirements
- Differences per country (priority)
- Voluntary programs
- Self-declaration
- Restriction on use of chemicals/substances
- Energy declaration
- Take-back
- Limitation on landfill, incineration
- "Green" taxation, levies on energy, and material use
- Enforcement tightened

approach. Market forces more and more drive the environmental improvements and take over from legislation. Essentially, regulations is "leveling the playing field" and offers therefore no opportunity for competitive advantage.

Technological trends are described in Table 6. Most (but not all) of the trends shown here are good for the environment. Per item the ecological effects of products in which modern technology are applied are mostly positive (less energy consumption, less resources use). However, if the number of products used increases steeply because they are cheaper, easier to operate, or offer more functionality, these effects can reverse in to negative ones. Similarly, ecoefficient transport can lead under circumstances to more transport and ecoefficient offices can lead to an expansion rather than to a reduction of office space. Moreover, a lot of high-tech products are operating in systems and are no stand alone. This brings up the issue that the impact of individual products on systems level performance should be checked as well.

Table 7 shows trends in availability of skills, techniques, and services. From this table it can be concluded that environmental care in products is becoming more and more nature. Some 5 years ago most companies had to

TABLE 5 Environmental Trends: Competitor Issues

- More and more companies develop environmental strategies, roadmaps, programs
- Competition is using green in marketing
- Environmental management and environmental systems (ISO 14000 and others) develop rapidly from a qualiter to a standard
- Competition is using green in market
- Asian pick up green issues on a fast track

TABLE 6 Environmental Trends: Technological Developments

- Technology offers huge environmental opportunities
  Miniaturization
  Digitalization
  Software
  Portability
- Function integration
- Ecoefficient offices
- Ecoefficient transport
- Global village (information exchange

rely on internal resources to make progress in green. Now many skills, techniques, and services are available from the marketplace or through the internet.

3 ENVIRONMENTAL STRATEGY, ROADMAP

An environmental *strategy* formulates all the actions to be taken to realize the vision in a given period of time; the results of the external and internal analysis are taken into account. Strategy formulation starts with a thorough analysis of the current environmental position (see also section 4.1.1) of the product line up, in technical, industrial, and commercial terms. Reference of this positioning are comparable competitors products. Once the question "where do we stand in environmental terms" has been answered, the directions go on to be formulated on various levels.

- *Corporate level.* Organization, responsibilities; programs, roadmaps; and monitoring, review of performance.
- *Division level ("ecobusiness/improvement").* Green product program; green/validation and communication; ISO certification; and monitoring, review of performance.

TABLE 7 Environmental Trends: Availability of Skills, Techniques, and Services

- Well-educated environmental specialists are available in the labor market
- More availability of design manuals, software
- Use of life cycle analysis and life cycle cost techniques is increasing
- More consultancy bureaus offer their services, quality of service is increasing
- Recycling companies offer their services, ecoefficiency of their operations is increasing

- *Business unit level ("green product management").* Application of green design manual; environmental benchmarking; formulation of green options and consolidating them in product concepts, technical targets (i.e., energy consumption, materials application, packaging/transportation, environmental relevant substances, and durability, recycling, disposal); industrial targets (i.e., supplier evaluation, utilities reduction, reduction of auxiliaries), and monitoring, review of performance.

Once the strategy is in place, formulation of roadmaps is straightforward. An example of roadmaps for the corporate and the divisional level is given in Table 8. As can be seen in this table, each roadmap item should have a target, an owner, and a time planning with milestones.

TABLE 8 Example of Roadmaps

		Environmental Strategy/Corporate Level		
Issue	Target	Owner	Time planning (milestones)	Remarks
1.1 Organization				
1.2 Ecoprogram				
1.3 Roadmaps (corporate)				
1.4 Performance indicators				
1.5 SWOT analysis update				
1.6 Create new strategies				

		Product Strategy/Business Unit Level		
Issue	Target	Owner	Time planning (milestones)	Remarks
2.1 Green flagships				
2.2 Percentage of ecodesigned products				
2.2 Instruments/validation Ecoindicator Other				
2.3 ISO 14001 certification for nonmanufacturing units (e.g., sales)				
2.4 Monitoring ecoprogram				

4 MANAGEMENT OF (ECO) PRODUCT CREATION PROCESSES

Ecodesign of products has had it roots in academia. It is therefore strongly based on science based methods like life cycle analysis (LCA). LCA has in the first instance a holistic approach, that is, it wants to take into account all environmental effects irrespective of their position in the life cycle or their origin. This is in contrast with the way in which ecodesign in companies has been developing. Today focus is primarily on working on environmental issues that can be influenced inside the company itself (the "internal" issues only). This difference in ecodesign in the initial steps is summarized in Table 9. In the present chapter the industry approach is followed

4.1 Embedding Ecodesign of Products in Business

The basic scheme to embed ecodesign of products in business is given in Figure 3. The processes to embed ecodesign into business are on three levels:

- A mainstream level consisting of green idea generation, product creation, and green communication/sales. These items are addressed below.
- A strategy level.
- The level of supporting tools.

The key element in the processes is the linking of the three phases in the mainstream. In the green idea generation stages, focus is exclusively on environment (Figure 4). To be fitted into a standard product creation process, these have to be assessed in business terms—that is, in terms of company,

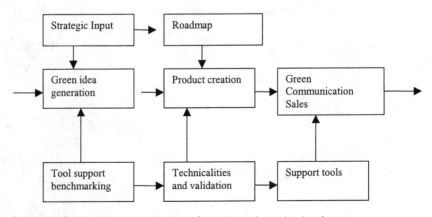

FIGURE 3   Embedding ecodesign of products into the business.

TABLE 9 Industry and Academia Approaches in the Initial Ecodesign Phase

	Step 1	Step 2	Step 3	Step 4	Step 5
Industry approach	Start with creative approach to environmental issues you can influence (benchmark, brainstorm).	Validate and prioritize according to ecodesign matrix (see Figure 4).	Check prioritized options against company, customer, and society benefits.	Check feasibility (physical, financial).	Implement in product creation.
Academia approach	Do LCA analysis, holistic approach.	Select internal and external improvement options.	Start stakeholder discussion.	Come to solutions.	Implement in production creation.

customer, and societal benefits and in terms of technical and financial feasibility. This is done by the so-called ecodesign matrix (Figure 4).

A next crucial step is the transition from product creation to green communication/sales. This is done according to the ecocommunication matrix, which shows many similarities with the ecodesign matrix. In fact, both have been set up in such a way that the same language can be used throughout the whole process.

4.1.1 Green Idea Generation

The basic scheme for the green idea generation process is given in Figure 5. Four elements can be distinguished:

- The input stages (benchmarking, supplier information)
- Brainstorm
- Green option classification
- Consolidation into product concepts

A generic environmental *benchmark method* has been developed by Delft University of Technology in the Netherlands [2]. The following steps have to be taken:

1. Definition of starting points and goals
2. Product system boundaries, functionality description and analysis, input–output diagram (input consumables, delivered functionality and waste)

Green options	Environmental benefit	Business Benefit	Customer Benefit	Societal Benefit	Feasibility Technical /financial
First option					
Second option					
Third option					
. . . .					
Nth option					

FIGURE 4   The ecodesign matrix.

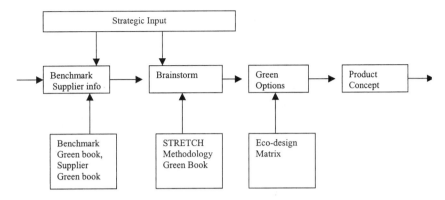

FIGURE 5   Basic scheme for the green idea generation process.

3. Energy analysis/measurement (energy sources, efficiency)
4. Embodiment: definition and analysis/measurement of physical product (including packaging), durability
5. Assembly/disassembly, end-of-life scenarios
6. Evaluation based on fact sheets generated in steps 2–5 and analysis of why facts are as they are, where do the differences come from
7. Generation of green options

A crucial element in green idea generation is that a comparison is made between the company's products and competitors products. To make a proper comparison, a product with the same or approximately the same functionality should be taken into account. An example of how benchmarking is carried out in practice is given in Chapter 19.

Also at Delft University of Technology a self audit tool for *suppliers* has been made available on the internet (www.tudelft.nl/research/dfs/ecoquest) to facilitate the distribution and availability throughout the world [3]. Ecoquest is organized into two parts, the Ecoquest site and the Ecoquest software tool (down loadable from the site). The site gives an introduction to ecodesign and to the industrial and environmental strategies of electronics companies. It also gives the environmental links for further information. The Ecoquest software tool requires the users to make a data sheet on their products and/or services. Both in the subsequent Product Review and Management Review parts, questions have to be answered. With help of the score obtained, a first roadmap for environmental improvement can be generated. Tests have shown that it takes the average supplier not more than 30 minutes to come to a meaningful result.

A useful method to do environmental *brainstorming* is the STRETCH (Strategic Environmental Challenge) [4]. The starting point is the identification of crucial during forces in the business. Basically this is identical to the process of general strategy making. In a second step, focus is put the environmental strategy as formulated according to sections 2 and 3. As a next step, environmental opportunities are brought in. These belong to two categories (Table 10). Facts to support environmental improvements in category 1A are provided by the supplier information and by the benchmarks. The category 1B and 2 elements are of a different nature; here the very structure and of the products business is to considered.

In the actual brainstorm it is essential that in the beginning the participants focus exclusively on the ideas how to realize big *environmental* gain. Experience has shown that bringing in too early elements of the ecodesign matrix (Figure 5) is killing a lot of green creativity.

The green options generated on basis of category 1B and 2 considerations generally score higher in an environmental scale than the ones based on category 1A. Mostly this applies to company, customer, and societal benefits as well. However; most of them are more difficult in terms of feasibility.

TABLE 10   Environmental Opportunities of STRETCH

Category 1	Category 2

Category 1A
Minimization of environmental impact through the life cycle
Minimization of energy use
Minimization of material
Minimization of environmentally relevant substances
Minimization of use of nonrenewable resources
Minimization of waste and emission
Efficient distribution and logistics
  Produce where you consume
  Direct distribution to consumer
Category 1B
Alternative functionality
Applying ecoefficient physical principles service concepts
	Increase of functionality
	Increase of intensity of use
	  Lease versus sell
	  Collective use durability of products
	  Longer life-time
	  Resell, reuse
	  Repairability, refurbishing
	  Technical upgrading
	Recyclability
	  Monomaterial, materials cascading
	  Design for disassembly

The essential task of the ecodesign matrix is to prioritize the green options generated by brainstorms (or in other ways) according to fitness to be incorporated in the business process. The scheme for doing so is shown in Figure 5. In the ecodesign matrix the generated green options are put in the left hand side. The environmental benefit of each option is first established for instance on basis of ecoindicator calculations [5]. The benefit analysis refers to the three important stakeholders in the business process: the company, the customer, and society as a whole. Benefits are considered in following categories:

- *Tangibles.* Cost or cost of ownership ($), less resources.
- *Intangibles.* Simpler to produce, easier to operate, fun, quality of life, better compliance.
- *Perceptions and emotions.* Better image, feel good, we make progress in green.

Feasibility has both technical and financial aspects. The technical side analysis physical and chemical feasibility (does nature allow us to do this?) and the amount of R&D needed to make the green option materialize. Also industrial aspects (changes in supplier base, different organization of production) are investigated. The financial feasibility looks at investment needed and possible pay-back times.

A completed ecodesign matrix is the perfect preparation for the next step: *product concept consolidation.* In this process, selected options from various brainstorms and feasibility studies (the environmental one as described above, a mechanical one, an electrical one, a software one, a commercial one, etc.) are brought together and prioritized further. Tradeoffs are being made in business terms. The ecodesign matrix is very helpful "to speak business language" for the green options, and this ensures that in the outcome the consolidation—a specification of the new product to be developed—environmental items are appropriately represented.

4.1.2 Green Product Creation

The basic scheme for green product creation is shown in Figure 6. It was described above how green has been integrated into the product concept and the specifications. What environmental aspects have been prioritized is strongly dependent on the functionality to be fulfilled and the business situation. It is therefore impossible to give here general rules how to tailor ecodesign optimally to a specific situation.

Proactive companies address these items in proprietary ecodesign manuals. Most of these are based on their experience and know-how and built up during the years. Chapters in such manuals are as follows:

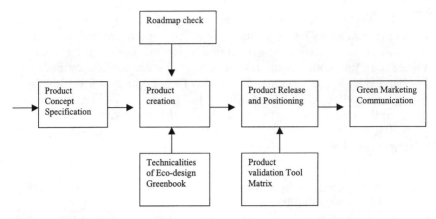

FIGURE 6   Green product creation.

- An introduction
- Explanation of the drivers for the company to be involved in green
- Presentation of vision, policy, strategy, and roadmap
- Overview of the organization of responsibilities and processes for ecodesign
- Overview of mandatory rules, directives, and guidelines
- List of environmentally released materials, components, and sub-assemblies

Ways to improve in the focal areas chapters are as follows:

- Energy consumption
- Materials application
- Packaging application
- Environmentally relevant substances
- Durability/recyclability
- Tool boxes for improvement metrics
- Environmental validation and product positioning

Environmental release and positioning forms also a integrated part of all products properties to be considered. The basis for this is already prepared with help of the ecodesign matrix, the product specification, and the tools applied in product creation. Note that this environmental release and positioning has a much wider scope than just doing environmental calculations such as LCA or ecoindicators. Also business, customer, and societal issues are addressed in a language that consists of $, kWh, kg, sec, m^3, %, and of descriptions of the intangibles (see also above and Chapter 19).

A next crucial step is the preparation of the green marketing and communication strategy. The basis for this strategy is the ecocommunication matrix as shown in Figure 7. Most data needed to fill this matrix can be derived from earlier steps in the ecodesign process (benchmarking, ecodesign matrix, product positioning). It should be noted that references can be made both to comparable product of competitors (if available) or to products of previous generations.

4.1.3 Exploitation of Results

A schematic overview of how exploitation of ecodesign results should be done in the marketplace is given in Figure 8. The core of this last part of integration of green into the business is green marketing and communication. This is of particular importance because the classical determinants of competition—price, delivery, and service—tend to become more and more equalized. "Green" can therefore be a tie breaker in customer decisions. Apart from customers themselves there are other stakeholders who directly or indirectly influence the companies—customer relationships. Other audiences for green communication are listed in Table 11.

Information needed for communication about green product attributes can be conveniently derived from the ecocommunication matrix (Figure 7). Especially when this environmental score in projected against performance of competitors and of own performance in previous years, this can be very powerful. Such score cards show the dynamics of progress gives a more transparent picture than the recently introduced environmental self declarations [6] which give rather a static picture a certain period of time.

The same objection holds for the so called ecolabels. These labels have the aim to inform the consumer whether a product is green by judging it against a set of criteria. Apart from lack transparency to the consumer (a

TABLE 11 Other Audiences for Green Communication

Audience	What do they primarily want
Employees	To contribute/work for a successful company
Authorities	
Local	Information
Regional	Compliance
National	Corporation
International	Data
Banks/shareholders	Performance, no scandals
Scientific/technical	Information
Green pressure groups	Suggestions to be taken

Focal area	Internal (company benefit)			Customer (customer benefit)			Other stakeholders (societal benefit)		
	T	InT	PXE	T	InT	PdFE	T	InT	PFE
Energy consumption									
Materials application									
Packaging/ transport									
Substances									
Durability/ Recyclability									
Manufacturing									
Life cycle perspective									

T = Tangible benefits
InT = Intangible benefits
PXE = Perceptions X Emotions

FIGURE 7 The ecocommunication matrix. T, tangible benefits; InT, intangible benefits; PXE, perceptions × emotions.

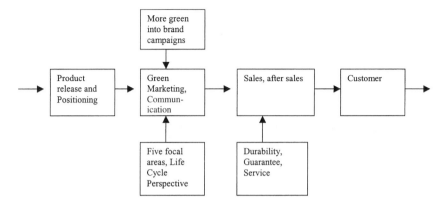

<small>Figure 8</small> Exploitation of ecodesign results.

variety of issues is brought together in an all or nothing label), ecolabels have a multitude of problems for producers. Worldwide there are now many programs of this kind, with different criteria (sometimes of subjective nature). There is a diversity of test procedures, sometimes high costs are involved, and obtaining the ecolabel involves lengthy procedures.

As an alternative for ecolabels, making of LCA scores has been proposed. This matter is still under hot debate. Main issues are subjectivity in the final valuation step. So from a scientific perspective LCA is unsuitable to make objective comparative statements (see ISO standard 14,042). Also, applicability and data uncertainly play a role in this debate.

From a business perspective the most important problem in using LCA in environmental communication is the "scientific" character of the method. This works in two ways:

• Intangibles and emotions cannot be described.
• LCA "language" is not understood by the average customer (and any other stakeholders).

Positive green products attributes need to be matched with a positive brand image as well. Market studies have shown that today in purchasing decisions about products of quality companies, brand image and specific attributes are about equally important. This also holds in the environmental communication field that showing to be a caring company is just as important as having good green score cards. Green image items to be considered are listed in Table 12.

To make appropriate communication strategies, the receiver of such messages also should be taken into account in more detail. Research [7] has shown that both in Europe and the United States, green attitudes of the buying public can be subdivided as follows:

- Approximately 25% strong green motivation: will pay higher prices for "demonstrated green."
- Approximately 50% positive for green but are not prepared to pay higher price ("neutral").
- Approximately 25% neutral or negative for green: perceive green products with same price as traditional products as products with lower performance.

From these results the following conclusions can be drawn:

1. Green as such does not sell; gains in the category "green motivated" will be undone by losses in the neutral/negative group.
2. When apart from green other customer benefits can be demonstrated (see §4 and 5), chances are much better; a least 75% of

TABLE 12   Green Brand Image Items

Leadership
  • Top management shows visible involvement in green effort
  • Environmental vision, policy, and strategy in place
  • Participation in World Business Council for Sustainable Development
  • Proactive in business associations
Programs put in place
  • Corporate environmental program
  • ISO 14001
  • Supplier requirements
  • Roadmaps/score cards
Documentation
  • Environmental reports
  • Brochures
  • Internet
  • Press releases/free publicity
  • Technical/scientific
Sponsorship of
  • Environmental research at universities
  • Nature conservation groups
  • Environmental-related events

customers will be interested, and chances to increase market share are good.
3. Special green products, that is, products that can command higher prices on the basis of green, should be customized to market segment, which is at best 25%.

In practice each company interested in green marketing and sales has to find out in what segment its present and potential customers are located and tailor its product strategy and communication policy toward the customer profile. This will make it necessary to dig deep into customers' feelings about environment. The feelings have a two-sided character. On one hand there is a strong feeling (which will become even stronger in the future) that society should take into account much more the environmental issues. On the other hand, presently a big majority is afraid that such efforts would require that they give up their present standard of living.

REFERENCES

1. The Brundtland Commission. "Our common future." Report of the World Commission on Environment and Development. Oxford University Press, Oxford, 1987.
2. A.J. Jansen and A.L.N. Stevels. The EPass Method, a systematic approach for Environmental Product Assessment. Proc. CARE Innovation 1998, pp. 313–320, Vienna, November 1998.
3. S. Brink, J.C. Diehl, and A.L.N. Stevels. Eco Quest: an Eco-design self audit tool for the suppliers in the electronics industry. Proc. IEEE Conference on Electronics and the Environment, Oak Brook, IL, 1998, pp. 129–132.
4. J.N. Cramer and A.L.N. Stevels. Strategic Environmental Planning within Philips Sound & Vision. J Environ Qual Manage pp. 91–102, 1997.
5. The Ecoindicator '95 Weighting Method. Novem/RIVM Editors. ISBN 90-72130-77-4. Ecoscan ®, a windows program for the calculation of one figure Eco scores available from Turtle Bay, info @turtlebay.nl. See also www.luna.nl/turtlebay
6. ECMA TC 38 committee. Technical Report TR70. Available from the ECMA office, Geneva, Switzerland. e-mail: helpdesk@ecma.ch and L. Mulder. Green Purchasing: Does it Make Sense. Proc. IEEE Conference on Electronics and the Environment, Oak Brook, IL, 1998, pp. 123–127.
7. J. Ottman. Green Marketing: An Opportunity for Innovation, 2nd edition. NTC/Contemporary Books.

25

Product Design and Product Recovery Logistics

V. Daniel R. Guide, Jr.
Duquesne University, Pittsburgh, Pennsylvania

Jonathan D. Linton
Polytechnic University, Brooklyn, New York

1 LOGISTICS DESIGN OF RECOVERABLE SYSTEMS

There is a common perception that design for the environment is only conducted for altruistic reasons; this is a misconception [1]. A rational firm will engage only in activities that increase the value of shareholder wealth. Fortunately, design for the environment is currently profitable in many circumstances as demonstrated by a number of firms [2,3], but trends in management practices, legislation, and technology are encouraging design for environment and a focus on the life cycle, as opposed to the sale, of a product. Since legislation, pollution prevention pays approaches, and ethics and value systems are considered elsewhere in this text, we only consider management practices and the technological drivers that are encouraging a life cycle approach to products. The design of a reverse logistics system and a recoverable product is important, since most life cycle costs are irreversibly established at the time of product design. Product recovery is currently more costly than product distribution, because only forward-distribution costs are currently considered in most business models. Furthermore, forward-distribution typically involves a single organization delivering to a number of locations, whereas reverse-distribution involves delivery to a single source with collection from a large number of customers in quantities and at times that are often unknown (Figure 1). The added cost and complexity of the reverse network is important, since between 18 and 45% of product costs are already associated with forward distribution—compared to about 5% for profit [4]. Having stated that reverse logistics is a challenge due to its cost and complexity, we consider the management and technological drivers that are encouraging the development of product recovery systems and then consider the factors that contribute to complexity of the recovery of product.

2 MANAGEMENT DRIVERS

There has been a gradual recognition in both academe and practice that firms are often able to identify benefits of their product that are not apparent to customers [5,6]. These manufacturers in many cases have moved away from a product-offering that sells a good for an agreed upon price to a combination of goods and services. The advantage to the customer is that they are purchasing the benefits of the product, a value that is better understood by the customer. The manufacturer by selling benefits more directly is able to gain a better understanding of the value of the goods and service

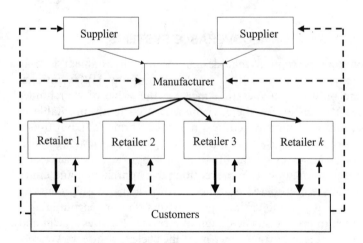

FIGURE 1   Forward (⟶) and reverse (--⟶) logistics flows.

package to specific customers. Consequently, it is possible to personalize the goods and services package and charge a price that is based on the value of the benefit received, as opposed to the cost of production. Furthermore, since the offering is more personalized and sophisticated, the likelihood of a competitor intervening declines, reducing the likelihood of having to compete on a price basis or losing established customers. Frequently, the manufacturer has a greater understanding of how to operate the product to offer the greatest possible benefit—this creates additional benefit that can be distributed to the customer and or supplier. There are not always benefits to moving closer to the customer. The important factors to determine whether there are benefits to being closer to the customer are the attractiveness of downstream business, the importance of customer relationships, and the power of the distribution channel.

2.1 Forces that Motivate Downstream Businesses

Attractiveness of the downstream business depends on the ratio of installed base to new products, life cycle economic cost of products, and difference in margin between upstream and downstream activities. Many products have reached the maturity stage of the product life cycle [7,8]. In such cases, a small fraction of the installed base is disposed of yearly, and product sales may serve to replace only this system fallout. Durables like locomotives, automobiles, kitchen appliances, and televisions are in this category. For these products, the number of products in operation is much greater in quantity than yearly sales. Consequently, any product that services the installed base has a much larger potential market. Warranties and maintenance programs are examples of products that are geared toward mature markets with a large installed base. Another factor that may make a downstream business attractive is the cost of a product over its entire life cycle. For many products, the purchase cost is a fraction of the actual cost of ownership. When this is the case, businesses that move downstream may tap into markets that are much larger than the manufacturing business they used to concentrate on. The ownership and operating costs of personal computers and automobiles are several times greater than the purchase price, and these are attractive industries for a downstream move by manufacturers, whereas nondurable goods tend to have low ownership and operating costs. The final factor that determines the attractiveness of a downstream market is the difference in margins between products and downstream product offerings. Downstream consulting and maintenance services frequently offer high margins. Alternatively, products are either treated as commodities or purchased on the basis of their initial price. In addition to the attractiveness of a downstream business, the importance of customer relationships must also be considered.

2.2 Customer Relationships

The importance of customer relationships relies on the magnitude of product-based differentiation, the market share of the top five customers, and the share of total profit earned from the top 20% of customers. As the importance of customer relationships increase, the attractiveness of the downstream business also increases. Substantial market share of top five customers increases the importance of customer relationships, since one or more unhappy customers changing to a different supplier could drastically affect sales volume and profitability. The greater the market share of a few customers, the greater the importance of customer relationships. Similarly, the greater the profit contribution of a few customers, the greater the importance of customer relationships. Finally, the magnitude of product-based differentiation is of great importance. If a product is differentiated from other products and this position is sustainable (either a result of patent protection or trade secrets), then the importance of customer relations is not so great since the customers will have difficulty replacing the product with alternative suppliers. However, if the product is a commodity, a strong relationship with the customer may prevent or reduce the frequency of customer defection to other suppliers. As customer relationships increase in importance, the attractiveness in being involved downstream or in other parts of the life cycle increases. Finally, the power of distribution channels also affects downstream decisions for firms.

2.3 Distribution Channel Strength

The power of distribution channels consists of the distribution and selling expenses as a percentage of the product's price, the degree of channel concentration, and the degree of channel innovation or multiplication. As the cost of sales and distribution increases, firm involvement with these functions become more attractive. If channel concentration is high, then there are a small number of distributors. High shares of sales in the hands of a small number of distributors allow the distributor substantial power over the producer. This power frequently results in reduced margins for the producer [9]. The presence of multiple channels or innovation in channels increases both the ease and the importance of manufacturers being involved in activities further downstream than manufacture of the product [10]. Increasing power in distribution channels, importance of relationships with the customer, and attractiveness of downstream business all drive manufacturers to become involved in activities beyond manufacture and during the life cycle of the product (Table 1). Firms that have pursued downstream activities have outperformed their competitors throughout the 1990s [11]. Downstream moves will be pursued because they are profitable for many reasons in many industries. Such moves provide information, infrastructure, and relationships with customers that facilitate product recovery throughout the life cycle of the good. In addition to these management trends driving product recovery systems, many technological advances are enabling and/or encouraging product recovery.

3   TECHNOLOGICAL DRIVERS

Advances in sensing, communications, and the rise of the internet and e-commerce all contribute to the attractiveness and range of possibilities available for recoverable product systems. The implications of these technologies are now briefly examined.

3.1   Sensing

There is increasing use of MEMS (Micro Electro Mechnical Systems) devices for sensing. These sensors are made using the same technologies that are used for transistor fabrication in microelectronics. Consequently, the sensors are inexpensive to manufacture, small in size, and light in weight. The combination of light weight, small size, and low cost opens a wide range of new applications to these sensors. MEMS technologies can be used today for a wide variety of sensing tasks. With time the development of nano-technologies is expected to further reduce size, weight, and cost while increasing flexibility [12]. Sensing applications that can effect life cycle management and return logistics includes sensors to detect wear, stress, vibration, and duration of use. Sensors can be used as part of a system to record the duration of use to facilitate refurbishment and to identify the timing of maintenance and the end of product life. Similarly, sensors that detect wear, and changes in thermal signature and vibration frequency can also be used to identify the need for maintenance or product retirement. How sensing is used depends, in part, on the use of communication.

3.2   Communications

The rapidly declining costs of computerization and wireless communication are creating new opportunities in product recovery. Advances in communications are important, since they allow for upgrading of product from a distance, identifying the location of a machine, and with proper sensing feedback on maintenance and wear that can indicate the ideal time for repair or product retirement. An awareness of location increases the ease of pick-

TABLE 1 Summary of Factors Affecting the Attractiveness of Downstream Products

Critical factor	Attractiveness of downstream business	Importance of customer relationships	Power of distribution channels
Unattractive	Small installed base. Major expenditure is on product acquisition. Low margin for postsales activities.	Large number of customers. Low sales volume and profitability from major customers. Commodity product—little or no opportunity for product differentiation.	Low cost of sales and/or distribution. Large number of distributors. Small number of distribution channels. Lack of innovation within distribution channels.
Attractive	Large installed base. Most expenditures occur after product acquisition. High margin of postsales activities	Small number of customers comprise a major share of sales and/or profit for a specific good. Product with substantial opportunity for product differentiation.	High cost of sales and/or distribution. Only a few distributors. Large number of distribution channels. Innovation within distribution channels is common.

up or offering information to the user about the most favorable return options.

3.3 The Internet

The rise of the Internet and electronic commerce also offers a technological trajectory that changes the options for recovery of product at its end of life [13]. In electronic commerce, the automation of service is the typical offering. Furthermore, the internet offers easy access from fixed locations and increasingly for mobile applications. This combination of ease of access and service automation is ideal for communication with the existing field population which allows for identifying the number of products, and possibly the location of these products, at different stages of their life. If buyback or recall of product with a certain length of operating time is desired, the internet can track and manage such a goal.

The use of the internet is not limited to futuristic sounding applications that have machines "speaking" to or interacting with other machines. The internet can be used for retrieval of product either through a vendor's website or a third-party website. In the case of a vendor's website, the vendor can offer product owners a specific package of goods or benefits in exchange for the return of a used product. These benefits can either be publicly stated on a website with instructions on how to return the product to obtain these benefits or can be negotiated through e-mail or an internet mediated discussion. This allows the firm and vendor to come to mutually acceptable terms for the return of the product and to have terms that can vary from customer to customer [14]. Another approach is to have the recovery of product handled through a metamediary [15]. A metamediary provides a multivendor, multiproduct marketplace. The metamediary also provides multiple services—such as payment settlement, fulfillment, quality assurance, and procurement management. The use of a metamediary is likely if recovery of products is managed by a number of independent firms—as opposed to the OEM (original equipment manufacturer). Other conditions that favor the use of metamediaries include large market size, fragmented supply chain, commodity, high information search costs, high product comparison costs, and high work flow costs [16]. The metamediaries provide benefits by offering one or more of the following services: matching buyers and sellers, reducing transaction costs, reducing search costs, ensuring anonymity of buyer and/or seller, and increasing the ease of communications [17]. Having considered the managerial and technical drivers of product recovery systems, the consideration of factors effecting product recovery and the reasons for product recovery is considered.

4 FACTORS AFFECTING PRODUCT RECOVERY

A variety of characteristics of a product and the product's market affects the ease and the preferred method of product recovery. Before considering the different types of product recovery methods and the factors that complicate product recovery, these important characteristics are briefly considered. Product characteristics such as size, ease of transport, acceptability of product downtime (repair), and what activities can be handled remotely affect product recovery activities.

4.1 Size

The weight and physical dimensions of the product are important considerations. Not only do these factors affect the cost of shipping, they also affect the ease with which a product is prepared for shipping and the ease with which a product is handled postshipping during recovery operations.

4.2 Ease of Transport

This characteristic can refer to the size of product, due to the effect product dimensions and weight have on the cost and options available for shipping. But ease of transport also considers whether there are restrictions associated with movement of the product. For example, moving products that are contaminated with radiation is exceedingly difficult. A more common example of this is laws restricting the export of waste (in an attempt to prevent illegal dumping) and regulations requiring special permitting or precaution for movement of postconsumer products. For example, in some jurisdictions the shipment of postconsumer electronics must be treated as hazardous waste movement (due to the high lead content), whereas new electronic products can be shipped in any method desired.

4.3 Product Downtime

In some applications, downtime is unacceptable. Consequently, the product may not be sent off-site for repair, refurbishment, or upgrade.

4.4 Remote Activities

An increasing number of products have value provided by information technology and computer programs. In these types of products, repairs, upgrades, and refurbishment may be handled through fixed or wireless telephony. This situation is common with direct digital control systems used in automation in office and industrial buildings. Alternatively, a kit or instructions may be provided to individuals at the customer's site for the purpose of repair, upgrade, or refurbishment of a product. This situation is common practice in telecommunications. Telephone service providers cannot allow for downtime; consequently, suppliers send them new products with instructions on how to remove the old component and replace it with the new component.

Having considered a number of factors that affect the recovery of a product, the different purposes for which products are recovered are considered.

5 REASONS FOR PRODUCT RECOVERY

There are a number of possible reasons for the recovery of product. Each reason for recovery typically has different considerations. Consequently, each reason for recovery and the associated considerations are addressed. We use the term "purpose" here in noneconomic terms. There are only two possibilities for why firms engage in product recovery: it increases shareholder wealth or it is mandated by legislative requirements. The primary purpose of our discussion here is to show that there is a variety of alternatives for recovered products.

5.1 Repair

In product repair, the customer may have to be provided with a replacement product either temporarily or while their product is being transported and repaired. Protection of the product's function and appearance are both important. If a new product is provided, prior to sending the broken product in for repair, then the shipping material for the replacement product can be used as return shipping for the broken product. This is a common practice for the repair of modular electronics products, such as hard drives in computers. If return shipping packaging cannot be provided, then it is important to provide sufficient instructions to ensure that the product is protected in transit. If a replacement product is not given, the duration of the repair and transport of the product is important to continued customer satisfaction. The challenges facing the manufacturer are ensuring that sufficient repair capacity and parts inventory is available to ensure rapid turn around of product, while minimizing the cost to provide this service. Forecasting future likely returns based on consideration of frequency of product failures, failure modes, and past sales data may be of assistance. Since there is little experience with forecasting product returns, the relative difficulty of this is unclear. Forecasting product failure is well understood for many systems [18]. However, predicting product returns in general is quite difficult since the decision to discard a product is a highly personal decision.

5.2 Upgrades

This option involves replacing product or parts of the product to either eliminate failure modes or improve functionality. Upgrades may differ in that some are planned to be at the customer's convenience, while others are at the company's convenience. To illustrate this difference, the practice of Nortel Networks is compared with an automotive recall. Nortel Networks [19] supplies telephone service providers with electronic equipment. Occasionally, an upgrade of these electronics parts is desirable to either add additional functionality or eliminate possible failure modes. Prior to upgrading the product, a functional replacement must be provided to ensure that product service is guaranteed through the repair service. Alternatively, upgrades at the company's convenience are common in the automotive industry. In this model, the customer can provide the product that requires upgrading, to the company. The company upgrades the product and then either returns the product to the customer or notifies the customer they can retrieve the product from the company. During the repair period, the customer is without the use of the product. An additional difference between these two forms of upgrade is in the second case the customer, who lacks a product during the upgrade process, is aware of the length of time required to modify the product.

5.3 Remanufacturing

We use the term remanufacturing to include terms such as refurbishment, overhaul, and rebuilding. In our experience the differences are slight, and using separate terms only serves to cloud the discussion. This option involves recapturing products after disposal and making changes to renew the product so that it can provide functions similar to those that it originally provided. Lund [20] identified 75 separate product types that are routinely remanufactured and developed seven criteria for remanufacturability:

1. The product is a durable good.
2. The product fails functionally.
3. The product is standardized and the parts are interchangeable.
4. The remaining value-added is high.
5. The cost to obtain the failed product is low compared to the remaining value-added.
6. The product technology is stable.

7. The consumer is aware that remanufactured products are available.

It is worthwhile to note that many of these criteria may be dominated by decisions made at the product design stage. Specifically, criteria 1, 3, and 6 may be addressed at the product design stage. Even though products may be classified as durable products in the economic sense, it is by no means an absolute standard for remanufacturability. Many products are implicitly designed for disposal, where the price of repairing a failed item exceeds the cost of a complete replacement. Videocassette recorders often employ an all-in-one board design that is economically impractical to replace in the event of a component or part failure and difficult to reuse at higher levels (e.g., value-added recovery). We argue that type of design decision is often made with insufficient information about true life cycle costs.

Criterion 3, products that are standardized and made with interchangeable parts, is also a conscious design decision that has a profound impact on whether a product may be remanufactured. Modular design has been the standard for expensive military assets, such as aircraft, for almost 30 years. Some of this design technology has filtered through to the commercial sector. One of the more notable consumer products in Europe is the smart™ car manufactured by MCC smart™ Gmbh, a sister company of DaimlerChrysler AG. This auto is designed for reuse and incorporates many modular design techniques for both exterior and interior components and parts.

Criterion 6, stable technology, may be greatly influenced by design choices. Products may be designed for technology upgrades, the smart™ being an excellent example of a product designed for continuous upgrades. Other examples may be found in telecommunications, industrial equipment, and construction.

Current survey findings [21,22] indicate that the majority of the products presently being remanufactured (value-added reuse) are commercial goods, and the majority of firms offering reused products are non-OEMs. However, consumer electronics represent one of the greatest growth markets, and consumer products in general, with sales volumes in the millions, offer the greatest potential sources of used products. The potential for reuse activities is limited by a lack of interaction between firms reusing products and firms designing the products. Additionally, few firms are designing products for incorporating remanufactured parts or components. This contributes to a lack of uses for recovered materials and may negatively influence the economic attractiveness of reuse activities. A recent discussion about the reuse of consumer electronic motors highlights many of the problems associated with the reuse of consumer goods [23].

5.4 Part and Component Recovery

In some cases a product that is not worth recovering contains many components or parts that have retained their value. Products that are assemblies of many components may have little value in their assembled form but great value in disassembled form. The best example is the automobile. Old automobiles are stripped of parts that have value either after they are remanufactured or as replacements for parts damaged by wear, age, or collision.

5.5 Material Recovery

The product in its current form may have no value, however, the materials that comprise the product are of value if they can be recovered. The recovery of materials (also termed recycling) is favored by much of the existing policy and legislation, although it tends to be the least attractive option for recovery. Recovery of materials is commonly used in situations where a valuable material can be obtained with little contamination. Examples include copper cables, steel carcasses of automobiles, and aluminum cans. Polymers are relatively undesirable for material recovery, since the cost of processing is high and the properties of recycled polymers are degraded, due to the tendency of the length of the polymer chains to be shorter than virgin material, after its reclamation.

5.6 Complicating Characteristics of Product Recovery

Seven major issues have been identified through the literature [24], case studies [25–28], and surveys [29,30] as being unique to product recovery. To be successful in product recovery these issues must be addressed. A variety of tactics has been developed to address these issues after the fact. However, an awareness of these issues at the design stage assists in designing the product to facilitate the maximization of its value through the product's entire life cycle by improving its recoverability throughout the entire life cycle. The seven issues that have been found to consistently complicate the recovery of product are as follows:

1. Uncertain timing and quantity of returns
2. The need to balance demand with returns
3. Need to disassemble the returned products
4. Uncertainty in material recovered from returned items
5. Requirement for a reverse logistics network
6. Complication of material matching restrictions
7. Problems of stochastic routings for material for repair and remanufacturing processes and highly variable processing times

Having identified the issues, they are now considered in depth.

5.6.1 Uncertain Timing and Quantity of Returns

Uncertainty in the timing and quantity of returns affects decisions regarding management of product recovery in terms of collection and processing of end-of-life materials. In the case of leasing operations, the timing of return is known. More frequently, the customer decides at what point in time the product is disposed of by the customer. Some researchers have considered the forecasting of return of reusable containers, but this is a special case since they are a close-loop system [31,32]. More frequently, planners must forecast the availability of postconsumer product. This is dependent on past sales and the lifespan and the mortality rate of product. The design of the product has a bearing on both the product lifespan and the mortality. During the design phase, the likely mortality function should be determined and it should be adjusted to reflect changes in parts and processes over time. To assist in validation of the mortality function, estimated failures can be compared with warranty data. This information is not only useful for understanding whether failure within the warranty period is greater or less than anticipated, but it can be used to gauge the accuracy of the predicted mortality function. However, it is important to note that field failure may not govern the availability of postconsumer product. If a product becomes obsolescent, large quantities of postconsumer product may become rapidly available. In such a case, remanufacturing may not be viable and the product may have to be substantially altered, used for parts, or only suitable for materials recycling. Understanding the likelihood of obsolescence as a function of time is a consideration that should be addressed during design. If a short product life, due to obsolescence, is anticipated features such as repairability, modularity, and durability are of little concern. But for products with short anticipated lives, the disposability or recoverable materials value is of greater concern. The use of lead, mercury, and other heavy metals that may require special disposal should be avoided, since a collection infrastructure with significant over-capacity will have to be available. High value components should be made easy to access and recover.

With improved sensor technology and expectations of smart appliances that are able to communicate using either the electrical lines or wireless technology, the potential to use predictive maintenance techniques to either warn a third party of impending failure or to identify products that are good candidates for trade-in or reprocessing. Although this concept may seem futuristic, photocopiers, automation, and computers in many cases currently fit this description. In the next few years, the range of products that fit this description are expected to increase greatly in terms of scope and number, thereby reducing the uncertainty over timing and quantity of returns.

5.6.2 The Need to Balance Demand with Returns

The need to balance return rates with demand and uncertainty in the timing and quantity in the timing and quantity of returns and inventory management is a challenge. Repairable inventory has been the subject of much research [33]. The repairable inventory problem is unique in recoverable systems, since the assumption of perfect correlation between returns and demands is reasonable in many cases. Upgrades of product can also involve this assumption. Linton and Johnston [34] provide an excellent case study and discussion of a product upgrade operation. However, many firms have little control over the quantity, quality, and timing of returned products.

Design of products that are intelligent and locatable can address the problem that balancing demands and returns can create. If it is possible to communicate with products to understand how close they are to their end of life and their location, it is possible to forecast the likely rate of future product return and the location of these products. Such a process is too expensive if it requires human intervention. However, this process can be automated and as a result relatively inexpensive. Alternatively, the sale of the benefit of a product as opposed to a product—such as an automotive or equipment lease or service agreements [35]—can provide a return date for the product itself. Otherwise, one is faced with attempting to balance returns with demands.

The knowledge in this field is reviewed to advise designers of the complications and in doing so convince at least some designers that there are advantages in designing products and product systems to better address this factor which complicates current practice. Firms must also balance inventories of cores and repairs or remanufactured end items with actual customer demand. Without careful control and management of inventories, stocks will grow uncontrollably. Firms frequently need to dispose of used product to avoid carrying excess inventory [36,37]. Consequently, a number of inventory control systems have been proposed [38–41]. Continuous-review models require constant monitoring and updating of inventory levels and ordering materials based on trigger points from set inventory levels. Decision variables typically involve setting threshold values to trigger replenishment quantities and determining the order quantities. Muckstadt and Isaac [42] developed the first published model that explicitly considers the problem of imperfect correlation between returns and demands. The model does not allow for disposals, but it does account for nonzero lead times. Push and pull strategies for joint production and inventory decisions for a mix of remanufactured and new items have been investigated in detail

[43–45]. Both strategies may consider disposal options, and findings indicate the choice of the appropriate strategy depends heavily on the cost-dominant relationship between remanufactured and new stocks. These studies also show that changes in the demand and return rates make updating the model's parameters necessary. This implies that as products go through the normal life cycle, managers must revise the decision rule for effective management and control of inventories.

Core (the term core is used to denote a discarded products that serves as the input to reuse activities) acquisition is an important activity in the supply chain for recoverable manufacturing firms, but there is little published on this subject [46]. Traditional manufacturing firms purchase only new parts and components and know the price, quantity, and the supplier base. Core acquisition is highly uncertain with respect to quality and quantity. Since the quality of incoming cores determines the amounts of materials and labor to restore the core to like-new condition, schemes to screen cores prior to acquisition may be particularly effective at reducing the variability. Enough cores must be obtained to meet demands and develop strategies to balance returns with demands. A firm may need to manufacture new items or purchase replacements when cores are unavailable. Firms must also integrate traditional demand management activities, such as demand forecasting, with core acquisition activities to balance return rates with demand rates. Firms must develop strategies for acquiring cores, since they may acquire cores from various sources, such as customers, core brokers and other third-party vendors, and OEM.

1. Firms may acquire cores directly from customers for remanufacture and return, in exchange for remanufactured units or through simple purchase. Charging a core deposit could help reduce the uncertainty of return quantities since a sale would generate a return but might not reduce timing uncertainty since demand rates would still be stochastic.

2. By acquiring cores from third-party vendors and broker firms, it is possible to acquire cores in larger lots and at a more constant rate. Remanufacturing can then be planned more easily and economies of scale can be achieved in acquisition and transportation. The disadvantage of this approach is the firm has little control over the age and condition of the returns.

3. OEM programs provide firms with economies of scale and a steady source of cores. Firms can expect greater homogeneity for the cores than those from third-party vendors and core brokers. Some firms are using innovative systems to ensure availability of cores. ReCellular, Inc., the major supplier of remanufactured cellular telephones in the world, has established working relationships with a number of cellular telephone manufacturers and service providers to improve the availability of

cores. The service providers return many used cellular telephones to ReCellular, which remanufactures and redistributes them. This provides service providers with a market for returned cellular telephones and provides ReCellular with a steady supply of cores. For example, ReCellular is the authorized service and remanufacturer center for Nokia in North America.

Firms need to develop formal core acquisition strategies to balance customer demand with returns and avoid holding excessive inventories of postconsumer product. Redesign of products and/or the sales life cycle to assist in the acquisition of postconsumer products could eliminate the challenge of matching returns with demands, which is currently a complicating factor in product recovery operations.

5.6.3 Disassembly

Returned items must be disassembled before the product may be restored to full use. The effects of disassembly operations impact a large number of areas, including production control, scheduling, shop floor control, and materials and resource planning. Disassembly is the first step in processing for remanufacturing and acts as a gateway for parts to the remanufacturing processes. Products are disassembled to the part level, assessed as to their remanufacturability, and acceptable parts are then routed to the necessary operations. Parts not meeting minimum remanufacturing standards may be used for spares or sold for scrap value. Purchasing requires information from disassembly to ensure that sufficient new parts are procured.

The process of disassembly is complicated since very few products are designed for disassembly. Products not designed for disassembly have less predictable material recovery rates, higher disassembly times, and generate more waste. Parts may be damaged during disassembly, especially on products not designed for disassembly, and this often increases material replacement rates. These factors significantly increase the variability inherent in reuse operations, and this increased variability makes production planning and control activities complex. Additionally, few firms have access to OEM specifications and must perform reverse engineering, and even the two thirds of the firms with access to OEM specifications reported significant amounts of reverse engineering activities [47].

All disassembly requires some expenses by the firm (i.e., labor, equipment, and overhead), and firms are often concerned with minimizing the associated costs. One approach to the recovery problem seeks to balance the costs of disassembly with the revenues from material recovered. This is not an easy task since many benefits may be quite difficult to quantify, such as improved corporate image, improved competitive position, and decreased liability from waste products. There are a numerous works on determining optimal disassembly strategies, and we show in Table 2 the relevant contributions. Each of these techniques is designed to aid the practicing engineer make informed choices about the economics surrounding the recovery problem.

Alternative approaches to the disassembly problem have been suggested. The most promising seems to be the development of metrics and tests that indicate whether a part/component is worth recovering for value-added reuse. Klausner et al. [48] discussed the development of an electronic data log to provide information about the suitability of a consumer electronic motor for reuse. Essential information about the conditions (i.e., temperatures, cumulative run time, speed, etc.) the motor has experienced are ported from the data log without having to first disassemble the motor from the product or disassemble the motor itself. The authors had not yet developed an accurate heuristic to determine the exact recoverability parameters of the go/no-go attribute, but the idea of simple tests to determine the overall quality of a returned product is appealing. Such a testing methodology would help reduce the unnecessary expense of disassembling a product that has little economic reuse potential (except for the material recycling value). Such testing systems help to reduce the variability in reuse systems, and these higher amounts of variation make reuse systems difficult to plan and control and more expensive to operate. Design engineers should be aware of the potential benefits of such a data log system and work to exploit existing sensor mechanisms on products to record such information.

TABLE 2 Techniques and Methods for Optimal Disassembly Decisions

Author	Methodology
Navin-Chandra (1994)	Expert system
Johnson and Wang (1995)	Genetic algorithms
de Ron and Penev (1995)	Dynamic programming and graph
Penev and de Ron (1996)	Theory
Lou et al. (1996)	AHP
Vujosevis et al. (1995)	Expert system
Chen et al. (1997)	Removability trees
Dutta and Woo (1996)	Parallel disassembly
Yokota and Brough (1992)	Expert system
Zussman et al. (1994)	Utility theory
Geiger and Zussman (1996)	Bayesian networks
Feldmann and Scheller (1994)	Expert system
Ari and Iwata (1993)	Expert system

However, part of the long-term solution requires that products developed in the future are designed with reuse in mind. There are several excellent sources with guidelines for new product development, and for specific design recommendations we refer the reader to Navin-Chandra's work [49], *Green Products by Design* [50] and *Design for Environment* [51]. However, these works tend to focus on material selection and other design criteria, and none consider logistics issues as part of the life cycle costs. As more products are designed for reuse and disassembly, there is a need for better and reliable feedback to exist between remanufacturing operations and designers, especially when OEMs do not have in-house remanufacturing.

5.6.4 Uncertainty in Materials Recovered

Materials recovery uncertainty complicates resource planning, purchasing, and inventory control since disassembly may provide variable yields of usable or repairable parts and components. Parts may be remanufactured, used for spares, sold to secondary markets, or recycled. This uncertainty makes inventory planning and control and purchasing more problematic. Until the product has been fully disassembled and the parts cleaned and inspected, their suitability for reuse or rebuilding as opposed to scrapping is not known. Firms should be able to forecast the recovery rates for parts to plan for new parts to replace those they cannot recover. Presumably, the development of electronic data logs and their acceptance will help reduce the uncertainty in materials requirements by making the prediction of recovery rates more certain. Additionally, product design should specify the maximum number of cycles a part or component may be reused before material recycling is mandated. Parts should also be clearly identified as reusable or recyclable to facilitate materials planning for remanufacturing and other value-added reuse.

Material recovery uncertainty is measured as the material recovery rate; the frequency that material recovered off a core unit is remanufacturable. Firms report a wide range of stability for most material recovery rates ranging from completely predictable to completely unpredictable. Products are equally likely to contain parts with known predictable recovery rates, as to contain parts with no known pattern of recovery.

Purchasing is complicated by the uncertain requirements resulting from material recovery uncertainty and short lead times. Further complications may include substitutable parts, proprietary parts or technology, out-of-production parts, and little or no support from the original manufacturer. The purchasing function is particularly important when the firm aims to provide a wide range of reworked products and sufficient shop capacity to handle the flow, to hold the inventory and work-in-process, and to use equipment and labor efficiently. A recent survey of remanufacturing firms shows that the primary causes of late deliveries of customer orders are a lack of availability of parts, long lead times for order delivery, or OEM parts being very highly priced [52].

5.6.5 Requirements for a Reverse Logistics Network

Reverse distribution is the task of recovering discarded products (cores); it may include packaging and shipping materials and backhauling them to a central collection point for either recycling or remanufacturing. Handling the mechanics of reverse distribution requires an understanding of state and federal laws in the United States, Europe, and Asia. Certain design decisions, such as the inclusion of hazardous or toxic materials, like lead or PCBs, can limit the flexibility in design of a logistics network.

Three key issues affect reverse logistics: the structure of the network, the planning for material flows, and the classification and routing of materials [54]. The collection of goods from the marketplace is a supply-driven flow rather than a demand driven flow. This flow is uncertain with respect to the quantity, timing, and condition of items. Thus as mentioned earlier, predicting the quantity of goods available may be difficult.

Recovering goods is further complicated by the fact that most logistics systems are not equipped to handle reverse product movement; returned goods often cannot be transported, stored, or handled in the same manner as outgoing goods; and reverse distribution costs may be several times higher than original distribution costs [55]. There are a number of reverse networks in operation in Europe, including OEMs like Digital and IBM and some third-party programs [56].

5.6.6 Materials Matching Requirements

When customers require repair and returns of the same items they originally bought, the firm must coordinate disassembly operations with repair and remanufacture operations and reassembly. This requirement may be customer driven, for example, when a customer turns in a unit to be remanufactured and then requests that same unit is returned to them. Xerox offers customer-driven returns as a part of customer service for copier remanufacturing. A unit may also be composed of a mix of common parts and components and serial number specific parts and components. This characteristic complicates resource planning, shop floor control, and materials management.

The other major impacts of this characteristic are on scheduling and information systems. To provide the same unit back to the customers, parts must be numbered, tagged, and tracked, and this places an additional burden on the information systems. The reassembly of a unit composed of matched parts may be easily delayed since a specific part number, not just a specific part type, may delay the order.

5.6.7 Routing Uncertainty and Processing Time Uncertainty

Stochastic routings are a reflection of the uncertain condition of units returned. A part will have a maximum set of processes that should be performed to restore the part to specifications. However, these routings represent a worst-case scenario, and the majority of parts will only require a subset of these processing steps. Highly variable processing times are also a function of the condition of the unit being returned. These additional forms of uncertainty make resource planning, scheduling, shop floor control, and materials management more difficult than in traditional manufacturing environments. Firms need to estimate the condition of the parts to be recovered from returned product to schedule for the work centers and plan capacity. Shifting bottlenecks are common because the materials recovered from disassembly vary from unit to unit, processing times vary, and routings vary. Also, the end products in a recoverable environment do not require all new parts, and remanufactured components do not necessarily require all the operations associated with newly manufactured component.

In remanufacturing operations, some tasks are known with certainty, (e.g., cleaning); however, other routings may be probabilistic and highly dependent on the age and condition of the part. Routing files are a list of all possible operations, and for planning purposes the likelihood of an operation being required is maintained. Not all parts will be required to route through the same set of operations or work centers; indeed, few of them would go through the same series of operations as a new part. This characteristic is cited as the single most complicating factor for scheduling and lot sizing decisions [57]. Order release mechanisms are limited to one-for-one core release, except for some common parts for which release may be delayed to provide a minimum batch size.

Remanufacturing lot sizing is complex, and there is no agreement on the best method. A standard lot size of one unit is commonly used in practice because of the unique requirements of remanufacturing; that is, unique routings for identical part types [58]. Cleaning operations represent a major portion of the processing time, with an average of 20% of total processing being spent in cleaning. Parts may return for multiple processing at the cleaning operation, and almost half of remanufacturing firms reported additional difficulties in cleaning due to part materials and sizes. Parts must be cleaned, tested, and evaluated before any decision is made as to remanufacturability. This late determination of a part's suitability further complicates purchasing and capacity planning due to the short planning horizon. Firms also report that the variability in a parts' condition creates problems in machine fixturing and set-ups. These additional sources of variability further complicate the accurate estimation of flow times.

The product design process can help in reducing these uncertainties in several ways. First, the development of quality metrics and tests that may be applied before products are disassembled will enable production planners to have crucial information earlier. Second, products designed for a maximum number of remanufacturing cycles will make the required manufacturing operations and processing times more predictable. Finally, an economic determination as to whether a part is to be recoverable (via value-added recovery) or disposable (recyclable) should be made in the design phase.

6 GENERALIZED DESIGN GUIDELINES

A number of considerations have been raised throughout the discussion of the types of postconsumer product recovery and the challenges that each presents. Each consideration offers advice or insight for new product design, and they are as follows:

1. *Selling benefit and retaining ownership.* If the benefit of a product is sold and the ownership is retained, it is possible for the seller to extract a large percentage of the value that each customer obtains from the product. This approach is followed by many firms, due to the high margins offered by this approach for the sale of certain products. However, this approach offers control over the product at its end of life and an in-depth knowledge of operating conditions, maintenance, and the duration of operating life to date.

2. *Remote repair or upgrade.* With the knowledge content of products increasing, due to the importance of software and advances in sensor technology, the ease of remote repair and upgrade improves. Design of products for remote repair and upgrade increases the life of the product. This approach allows the OEM to sell upgrades, warranty, and repair services that cost little to fulfill but offer great value to the customer. Consequently, these products offer a high margin.

3. *Prediction of failure.* During the design and testing of a product, decisions are made that effect the failure rate of the product. The failure rate is rarely calculated and considered by designers and manufacturers. The lack of this information complicates not only postconsumer product recovery but the understanding of sales and marketing regarding the likely rate of product replacement and expectations about how much time will pass prior to product obsolescence. By developing knowledge about failure rate, during the design and testing of a product, a firm gains a knowledge advantage about their products. This knowledge advantage can not only be useful for

planning product recovery activities, but provides an advantage over third parties that attempt to recover the OEM's products.

4. *Modify products so that age and probable remaining use is easily predictable*. With minor design changes, especially in electronic products, it is possible to maintain information regarding total number of operating hours and operating conditions. Easy access to this information is important for making a quick assessment of the operating life remaining in the product and its components. This is analogous to checking the odometer on a used automobile.

5. *Modularity for parts that are likely to have great recovery value and modularity so obsolescent parts can be replaced easily*. Modularity assists in recapture of product value. There are two separate situations where modularity adds value. A design should allow for the easy replacement of components that have short lives or become obsolete quickly. The rapid onset of either obsolescence or failure may be unavoidable, but modularity in design of parts of the product that retain value even when the product does not is desirable. Such a design allows for the recovery of valuable parts and/or materials.

6. *Consideration of ease of disposal at the time of design*. Certain products are difficult to dispose of due to the presence of undesirable components or substances. Examples include PCBs in capacitors in fluorescent light ballasts, mercury switches in automobile dashboards and other equipment, and lead in cathode ray tubes and other electronic products. Products can be designed to either avoid or reduce the quantity of undesirable materials used. The design change improves the flexibility in both disposal and recovery options. Presence of undesirable substances in a product can lead to special transport, handling, and disposal requirements. These requirements can result in increases of the life cycle cost of a product by elevating the cost of postconsumer options.

7 IDENTIFY ISSUES THAT MUST STILL BE CONSIDERED

Through consideration of the current challenges and knowledge relating to reverse product flow and recoverable systems, it is possible to better position a product for recovery as a postconsumer product. However, there are two major areas that further work and knowledge are required. Much attention has been given to forward flows of product. But there is a need for a much better understanding of the efficient operation of reverse flow networks. With time it will be more apparent which product characteristics make what type of networks advisable. In addition, there is a need to experiment with different offers and procedures to allow for controlled retrieval of postconsumer products. Voluntary and legislated systems dominate post-consumer product retrieval. The consideration of contractual and monetary incentives is still in its infancy.

8 CONCLUDING STATEMENTS

Changes in design of a product can result in the elimination or reduction of factors that complicate the retrieval and reprocessing of products. Consider the following:

- Uncertain timing and quantity of returns
- The need to balance demand with returns
- Need to disassemble the returned products
- Uncertainty in material recovered from returned items
- Requirement for a reverse logistics network
- Complication of material matching restrictions
- Problems of stochastic routings for material for repair and remanufacturing processes and highly variable processing times

These seven complicating factors act as a barrier, in some cases to product retrieval and recovery operations and in other cases they reduce the profitability of such operations. Even with these factors in existence recovery activities like remanufacturing can be highly profitable [59–61]. By consideration of these factors a product is designed taking the entire life cycle, including postconsumer disposal, into account. As a result, the cost and ease of such life cycle activities like repair, upgrades, remanufacturing, and recycling become simpler to coordinate and more cost effective—without having adverse effects on the product's initial value proposition.

REFERENCES

1. M Porter, C Van der Linde. Green and competitive: ending the stalemate. Harvard Business Rev 73(5):120–134, 1995.
2. VDR Guide Jr. Production planning and control for remanufacturing: industry practice and research needs. J Oper Manage 18:467–483, 2000.
3. VDR Guide Jr, V Jayaraman, R Srivastava, WC Benton. Supply chain management for recoverable manufacturing systems. Interfaces 30(3):125–142, 2000.
4. B LaLonde, J Grabner, J Robeson. Integrated distribution systems: a management perspective. In: P.M Van Buijtenen, ed. Business Logistics. The Hague: Martinus-Nijhof, 1976.
5. M Michaelis, JF Coates. Creating integrated performance systems: the business of the future. Technol Anal Strateg Manage 6(2):245–50, 1994.
6. R Wise, P Baumgartner. Go downstream: the new profit imperative in manufacturing. Harvard Business Rev 77(5):133–141, 1999.

7.   T Levitt. Exploit the product life cycle. Harvard Business Rev 43(6):93–100, 1965.

8.   HB Thorelli, SC Burnett. The nature of product life cycles for industrial goods businesses. J Market 45(4):97–108, 1981.

9.   M Bensaous. Portfolio of buyer-supplier relationships. Sloan Manage Rev 41(2):35–44, 1999.

10. C Savasakan, LN Van Wassenhove, S Bhattacharya. Channel choice and coordination in a remanufacturing environment. Working Paper 99/14/TM, INSEAD, Fontainebleau, France, 1999.

11. R Wise, P Baumgartner. Go downstream: the new profit imperative in manufacturing. Harvard Business Review 77(5):133–141, 1999.

12. R David. Will the real nanotech, Please stand up? Technol Rev 2: 47–62, 1999.

13. A Kokkinaki, R Dekker, JAEE van Nunen, C Pappis. An exploratory study on electric commerce for reverse logistics. Econometric Institute Report EI-9950/A, Erasmus University Rotterdam, The Netherlands, 1999.

14. C Bayers. Capitalist econstruction. Wired Magazine 8(3):210–219, 2000.

15. S Ehrens, P Zapf. The Internet Business-to-Business Report. Equity Research Bear Stearns, New York, 1999.

16. S Ehrens, P Zapf. The Internet Business-to-Business Report. Equity Research Bear Stearns, New York, 1999.

17. S Ehrens, P Zapf. The Internet Business-to-Business Report. Equity Research Bear Stearns, New York, 1999.

18. DJ Klinger, Y Nakader, MA Menendez. AT&T Reliability Manual. New York: Van Nostrand Reinhold, 1990.

19. J Linton, D Johnston. A decision support system for the planning of remanufacturing at Nortel. Interfaces, forthcoming.

20. R Lund. Remanufacturing: an American resource. Proceedings of the Fifth International Congress for Environmentally Conscious Design and Manufacturing, Rochester Institute of Technology, Rochester, New York, 1998.

21. VDR Guide Jr. Production planning and control for remanufacturing: industry practice and research needs. J Operat Manage 18:467–483, 2000.

22. N Nasr, C Hughson, E Varel, R Bauer. State-of-the-art assessment of remanufacturing technology. National Center for Remanufacturing and Resource Recovery, Rochester Institute of Technology, Rochester, New York, 1998.

23. M Klausner, W Grimm, C Hendrickson. Reuse of electric motors in consumer products. J Indust Ecol 2(2):89–102, 1999.

24. VDR Guide Jr, V Jayaraman, R Srivastava, WC Benton. Supply chain management for recoverable manufacturing systems. Interfaces 30(3):125–142, 2000.

25. M Thierry, M Salomon, JAEE Van Nunen, LN Van Wassenhove. Strategic issues in product recovery management. Calif Manage Rev 37:114–135, 1995.

26. HM Driesch, JE van Oyen, SDP Flapper. Control of Daimler-Benz product recovery options. Logistik auf Umweltkurs: Chancen und Herausforderungen. Magdeburg, Germany: Otto-Von-Guericke-Universitat Magdeburg, 1997, pp 157–165.

27. G Ferrer. The economics of personal computer remanufacturing. Resourc Conserv Recycl 21:79–108, 1997.

28. H. Krikke. Business case Oce: reverse logistic network re-design for copiers. OR Spektrum 21:381–409, 1999.

29. VDR Guide Jr. Production planning and control for remanufacturing: industry practice and research needs. J Operat Manage 18:467–483, 2000.

30. N Nasr, C Hughson, E Varel, R Bauer. State-of-the-art assessment of remanufacturing technology. National Center for Remanufacturing and Resource Recovery, Rochester Institute of Technology, Rochester, New York, 1998.

31. TN Goh, N Varaprasad. A statistical methodology for the analysis of the life-cycle of reusable containers. IIE Trans 18:42–47, 1986.

32. P Kelle, EA Silver. Forecasting the returns of reusable containers. J Operat Manage 8:17–35, 1989.

33. VDR Guide Jr, R. Srivastava. Repairable inventory theory: models and applications. Eur J Operational Res 102(1):1–20, 1997.

34. J Linton, D Johnston. A decision support system for the planning of remanufacturing at Nortel. Interfaces, forthcoming, 2000.

35. M Michaelis, JF Coates. Creating integrated performance systems: the business of the future. Technol Anal Strat Manage 6(2):245–250, 1994.

36. E van der Laan. The effects of remanufacturing on inventory control. PhD Series in General Management 28, Rotterdam School of Management, Erasmus University, Rotterdam, The Netherlands, 1997.

37. E van der Laan, M Salomon, R Dekker. Production remanufacturing and disposal: a numerical comparison of alternative control strategies. Int J Produc Econ 45(1-3):489–498, 1996.

38. E van der Laan. The effects of remanufacturing on inventory control. PhD Series in General Management 28, Rotterdam School of Management, Erasmus University, Rotterdam, The Netherlands, 1997.

39. K Inderfurth. Modeling period review control for a stochastic product recovery problem with remanufacturing and procurement leadtimes. Preprint No. 2, Fakultät für Wirtschaftswissenschaft, Otto-Von-Guericke-Universität Magdeburg, Germany, 1996.

40. K Inderfurth. Simple optimal replenishment and disposal policies for a product recovery systems with leadtimes. OR Spektrum 19(2):111–122, 1997.

41. E van der Laan, M Salomon, R Dekker, LN Van Wassenhove. Inventory control in hybrid systems with remanufacturing. Manage Sci 45:733–747, 1999.

42. J Muckstadt, M Isaac. An analysis of single-item inventory systems with returns. Naval Res Logist Q 28:237–254, 1981.

43. M Salomon, E van der Laan, R Dekker, M Thierry, A Ridder. Product remanufacturing and its effects on production and inventory control. Management Report Series No. 172, Rotterdam School of Management, Erasmus University Rotterdam, The Netherlands, 1994.

44. E van der Laan, M Salomon. Production planning and inventory control with remanufacturing and disposal. Eur J Operat Res 102:264–278, 1997.

45. E van der Laan, M Salomon, R Dekker. Production remanufacturing and disposal: a numerical comparison of alternative control strategies. Int J Product Econ 45:489–498, 1996.

46. VDR Guide Jr, V Jayaraman. Product acquisition management: current industry practice and a proposed framework. Int J Product Res, forthcoming, 2000.
47. N Nasr, C Hughson, E Varel, R Bauer. State-of-the-art assessment of remanufacturing technology. National Center for Remanufacturing and Resource Recovery, Rochester Institute of Technology, Rochester, New York, 1998.
48. M Klausner, W Grimm, C Hendrickson. Reuse of electric motors in consumer products. J Indust Ecoly 2(2):89–102, 1999.
49. D Navin-Chandra. The recovery problem in product design. J Eng Design 5(1):65–86, 1994.
50. U.S. Office of Technology Assessment. Green Products by Design: Choices for a Cleaner Environment. Washington, DC: Congress of the United States, 1992.
51. TE Graedel, BR Allenby. Design for environment. Upper Saddle, NJ: Prentice-Hall, 1996.
52. VDR Guide Jr. Production planning and control for remanufacturing: industry practice and research needs. J Operat Manage 18:467–483, 2000.
53. VDR Guide Jr, V Jayaraman, R Srivastava, WC Benton. Supply chain management for recoverable manufacturing systems. Interfaces 30(3):125–142, 2000.
54. J Sarkis, N Darnall, G Nehman, J Priest. The role of supply chain management within the industrial ecosystem. Proceedings of the 1995 IEEE International Symposium on Electronics and the Environment, Orlando, Florida, 1995, pp. 229–234.
55. J Sarkis, N Darnall, G Nehman, J Priest. The role of supply chain management within the industrial ecosystem. Proceedings of the 1995 IEEE International Symposium on Electronics and the Environment, Orlando, Florida, 1995, pp. 229–234.
56. T Cooper. The re-use of consumer durables in the UK: obstacles and opportunities. In: SDP Flapper, A de Ron, eds. Proceedings of the First International Working Seminar on Reuse. Eindhoven University of Technology, Eindhoven, The Netherlands, 1996, pp. 53–62.
57. VDR Guide Jr. Production planning and control for remanufacturing: industry practice and research needs. J Operat Manage 18:467–483, 2000.
58. VDR Guide Jr. Production planning and control for remanufacturing: industry practice and research needs. J Operat Manage 18:467–483, 2000.
59. R. Lund. Remanufacturing: an American resource. Proceedings of the Fifth International Congress for Environmentally Conscious Design and Manufacturing, Rochester Institute of Technology, Rochester, New York, 1998.
60. N Nasr, C Hughson, E Varel, R Bauer. State-of-the-art assessment of remanufacturing technology. National Center for Remanufacturing and Resource Recovery, Rochester Institute of Technology, Rochester, New York, 1998.
61. VDR Guide Jr. Production planning and control for remanufacturing: industry practice and research needs. J Operat Manage 18:467–483, 2000.

26

Decision Tools for the Design for Environment

Jukka Kaipainen and Eero Ristolainen
Tampere University of Technology, Tampere, Finland

1 INTRODUCTION

Environmental impacts of a product are consequences of raw material consumption and emissions to soil, water, and air during its life cycle. Decisions that are made in product design affect either directly or indirectly all phases of the life cycle and thus also their environmental impacts. Designers are particularly well placed to contribute actively to the production of environmentally sound products [1]. They have to be aware of the consequences of their decisions to find environmentally preferable solutions in product design. They must be able to compare different material, process, and component alternatives in a fast and reliable way, so that environmental issues can be considered among other objectives of product design.

The less harmful the environmental impacts of a product are during its life cycle, the better are its environmental quality and performance. The ability to evaluate environmental performance in objective measurable terms is a key component of an effective design for environment (DFE) capability [2]. Measuring can be done in two ways. Assessment of environmental quality can be based on the physical environmental properties of a product, such as its number of parts, or the environmental impacts of the product's parts and life cycle phases. The latter is more difficult, but it must be done to focus on the right things in product design.

An organization's environmental experts can identify the most significant sources of impacts by means of software tools and provide advice and checklists for designers as the next step, thus moving the issue of DFE from theory into practice. As industrial products are rarely manufactured from raw materials to finished products within one organization, environmental cooperation with suppliers is very important. According to life cycle thinking, manufacturers are responsible for their products from cradle to grave.

This chapter reviews some of the tools that can be used both in environmental cooperation within the supply chain and DFE. Their applicability in organizations of different sizes, producing varying ranges of products, is discussed and the utilization and reliability of the results assessed from the point of view of life cycle thinking. Table 3 presents a summary of the features of the tools and gives sources of additional information.

2 TOOLBOX

Different tools can be used to support the manufacturer in making the right decisions in DFE. Figure 1 illustrates how the DFE tools can be grouped. In this example the tools are divided into two groups—nonsoftware and software tools.

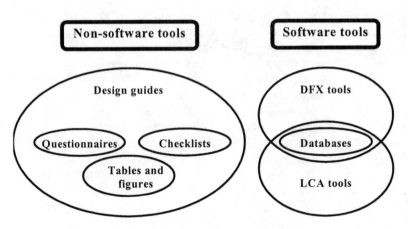

FIGURE 1   Design for environment toolbox.

Nonsoftware tools comprise questionnaires, checklists, tables, figures, or design guides. Questionnaires are used in environmental cooperation within the supply chain. For example, data sheets of electronic components do not usually give any information on material contents of the components, and issues like this must be determined by questionnaires. Companies can create checklists to help designers for example in placing components on printed wiring boards in the right way. Tables and figures can be used to work out or visualize whether or not the environmental design objectives were achieved in a particular product concept. Design guides are manuals that can be used in environmental training and as a basis when planning how environmental issues could be taken into consideration in practise.

Software tools are divided into DFE tools, databases, and life cycle assessment (LCA) tools. Databases can be either tools of their own, like component databases, or they can be integrated with larger LCA or DFE tools. LCA tools are not reviewed in this article because LCA and its tools have their own chapter in this book. DFE tools can be used in modeling for example disassemblability of a product or in modeling the overall environmental quality of a product, like the two software tools reviewed in this article. They are meant to be used not only by environmental experts but also by product designers. The following chapters give a more detailed view of different tools of DFE.

2.1 Nonsoftware Tools

2.1.1 Questionnaires

Questionnaires are used in environmental cooperation with supply chain partners. Their usage has increased along with implementation of certified environmental management systems. Questionnaires can be used for getting environmental information about products of suppliers or their environmental management. The grade of environmental management and the environmental quality of the products are then used among other decision criteria when choosing suppliers.

EcoPurchaser^{TM} [3] developed by the Swedish research institute IVF is an example of a questionnaire tool. It consists of a large questionnaire and a spreadsheet application for evaluating the answers. The multiple-choice questionnaire is divided into two sections which cover aspects of environmental management of an organization and environmental features of a product. When a supplier returns the questionnaire, answers are fed into spreadsheet matrices. The evaluation method calculates two result values, one for the grade of environmental management (company score) and the other for the product (product score). These can be used to make comparisons between alternative suppliers and their products. Additional information about the tool can be found in the summary table in Section 3.

2.1.2 Checklists

Checklists are used for dividing general environmental design objectives into smaller units to turn them into practical tasks for designers. These general objectives can be, for example, low energy consumption, minimizing use of hazardous materials, or good recyclability of a product.

Checklists can be created both for mechanics and electronics designers. They should consist of simple practical advice on how to improve the aspects of environmental quality of a product without being an environmental expert. Designers have to be supported by practical advice on what to do and what not to do. It is important that checklists do not give too general advice but focus on the very things that their users can contribute. Checklists are at their best when they are built "in house" and focused on the special features of the company's own products. Checklists have to be unambiguous, updatable, and simple, so they could really be used in practical everyday design work. For example, the following checklist could be used in printed wiring board layout design to minimize the use of environmentally hazardous or scarce materials:

- Do not use gold plated connectors.
- Choose a board material that does not contain brominated flame retardants.
- Do not use electrolytic capacitors that have a volume of over two cubic centimeters.
- Do not use solder that contains lead.
- Do not use components that contain lead or mercury.

This kind of a checklist is not too general in scope, and it is focused on things that usually can be taken into consideration by a layout designer. However, it must be noted that it is not the designer's task alone to make it possible to use environmentally more preferable alternatives, like lead-free components. These issues must be taken into consideration in other departments of the organization, like the purchase department, as well. The designer can only work within limits set by management operations.

2.1.3 Tables and Figures

Progress in environmental design can be monitored by tables and figures. They can be used for measuring and clarifying success in tasks that have been set by checklists. Tables and figures can give an overview of the whole product. Table 1 gives an example.

TABLE 1   Example of Estimating Environmental Quality of an Electronic Product

	Electronics	Mechanics
Recyclability	4	1
Repairability	2	3
Hazardous/scarce materials	3	3
Weight	2	4
Energy use	3	2

Some general environmental objectives were set for the product, given in the first column of Table 1. Concerning each general aim, checklists were created both for electronics and mechanics designers of the product. Numbers in the other colums show how well the goals of each checklist were achieved in this particular case. For example, it can be seen from Table 1 that three of five goals in the checklist that was created to minimize the use of hazardous or scarce materials in the electronic parts of the product were achieved. Behind each other number there lies a checklist of its own.

Numbers in the table can also be presented as a "spider web" graph, which is more explanatory than a tabular presentation. Figure 2 presents the numbers of Table 1 in this way. The outermost circle corresponds to score one and the innermost circle to score five in the table.

It can be seen from the graph that the most difficult tasks for designers were given in the checklists that were used for improving recyclability of mechanical parts and repairability and weight of electronic parts. Obstacles to success should be first examined on the basis of these checklists before considering other improvements.

2.1.4   Design Guides

There are many environmental design guides available for electronic products. They are published by both academic institutions and companies. They give basic information on how to take environmental aspects into account in product design and management. Guides can be used as educational material and as a general information source. However, each company that decides to move DFE from theory into practice has to create its own methods of how to organize training, environmental assessment of products, and DFE. Examples of guides are *Environmentally Oriented Product Design* [4] and *Handbook for Design of Environmentally Compatible Electronic Products* [1].

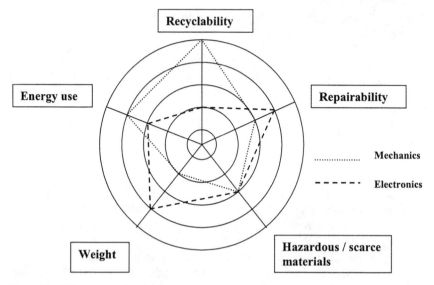

FIGURE 2 "Spider web" graph.

2.2 Software Tools

2.2.1 DFE Software Tools in Assessment of Environmental Impacts and in Product Design

DFE software tools are used in product design to find out the environmental impacts of design alternatives and their importance as a part of the whole product's environmental quality. Essential in their usability and reliability is comprehensiveness of their databases, so that the tools can be used for assessing and comparing most practical design alternatives in both electrical and mechanical design of products. The databases should include a thorough list of different materials, components, manufacturing processes, transportation methods, end-of-life treatments, and their environmental impacts, so that the user can build a reliable model of environmental impacts over the whole life cycle of a product.

Software should also be usable both for environmental experts and designers and able to communicate with other software tools used in product design. Present DFE tools can fulfill most of these expectations at some level, but none of them is compatible with other software tools that are used in mechanic and electronic design at the moment.

The main difference between LCA and DFE tools is that the latter are intended to be used without collecting inventory data. All input information for modeling should be possible to find in the company's existing databases or by measuring simple attributes such as weight or surface area of a product or its parts under assessment. In this way the most time- and resource-consuming phase, inventory analysis, of LCA can be avoided, and estimations about environmental impacts of products can be made using only existing information in databases of the tools.

The sections below present a closer look at two software tools, EcoScan®2.0 and EIME™. These DFE tools are often referred to in scientific articles at environmental design conferences, such as ISEE [5,6], and are used in many companies in Europe and the United States. The tools have a lot in common but they also have substantial differences in price, calculation methods of environmental impacts, comprehensiveness of databases, operating system requirements and form of results. However, both can be used for modeling environmental impacts of products and for finding out which parts, manufacturing processes, and phases of life cycle cause most of the environmental burden of a product under assessment.

When evaluating reliability of results of environmental assessments, one must be familiar with the theory of LCA. Figure 3 illustrates formation of environmental impacts and the phases of LCA. A product under assess-

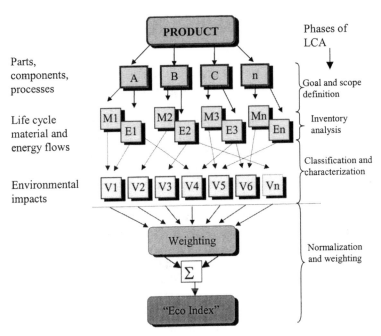

FIGURE 3 LCA flow from product to environmental index.

ment can be divided into parts, components, and their manufacturing, transportation, and handling processes over the whole life cycle from raw material production to end-of-life handling. Product system boundaries are chosen in the first phase of LCA. This is done by choosing which unitary processes of the life cycle process tree (e.g., transportation of a part from supplier to factory) are taken into account. The next phase of the LCA is inventory analysis. Each of the unitary processes has a unique combination of inputs and outputs, so-called environmental exchanges. These material and energy flows entering and leaving unitary processes are energy, raw materials, emissions to soil, air and water, waste and end products of unitary processes. Inventory analysis thus gives results as quantitative values of material and energy flows between the product system and the environment.

The rest of Figure 3 is the impact assessment phase of the LCA. The first step of impact assessment is classification, in which the environmental exchanges are allocated to chosen environmental impacts. For example, CO_2 emissions have a direct influence on global warming but not on euthrophication. The second step is called characterization. It includes multiplication of each environmental exchange with a factor that expresses how much a particular exchange contributes to each environmental impact. For example, the factor of CO_2 per mass unit in global warming for 100 years is one, whereas the factor of methane for the same impact is about 25 [7]. After these multiplications, exchanges that contribute to the same impact can be summed. The result is a numerical value for the magnitude of the impact.

The next step of impact assessment is normalization. It has two objects: to provide an impression of the relative magnitudes of the environmental impact potentials and to present the results in a form suitable for subsequent weighting [7], which is the last step of impact assessment.

In weighting, the impacts are combined into a single figure that represents the total environmental burden caused during the life cycle of the product. Weighting methods are based on political and subjective decisions on importance of different environmental impacts. A number of different methods have been developed to do this in Europe and the United States.

Subjective decisions thus have to be made in each phase of the LCA, and the further the process is continued the more uncertain the results are. There is no unambiguous method of how to set boundaries to the product system. In inventory analysis all the environmental exchanges cannot be measured accurately or even taken into account at all. However, the most subjective phase is impact assessment. Many environmental exchanges must be omitted in characterization because it is not fully known which substances contribute to chosen environmental impacts. Multiplication values used in characterization are subject to change as knowledge about formation of impacts develops, and many different valuation methods are used.

Limitations of the LCA method cause difficulties in interpretation of results of environmental assessments. Due to these limitations it is also difficult to compare results of LCAs that are made independently, even if the product under analysis is the same. Different system boundaries, sources of information and valuation methods in all probability lead to different results. When the LCA method and its limitations are known, it is possible to critically evaluate the reliability of software DFE tools and their results. A review of two of these tools is presented.

2.2.2 EcoScan 2.0

EcoScan 2.0 is a PC software tool that can be used to calculate single numerical values for the environmental burden of products and for finding out which parts and processes contribute the most to the burden.

Product Modeling. Products are modeled in EcoScan 2.0 by picking up and combining database units that are materials, manufacturing processes, transportation methods, and end-of-life treatments. For example, if there are 50 g of aluminum compound in a cover part of a product, a database unit that is as close to the real material as possible is picked up from a database and the weight of the compound is given as a parameter for the part. Manufacturing processes (e.g., extrusion of aluminum) and transportation methods (16-ton truck) can be modeled correspondingly. The software then calculates an environmental index for each database unit in the product model. These are summed up to form indices for parts, subassemblies, life cycle phases, and finally the entire life cycle. Numerical results can also be viewed as pie or bar charts. These show how the overall environmental burden is distributed to life cycle phases or to subassemblies, parts, processes, and materials in manufacturing.

It is easy to build up the model of a product. However, it must be noted that only one of the available databases can be used for each model, because the databases are created from different sources even if their valuation methods might be the same.

Comprehensiveness of Databases. The wider the selection of materials, processes, and transportation methods in a database is, the wider the range of products and product systems that can be modeled with it becomes. There are two large databases in EcoScan 2.0. These are EcoScan Eco-Indicator '97 (ES97) and IdeMat '96. The former is composed by Philips and the latter by the Delft University of Technology [8].

Both databases include, for example, a comprehensive list of metals, plastics, and most common manufacturing processes of industrial products.

Even though ES97 is created by Philips, it is not specifically tailored for modeling electronic products. It suits best for modeling machines and equipment whose environmental burden does not originate from electrical parts but from material consumption and manufacturing of mechanical parts. Idemat '96 can be used to build more accurate models if the exact material composition is known, because it includes, for example, 23 aluminum compound database units, whereas ES97 includes only two.

Both databases can be used for modeling mechanical parts and some of their manufacturing processes, but because neither of them includes units for electronic components, these have to be modeled only according to their material content. Thus, the environmental burden caused by electronic parts becomes underestimated because their manufacturing processes cannot be modeled by the software. It is impossible to build a reliable model, for example, for printed wiring boards or any electronic components by EcoScan 2.0.

User Interface.   The user interface of the software looks familiar to most users of other Windows® applications. Figure 4 shows the user interface of EcoScan 2.0. There is a tree structure of an example product in the

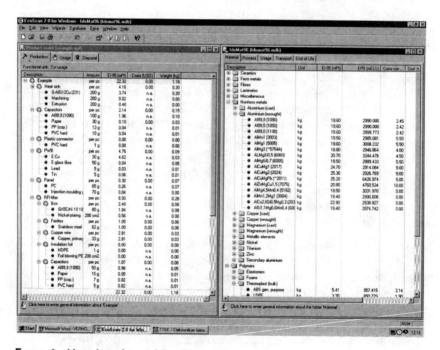

FIGURE 4   User interface of EcoScan 2.0.

window on the left. The structure consists of parts and their manufacturing processes. The window has separate cards for manufacturing, use, and disposal phases of the life cycle. The analyzed product is a frequency converter whose material content and manufacturing processes were known. Numbers and part list have been altered, so any similarities with any real product are incidental. There is a database window on the right. The IdeMat '96 database is open. The database comprises cards for materials, processes, usage, transport, and end of life.

Methods of Calculating Environmental Impacts. The software uses two calculation methods for environmental impacts. Both methods combine the impacts into a single numerical value that describes the magnitude of the total environmental burden. The ES97 database uses an environmental impact assessment method, called Eco Indicator 95 (EI95). The IdeMat '96 database uses also another method called EPS, so most of the database units in the IdeMat '96 have two different environmental indices. Characterization and valuation steps of the methods are different, so their values are not comparable, and the results may be contradictory. For example, according to the EI95 method, use of PVC plastic is more harmful to the environment than that of LDPE, whereas the EPS method gives a converse valuation for these materials. The EI95 method is described on a general level in Help windows of EcoScan 2.0, but the units of the databases are not transparent. Thus users cannot see system boundaries or inventory data behind the environmental indices of the database units. The software does not include any background information about the EPS method, but it can be found in literature [9].

Results and Their Utilization. It is easy to view the results graphically in EcoScan 2.0 and see how the environmental burden is distributed within the product concept in the manufacturing and other phases of its life cycle. Figure 5 shows how the environmental burden of manufacturing was distributed within the example product which was presented in Figure 4.

As Figure 5 shows, the three most significant sources of environmental burden were RFI filter, heatsink, and printed wiring board (PWB). These are therefore the parts in the product concept where the possibilities of improvement should be analyzed most carefully. The environmental burden of the PWB was probably underestimated in relation to the other two main contributors in this case because the model was built up only according to the material content of the board. It was not possible to model the manufacturing of PWB because suitable database units did not exist. It was possible to include some manufacturing processes in the models of the RFI filter and heatsink, so their model was more accurate.

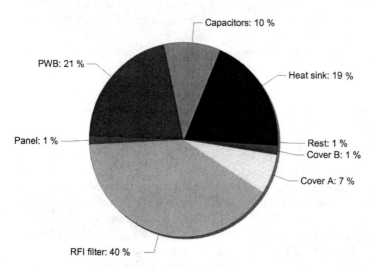

Figure 5 Distribution of environmental burden of manufacturing in the Example case according to the EI95 method.

Databases of EcoScan 2.0 are kind of checklists in themselves, because environmental indices of database units are visible. It is very easy, for example, to make comparisons between different material alternatives if the user accepts the impact assessment methods of the software. Additional information on EcoScan 2.0 can be found in Table 3.

2.2.3 EIME

EIME (Environmental Information and Management Explorer) is a DFE software tool developed by Ecobilan Group. Its database is the only commercial one available that is tailored for modeling electr(on)ic products. Database units are called modules; they are models of components, processes, or materials. There are about 40 component modules (e.g., LED), about 40 process modules (wave soldering), and about 100 material modules (ABS plastic) in the database. Product models are created in a similar way to EcoScan 2.0, that is, by picking the modules from the database into a product tree. The software calculates environmental impacts as a combination of impacts of the modules used in the product model. Impact assessment does not include weighting of impacts, so results of environmental assessment are presented as numerical values of 11 different impact indicators, for example, of global warming.

User Interfaces and Product Modeling. There are two user interfaces in EIME called Expert and Designer. As their names indicate, they are meant to be used either by environmental experts of a company or product designers. Users of Expert can create user profiles for the interfaces, start assessment projects, create design guidelines and checklists for designers, and revise the module database and calculation methods of environmental impacts. Thus, users of Expert are specialists in environmental issues, and they maintain the software and support users of Designer.

The Designer interface is used for building up a model of a product, for adding information on its energy consumption and transportation, and for calculating environmental impacts of the manufacturing, distribution, and use phases. There are three modes in Designer. The first of these is Design Mode, in which the model of a product is built. The second is Product Data Mode, which is used for adding energy consumption and transportation data. The Evaluation/Compare Mode is used for calculating results and making comparisons between analyzed product concepts. Figure 6 presents the Design Mode window of the Designer interface.

In Design Mode the product under assessment is divided into sub-assemblies and parts. These are built up of modules that are picked from the database. There is a tree structure of an example product in the upper left window of Figure 6. The product under assessment is the same as in the case of EcoScan 2.0. The next window to the right shows the modules that were used in modeling of the printed wiring board. Percentages shown on each module describe their contribution to a selected environmental impact, which is raw material depletion in this case. The lower left window shows a list of privileged users and their product assessment projects. The last of the four windows shows the database that the modules were picked from.

Comprehensiveness of the Database. There are about 40 component modules (e.g., LED), about 40 process modules (wave soldering), and about 100 material modules (stainless steel) in the database. The database of EIME is customized for estimating environmental impacts of electrical and electronic products. It is the only software available on the market whose database includes completed models for electronic components, special materials, and industry-specific manufacturing processes. The database is comprehensive enough to be used for modeling most mechanical and electronic parts and for finding out the "hot spots" in products. In many reported LCAs for electronic products such as TV sets [10], electronic parts have been left outside the product system boundaries or they have been modeled only according to their material content due to difficulties in

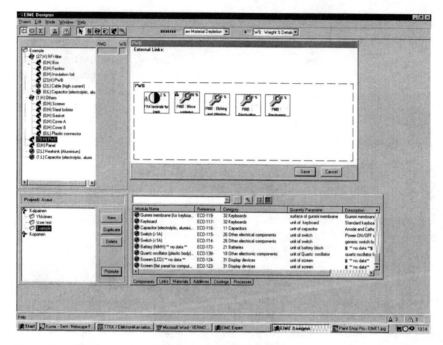

FIGURE 6   Design mode window of EIME™ Designer.

getting the inventory data from suppliers of these parts. However, the EIME database is not even nearly comprehensive enough to have an accurate module for all electronic components. For example, there are only two integrated circuit (IC) modules in the database, one for silicon ICs and one for GaAs ICs. Thus, all the ICs in products have to be modeled by these two alternatives.

Methods of Calculating Environmental Impacts.   Users of the Expert interface are able to see and modify inventory data of each module. These data are partly collected from processes of companies that took part in the development of the software and partly from literature and other public databases. EIME calculates the results in 11 impact catagories using the inventory data of modules that have been used in the model. The impact categories are the following:

- Raw material depletion (RMD);
- Energy depletion (ED);
- Water depletion (WD);
- Global warming (GW);

- Ozone depletion (OD);
- Air toxicity (AT);
- Photochemical ozone creation (POC);
- Air acidification (AA);
- Water toxicity (WT);
- Water euthrophication (WE);
- Hazardous waste production (HWP).

Each category has its own calculation method for classification and characterization. Formulas are presented in the manuals of the software, and environmental exchanges and characterization factors related to impacts are modifiable by users of Expert. Calculations do not include normalization and weighting phases of impact assessment, so if these are needed, users have to perform them themselves with data from literature or other software.

Results and Their Utilization. EIME presents results of environmental assessment as numerical values in each of the impact categories. As Figure 6 showed, distribution of impacts of manufacturing among parts and processes can be viewed, and thus the environmental hot spots of the product concept can be identified. Possibilities for environmental improvement should be investigated first with respect to these parts or processes. Table 2 shows how the environmental impact categories were distributed within the product concept model in the Example 1 case.

TABLE 2 Distribution of Impacts Within Example 1

	Heatsink	Capacitors	PWB	Panel	RFI filter	Others
Raw material depletion	2	1	70	0	27	1
Energy depletion	27	13	7	7	16	30
Water depletion	47	20	5	1	18	8
Ozone depletion	52	21	2	1	19	5
Photochemical ozone creation	20	12	17	4	19	28
Air acidification	12	46	4	2	28	8
Water toxicity	15	7	46	3	20	9
Water euthrophication	2	7	69	1	24	3
Hazardous waste production	0	1	30	13	14	41
Air toxicity	12	46	4	2	28	8
Global warming	31	14	7	7	16	26

The heatsink is a significant source of impact in the categories of energy and water depletion, ozone depletion, and global warming. Capacitors are sources of air toxicity and acidification. Manufacturing of PWB causes raw material depletion, water toxicity, euthrophication, and hazardous waste production. Photochemical ozone creation is quite evenly shared with all main parts of the product.

Results of two or more product assessments can also be compared. Relative result values of assessments can be presented as a spider web graph (Figure 2). This can be used for example in monitoring the development between product generations. Figure 7 presents a comparison between two example cases in the Evaluation/Compare Mode window of Designer.

The area of the PWB of Example 2 was decreased by 10% and the weight of the heat sink by 20% compared with Example 1. The number of capacitors was increased from six to eight correspondingly.

The step between the circles in the graph in Figure 7 is 10%. It can be seen from the figure that the changes in the model resulted in improvements in all but two impact categories, air acidification and air toxicity. This was

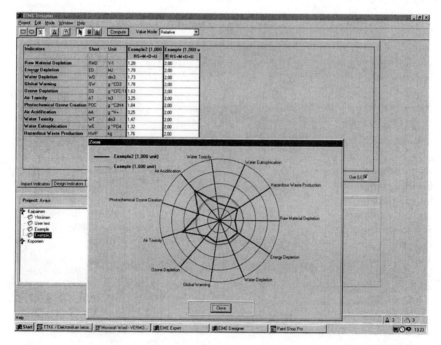

FIGURE 7 Spider web graph for comparison of two cases.

to be expected because it was shown in Table 2 that the capacitors were the biggest source of impact in these two categories.

As shown above, the results are not as easy to interpret as in EcoScan 2.0, but on the other hand they are not as subjective, because the phases of normalization and valuation are not done. However, these phases can be done the by user if usable normalization values for categories are available and she or he wants to valuate the normalized results by a chosen valuation method. Normalization factors for the most commonly used impact catagories, such as ozone depletion and global warming, can be searched for in literature, for example in [7], but they are not available for all the 11 categories.

In addition to calculation of environmental impacts, product modeling with EIME also gives other results. These are, for example, number of parts in the product, weight of the product and its packaging, and portion of recyclable material in the product.

Warnings and "To-Dos." A unique feature in EIME is that users of Expert can create warnings and design instructions for users of Designer. Modules of the database can have their own checklists which appear on screen in Designer when the module is used. For example, PWB material modules can be linked to a checklist concerning preferable material alternatives or to a larger checklist concerning also other environmental aspects of PWB design. An example of a warning could be that when a designer is planning to use a plastic part that consists of ABS and PC polymers and he or she picks these modules from the database into the same part, he or she is warned that compound polymers are hard to recycle.

Users of the Expert interface of the software thus can counsel users of Designer in how to take environmental issues into consideration both in mechanical and electronic design of products by bringing company specific design instructions available to them by the aid of the software. In this way it is easier to maintain and update the checklists because information can be given right where it is needed and it is always available to designers.

3 SUMMARY

Even though designers play a key role in defining the environmental quality of industrial products, they can achieve good results only if environmental issues are considered also in the management operations of a company. Life cycle thinking means environmental cooperation with suppliers and customers. EcoPurchaser is a suitable tool for developing this cooperation. Results can be used for example for supporting implementation and use of other DFE tools.

Both DFE software tools described in this article have to be used as separate tools along with other design software. The databases of EcoScan 2.0 have a larger number of units than the database of EIME. EcoScan 2.0 is also easier to use and cheaper to buy and maintain than EIME, but reliability of results is poor because all environmental impacts are combined into single environmental indices and calculations behind the database units are not transparent.

A particular strength of the database of EIME is that it includes models for electronic components and manufacturing processes of electronic products. In addition, the database and calculation formulas are modifiable by users of Expert, and it is possible for an expert to create company specific checklists and make them available to designers by means of the software. This enhances the usability and flexibility of the software. Avoiding collection of inventory data for electronic parts and flexibility of the software have been seen as the best features of the software in earlier articles as well [6]. However, the software makes great demands on hardware, and its maintenance takes up more resources than that of EcoScan 2.0. Another disadvantage is that is not possible to model or compare end-of-life options (reuse, recycling, landfilling) of a product using the software. It is also more difficult to interpret the results of environmental assessment of EIME than those of EcoScan 2.0, because they are presented in 11 impact categories. Thus EIME is more suitable to be used as a tool for environmental experts in larger organizations. EcoScan 2.0 is easier to adopt in smaller organizations and by designers because of lighter hardware requirements and easier use. Finally, the most important features of EcoPurchaser, EcoScan 2.0, and EIME are summarized in Table 3.

TABLE 3 Summary of the Properties of the DFE Tools

Tool	Description	Used for	Easiness of implementation
EcoPurchaser	Questionnaire tool. Consists of a questionnaire and a spreadsheet evaluation method for the answers.	Evaluating the environmental performance of supply chain companies and their products.	Easy. Questions and their weighting must be chosen according to user company's environmental policy. Good manual.
EcoScan 2.0	PC software for evaluation of environmental impacts of products according to EI95 or EPS methods.	Compiling assessment or comparison of environmental impact of industrial products.	Easy. A typical Windows interface with a built-in tutorial.
EIME	Unix and PC software with two user interfaces for both environmental experts and designers	Rough LCAs of electronic products including possibility to integrate checklists for designers.	Difficult. Great demands on hardware. Imperfect installation instructions, but well supported by vendor. Requires an environmental expert. Good manuals.

Tool	Inputs	Outputs	Modifiability
EcoPurchaser	Users can modify the questionnaire as they wish. Inputs for the evaluation procedure are the answers obtained.	Company and product scores describing the level of environmental performance of supply chain partners.	Fully modifiable. Own questions and weighting factors can be used.

(continued)

TABLE 3 (continued)

	Inputs	Outputs	Modifiability
EcoScan 2.0	Materials, manufacturing processes, surface areas, lifetime, energy consumption, transportation methods. Data about associated costs can be fed in, if known.	Two different single-value scoring methods (EPS and EI95) for environmental impacts. Bar or pie chart presentation of weight, cost and impacts for the whole product or its parts.	Databases not transparent, but own values for environmental indices can be used among the built-in values.
EIME	A mode of a product is made using built-in component, process, and material modules. Also life-time, energy use, transportation methods, and EOL treatment can be fed in.	Environmental impacts (11 categories) caused by the whole product and its parts, comparisons of products by "spider web" graph, warnings about banned materials, and "to-dos" for designers. Also amount of different materials used, product weight, recycled content.	Databases and modules fully modifiable by Expert. Own modules can be created and existing ones can be tailored. User can also modify environmental impact calculation methods by adding and re-valuing material flows.
	Price (one user license, subject to change)	Contact	Information in World Wide Web
EcoPurchaser	2200 SEK (280 USD). Same for universities and companies	IVF Attn. Annika Westergren Argongatan 30 SE-431 53 Mölndal Sweden	http://www.ivf.se/elektronik/Ep/Eco/Ecomain.htm

			Additional comments
EcoScan 2.0	815 USD including all three databases.	E-mail: ecoscan@ind.tno.nl TNO Industrial Technology Sustainable Product Innovation P.O. Box 5073 2600 GB Delft	http://www.ecoscan.nl
EIME	7500 USD including server software, Designer and Expert client software, and database.	Email: info@ecobilan.com Phone: +33(0)153 282 378 Fax: +33(0)153 782 379 Ecobilan S.A. and Ecoaudit Immeuble le Barjac 1 Boulevard Victor 75015 Paris	http://www.ecobalance.com/ software/eime/eime_ovr.htm
	Potential users	Software requirements	Additional comments
EcoPurchaser	Quality and environmental managers, anyone developing supply chain management.	The three evaluation matrices are Excel applications	Usable tool for environmental supply chain management. Could be used as a model and an idea source in building up a customized tool. Questionnaire does not have to be restricted to environmental questions.

(continued)

TABLE 3 (continued)

	Potential users	Software requirements	Additional comments
EcoScan 2.0	Mechanics designers, environmental managers	Minimum 486-66MHz or higher PC, 8 M RAM, 8M free hard disc space, SVGA 800*600 monitor, Win 95.	Simpler and faster to use than EIME, presents the results as single number is cheaper by a factor of > 10, but lacks in reliability and comprehensiveness of databases and results.
EIME	Electronics designers, mechanics designers, environmental expert as software and database supporter.	Tool is built on a client/server architecture. Requires a Sun SPARC (Solaris 2.5) as server platform and PCs (Windows 95/NT) for clients. The database is managed under ORACLE and the client/server communication utilizes ORBIX.	The "heaviest" one of these DFE tools in many respects, e.g. purchase price and work required to use and maintain the tool. Databases for quick LCA are the best ones available for electronic products. Gives quite similar results to fully executed LCA, but saves a lot of time and effort in comparison. Introduction of the software in companies requires serious commitment to improving the environmental performance of the products, as the tool is quite expensive to buy and maintain.

EOL, end of life.

REFERENCES

1. C.G. Bergendahl, et al. 1995. Handbook for Design of Environmentally Compatible Electronic Products. An aid for designers. Edition 1. IVF Research Publication 95851.
2. J. Fiksel. Design for Environment. Creating Eco-Efficient Products and Processes. McGraw-Hill.
3. C.G. Bergendahl and T. Segerberg. 1998. EcoPurchaser™1.0. A tool for environmentally conscious procurement of electronic components and electronic products. IVF Research Publication 98803.
4. A. Kärnä. 1997. Environmentally Oriented Product Design. SET. Available from http://www.electroind.fi/ymparisto/eng_guide.htm
5. S. Criel, et al. Metrics for telecom products: essential part of an ecodesign methodology. Proceedings of International Symposium on Electronics & the Environment, May 11–13, 1999, Danvers, Massachusetts, IEEE. pp. 25–30.
6. D. M. Timmons. 1999. Building an Eco-Design toolkit at Kodak. Proceedings of International Symposium on Electronics & the Environment, May 11–13, 1999, Danvers, Massachusetts. IEEE, pp. 122–127.
7. M. Hauschild and H. Wenzel. 1998. Environmental Assessment of Products. Volume 2: Scientific Background. Chapman & Hall.
8. http://utopia.tno.nl/instit/indus/development/ecoscan/database/generic.html
9. B. Steen. 1996. EPS-Default Valuation of Environmental Impacts from Emissions and Use of Resources. Version 1996. Report AFR 111 from Swedish Environmental Protection Agency.
10. Wenzel and Hauschild. 1998. Environmental Assessment of Products. Volume 1: Methodology, Tools and Case Studies in Product Development. Chapman & Hall.

Index